MECHANICS OF FLEXIBLE FIBRE ASSEMBLIES

NATO ADVANCED STUDY INSTITUTES SERIES

Proceedings of the Advanced Study Institute Programme, which aims at the dissemination of advanced knowledge and the formation of contacts among scientists from different countries.

The series is published by an international board of publishers in conjunction with NATO Scientific Affairs Division

A	Life Sciences	Plenum Publishing Corporation
B	Physics	London and New York
C	Mathematical and Physical Sciences	D. Reidel Publishing Company Dordrecht and Boston
D	Behavioural and Social Sciences	Sijthoff & Noordhoff International Publishers B.V.
E	Applied Sciences	Alphen aan den Rijn, The Netherlands and Germantown, Maryland, USA

Series E: Applied Sciences - No. 38

MECHANICS OF FLEXIBLE FIBRE ASSEMBLIES

edited by

JOHN W.S. HEARLE
Professor of Textile Technology
The University of Manchester Institute
of Science and Technology (UMIST)

JOHN J. THWAITES
Lecturer in Engineering
Fellow of Gonville and Caius College
Cambridge, U.K.

JAFARGHOLI AMIRBAYAT
Honorary Research Fellow
The University of Manchester Institute
of Science and Technology (UMIST)

Springer-Science+Business Media, B.V.

Proceedings of the NATO Advanced Study Institute on
Mechanics of Flexible Fibre Assemblies
Kilini, Greece
August 19 - September 2, 1979

ISBN 978-94-011-9776-2   ISBN 978-94-011-9774-8 (eBook)
DOI 10.1007/978-94-011-9774-8

Copyright © Springer Science+Business Media Dordrecht 1980

Originally published by Sijthoff & Noordhoff International Publishers B.V., Alphen aan den Rijn, The Netherlands in 1980.

All rights reserved. No part of this book may be reproduced, stored in a retrieval system, or transmitted, in any form or by any means, electronic, mechanical, photocopying, recording, or otherwise, without the prior permission of the copyright owner.

PREFACE

Continued and systematic analysis of the mechanics of flexible
fibre assemblies dates from about 1945, although the growth of
research into textiles after 1920 had included studies of fabric
structure and the measurement of mechanical properties. The
subject is thus a young one, although this NATO Advanced Study
Institute is a sign of developing maturity.

However there is an earlier tradition. Relevant, even if
somewhat loosely connected, quotations can be found in the works
of the engineers of the ancient civilisations, recurring during
the Renaissance with Leonardo da Vinci and Galileo. But the
glorious libk is with Euler and the Bernoulli family, with
their theories of the mechanics of flexible slender rods.
While mathematicians have admired the beauty of this work, the
invention of elliptic integrals, and the grace of the different
classes of planar elastica, it is in the technology of textile
materials, composed of flexible fibres and yarns, that the subject
has found its more direct application. All this, and much more
such as Max Born's doctoral thesis, was brought to our attention
in a delightful discourse by Milos Konopasek, who is not only
fascinated by the mathematics of Euler and the modern movement of
the solutions of bending curves from two dimensions into three
by the use of the computer, but also feels a personal link
through having lived and studied within sight of the scene of
Euler's triumphs in St. Petersburg.

Although the mechanics of flexible fibre assemblies has developed
as a subject over the last 30 years, this advanced study institute
was the first occasion on which a major group could concentrate
solely on mechanics. Usually the subject is discussed at meet-
ings where the majority of the audience are interested in other
aspects of textile technology and fibre science. A great
virtue was the interdisciplinary link, which brought the subject
into contact with the main stream of engineering mechanics and
with the applications in civil engineering. Those of us who were

to be in Kilini for the last two weeks of August 1979 learnt a lot in a highly stimulating atmosphere, with discussion continuing into the warm water of the Ionian sea and among the stones of ancient Delphi. I hope that this publication will bring the interest and engineering relevance of the subject to many others.

As Director, I would like to express my thanks to all who contributed to the great success of the ASI: to the NATO Science Committee for their generous grant; to the members of the Advisory Committee; to all the lecturers, discussion leaders, rapporteurs and speakers; to the Greek National Tourist Organisation for providing an excellent location, and to their staff for contributing to our comfort, enjoyment, and efficiency; to my Associate Directors, Denney Freeston, unfortunately unable to be present but very helpful with contacts across the Atlantic; John Thwaites, who took on responsibility for the discussion sessions as well as guiding the choice of programme, and Stelios Arghyros, who gave unstinting help in making all the arrangements in Greece; to Jafar Amirbayat, who displayed an incredible range of talents in dealing with every detail of the organisation, before, during and after the event; to Pericles Arghyros and Adrian Hearle for their services during the meeting; to the European Research Office of the United States Army for an additional grant; to the Hellenic Cotton Board for hospitality; to many industrial companies and other organisations who provided an invisible subsidy by meeting the costs of speakers; and, not least, to Pat Bardsley and Marion Mellor in my office.

J.W.S. Hearle
Manchester
November 1979

TABLE OF CONTENTS

Preface   V

J.W.S. Hearle
    The formation of textile structures   1

C.P. Buckley
    Mechanical properties of fibres   35

J.W.S. Hearle
    Mechanics of dense fibre assemblies   51

J.J. Thwaites
    A continuum model for yarn mechanics   87

G. Carnaby
    The compression of fibrous assemblies, with applications
    to yarn mechanics   99

L. Phoenix
    Statistical models for the tensile strength of yarns
    and cables   113

G. Leaf
    Woven fabric tensile mechanics   143

S. Moghe
    From fibres to woven fabrics   159

R.N. Hepworth
    The mechanics of a model of plain weft-knitting   175

P. Grosberg
    The bending of yarns and plain woven fabrics   197

J. Skelton
    Shear of woven fabrics   211

S. de Jong and R. Postle
    Energy optimisation methods in fabric mechanics   227

J. Skelton
    Tearing behavior of woven fabrics   243

M. Konopasek
    Classical elastica theory and its generalizations.   255
    Computational aspects of large deflection analysis of
    slender bodies   275
    Textile applications of slender body mechanics   293

D.W. Lloyd
    The analysis of complex fabric deformations   311

C.M. Leech
    The dynamics of flexible filament assemblies   343

B.M. Chapman
    Viscoelastic, frictional and structural effects in
    fabric wrinkling   391

S. Kawabata
    Examination of effect of basic mechanical properties
    of fabrics on fabric hand   405

D.H. Page, R.S. Seth and J.H. De Grâce
    The elastic modulus of paper - The controlling
    mechanism   419

P.S. Theocaris and G.C. Papanicolaou
    An introduction to the mechanics of fibre-reinforced
    composites   433

J. Skelton
    Mechanical properties of coated fabrics   461

S.K. Clark
    The role of textiles in pneumatic tires   471

R. Blum
    Mechanics of fabrics in pneumatic structures   495

J.P. Giroud
    Behaviour of geotextiles   513

TUTORIAL DISCUSSIONS

    I   Yarns etc   535

    II   Non-wovens   543

    III Variability   549

    IV   1-d strands   557

    V   Woven/knitted fabrics   563

    VI   Complex fabric deformations   569

    VII Characterisation of structure   575

    VIII Dynamics   579

    IX   Composites etc   587

    X   Stress concentration etc   593

    XI   Engineering   597

    XII Civil Engineering   601

    XIII Clothing   609

    XIV Computation   623

    XV   Product life   631

    XIV Communication   641

APPENDIX : Some textile quantities   645

List of participants   647

NATO ADVANCED STUDY INSTITUTE ON

THE MECHANICS OF FLEXIBLE FIBRE ASSEMBLIES

Kilini, Greece, August 19 - September 2, 1979

Director : Professor J.W.S. Hearle, Dept. of Textile Technology,
U.M.I.S.T., Manchester, United Kingdom

Associate Directors :   Dr S. Arghyros, Athens, Greece
                        Professor W.D. Freeston, Atlanta, U.S.A.
                        Mr J.J. Thwaites, Cambridge, U.K.

Assistant to the Director :   Dr J. Amirbayat, U.M.I.S.T., Manchester.

Advisory Committee:

Professor S. Backer, M.I.T.
Professor P. Grosberg, Leeds
Mr V.E. Hansen, U.S. Army, Natick
Professor K.V. Harten, Delft
Professor J. Lunenschloss, Aachen
Mr M. Mackeprang, Lyngby
Dr T. Moynehan, M.O.D. Colchester
Professor K. Slater, Guelph, Canada
Dr M.Sotton, I.T.F., Paris
Dr R.Wilfong, du Pont, Wilmington

Additional sponsorship by the European Research Office, United States Army.

THE FORMATION OF TEXTILE STRUCTURES

J.W.S. Hearle

Department of Textile Technology, University of
Manchester Institute of Science and Technology,
Manchester, England.

ABSTRACT. The essential characteristics of fibre assemblies,
which combine flexibility with strength, are discussed. A
brief account of the methods of making yarns and fabrics is given,
together with a description of the structural forms and some
discussion of their fundamental representation.

1. INTRODUCTION

1.1 Characteristics of fibre assemblies

About ten thousand years ago, primitive man started to make
flexible fibre assemblies in order to protect himself from the
environment. By the time of the ancient civilisations,
sophisticated textiles were being made of wool, flax, cotton,
silk and other fibres. Apart from clothing and bedding, textiles
were also used for engineering purposes as ropes and cables,
filter media, reinforcement, and tented buildings. A high
expertise in craft developed in the manufacture of what are very
complex systems in the mechanics both of the internal response
of the fibres and the assemblies and of the external response of
fabricated structures. For a variety of reasons, discussed
elsewhere [1], the analysis of the mechanics lagged behind that
in other branches of engineering. The introduction in recent
years of a wide variety of man-made fibres and of more demanding
applications, together with changing social patterns, has now
made the study more important; and the development of computation
makes a more useful contribution possible. But in order to achieve
this, we must be able to measure mechanical properties, to
characterise behaviour, to describe structure, and to analyse

mechanical interactions - all in sensible and productive ways. Fortunately such pioneers as F.T. Peirce and Walter Hamburger established the subject during the second quarter of this century, and there is now a body of knowledge to cover in an advanced study institute. An account of earlier studies has been published by Hearle, Grosberg and Backer [2].

The main purpose of this introductory chapter is to review, for those who are not textile specialists, the practical ways in which fibres are assembled into useful materials by the textile industry.

Textiles are solid structural materials, but little of direct relevance to textile behaviour will be found in any standard text-book on the mechanics of materials. Table I lists some

TABLE I

IMPORTANT DISTINCTIONS

common engineering materials	*intermediate categories*	*textile materials*
RIGID	flexible	FLEXIBLE
HOMOGENEOUS	solid	DISCONTINUOUS
HARD	sheets	SOFT
IMPERVIOUS		POROUS
SMOOTH	soft, loose fillings	TEXTURED
STRONG	of no strength	STRONG

features which distinguish textile materials from those usually studied in engineering mechanics. In ordinary engineering the development of discontinuities, of porosity, of buckling, of long range displacement, of surface roughness, or of a soft elastic yielding under transverse pressure are often taken as signs of failure of the material, and the need for mechanical analysis ceases with the onset of these phenomena: but in textiles their manifestation signals the value of the materials, and the beginning of the region where the mechanical analysis is of most interest.

Textiles are able to combine flexibility and texture with strength because, following the invention of primitive man, they are made by assembling together long, fine fibres in ways which maintain and enhance the inherent fibre flexibility under small forces, but cause the structure to lock together and resist large forces

without failure. Both high quality fibres and complicated means
of assembly are needed in order to give high quality materials.
If cheaper wood-pulp fibres, assembled at high-speed from a
slurry, are used, then we get products, paper and board, which
are excellent for certain purposes, but do not display good
textile qualities. In addition to such obvious advantages/
disadvantages as easy tearing, easy creasing, low wet strength,
smooth surface, and dense packing, paper differs from textiles
in not being able to accept double curvature. It may be flexible
in uniaxial bending, but it resists bending in two directions
simultaneously: if forced to do so, paper will form point
singularities, rather than regions of smooth double curvature.
The fact that textiles will take up double curvature is what
allows them to drape gracefully, and to buckle comfortably as
the body moves.

In classical engineering materials, flexibility is achieved only
when the material becomes thin enough (or too thin!). In flexible
fibre assemblies, it is achieved, even in thick fabrics, by
allowing individual fibres within the structure to slide over one
another, without loss of integrity of the structure. The rule
that bending and twisting rigidity increase as the square of
cross-sectional area is beaten. This is one mechanical principle
of fibre utilisation in industrial products. But there is another
mechanical principle associated with fibres, which also has an
old history although its application has recently increased
enormously: this is the principle that extremely strong, stiff,
light-weight materials can be made by bonding fibres together in
a solid matrix. The mechanics of fibre-reinforced materials has
progressed rapidly in the last 20 years, although major problems
remain unsolved. Rigid fibre assemblies will be covered briefly
as a related material, in one chapter in this advanced study
institute, but the emphasis otherwise is on flexible assemblies.

1.2 Fibres

In the jargon of applied mechanics, fibres are fine slender rods -
microscopic in lateral dimensions and macroscopic in axial dimension.
The dimensions of a typical, though rather fine and short, fibre
are conveniently remembered though the three "ones" listed in
Table II, which also includes some derived dimensions. The
complexity of the total system is then indicated in Table III.
The job of the textile industry is to put these millions of fibres
in the right place and the right configuration. If we say,
reasonably, that a full mathematical analysis would have to be
based on short fibre elements with lengths comparable to their
diameter, namely about 10μm, then we have 10^{11} fibre elements in
the typical fabric specimen. The job of the theoretician is to
find ways of handling the mechanics of such large systems.

TABLE II

DIMENSIONS OF A "TYPICAL" FIBRE

fineness	1 decitex (0.1g/km)	up to ~ 20 dtex (then bristles)
length	1 cm	up to ~ 20 cm (then cont. fil.)
density	1g/cm^3	0.9 to 1.5
- - - - - -	- - - - - - - - - - - -	- - - - - - - - - - - -
mass	1 µg	
number of fibres in 1 kg	1000 million	(> 1 million in any product)
diameter	~ 10 µm	
length/width	1000:1	and higher

TABLE III

FIBRES IN FABRICS

fibre	1 m^2 of fabrics
1 µg	100 g
	10^8 fibres

1.3 Textile operations

Most textile fabrics are flexible, planar sheets, and the traditional manufacturing process is shown in the centre line of Table IV. One-dimensional fibres are taken from a three-dimensional array and assembled into one-dimensional yarns, capable of being interlaced into two-dimensional fabrics. The fabrics are then cut and sewn into garments or other final products. Variants of this sequence consist of the use of parallel continuous filaments as raw material; the manufacture of nonwoven fabrics from fibre webs; the assembly of yarns into cords and ropes, utilised as one-dimensional final products; the production of non-planar shaped sheets by knitting or braiding; and, to a very small extent, the production of three dimensional structures.

1.4 Quantities and units

Partly for historical reasons and partly because it is almost always more useful to characterise fibres and fibre assemblies by weight (strictly by mass) rather than by linear dimensions, the quantities and units used in textile technology often differ from those in general engineering mechanics.

The basic characterisation of a fabric will be in terms of *(a)* its fibre content *(b)* its construction (e.g. plain weave) *(c)* its "weight", strictly mass per unit area, expressed in g/m^2 or oz/sq. yd. Yarns are characterised by *(a)* fibre content *(b)* yarn type *(c) either* the yarn count, which is the length of unit mass,

TABLE IV

TEXTILE MANUFACTURING SEQUENCES

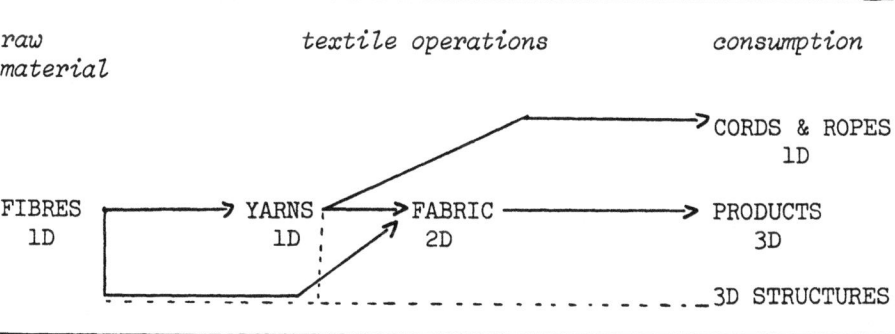

expressed in a variety of traditional units, such as the number of 840 yard hanks per lb (cotton count) 560 yd/lb (worsted count), km/kg (metric number) etc., *or*, preferably, by the linear density, namely mass per unit length, with the strict SI unit of kg/m, the recommended subsidiary SI unit of g/km with the name tex, and the historical unit of g/9000m with the name denier. Fibre "fineness" would be expressed in tex (or the widely used submultiple decitex) or denier. Some important textile units are listed in the Appendix.

Turning to mechanics, we note that instead of dividing force by area of cross-section to obtain stress, the usual practice is to divide by linear density to obtain specific stress* (and modulus, strength etc). There is a dimensional identity of:

force/linear density - specific stress

stress/density - as often used by engineers, concerned with material weight

energy/mass

velocity^2 - wave velocity = (specific modulus)$^{\frac{1}{2}}$

length x *(force/mass)* -"breaking length" units

Unfortunately the choice between these different forms, combined with choices between c g s, m k s and Imperial units, gravitational and inertial measures of force, and various traditional textile units has led to a terrible diversity. A conversion chart is included in the Appendix.

In theoretical analysis (or computation) it is best to formulate the problem in any consistent set of units, and then at the end to provide the conversion from a given consistent set, such as strict SI, to the required practical units. In many instances, particularly when energy methods are used, it is most convenient to follow chemical rather than applied mechanics practice and to work in terms of mass elements and energy per unit mass (specific stress) rather than volume elements and energy per unit volume (stress). Where geometry forces it, we use the relation:

$$\text{force} = \frac{\text{specific stress x area}}{\text{specific volume}} \text{ or specific stress x area x density}$$

---

*There is an American usage which terms this quantity the tenacity, but the British usage is to define tenacity as specific stress at break (i.e. a measure of strength).

Specific stress is naturally defined for one-dimensional elements, or, through the above relation, for solid materials. For two-dimensional sheet materials, the "natural" quantity to use in many applications is the force per unit width. A useful relation, normalizing fabric properties and bringing them into relation with fibre properties, is then:

$$\text{specific stress} = \frac{\text{force/width}}{\text{mass/area}}$$

In particular:

$$\text{specific stress in N/tex} = \frac{\text{force/width in N/mm}}{\text{fabric "weight" in g/m}^2}$$

2. FIBRES

2.1 Raw materials for the industry

The traditional raw materials of the textile industry were fibres or filaments found in nature and then cultivated through agriculture. In this century, man-made fibres, either regenerated from natural products or synthetised from simple chemicals, have been introduced, with such success that polyester now challenges cotton as the cheap general-purpose fibre.

For general textile purposes, an extensibility in the range of 10 to 50% is required. For this and other reasons, all the widely used commercial fibres are partially oriented, partially crystalline, linear polymers of six chemical types: cellulose or cellulose derivatives (cotton, rayon, flax etc.), keratin (wool), polyamide (nylon 6 and 66), polyester, acrylic, and polyolefin (polypropylene and polyethylene). Some other polymers are used to a limited extent, usually in high-price fibres with special properties. For some industrial purposes, high-strength fibres with low extensibility are required (for example: glass, carbon Kevlar), and for other purposes elastomeric fibres with high extensibility are used. The peculiar rheological properties of fibres, and the influence of temperature and moisture, are reviewed in the next chapter.

The ideal fibre (in mathematical and not utilitarian terms) would be a uniform circular cylinder, internally homogeneous and continuous. Real fibres have a variety of forms of cross-sectional shape, vary along their length, may contain internal voids, usually possess some heterogeneity of structure, and may be deliberately multi-component, either sheath-core or side-by-side. Furthermore the fibres supplied to the textile industry take a variety of forms.

2.2 Staple fibres

Natural textile fibres (except silk) are short in length (about 1 cm to 20 cm) and are supplied baled in an irregular three-dimensional array. Any of the man-made fibres may be cut to length and similarly supplied.

2.3 Continuous filament yarns

In man-made fibre production, the required number of filaments (from one to a few thousand) can be extruded together from a spinnerette and wound up as a single continuous monofilament or multifilament yarn. Except for accidental breakges there will be no free ends, and the filament length is as long as the package being produced. The only traditional yarns, which were effectively continuous filament, were made by reeling together individual silk filaments, each about 1 km in length.

2.4 Continuous filament tow

An alternative form of supply for man-made staple fibre yarns is tow, consisting of hundreds of thousands of filaments extruded together. This is supplied boxed in a long continuous-filament rope-like form, which can then be converted to a short staple fibre assembly by breaking or cutting, without destroying the parallel arrangement.

2.5 Film

For some purposes, it is more economical to extrude polymer film instead of fine fibres. This film can then be cut into narrow ribbons, which may be regarded as monofilament yarns with a highly asymmetric cross-section, and woven into fabric. It is also possible to break up the split film by fibrillation into interconnected elements, and so give a more typical fibrous texture.

2.6 Fibrous networks

There are other ways in which fibrous networks can be produced. For example the polyethylene spunbonded fabric, Tyvek, is directly produced as a sheet material with a structure which is a continuous net of fibrous elements.

3. YARNS

3.1 Continuous filament yarns

Table V gives a classification of yarn types. No more need be said about the simplest forms of monofilament and multifilament yarns, since yarn manufacturing is then the same as fibre manufacturing. However a parallel multifilament assembly has no cohesion or real identity: the filaments will easily spread away from one another, and can be split into groups and recombined in other ways.

The traditional method of giving coherence to a continuous filament yarn was to twist it, typically by about $\frac{1}{2}$ turn per inch. Each yarn then holds together as a separate bundle. For special purposes, notably to avoid fatigue in tyres, high-twist continuous filament yarns are produced.

TABLE V

TYPES OF YARN

monofilament	
parallel multifilament	[staple roving]
coherent multifilament	spun yarn (twisted etc)

one component	*or*	blended
single	*or*	plied, corded
simple	*or*	textured
simple	*or*	fancy

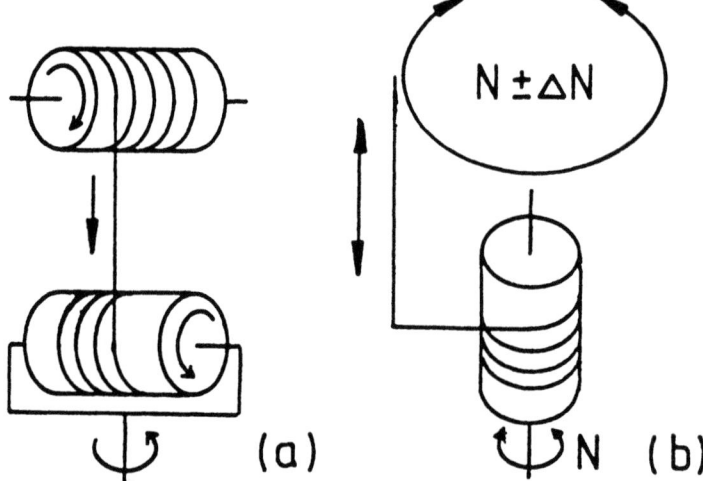

Figure 1 (a) Basic principle of yarn twisting method
(b) Schematic view of usual arrangement.

The basic method of continuous-filament yarn twisting is illustrated in figure 1(a) with the axes of rotation for twist insertion and axial motion perpendicular. In practice they are usually coaxial, as in figure 1(b), with a high rate of yarn rotation N causing twisting and the different rate of package rotation $(N-\Delta N)$ allowing axial motion $R\Delta N$. In uptwisting the twisting rotation is at the supply package; and in downtwisting at the take-up package.

Twisting is expensive, and gives a limit in possible package size by the requirement to rotate the whole package at high speed. Coherence is now more commonly achieved by interlacing. The yarn is passed through a turbulent air-jet which causes the fibre paths to cross over one another. In topological terms, this is a false interlacing (or false irregular braiding) since all the crossings must be paired as in figure 2. However, with a sufficient density of interlacing the yarn does not disentangle: if a pin is pushed through the yarn, and then moved along, it is eventually stopped when the filaments form a "knot".

3.2 Staple fibre yarns (or "spun" yarns).

The early stages of yarn manufacturing from short fibres consist of processes intended to produce relatively thick strands of more-or-less parallel fibres, known as slivers or rovings*. These

---

*Note a confusion in terminology. Glass rovings are thick bundles of parallel continuous filaments.

Figure 2. A single interlacing pair of crossings. An interlaced yarn consists of a large collection of such pairs.

have no strength since the fibres slide past one another under tension: in other words, they can be drafted into finer strands. The final stage of manufacture then gives coherence and strength to the assembly. Most yarns are held together by frictional forces between the fibres, and figure 3 illustrates three ways - twisting, interlacing, and wrapping - in which the necessary normal forces can be generated. In varying degrees, these features, or combinations of them, are found in commercial yarns.

The commonest method is twisting, and it was the traditional spinning whorl that gave the name to the whole operation. In ring-twisting, figure 4, the fine strand is fed through nip rolls and then passes down to a guide (the "traveller") which is rotating at high speed on a ring and so inserting twist. The traveller is, in fact, dragged round by the high-speed rotation of the take-up package, thus generating the wind-up tension and leading to the slight difference in rotational speed as the yarn

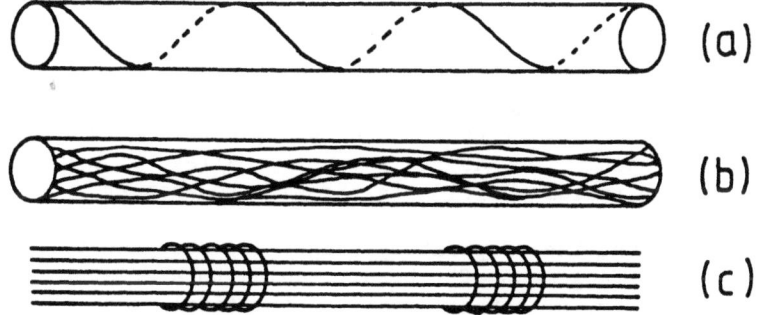

Figure 3. Three modes of generating transverse forces to hold staple yarns together:(a) twisting; (b) interlacing; (c) wrapping.

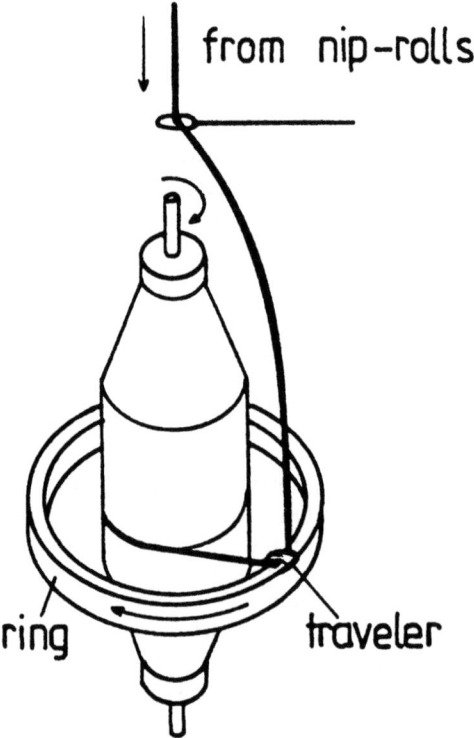

Figure 4. Ring twisting.

fed by the nip rolls is wound on to the package. Subject to the irregularities of the supply strand, the yarn is twisted into a fairly regular geometry. To a first approximation, each elementary cylindrical bundle of fibres can be regarded as twisted about the yarn axis. Such a structure would be of little use since fibres on the outside could be peeled off the yarn surface; and under tension, they would not be gripped, and so could not grip the underlying fibres and generate a self-locking structure. Fortunately, for several reasons, notably tension differences between fibres following the straight path at the core and those following the longer helical path on the surface, there is a measure of interlacing - or, as it is usually called, of migration of fibres from one radial position in the yarn to another.

In woollen spun yarn, the preparatory stages are designed to give a more irregular arrangement of fibres, leading to a greater degree of interlacing, or entanglement, in the final yarn, thus allowing lower twists to be used.

Ring twisting is limited by the need to rotate the whole yarn package, and, in recent years, various forms of open-end spinning

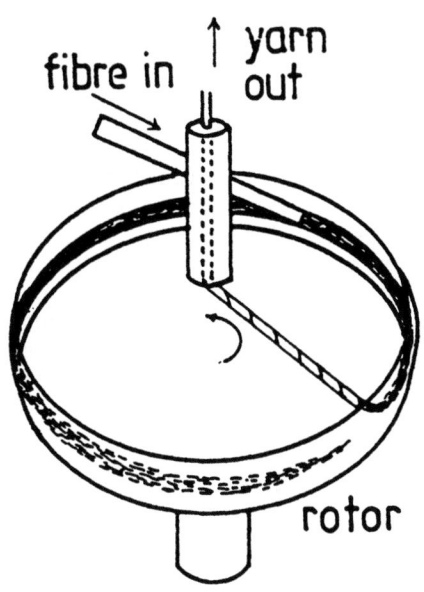

Figure 5. Rotor spinning.

have been introduced. Here the fibres are twisted into the yarn at a free growing end. The commonest method is rotor spinning illustrated in figure 5. The fibres are fed into a circumferential groove in a drum rotating at high speed. The free end of the yarn, which is being pulled away to the wind-up package, continually collects the fibres from the groove, and these are twisted in by the rotation of the system. Open-end yarns have a less simple twist structure, and some fibre wrapping also occurs.

Wrapping was tried as the dominant mechanism in fasciated yarns, but these are no longer produced. There are some new yarn types involving the wrapping of staple fibre strands with fine continuous filament yarns.

An as alternative to frictional coherence, the fibres in low-twist yarns can be stuck together by an adhesive binder. This will destroy the advantages of fibre flexibility, by causing the yarn to act as a more-or-less solid rod, but will give the strength needed to weave fabric. The adhesive can then be removed, since the fabric interlacing will give rise to the necessary normal forces.

3.3 Textured yarns

Staple fibre yarns, because of the stochastic arrangement of individual fibres, of irregularities along the length, and of the hairy surface of projecting fibre ends inevitably give a characteristically rough and bulky texture to fabrics. On the other hand, regularly packaged continuous filaments give dense, smooth, lustrous, silky fabrics – ideal for some purposes but not for many other uses. Consequently a yarn texturing industry has arisen in recent years, intended to give in varying degrees bulk, stretch and texture to continuous filament yarns. To some extent the same principles may be used to intensify effects in staple fibre yarns. The first method described below is purely mechanical and may be applied to any continuous filament yarns; the others depend on particular thermal responses, either shrinkage or heat-setting.

Air-jet texturing. If a yarn is overfed into a turbulent air-jet, the excess length is forced into projecting loops which are trapped in an interlaced yarn core. This provides bulk and texture on the yarn surface.

Differential shrinkage between fibres. If a blend of shrinking and non-shrinking fibres is formed into a yarn, then activation of shrinkage leads to a reduction of yarn length, and the non-shrink fibres can only respond by buckling into bulky surface layers on the yarn.

Differential shrinkage within fibres. If one constituent of a side-by-side bicomponent fibre wants to shrink, on some activating treatment, more than the other constituent, the fibre will curve like a heated bimetallic strip. If completely free, a short length would bend into the arc of a circle; but a long length, with some restriction on axial contraction, would form a helix. However, a straight strand can only go into bent coils if the ends are allowed to rotate relative to one another. When this is not possible, as in a continuous filament the only way of bending is to form alternating right-handed (Z) and left-handed (S) helices with rotation at the reversals linking the helical sections. Yarns of bicomponent fibres, buckling into alternating helices, have high bulk at a moderate degree of stretch.

Set forms. There are a variety of ways of forcing filaments into contorted forms and then setting these by heat. The most important are stuffer-box methods, forcing the filaments into a

restricted space, and jet-screen texturing, disturbing the
filaments by turbulence and then blasting the yarn on to a screen;
but there are also gear-crimping, knit-set-deknit, and other
methods.

Unbalanced forces. An alternative procedure, is to generate
stress in the filament in such a way that it cannot be directly
removed, but which can be reduced by some other form of buckling.
A simple example is a twisted monofilament, which will relieve
torque by snarling in one of the ways show in figure 6: the usual,
normal snarls would give bulk and texture, and their removal under
tension would provide stretch. Twisting of a multifilament yarn
will not produce the same result (though it can be used to give
crepe effects in fabrics) because the filaments will all be twisted
together into a bundle and will not be free to snarl individually.
Instead it is necessary to twist, heat-set, and then untwist:
this gives a twist-free bundle of twist-lively filaments. Furthermore,
the whole process can be carried out continuously by passing a
yarn under controlled tension through a heater and on to a
rotating spindle or friction twister: this so-called false-twisting
generates twist upstream of the spindle on the heater, but the
twist is lost on leaving the spindle downstream. It is, of course,
impossible to insert real twist continuously in the final yarn
without rotating one package relative to the other.

There is a further difference from the twisted monofilament in
that the filaments are set in a helical path in the yarn. We are
therefore concerned with an untwisted helix (actually a helix of

Figure 6. Two forms of snarling which reduce torsional energy in
twisted monofilaments: (a) normal snarl; (b) cylindrical
snarl.

varying radius, due to the filament migration between yarn core and yarn surface). If an untwisted helix is pulled straight and then allowed to contract, it first of all buckles into helical sections of alternating sense and then collapses into tight snarls. In contrast to the bicomponent fibre, the helices are different in their parameters as well as well as in sense. Yarns made by this method on a single-heater machine are torque-stretch yarns with an elongation from the fully contracted state of several hundred per cent.

Double-heater machines are used to give a second heating of the yarn under a small contraction (about 15%) from the fully extended state. This stabilises the alternating helical crimp, giving a yarn with high bulk but low stretch.

3.4 Other yarns

Any of the single yarns can be twisted together to make multi-ply yarns. Usually twisting is in the reverse direction in order to give a balanced-torque yarn. For reasons of packing, preference is given to two-ply, three-ply and seven-ply yarns.

There is one special case of two-ply yarns which is of particular mechanical interest. If two twisted yarns are brought together into frictional contact, then as they untwist, they will roll round one another to give a plied structure. This is exploited in self-twist yarns, where a pair of single yarns are alternately twisted in opposite directions and then brought together, so that they form a two-ply yarn with sections of alternating twist. This does, surprisingly, give a stable yarn structure – essentially because it is an energy minimum under the constraint that slippage at the contact points is not possible.

Another variant that is always possible is to make yarns of more than one fibre type, either intimately blended or segregated into zones. This category may be regarded as including core-spun yarns, with a continuous filament yarn at the centre and a sheath of staple fibres, or other such composite structures. One example, the Bobtex yarn, is produced in a very fast process by feeding a continuous filament yarn through an extrusion of molten polymer, and then causing a layer of staple fibres to be stuck on to the yarn surface.

Finally the irregularity, which is often a nuisance and a source of weakness, may be exploited in the production of fancy yarns for decorative purposes.

4. FABRICS FROM INTERLACED YARNS

4.1 Weaving

The commonest form of fabric manufacturing is weaving, in which one set of yarns - the warp - passes lengthwise through the loom, while another set - the weft or filling - is sent across to interlace with the warp. In principle, the weft can be a single length of yarn, which unwinds from a package carried by a shuttle, which is being banged backwards and forwards across the loom. This gives a loop of yarn at the selvedge: a reversal in direction of a continuous strand. In modern shuttleless looms, separate pieces of weft yarn are carried across the loom, and this leaves free ends at the selvedge, either sticking out in a fringe or tucked in by another mechanism. After each passage, the weft yarn is beaten up into the body of the cloth, and the warp threads are raised or lowered into the right position for interlacing on the next passage.

The variants of woven fabric structure form a major study in themselves. A simple account is given by Taylor [3], and fuller treatments by Grosicki [4] and Robinson and Marks [5]. In the brief comments here, we will first consider the situation where the warp and weft yarns are roughly similar in type and spacing: so that the same diagrams can be used to represent topology and geometry.

The simplest form is the plain weave in which crossings alternate between warp and weft on top, as illustrated in figure 7. If the warp and weft yarns are identical and equally spaced, the plain weave will have axes and planes of symmetry perpendicular to the plane of the fabric: in topological terms, the weave itself has this symmetry even when the yarn geometry is different. Figure 7 (c) shows the two types of repeat cells centred on axes of symmetry: it may be noted that they are composed of four sub-cells, which are identical except that one pair are mirror images of the other pair. There is also a central plane which splits the fabric thickness into two identical halves, although it is not a plane of symmetry since the two halves are displaced relative to one another.

More complicated weaves derive from other patterns of interlacing. For example in basket weaves, figure 8 (a), several neighbouring ends follow the same alternation: topologically, they may be regarded as plain weaves of multiple yarns. In other weaves some of the symmetry is lost. For example, in twill weaves, figure 8 (b), diagonal lines develop and the fabric is naturally skew without planes of symmetry; and in satin weaves, figure 8 (c), there is a marked difference between the face and back of the fabric.

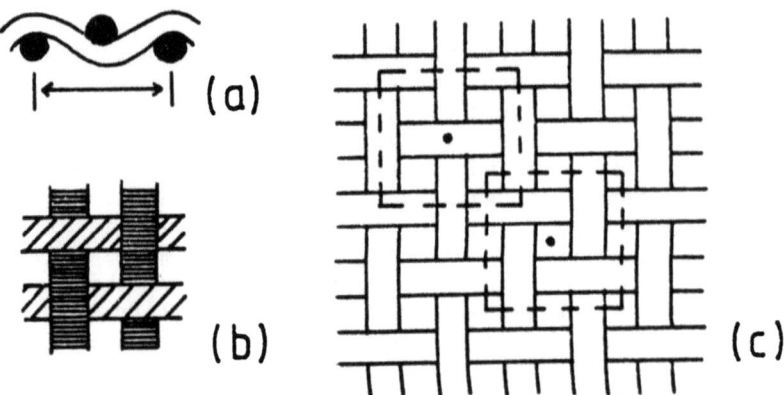

Figure 7. Plain weave. (a) (b) Pattern of interlacing. (c) Two types of repeat cells.

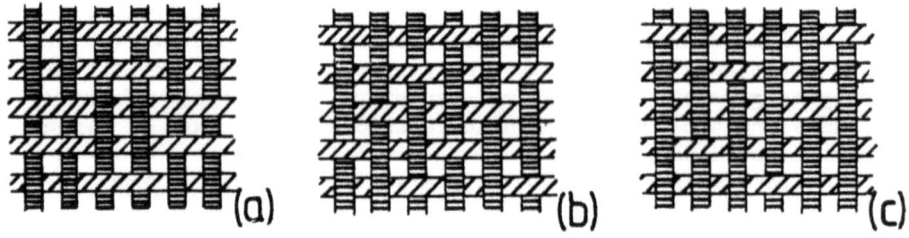

Figure 8. Three other weaves: (a) hopsack; (b) 2 x 1 twill; (c) satin.

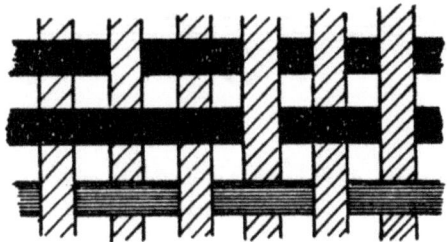

Figure 9. (a) Velvet weave, separating into two layers.
(b) Transformation showing the basic single layer weave topology.

Some weaves lead naturally to a successive displacement perpendicular to the fabric plane, and can be used to give ridged or cellular (honeycomb) effects in fabric. Other weaves separate into two or more layers linked by cross threads, as in figure 9: they can, in fact be regarded as two separate sets of warp and weft yarns forming the top and bottom fabrics, and then a third set interlacing with both. The resulting material may be used as a double (or higher multiple) fabric, or alternatively cut in two to give a velvet or other pile fabric.

Within a particular weave structure, major differences in fabric can be obtained by the use of different yarns, tensions, and spacings: the fabrics are toplogically identical but geometrically different. For example, figure 10 is a plain weave, but its geometry consists of: (a) dominant load-bearing warp A of strong yarns lying straight along the length of the fabric; (b) weft yarns B lying above and below the main warp; and (c) a light binder warp C holding the assembly together. Fabrics designed on this principle, though not necessarily with quite as simple a

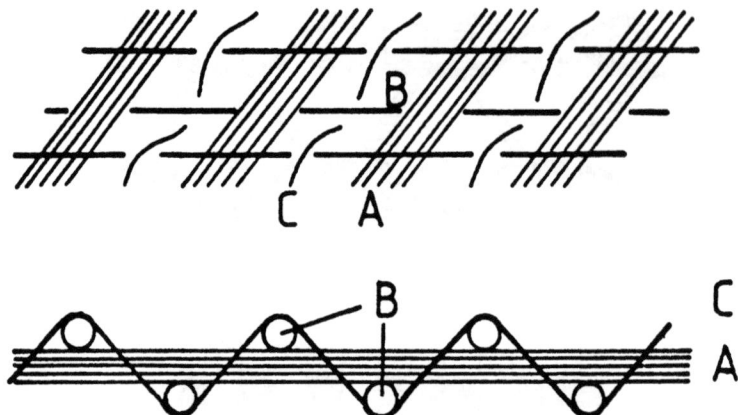

Figure 10. A plain weave arranged to give a strong warp A lying straight in the fabric, held in position by the weft B and secondary warp C.

weave, are used in heavy-duty conveyor belts. Other geometrical variants come from major differences in lengths between crossovers. For example, in a terry weave, figure 11, loops are forced out of the fabric.

4.2 Weave representation

Traditionally, woven fabrics are described by a series of common names: afgalaine, barathea, batiste, bedford cord, buckram etc. Furthermore this nomenclature mixes up differences in topology, geometry, fibre and yarn characteristics, fabric weight, and fabric use. A more rational representation is required.

We note from figures 7 and 8 that the weave is defined by whether the warp or the weft thread is on top at any particular crossover. A woven fabric is thus in mathematical terms a binary matrix. The plain weave is a repeat of the simple 2 x 2 matrix:

$$\begin{matrix} 0 & 1 \\ 1 & 0 \end{matrix}$$

Other weave repeats will be larger matrices. The number of possibilities becomes very large. At 1 mm thread spacing, a square metre corresponds to 2 raised to the power 10^6 different matrices, ranging from zero interlacing, with one set of threads

Figure 11. Terry weave with projecting yarn loops.

on top of the other, to the maximum interlacing of the plain weave. It is this diversity which enables complicated pictures or letters to be represented in woven fabrics. A scientific approach to the subject requires more work to classify the matrices into different types, to specify which generate new weaves and are not just multiples of simpler weaves, which are mere transpositions or rotations, which generate true interlacing, what are the relations to fabric characteristics such as skewness, float length, separation into layers and so on, and how the system may be most conveniently related to the needs of other calculations, design, and manufacturing operations, and handled economically in computing.

Traditionally the fabric geometry is defined by the thread spacing in warp and weft, and by the yarn crimp (which is the ratio of the actual yarn length to the straight line distance in the plane of the fabric). However, these quantities depend on the mechanical state of the fabric - pulling in one direction will reduce crimp one way, increase it the other way, and change the thread spacings. Nordhammer recommended the modular length, namely the actual yarn lengths between cross-overs, as a more rational representation: however even this varies when tensions are large enough to stretch yarns.

The quantity which is invariant, unless there is a chemical attack on the fibres or a slippage at crossover points, is the mass between crossovers (or in a repeat). This should be the starting point of a rational representation of geometry. Figure 12 indicates the sequence of factors to be considered in determining an actual geometry, starting from what may be termed the basic fabric,

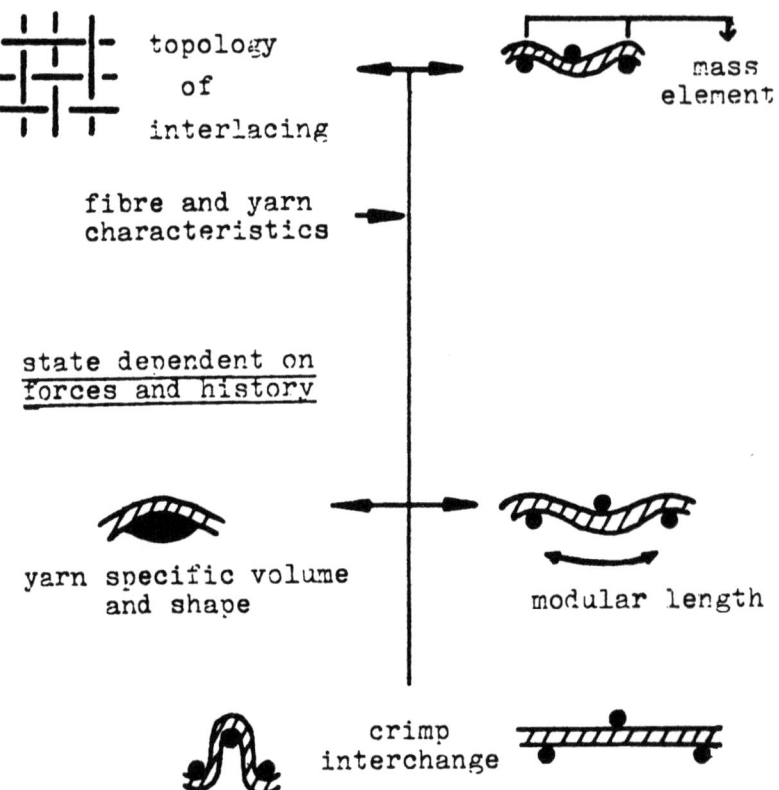

Figure 12. Sequence of factors defining woven fabric geometry.

defined by the invariant interlacing of mass elements of yarn, and going on to the state of the fabric, determined by its mechanical (or thermo-mechanical) history and the current applied forces.

4.3 Knitted fabrics

Woven fabrics are made of two sets of yarns, and free ends must be available in one set to interlace with the other set. In contrast to this, knitted fabrics are made from a single set of yarns, as in figure 13, which shows the simplest plain knit structure, and free ends are not required. Neighbouring yarns in the set are interlaced by forming knitted loops. The rows of stitches across the fabric are known as courses, and those along the fabric as wales. In practice, in weft knitting, figure 14

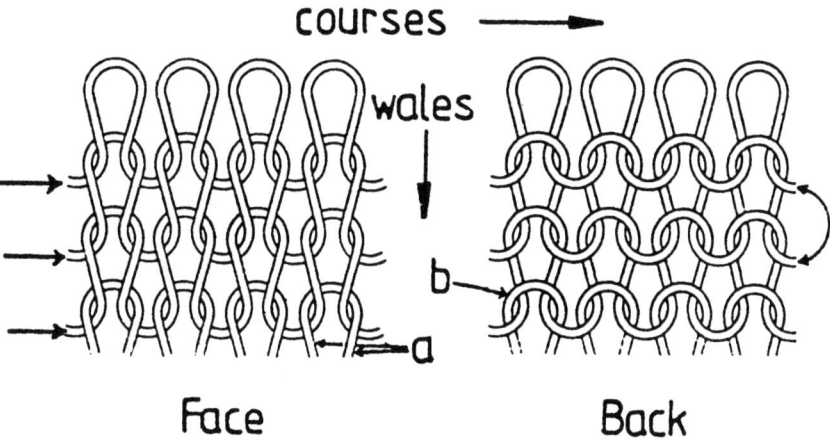

Figure 13. Plain knit.

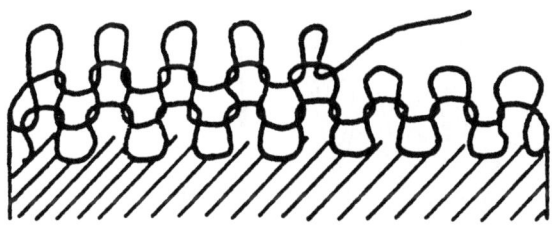

Figure 14. Weft knit fabric growing by addition of single end of yarn.

a single end of yarn is fed in and knitted by needles into the growing fabric along the line of the courses. The knitting yarn may go backwards and forwards in a flat fabric, or round-and-round in circular knitting, where multiple feeds may follow one another.

We note that the plain knit fabric, figure 13, is not symmetrical about the centre plane. The arms of the loops, a, are on the face of the fabric, and the tops of the loops, b, are on the back of the fabric. One consequence of this is that the minimum energy state of an ideal elastic plain knit would not be in a plane, but would be curled up. A variety of other knit structures, such as rib and purl, can be obtained by alternating the stitches between face and back. Other structural variations come by knitting yarns together, by alternating knitting and floating, by jumping courses, by transferring stitches diagonally, and so on.

In weft knitting, the choice of structures is constrained by the fact that the fabric is growing in a direction perpendicular to the courses. In warp knitting a large parallel set of yarns is fed from a beam, and these are then interlaced side-by-side at the needle bars. The yarns follow the wale direction along the fabric. Interlacing can be to right or left and a simple warp-knit construction is shown in figure 15. Variants in the details of interlacing and the extent of sideways displacement enable a wide range of warp-knit structures to be formed.

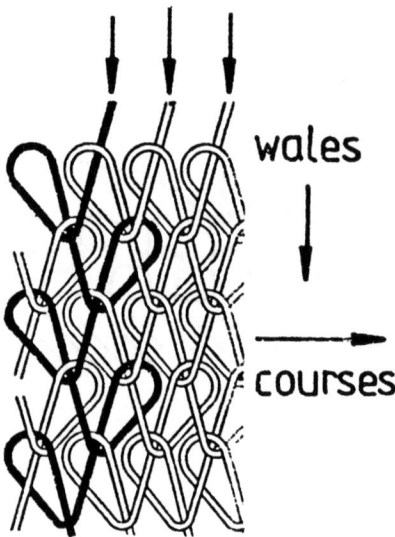

Figure 15. A simple warp-knit construction (half tricot).

Further variants in both warp and weft knitting can be obtained by laying in yarns so that they are trapped within the knitted structure. An extreme example of this would be Malimo fabrics where a set of weft yarns is laid across a set of warp yarns and then a third set is knitted round these to hold the fabric together.

Taylor [3] gives a simple account of knit structures and more detailed treatments are given by Gallemaert [6] and Paling [7].

4.4 Comparative features of interlacing

The simplest model of both plain weave and plain knit fabrics is a rectangular grid of yarns, as in figure 16(a). However they differ in the yarn path through the grid and in the form of interlacing, as shown in figures 16(b) and (c). This gives an interesting paradox. It is easier to make denser fabrics by weaving than by knitting, and so knitted fabrics are generally regarded as less stable and more easily distorted; but, for open structures with a given density of cross-overs, the knitted fabrics are more stable.

The curvature of yarn paths in woven fabrics arises only from out-of-plane displacement, but in knitted fabrics it arises from the in-plane yarn path. Figure 17(a) shows that in woven fabrics, the yarn curvature becomes proportional to t/s^2 where t is the yarn thickness and s is the thread spacing, and will thus tend rapidly to zero in open fabrics with large thread spacing and small yarn thickness. But in knitted fabrics, figure 17 (b), the curvature becomes proportional to $1/s$, merely changing in proportion to the scale of the fabric. In very open woven fabrics, figure 17 (c), the yarn paths become almost straight and there can

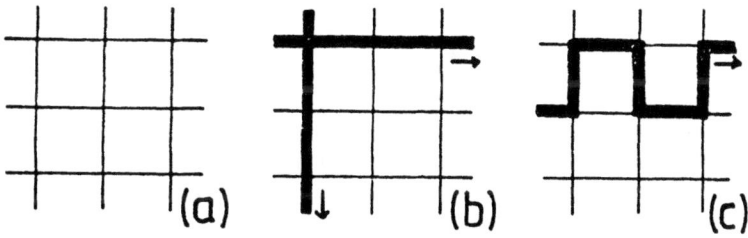

Figure 16. (a) Biaxial grid
(b) Path of warp & weft yarns in woven fabric
(c) Path of yarn in weft knit fabric.

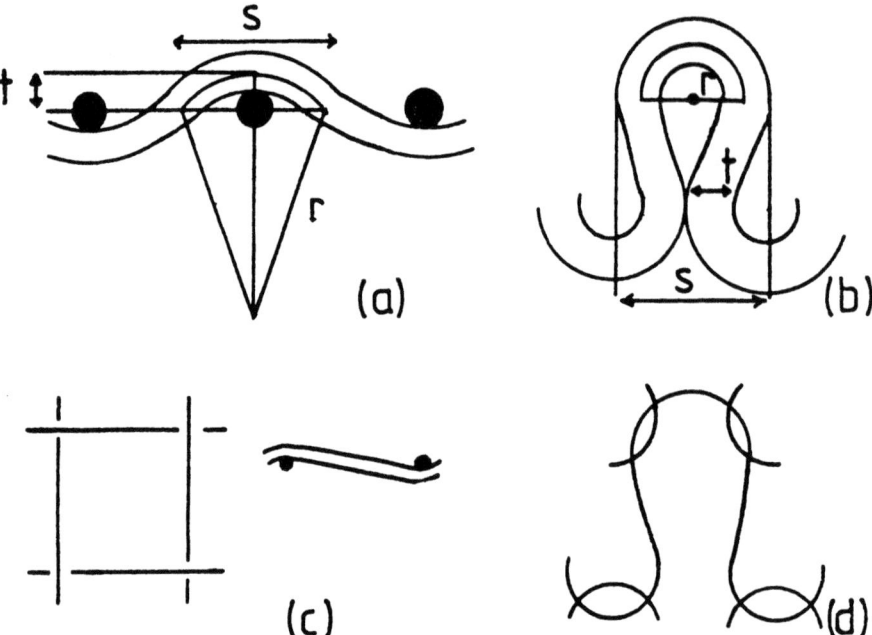

Figure 17. Curvature of yarn in fabrics. (a) In thickness of woven fabric. By Pythagoras' theorem, the radius of curvature r is given by:
$$r^2 = (s/2)^2 + (r-t)^2$$
$$= s^2/4 + r^2 - 2rt + t^2$$

Hence:
$$r = (s^2/4 + t^2)/2t \to s^2/8t \text{ if } s \gg t.$$

(b) In plane of knit fabric.
$$r = \tfrac{1}{2}(s-t) \to s/2 \text{ if } s \gg t.$$
(c) Open woven fabric.
(d) Open knit fabric.

be little extension due to crimp interchange, but in knitted fabrics, figure 17 (d) the constant geometric similarity gives a constant possibility of extension due to yarn straightening. The low curvature in open woven fabrics, means that the normal forces at crossovers will be small, and hence there will be little frictional resistance to slip; and the changed yarn path will cause little elastic resistance when the curvatures are very small. Both effects will be stronger with the curvature in the yarn paths in knitted fabrics.

However very open structures will inevitably be unstable. There are two ways of overcoming this. One is to tension the threads in woven fabric, as in tennis rackets. The other is to give a more complicated pattern of interlacing at the cross-overs of the grid. The woven interlacing is the simplest and least stable; the knitted involves rather more interaction; forms of interlacing used in lace-making involve more wraps of one yarn round another; and in nets a tight knot is formed at each cross over.

Another form of variation of yarn interlacing in fabrics is to depart from the basic biaxial grid of figure 16. Indeed in the warp knit structures and to a lesser extent in some weft knits, the grid is no longer recognisable. In leno weave, figure 18, some warp threads are moved sideways: the more complicated interlacing leads to stable fabrics with an open, cellular structure.

Another possibility is to make multiaxial fabrics, and there is an interest at present in triaxial fabrics where a higher degree of isotropy in properties is required. Figure 19 shows what would be a triaxial plain weave. We note that it can be modelled as triangular spaces with triple interlacing at each corner, or as a mixture of triangular and hexagonal spaces with simple interlacing at each cross over.

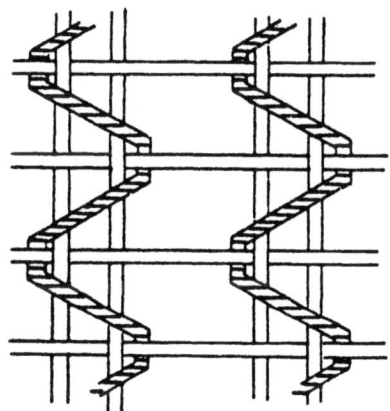

Figure 18. Leno weave.

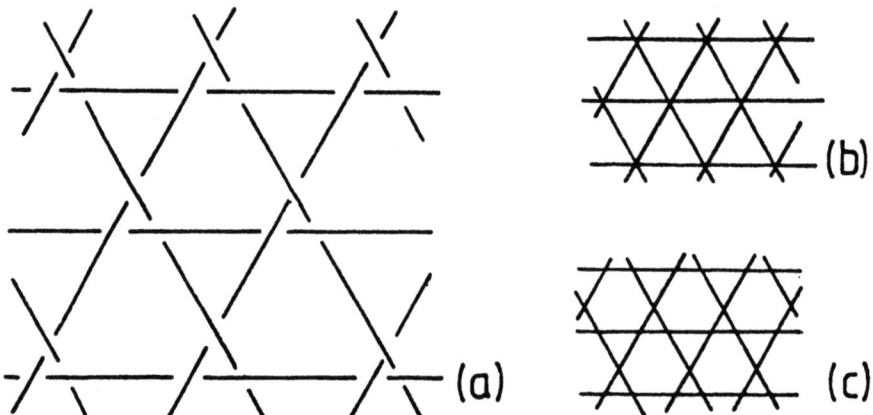

Figure 19. Triaxial weave: (a) plain weave; (b) extreme set of triangles when triple crossings reduce to points; (c) extreme with equally spaced crossings giving equilateral hexagons and triangles.

5. OTHER TOPICS

5.1 One-dimensional products

Most one-dimensional fibre assemblies are the intermediate product - yarns - which are interlaced into fabrics. But there is some utilisation of one-dimensional structures in threads, cords, ropes, cables, ribbons, and webbings, and in hoses and other tubes. In its simplest form, a rope might consist of a parallel collection of yarns or filaments; but special structures are needed in order to give coherence, to avoid damage from lateral forces, to give load-sharing and axial strength, to allow flexibility, and to free the system from torque or a tendency to untwist. One mode of assembly is by a hierarchy of twisting. For example, in a common rope construction, a collection of single or multi-ply yarns are twisted into strands, and then three or four of these strands are twisted together in the reverse direction to give the rope.

Another method is plaiting or braiding. This is a form of construction analagous to weaving in its simple interlacing by yarn crossovers; but the two sets of threads are following helices in opposite senses around and along the braid. Figure 20 (a) shows a typical braid; it is manufactured by pulling yarn off packages which are following tracks in opposite directions along circular interlacing paths, as in figure 20 (b). With a large number of supply packages, the braid will form a tube, which will jam tight with a hollow core; but when made of a few strands the braid collapses into a "solid"

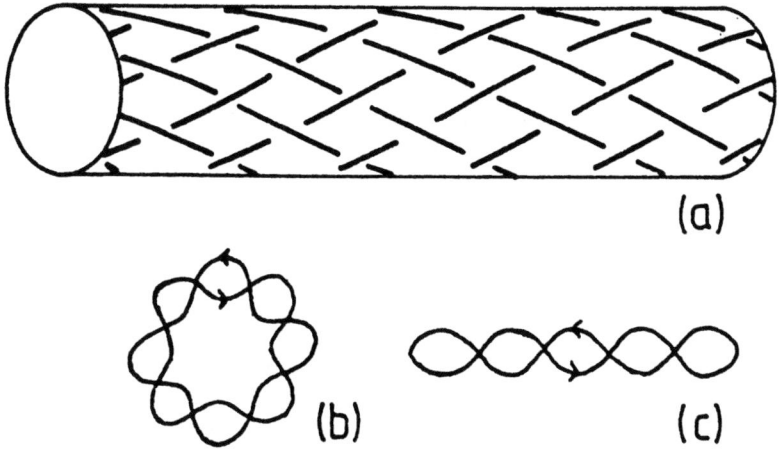

Figure 20. Braids: (a) typical circular braid; (b) interlacing pattern for circular braid; (c) interlacing pattern for flat braid.

structure. A typical square braided rope would consist of eight strands, interlacing in pairs, so that two pairs go round one way and two pairs in the other sense.

Other commercial ropes consist of a solid braid inside a hollow braid, or a parallel yarn core inside a hollow braid.

Tubes can be made by braiding, circular knitting, circular weaving, or flat weaving of two separate fabrics with the weft changing from one to the other in successive passages. Narrow ribbons and webbings can be made by weaving or knitting in the usual way. Alternatively flat braided webbings can be made by causing the supply packages to interlace as shown in figure 20 (c).

5.2 Fibres to fabrics

The advantage of interlacing yarns to form fabrics is that this is a well-controlled process with a definable geometry capable of specification by a fabric designer, in contrast to the stochastic arrangement of fibres within yarns. The disadvantage is the cost of the separate processes.

A cheaper route to fabric formation is to produce a stochastic arrangement of fibres in a sheet and then to cause this to cohere and achieve strength. One such route is paper-making. Short, cheap, wood-pulp fibres are dispersed in a slurry in water, which is then allowed to drain away, leaving a fibre web which can be dried. Hydrogen bonding between the cellulosic fibres, which have been beaten to give fine fibrils making close contact over large total areas, then causes the paper to hold together. However, such a tightly bonded sheet, while it has the right properties for the uses of paper, is not a good textile product.

There are a number of ways of making "nonwoven" fabrics from textile fibres.

The first operation is to form a web of fibres. With staple fibres, a traditional card web may be either parallel-laid in multiple layers, giving a highly anisotropic web, or cross-laid by folding backwards and forwards at an angle across the length of the material. Alternatively special aerodynamic machines may be used, giving so-called random-laid webs. Continuous filaments may be directly laid down on a moving belt following their extrusion. These webs will be loose assemblies of fibres.

Bonded-fibre (from short fibre) and spun-bonded (from continuous filament) fabrics are adhesively bonded, either by the application of a binder material or by self-bonding. The simplest form would be a more-or-less uniform web, more-or-less uniformly bonded, as illustrated in figure 21 (a). Other fabrics have holes blown in them to give a patterned web, or are printed-bonded to give a pattern of discontinuous bonding. Alternatively coherence may come from fibre entanglement. This is achieved in traditional wool and hair felts by agitation in the wet state, since these fibres have peculiar frictional properties, with a higher resistance in one direction than the other, and so they tend to move preferentially to interlace with one another. A universal method is needling, in which needles are punched through the fabric transferring some fibres from paths in the plane of the web to paths through the fabric, as shown in figure 21 (b) and (c). A more complicated rearrangement is achieved in spunlaced fabrics by fluid jets.

Finally, fabrics can be formed by using a limited amount of yarn to stitch or knit through the fibre web. These are known as stitch-bonded fabrics.

5.3 Fabric finishing

A simple-minded view of the situation would suggest that fabric structure and properties are determined by the fibre properties and the fibre arrangement resulting from yarn and fabric manufacturing. However many changes can be brought about by mechanical, thermo-mechanical, and chemical changes imposed after the fabric is formed. This is the province of the textile finishing industry.

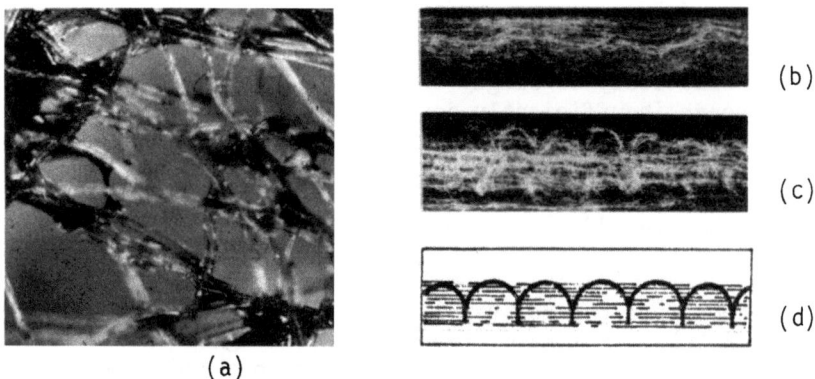

Figure 21. (a) Bonded fibre fabric
(b) Web before needling
(c) Needled fabric
(d) Vertical tufts through web.

It is not possible to go into detail here, other than to note that finishing treatments may alter fibre properties and dimensions, will certainly set fibres into new stress-free shapes, will alter interactions between fibres, may change fabric dimensions, and may alter fibre arrangements in the assemblies.

5.4 Fabrics to product

Although the textile industry may think of the fabric as the final product, it is, usually only an intermediate for the production of the consumer's requirement. Fabrics therefore have to be cut to shapes and joined together. There is much interesting geometry in cutting patterns, and, because of fabric anisotropy, this may also relate to the mechanical performance of products. Joining certainly involves mechanics in seam strength and other ways. In the traditional technology, there are many forms of overlap at the junctions, and there are many ways of stitching the seams together with sewing thread. Nowadays there is also an interest in other joining methods such as adhesive bonding or rivetting.

5.5 Three-dimensional fabrics

Most fabrics are made as flat sheets, or to a lesser extent as uniform circular tubes. However, it is possible in knitting, and some other forms of interlacing, to change the numbers of stitches and so shape the fabric into curved surfaces. Thus the heel of a sock can be knitted in the required shape (alternatively the planar form may be retained, but the fabric width varied). In another twist of the process, Milos Konopasek has had a continuous Mobius strip made by knitting. Other ways of making curved surfaces are to mould or shear flat fabrics into shape after manufacture. The study of singly or doubly curved sheets is an important, though insufficiently developed, part of fabric mechanics.

Finally, although the commercial interest is very small, it should be mentioned that in addition to one-dimensional strands, with diameters much less than their length, and two-dimensional fabrics, with thickness much less than their length and width, it is possible to construct assemblies with all three dimensions comparable. A bulky filling is one example, though it lacks coherence and strength; and so are the various packages used in manufacturing operations. More interesting would be multiaxial interlaced structures, where the axes are not coplanar. For example, there could be a multiplication in all three dimensions of the interlacing shown in figure 22 (a). It is interesting to note that, in contrast to two-dimensional interlacing and as hinted in figure 22 (b), such a structure would be coherent, even though all the strands were straight, and no single sheet of strands was interlaced with any other single sheet. Three-dimensional fibre assemblies could well be valuable in the manufacture of rigid composites, or as filter media; and there might be uses, even apart from the loose fillings, where a degree of easy deformability was utilised.

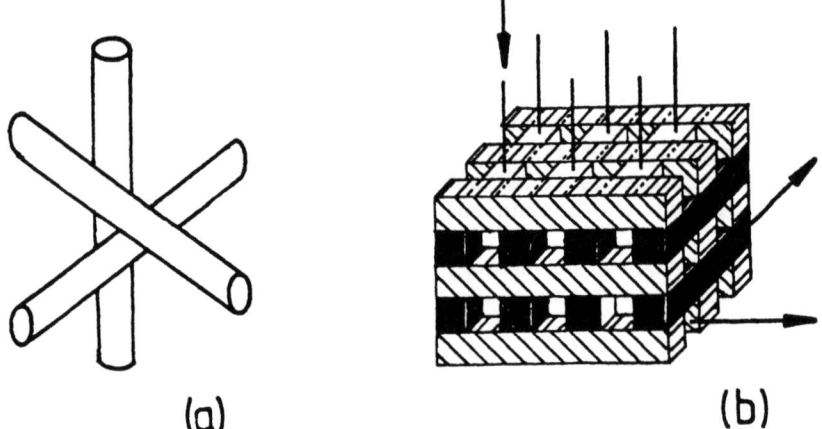

Figure 22. (a) Triple crossing in three dimensions
(b) Indication of simple interlacing in three dimensional structures.

REFERENCES

1. J.W.S. Hearle, P. Grosberg and S. Backer. Structural Mechanics of Fibres, Yarns and Fabrics, Volume II, Wiley-Interscience, in press.

2. J.W.S. Hearle, P. Grosberg and S. Backer. Structural Mechanics of Fibres, Yarns and Fabrics. Volume I. Wiley-Interscience, 1969.

3. M.A. Taylor. Technology of Textile Properties. Forbes Publication Ltd., 1972.

4. Z.J. Grosicki. Watson's Textile Design and Colour; Watson's Advanced Textile Design. Borough Green, 1975 and 1977.

5. A.T.C. Robinson and R. Marks. Woven Cloth Construction. Textile Institute, 1973.

6. L. Gallemaert. Initiation a la bonneterie. Editinos La Mallie. 1972.

7. D.F. Paling. Warp-knitting Technology. Columbine, 1965.

REVIEW OF THE MECHANICAL PROPERTIES OF FIBRES

C.P. Buckley

Department of Textile Technology, UMIST,
Manchester, U.K.

ABSTRACT. An introduction is given to the mechanical behaviour of fibres. The special problems encountered with these materials are outlined, together with the principles of methods available for dealing with them. Specific aspects covered are: effects of time and environment, anisotropy, and twisting and bending of fibres.

1. INTRODUCTION

A wide variety of materials occur naturally as fibres, or are manufactured in this form. Here we concentrate on those most widely used for flexible assemblies such as yarns and fabrics. They fall into three categories: natural fibres, man-made linear-polymer fibres, man-made nonpolymeric fibres. General information on the various fibre types is given in the books by Cook [1], Mathews [2] and Moncrieff [3].

For comparing their mechanical behaviour, consider the axial tensile test for a fibre of cross-section area A, in which the force F required to extend the fibre by a tensile strain ε is determined, at either constant loading-rate $\dot{F}$ or strain-rate $\dot{\varepsilon}$. Whereas in other applications the objective measure of force used is usually the nominal stress σ (=F/A), in the case of fibres it is often convenient to introduce the fibre linear density ρ_ℓ (related to A and the fibre density ρ by $\rho_\ell = \rho A$)* to express the

---

*Although the consistent SI unit for ρ_ℓ is kg/m, the accepted practical unit is tex (1 tex = 1g/km = 10^{-6} kg/m).

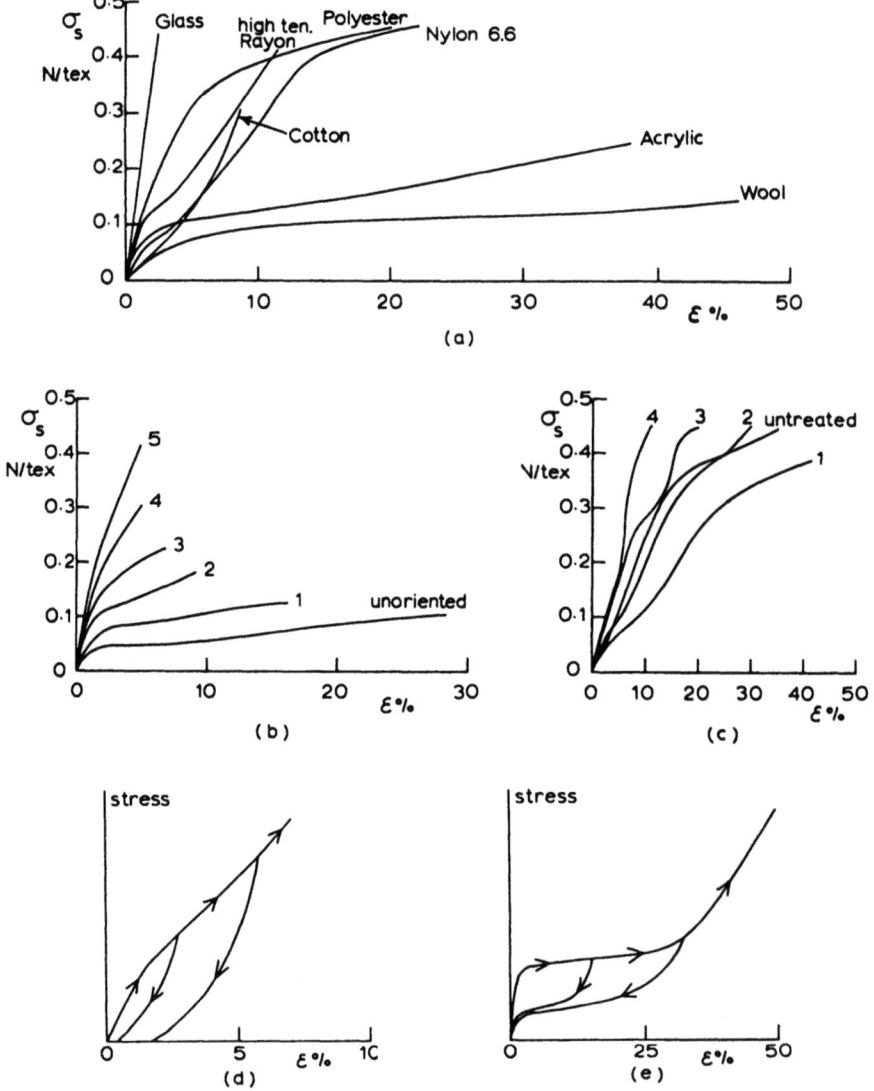

Fig. 1 Examples of fibre tensile specific stress-strain curves.
(a) Comparison of various fibre types, from Ford [4].
(b) Secondary cellulose acetate fibres with molecular orientation increasing from (1) to (5), from Work [5].
(c) Nylon 6.6 fibres heat-treated for 1.5s under various temperature/tension conditions (1) 200°C/0, (2) 200°C/0.0126 N/tex, (3) 200°C/0.0378 N/tex, (4) 200°C/0.0945 N/tex, from Hearle et al [6].
(d) Schematic loading/unloading paths for a polyester fibre.
(e) As (d) but for a wet wool fibre.

force as specific stress $\sigma_s = F/\rho_\ell$ (= σ/ρ). Curves such as those in Figures 1(a) - 1(c) are then obtained.

Useful parameters obtained from stress-strain curves are the "initial" axial tensile modulus E (= $d\sigma/d\varepsilon$), the stress at failure σ_f (= "strength" = "tenacity") and the strain at failure ε_f. The former two in specific units are E/ρ and σ_f/ρ. Some typical values, gathered from many sources, are collected in Table 1. Morton and Hearle [7] provide a general survey of the mechanical properties of fibres, while detailed data may be found in the handbooks by Ford [4] and Rae and Bruce [8]. Earlier work was reviewed by Meredith [9].

Figure 1(a) illustrates the wide range of mechanical behaviour encountered with fibres. Indeed even for a given fibre type its properties may vary widely, according to the degree of molecular orientation introduced during manufacture, and the effects of any subsequent thermomechanical treatment. Two examples are given in Figures 1(b) and (c). It is also clear from Figures 1(a) - 1(c) that for many fibre types we are dealing with awkward non-linear stress-strain relations, involving large strains. Moreover, the behaviour is often highly inelastic, as demonstrated by the hysteresis shown in Figures 1(d) and 1(e), where the loading sequence has been immediately followed by unloading. Data such as those given in Figures 1(a) - 1(c) and Table 1 must always be used with caution, therefore, since they depend greatly on the choice of load-time sequence used to obtain them. A similar caution applies also to environmental parameters such as temperature and relative humidity* (RH). Not only their current values but also their past history may have large effects on measured mechanical properties. The situation is further complicated by the facts that fibres are usually highly anisotropic, and are often subjected to inhomogeneous stress states such as twisting and bending. We consider each complication in turn.

2. EFFECTS OF TIME AND ENVIRONMENT

Most fibres including the natural fibres are polymeric, and polymers are notoriously time-dependent in their mechanical properties and hence are said to be "viscoelastic". This is clear, for example, in the creep (monotonically increasing strain) of a fibre under constant stress, or the stress relaxation that occurs in a fibre held at constant strain (see Figure 2). It is obvious that the stress-strain relationship is then not unique and is inadequate for describing mechanical response. When strains are sufficiently small, however, the behaviour is linear viscoelastic. There are then linear relations between stress and strain so that

---

*Data given in Figures 1(a) -1(c) and Table 1 refer to a standard atmosphere of 20°C, 65% R.H.

Fibre type	ρ kg/m^3 × 10^{-3}	E GN/m^2	10^{-6}E/ρ N/tex	σ_f GN/m^2	10$^{-6}\sigma_f/\rho$ N/tex	ε_f %
Wool	1.31	3.3	2.5	0.16	0.12	35
Cotton	1.35	6.8	5.0	0.55	0.40	7
Rayon (high ten)	1.51	11	7.0	0.45	0.30	15
Nylon 6.6 (h.t.)	1.14	4.6	4.0	0.68	0.60	17
Polyester (h.t.)	1.38	14	10.0	0.83	0.60	10
Graphite	2.00	400	200	2.0	1.0	0.5
Kevlar 49	1.54	150	97	2.4	1.6	2
E-glass	2.50	63	25	1.7	0.68	2
Steel wire	7.87	240	30	2.0	0.25	8

Table 1. Typical values of fibre mechanical properties.

at time t after loading:

$$\varepsilon(t,\sigma_o) = D(t)\,\sigma_o, \quad \sigma(t,\varepsilon_o) = E(t)\,\varepsilon_o \qquad (2.1)$$

for creep and stress relaxation respectively, where functions D(t) and E(t) are the "creep compliance" and "stress relaxation

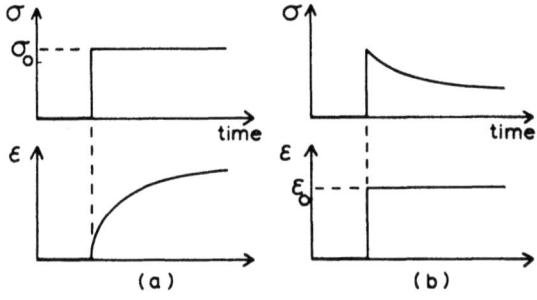

Fig. 2. Schematic diagram of (a) creep and (b) stress relaxation.

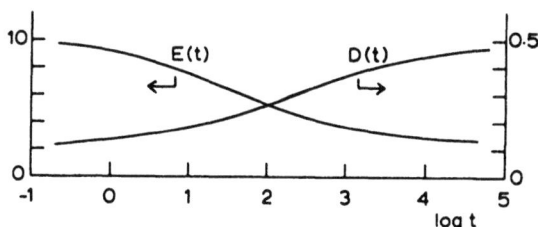

Fig. 3. Schematic diagram of response functions E(t) and D(t) versus logt for a fibre with $E_U = 10$, $E_R = 2$.

modulus" respectively. Moreover, the strain response to an arbitrary stress history $\sigma(u)$ is given by linear superposition of the separate responses to each part of the history. Thus at time t the strain may be expressed as the sum of the strains resulting from all increments of stress at times up to and including t:

$$\varepsilon(t) = \int_{u=-\infty}^{u=t} D(t-u)\, d\sigma(u) = \int_{-\infty}^{t} D(t-u)\, \frac{d\sigma}{du}\, du \qquad (2.2)$$

and by analogy the stress is expressed by

$$\sigma(t) = \int_{u=-\infty}^{u=t} E(t-u)\, d\varepsilon(u) = \int_{-\infty}^{t} E(t-u)\, \frac{d\varepsilon}{du}\, du. \qquad (2.3)$$

(2.2) and (2.3) are expressions of the Boltzmann Superposition Principle.

Let p be the Laplace transform operator and a bar placed above denote transform with respect to time. (2.2) and (2.3) are then compactly expressed as their Laplace transforms:

$$\bar{\varepsilon}(p) = p\bar{D}(p)\bar{\sigma}(p), \quad \bar{\sigma}(p) = p\bar{E}(p)\bar{\varepsilon}(p) \qquad (2.4)$$

and hence D(t) and E(t) are related through

$$\bar{D}(p)\bar{E}(p) = 1/p^2, \quad \text{or} \quad \int_0^t D(u)E(t-u)du = t. \qquad (2.5)$$

Viscoelastic behaviour is usually concentrated within a few "relaxation" regions of time-scale. Within any one relaxation region D(logt) and E(logt) have sigmoidal form, with the inflection in E occurring before that in D (see Figure 3). In the limits of short and long times the behaviour is linear elastic, with moduli $E = E_U = 1/D_U$, $E = E_R = 1/D_R$ respectively. This behaviour corresponds to recoverable or "primary" creep. An exception will arise, however, if nonrecoverable flow or "secondary" creep is superposed, when an extra term t/η must be added to $D(t)$ (η is a viscosity).

It is clear from Figures 2 and 3 that, qualitatively at least, $D(t)$ and $E(t)$ have approximately exponential forms

$$D(t) = D_U + (D_R - D_U)[1 - \exp(-t/\tau_\sigma)] \qquad (2.6)$$

$$E(t) = E_R + (E_U - E_R)\exp(-t/\tau_\varepsilon) \qquad (2.7)$$

where τ_σ and τ_ε are time constants governing creep and stress relaxation respectively. In order to satisfy (2.5) they must be related through

$$\tau_\varepsilon/\tau_\sigma = D_U/D_R = E_R/E_U. \qquad (2.8)$$

To fit experimental data quantitatively, however, it is necessary to generalise (2.6) and (2.7) by introducing continuous spectra of time constants: the retardation spectrum ϕ_σ and relaxation spectrum ϕ_ε

$$D(t) = D_U + (D_R - D_U) \int_{-\infty}^{\infty} \phi_\sigma(\ln \tau)[1 - \exp(-t/\tau)]\, d\ln \tau \qquad (2.9)$$

$$E(t) = E_R + (E_U - E_R) \int_{-\infty}^{\infty} \phi_\varepsilon(\ln \tau) \exp(-t/\tau)\, d\ln \tau \qquad (2.10)$$

where $\int_{-\infty}^{\infty} \phi_\sigma(\ln \tau)\, d\ln \tau = \int_{-\infty}^{\infty} \phi_\varepsilon(\ln \tau)\, d\ln \tau = 1.$ (2.11)

ϕ_σ and ϕ_ε are of course related through the need to satisfy (2.5).

The response functions $D(t)$ and $E(t)$ are usually strongly dependent on temperature. The major effect is to alter the timescale of response, i.e. to alter the spectra ϕ_σ, ϕ_ε. Frequently it is found to be a reasonable approximation that they simply shift along the $\ln \tau$ axis without change in shape. This means that each time constant at temperature T, say τ_T, is in the same ratio a_T to its value at temperature T_o:

$$\tau_T/\tau_{T_o} = a_T. \qquad (2.12)$$

In the case of glass transitions of amorphous materials a_T is given by the Williams-Landel-Ferry equation

$$a_T = \exp\left[-C_1(T-T_o)/(C_2 + T - T_o)\right] \qquad (2.13)$$

where C_1 and C_2 are material constants depending on the value chosen for the reference temperature T_o. For other relaxations the Arrhenius equation is found to be obeyed

$$a_T = \exp\left[\frac{\Delta H}{k}\left(\frac{1}{T} - \frac{1}{T_o}\right)\right] \qquad (2.14)$$

where ΔH is an activation energy and k is Boltzmann's constant. An ideal material for which this shift of $\phi_\sigma(\ln \tau)$ and $\phi_\varepsilon(\ln \tau)$ is the only effect of temperature on its viscoelastic behaviour is termed "thermorheologically simple". For such a material an interval of time Δt at temperature T is clearly entirely equivalent to an interval $\Delta t/a_T$ at temperature T_o. Comparing the creep and stress relaxation functions at temperature T, D_T and E_T, with those at temperature T_o we hence find they are related through

$$D_T(t) = D_{T_o}(t/a_T), \quad E_T(t) = E_{T_o}(t/a_T). \qquad (2.15)$$

The discussion above refers to isothermal situations. If however temperature is varying during deformation, the Superposition Principle still applies to a linear viscoelastic material but in the modified form

$$\left.\begin{aligned}\varepsilon(t) &= \int_{-\infty}^{t} D(t-u, u) \frac{d\sigma}{du} du \\ \sigma(t) &= \int_{-\infty}^{t} E(t-u, u) \frac{d\varepsilon}{du} du\end{aligned}\right\} \qquad (2.16)$$

where functions D and E now depend on the time of loading u, in addition to the elapsed time since loading $t-u$, and are functionals of the temperature-time sequence. For a thermorheologically simple material (2.16) can be reduced to

$$\varepsilon(t) = \int_{-\infty}^{t} D_{T_o}(\xi) \frac{d\sigma}{du} du, \quad \sigma(t) = \int_{-\infty}^{t} E_{T_o}(\xi) \frac{d\varepsilon}{du} du \qquad (2.17)$$

by introducing an "effective" elapsed time ξ given by

$$\xi = \int_{u}^{t} [a_T(t')]^{-1} dt' \qquad (2.18)$$

for any particular temperature-time sequence $T(t')$.

Changes of temperature may in addition cause structure

changes involving concurrent property changes. Moreover, these changes may take place gradually with time at constant temperature (e.g. chemical degradation, crystallisation, or compacting of an amorphous material near the glass transition). Such effects are sometimes known as "ageing". When ageing occurs over a time-scale comparable with that of the loading sequence, the Superposition Principle can only be applied in its modified form (2.16). In particular cases, however, when ageing involves merely a change in the time-scale of viscoelastic behaviour, it will be possible during ageing to invoke equations analogous to (2.17) and (2.18).

Some fibres are highly hydrophilic, and therefore sensitive to the relative humidity (RH) of their environment. Loosely, moisture acts as a plasticiser making the fibre soft. Raising RH, as with raising temperature, usually reduces the time-scale of response and again it may be possible to invoke relations analogous to (2.17) and (2.18). In other instances, however, D_U and D_R are also affected. Ageing effects are sometimes also seen following changes in RH.

At larger strains* viscoelastic behaviour becomes measurably non-linear and (2.1) - (2.5) are no longer valid, although they may give useful approximate predictions. With increasing strain the fibre is progressively damaged, and will show incomplete recovery on removal of the load. There is not yet any general theory of non-linear viscoelasticity which is also practically useful. An empirical approach must be used in any particular case.

Viscoelasticity of polymers is introduced in books by Ward [10] and Arridge [11], and is discussed in greater detail by Ferry [12] and McCrum, Read and Williams [13].

3. ANISOTROPY

The remarks given above apply also to modes of deformation other than axial tension. Almost all fibres, however, are anisotropic. To express this we recall that under general loading conditions stress and strain are second rank tensors σ_{ij}, ε_{ij}. A linear elastic material therefore obeys a generalised Hooke's law

$$\varepsilon_{ij} = s_{ijkl}\sigma_{kl}, \quad \sigma_{ij} = c_{ijkl}\varepsilon_{kl} \quad (i,j,k,l = 1,2,3) \quad (3.1)$$

where s_{ijkl}, c_{ijkl} are fourth rank tensors of compliance and stiffness respectively. The symmetries of σ_{ij} and ε_{ij} reduce the

---

* Typically beyond strains of 0.001 to 0.01.

maximum number of independent terms in s_{ijkl} and c_{ijkl} to 36.

Equations (3.1) are the most useful expression of Hooke's law in problems involving any rotation of axes, since one can take advantage of s_{ijkl}, c_{ijkl} transforming according to the rule for fourth rank tensors. In other circumstances, however, it is often convenient to use a contracted matrix notation (without tensorial character) in which (3.1) are replaced by

$$\underset{\sim}{\varepsilon} = \underset{\sim}{S}\,\underset{\sim}{\sigma}, \qquad \underset{\sim}{\sigma} = \underset{\sim}{C}\,\underset{\sim}{\varepsilon} \qquad (\underset{\sim}{S} = \underset{\sim}{C}^{-1}) \qquad (3.2)$$

where $\underset{\sim}{\varepsilon}$ and $\underset{\sim}{\sigma}$ are column matrices with transposes

$$\begin{aligned}\underset{\sim}{\varepsilon}^T &= [\varepsilon_1\ \varepsilon_2\ \varepsilon_3\ \varepsilon_4\ \varepsilon_5\ \varepsilon_6] = [\varepsilon_{11}\ \varepsilon_{22}\ \varepsilon_{33}\ 2\varepsilon_{23}\ 2\varepsilon_{13}\ 2\varepsilon_{12}] \\ \underset{\sim}{\sigma}^T &= [\sigma_1\ \sigma_2\ \sigma_3\ \sigma_4\ \sigma_5\ \sigma_6] = [\sigma_{11}\ \sigma_{22}\ \sigma_{33}\ \sigma_{23}\ \sigma_{13}\ \sigma_{12}]\end{aligned} \qquad (3.3)$$

and $\underset{\sim}{S}$ and $\underset{\sim}{C}$ are 6 x 6 matrices of compliance and stiffness. (Note the factors of 2 introduced to convert tensor shear strain in ε_{ij} to "engineering" shear strain in $\underset{\sim}{\varepsilon}$). Uniqueness of the strain energy function for an elastic material ensures that S and C are symmetric, and therefore have at most 21 independent elements.

This number is usually further reduced by the material itself having some symmetry. This will be at least the local symmetry of the growth process for a natural fibre, or the local symmetry of the extrusion and subsequent manufacturing process for a synthetic fibre. Apart from a few exceptions, the material of a fibre can usually be assumed to have transverse isotropy perpendicular to the fibre axis. The natural co-ordinate system to choose within a fibre is the cylindrical system (r, θ, z) centred on the fibre axis. With respect to this system $\underset{\sim}{S}$ and $\underset{\sim}{C}$ then reduce to

$$\begin{bmatrix} s_{11} & s_{12} & s_{13} & 0 & 0 & 0 \\ s_{12} & s_{11} & s_{13} & 0 & 0 & 0 \\ s_{13} & s_{13} & s_{33} & 0 & 0 & 0 \\ 0 & 0 & 0 & s_{44} & 0 & 0 \\ 0 & 0 & 0 & 0 & s_{44} & 0 \\ 0 & 0 & 0 & 0 & 0 & 2(s_{11}-s_{12}) \end{bmatrix} \begin{bmatrix} c_{11} & c_{12} & c_{13} & 0 & 0 & 0 \\ c_{12} & c_{11} & c_{13} & 0 & 0 & 0 \\ c_{13} & c_{13} & c_{33} & 0 & 0 & 0 \\ 0 & 0 & 0 & c_{44} & 0 & 0 \\ 0 & 0 & 0 & 0 & c_{44} & 0 \\ 0 & 0 & 0 & 0 & 0 & \tfrac{1}{2}(c_{11}-c_{12}) \end{bmatrix}$$

$$(3.4)$$

each with only 5 independent elements. The relation to the familiar axial tensile modulus E and shear modulus G is through

$$E = 1/s_{33}, \quad G = c_{44} = 1/s_{44}. \tag{3.5}$$

A point of frequent interest in the mechanics of fibre assemblies is the extent to which the fibres can be assumed incompressible. From (3.2) and (3.4), for a fibre with transverse isotropy, the dilation Δ is given by

$$\Delta = \sum_{q=1}^{3} \varepsilon_q = (s_{11} + s_{12} + s_{13})(\sigma_1 + \sigma_2) + (s_{33} + 2s_{13})\sigma_3 \tag{3.6}$$

and the compressibility B can be expressed as the sum of two parts

$$B = \frac{2}{3} B_1 + \frac{1}{3} B_2 \tag{3.7}$$

where

$$B_1 = 3(s_{11} + s_{12} + s_{13}), \quad B_2 = 3(s_{33} + 2s_{13}) \tag{3.8}$$

Some typical values of s_{pq}, together with B_1 and B_2 are collected in Table 2, where time effects are ignored.

For an anisotropic linear viscoelastic fibre (2.2) and (2.3) should be generalised to

$$\underline{\varepsilon}(t) = \int_{-\infty}^{t} \underline{S}(t-u) \frac{d\underline{\sigma}}{du} du, \quad \underline{\sigma}(t) = \int_{-\infty}^{t} \underline{C}(t-u) \frac{d\underline{\varepsilon}}{du} du. \tag{3.9}$$

In this case there is no rigorous proof that $\underline{S}$ and $\underline{C}$ are symmetric, although this is frequently assumed in practice.

For further discussion of anisotropic polymers the reader should consult Ward [10] and Arridge [11]. In the following section we shall make the assumption of transverse isotropy, so that $\underline{S}$ and $\underline{C}$ are as given in (3.4).

4. GEOMETRICAL FACTORS

In practical applications it often happens that fibres are subjected to twisting and bending. In both cases the strain field is inhomogeneous, and the fibre geometry strongly affects its overall behaviour.

Fibre type	s_{11}	s_{12}	s_{13}	s_{33}	s_{44}	B_1	B_2
Polyethylene	24	-12	-5.1	11	34	21	2.4
Polypropylene	19	-13	-2.8	6.7	18	9.6	3.3
Polyester	16	-5.8	-0.31	0.71	14	29.7	0.27
Nylon	7.3	-1.9	-1.1	2.4	15	12.9	0.60

Table 2. Compliances of some man-made polymeric fibres quoted by Ward [10]. Unit of compliance: 10^{-1} m^2/GN.

Consider first the case of a straight fibre which is given a uniform angular twist per unit length of ψ about the z-axis, by a torque applied at its ends. ψ may vary with time. This problem has no general analytical solution applicable to any cross-sectional shape. It is treated in texts on elasticity; see for example Timoshenko and Goodier [14]. What is true quite generally, however, is that a particle at (r, θ, z) is displaced circumferentially by $r\psi(z + c)$ due to rotation of the cross-section, and longitudinally by $\psi\omega(r, \theta)$ due to warping of the cross-section, where c is a constant and ω a function of r and θ only. The strains arising at time t are then

$$\varepsilon_4(t) = (r + \frac{1}{r} \frac{\partial \omega}{\partial \theta}) \psi(t)$$

$$\varepsilon_5(t) = \frac{\partial \omega}{\partial r} \psi(t).$$

(4.1)

The axial torque Q is generated by the shear stress σ_4, from

$$Q(t) = \iint \sigma_4(t,r) \, r^2 \, dr d\theta \qquad (4.2)$$

where the double integral is taken over the fibre cross-section. Q is related to ψ by linking (4.1) with (4.2) through the constitutive equation. As an example, consider a linear viscoelastic fibre, obeying the analogue of (2.3) for axial shear deformation. Combining this with (4.1) and (4.2) yields

$$Q(t) = J \int_{-\infty}^{t} G(t-u) \frac{d\psi}{du} du \qquad (4.3)$$

where

$$J = \iint (r^3 + r \frac{\partial \omega}{\partial \theta}) \, dr \, d\theta \qquad (4.4)$$

and $G(t)$ is the axial shear stress relaxation modulus. J is a geometrical parameter which depends on both the cross-section shape and the fineness of the fibre. In the case of a fibre with the shape of a solid or hollow circular cylinder, warping of the cross-section is absent i.e. $\omega = 0$ everywhere. J is then simply the polar second moment of area of the cross-section. Introducing a dimensionless shape parameter $\lambda = 2\pi J/A^2$, of value unity for a solid circular cross-section, we may always write J as

$$J = \lambda \rho_\ell^2 / 2\pi \rho^2 . \qquad (4.5)$$

In the special case of elastic behaviour (4.3) reduces to the well-known relation

$$Q(t) = JG\psi(t). \qquad (4.6)$$

More generally, (4.3) is required and it may be more conveniently expressed as its Laplace transform

$$\bar{Q}(p) = pJ\bar{G}(p) \bar{\psi}(p). \qquad (4.7)$$

The analysis of bending follows similar lines. It is, however, convenient to switch to rectangular Cartesian co-ordinates (x,y,z), with Oz again axial, but Ox and Oy parallel and perpendicular to the plane of bending; see Figure 4. Axes Oy, Oz define the neutral plane where $\varepsilon_3 = 0$. Suppose an initially straight fibre is bent to a radius of curvature R by pure couples applied at its ends. Cross-sections can then be assumed to remain plane and at position (x, y) in the cross-section the axial tensile strain is given by

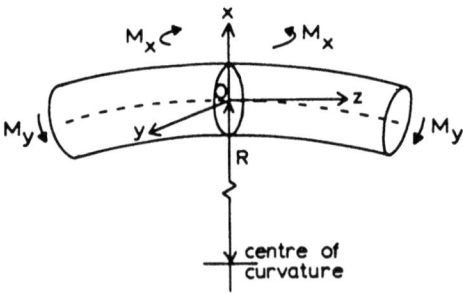

Fig. 4. Definition of axes in a bent fibre.

$$\varepsilon_3(t) = x/R(t). \tag{4.8}$$

The only non-zero stress will be σ_3, producing no net axial thrust and moments M_x and M_y about x and y-axes respectively, where

$$0 = \iint \sigma_3(t,x) \, dx \, dy$$

$$M_x(t) = \iint \sigma_3(t,x) \, y \, dx \, dy \tag{4.9}$$

$$M_y(t) = \iint \sigma_3(t,x) \, x \, dx \, dy$$

and double integrals are again taken over the fibre cross-section.

Again illustrating the procedure by assuming linear viscoelasticity, we may combine (2.3) with (4.8) and (4.9) to obtain

$$0 = \iint x \int_{-\infty}^{t} E(t-u) \frac{d}{du}\left(\frac{1}{R}\right) du \, dx \, dy \tag{4.10}$$

$$M_x(t) = I_{xy} \int_{-\infty}^{t} E(t-u) \frac{d}{du}\left(\frac{1}{R}\right) du \tag{4.11}$$

$$M_y(t) = I_{xx} \int_{-\infty}^{t} E(t-u) \frac{d}{du}\left(\frac{1}{R}\right) du \tag{4.12}$$

where $I_{xy} = \iint xy \, dx \, dy$, $I_{xx} = \iint x^2 \, dx \, dy$. (4.10) defines the position of the neutral plane: clearly, the y-axis is a centroidal axis of the cross-section. I_{xy} and I_{xx} are two second moments of area of the cross-section. I_{xy} is identically zero if the cross-section shape is symmetrical about either the x or y-axis. I_{xx}, similarly to J, depends on both the cross-section shape and the fibre fineness. Introducing a dimensionless shape parameter $\mu = 4\pi I_{xx}/A^2$, of value unity for a solid circular cross-section, we may write I_{xx} as

$$I_{xx} = \mu \rho_\ell^2 / 4\pi \rho^2 . \tag{4.13}$$

Elastic behaviour reduces (4.12) to the Euler-Bernoulli relation

$$M_y(t) = I_{xx} E/R(t) \tag{4.14}$$

while the Laplace transform of (4.12) becomes

$$\overline{M}_y(p) = p\, I_{xx}\, \overline{E}(p)\, \overline{R^{-1}}(p). \tag{4.15}$$

In cases of twisting and bending involving large deformations, where (2.2) and (2.3) are not obeyed, empirical methods may be necessary. To evaluate the integrals in (4.2) (or (4.9)) it is necessary to know σ_4 (or σ_3) for the particular history of ψ (or R) for each point in the cross-section. In particular cases this information may be available. On the other hand, it will be adequate in other cases to treat the torque-twist and moment-curvature relations as functions to be determined empirically. This approach may also be necessary if the fibre cannot be assumed homogeneous. Most fibre formation processes lead to fibres with properties independent of θ and z to reasonable precision; but assuming independence of r is often more difficult to justify. In addition, there is a possibility of heterogeneities such as voids and foreign inclusions (e.g. particles of delustrant).

Further complications arise if the fibre has initial curvature or twist. Any initial curvature will cause load-extension relations, for example, to contain geometrical non-linearity in addition to other effects discussed in this paper. Initial twist in the fibre reduces its symmetry and so increases the number of non-zero elements in $\underline{S}$ and $\underline{C}$, and as a result introduces coupling between the deformation modes of axial tension and torsion.

REFERENCES

1. J.G. Cook, <u>Handbook of Textile Fibres</u>, 4th Ed. Merrow, Watford, 1968.
2. J.M. Mathews, <u>Textile Fibres</u>, 6th Ed. (H.R. Mauersberger, editor), Wiley, New York, 1954.
3. R.W. Moncrieff, <u>Man-Made Fibres</u>, 6th Ed., Butterworths, London, 1975.
4. J.E. Ford, <u>Fibre Data Summaries</u>, The Cotton, Silk and Man-Made Fibres Research Association, Manchester, 1966.
5. R.W. Work, Text.Res.J., <u>19</u> (1949) 381.
6. J.W.S. Hearle, P.K. Sen Gupta and A. Mathews, <u>Fibre Sci.Technol.</u>, <u>3</u> (1971) 167.
7. W.E. Morton and J.W.S. Hearle, <u>Physical Properties of Textile Fibres</u>, 2nd Ed., Heinemann, London, 1975.
8. A. Rae and R. Bruce, <u>The Wira Textile Data Book</u>, WIRA, Leeds, 1973.
9. R. Meredith (editor), <u>The Mechanical Properties of Textile Fibres</u>, North Holland, Amsterdam, 1956.
10. I.M. Ward, <u>Mechanical Properties of Solid Polymers</u>, Wiley-Interscience, London, 1971.

11. R.G.C. Arridge, <u>Mechanics of Polymers</u>, Clarendon Press, Oxford, 1975.
12. J.D. Ferry, <u>Viscoelastic Properties of Polymers</u>, 2nd Ed., Wiley, New York, 1971.
13. N.G. McCrum, B.E. Read and G. Williams, <u>Anelastic and Dielectric Effects in Polymeric Solids</u>, Wiley, London, 1967.
14. S. Timoshenko and J.N. Goodier, <u>Theory of Elasticity</u>, 2nd Ed., McGraw-Hill, New York, 1951.

THE MECHANICS OF DENSE FIBRE ASSEMBLIES

J. W. S. Hearle

Department of Textile Technology, University of
Manchester Institute of Science & Technology,
Manchester 1. England.

ABSTRACT. Dense fibre assemblies are defined as those in which
the mechanics is dominated by fibre deformation predicted from
assembly deformation. Strain relations are derived for various
systems. Then general energy relations for the prediction of
system response are developed, and applied to twisted yarns and
nonwoven fabrics. The influences of compacting of the structure
and straightening of fibres and of slippage at fibre ends are
discussed.

1. INTRODUCTION

1.1 Characteristics of dense structures

It is necessary to start with an explanation of the title of this
chapter. Originally it was planned as the mechanics of twisted
yarns and nonwoven fabrics. But it is better to examine the
features of the structural mechanics which lead to this association.
Operationally what is important is that these are systems in
which:

> *(i) a statement of the assembly deformation enables a
> reasonable prediction of the fibre deformation to be
> made on the basis of a knowledge of the assembly geometry,
> without a need to invoke mechanics; (ii) most of the
> energy of deformation goes into fibre deformation.*

For uniaxial extension, the fibre stress-strain curve is the first
approximation to the stress-strain curve of the assembly, but it
is then modified by fibre orientation, by slippage at fibre ends,

and by limited compacting of the structure and straightening of
fibre paths. In the development of the methods, some modifications
of the strict criteria are found to be possible.

The structural features which lead to this situation are:

  (a) *relatively dense packing of fibres in the assembly;*
  (b) *low curvature of fibre axial paths;*
  (c) *absence of major change in geometric form of the*
    assembly during deformation;
  (d) *limited relative motion between neighbouring fibre*
    elements.

We thus exclude such structures as: very open fillings, with
deformation dominated by major compacting of the structure
(though Carnaby later provides a link to this sort of material);
low twist strands of staple fibres, such as slivers and rovings,
which can be drafted through sliding of complete fibres past one
another; bulked yarns in which the important mechanisms are
straightening of high fibre crimp or pulling-out of snarls;
knitted fabrics where straightening of loops is the major effect;
woven fabrics with crimp interchange in extension and a trellis
action in shear.

However, except in special cases, we do not carry the criteria to
the limit, such as would be found in a solid fibre-reinforced
composite, which, until some internal failure occurs, does act as
a true, but inhomogeneous, continuum with no relative motion of
the component parts. A limited degree of relative sliding under
low or moderate forces is an important characteristic of textile
structures, if they are not to act as solid rods or sheets,
although they must hold together under large forces. In analysis,
the material is divided into elements, which are treated as part
of a continuum, but at the microscopic level continuity between
fibres is not required.

Within the category of dense structures the main studies have
been on simple twisted yarns and on nonwoven fabrics. The
cylindrical helical geometry of the idealised twisted yarn is
quasi-planar, since the cylindrical shells can be "opened-out"
flat, and the nonwoven fabrics can be treated as planar sheets
of variously oriented fibre elements. In principle, the methods
can be extended to three-dimensional assemblies. Furthermore
many of the excluded structures, which deform in other ways under
small loads, come into the category of dense structures in the
limiting situation under high loads: for example, after crimp
interchange has been completed in a woven fabric, further
extension is due to fibre extension, and, if the geometry in this
state is known, the methods described in this chapter can be used.

As criteria for the possibility of direct application of the methods described in this chapter, we should also add two features of the type of deformation:

(e) the applied forces must be large enough;
(f) the deformation of the whole piece of material must be geometrically defined by the boundary conditions.

The first of these was implicit in the comment at the end of the previous paragraph. Another way of putting it is to say that deformations under small forces, such as decrimping at the beginning of yarn extension, are either negligible or treated separately.

The second criterion clearly applies when we pull a string: its path is a straight line. Fabrics under tension will be planar sheets. But under compression, which is a load-controlled boundary condition, or on reduction of the distance between the ends of a string or the edges of a sheet, which is a geometrical boundary condition, the material will buckle. Its path is then unknown, not only in magnitude of deformation, but also in form. A string may buckle in a planar elastica or in some more complicated three-dimensional path. Fabric buckling is even more complex. The overall material deformation problem has to be solved as well as the internal mechanics.

Fortunately there is a simplification which is often possible. Many yarns, and other one-dimensional strands, are much easier to bend than they are to extend. Consequently under any appreciable tension, the geometric path is known: either straight, or curved in the shortest path round constraints. Conversely, in any situation in which the yarn is buckled, the tensile forces must be negligible (axial compressive forces can never be large, except over very short lengths or in laterally constrained situations). Thus some separation of effects and methods is possible. These arguments will not apply to stretch yarns with a very low initial modulus of extension; and the analagous situation in fabrics is also more complicated because there are often some easy modes of in-plane deformation - indeed these are required in double curvature.

1.2 Approaches to the problem of dense structures

As a simple example of a dense structure problem we can consider the extension of a twisted yarn with an idealised helical structure. This was first analysed by Gegauff [1] in 1907, and much of the theoretical and experimental work has been reviewed by Hearle [2]. Although we like to think of force causing deformation, the strategy to follow in analysis is the reverse one shown in figure 1.

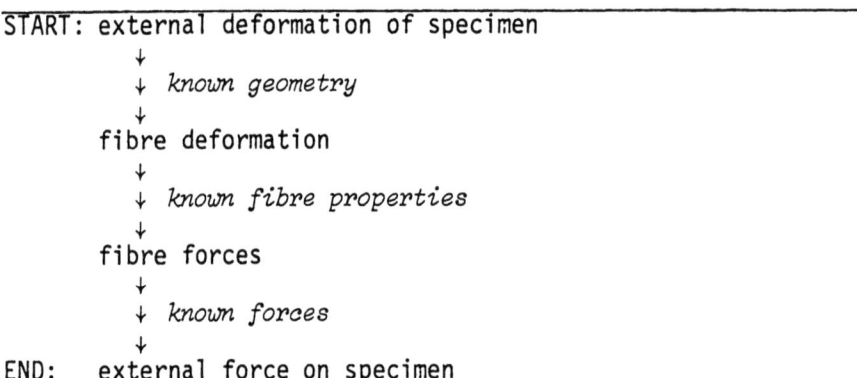

Figure 1. Strategy in analysis

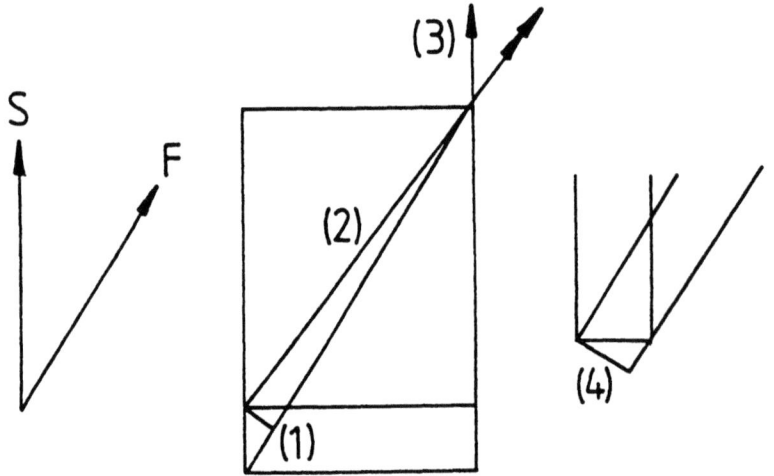

Figure 2. Simple prediction of structure modulus. The factor $\cos \theta$ comes in four times between fibre element F and structural element S: (1) ratio of elongations; (2) inverse ratio of lengths; (3) component of force; (4) inverse component of area. Hence conversion factor = $\cos^4 \theta$.

The early workers limited consideration to taking the components of fibre tension which did lead to the simple but very useful $\cos^4\theta$ (=$\cos^2\alpha$, for yarns) relation shown in figure 2. Since then a number of others, including myself in 1958 and more recently Jones [3], Konopasek [4], Huang [5] and Neckar [6], have provided more complete analyses of the forces or forces and moments on fibres in yarns.

These approaches all require several pages of mathematics - some of it interesting, but a good deal of messy algebra - and have been difficult or impossible to apply to the more realistic situation of large strains and non-linear fibre properties.

The force methods, or the continuum stress analysis, to be described by Thwaites, may be the best methods to use when a full statement of the internal forces is required. However, the overall prediction of properties is much more conveniently and powerfully treated by the energy methods which will be treated in this chapter.

A similar development from force methods to energy methods has occurred in the study of nonwoven fabrics, as described by Hearle [12]

2. STRAIN RELATION

2.1 The simplest approximation

As indicated in figure 1, it is first necessary to relate the external deformation to the fibre strain. If we consider a fibre lying at an angle θ to the direction of a uniaxial extension, without lateral contraction, then figure 2 shows that:

fibre strain/assembly strain = $\cos^2\theta$

More exact expressions can be derived for particular circumstances, and a few examples will be given here.

2.2 Three-dimensional, orthogonal

For orthogonal deformation of a three-dimensional assembly, we consider a fibre element of length l oriented with direction cosines C_x, C_y, C_z subject to strains e_x, e_y, e_z, as shown in figure 3.

In the strained state, we have:

$$(l+\delta l)^2 = (x+\delta x)^2 + (y+\delta y)^2 + (z+\delta z)^2 \qquad (2)$$

$$(1+\tfrac{\delta l}{l})^2 = (1+\tfrac{\delta x}{x})^2 \tfrac{x^2}{l^2} + (1+\tfrac{\delta y}{y})^2 \tfrac{y^2}{l^2} + (1+\tfrac{\delta z}{l^2})^2 \tfrac{z^2}{l^2} \tag{3}$$

$$(1+e_j)^2 = (1+e_x)^2 c_x^2 + (1+e_y)^2 c_y^2 + (1+e_z)^2 c_z^2 \tag{4}$$

where e_j is the fibre strain.

2.3 Planar with shear

Figure 4 shows a planar element with a fibre at an angle θ to the y-direction, taken as the direction along the length of yarn or fabric, subject to biaxial strain, e_L and e_T, and shear β, defined as the angular shift of lines in the long direction (originally y-) with the cross direction (x-) remaining unaltered. We then have in the triangle ABC:

$$(l+\delta l)^2 = (y+\delta y)^2 + [x+\delta x+(y+\delta y)\tan\beta]^2 \tag{5}$$

$$(1+\tfrac{\delta l}{l})^2 = (1+\tfrac{\delta y}{y})^2 \tfrac{y^2}{l^2} + [1+\tfrac{\delta x}{x} + \tfrac{y}{x}(1+\tfrac{\delta y}{y})\tan\beta]^2 \tfrac{x^2}{l^2} \tag{6}$$

$$(1+e_j)^2 = (1+e_L)^2 \cos^2\theta + [1+e_T+(1+e_L)\cot\theta\tan\beta]^2 \sin^2\theta \tag{7}$$

We also see from figure 4 that there is a shear of the fibre direction, given by the angle γ: a similar shear is present in figure 3, since the change in direction results from axial strains as well as shear strains. Introducing the angle α between the final fibre direction and the x-direction, we note that:

$$\gamma = \pi/2 - \theta - \alpha \tag{8}$$

$$\tan\alpha = \frac{y+\delta y}{x+\delta x+(y+\delta y)\tan\beta}$$

$$= \frac{1+\delta y/y}{(1+\delta x/x)(x/y)+(1+\delta y/y)\tan\beta}$$

$$= \frac{1+e_L}{(1+e_T)\tan\theta+(1+e_L)\tan\beta} \tag{9}$$

One could now write out an explicit expression for γ in terms of the geometry θ, and the deformations e_L, e_T and β but this is unnecessary since equations (8) and (9) are in a convenient form for incorporation in a computer program.

We should note that large strains do give rise to alternative definitions, and the forms given here should be examined in the light of modern treatments of large strains.

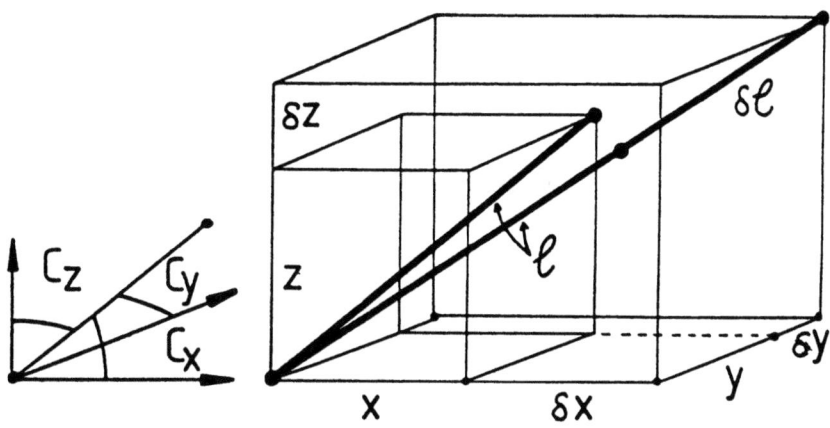

Figure 3. Fibre strain in a three-dimensional element

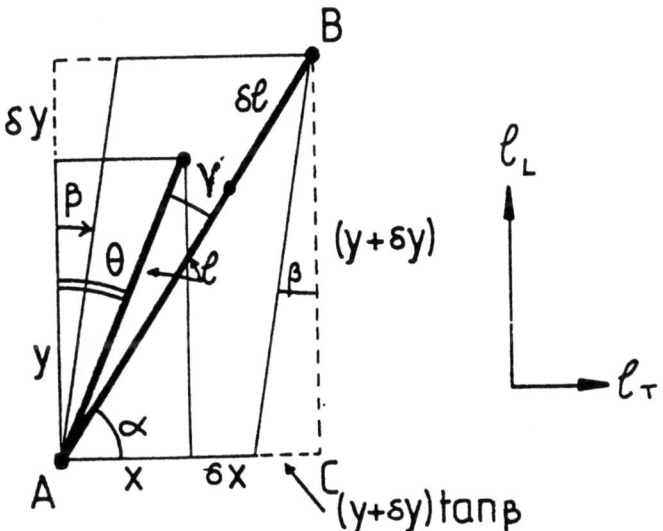

Figure 4. Fibre strain in plane, with shear

2.4 Planar, orthogonal, with curl

If the fibre element is curved, then it must be straightened before fibre extension occurs. Figure 5 shows the appropriate element. The curl factor c of the fibre is defined by the equation:

$$l_f = c \, l_o \tag{10}$$

where l_f is the actual fibre length and l_o is the straight length of the diagonal in the element.

We then have:

$$(l_f + \delta l_f)^2 = (y + \delta y)^2 + (x + \delta x)^2 \tag{11}$$

$$(1 + \frac{\delta l_f}{l_f})^2 = (1 + \frac{\delta y}{y})^2 \frac{y^2}{c^2 l_o^2} + (1 + \frac{\delta x}{x})^2 \frac{x^2}{c^2 l_o^2} \tag{12}$$

$$(1 + e_j)^2 = (1/c_j^2) \, [(1 + e_L)^2 \cos^2 \theta_j + (1 + e_T)^2 \sin^2 \theta_j] \tag{13}$$

where c_j and θ_j are the curl factor and orientation of the particular fibre.

2.5 Planar, no-shear, no-curl

For the commonest case, without system shear or fibre curl, equations (4), (7) and (13) all reduce to the explicit form:

$$e_j = [(1 + e_L)^2 \cos^2 \theta + (1 + e_T)^2 \sin^2 \theta]^{\frac{1}{2}} - 1 \tag{14}$$

For uniaxial extension, we put:

$$e_T = - \sigma \, e_L \tag{15}$$

where σ is a Poisson's ratio.

For constant volume deformation, the Poisson's ratio is $\frac{1}{2}$: this is not strictly true at large strains, and more exact expressions can be introduced, but is usually an adequate approximation. We thus get the expression:

$$(1 + e_j)^2 = (1 + e_L)^2 \cos^2 \theta + (1 - \sigma \, e_T)^2 \sin^2 \theta \tag{16}$$

$$(1 + e_j)^2 = (1 + e_L) \cos^2 \theta + (1 - \tfrac{1}{2} e_L)^2 \sin^2 \theta \tag{17}$$

For the uniaxial extension of the three-dimensional system shown in figure 3, we put $e_L = e_z$. If the material has some symmetry about the z-axis (orthotropy is a sufficient, but not a necessary

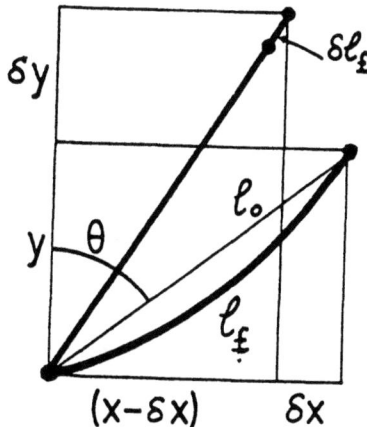

Figure 5. Fibre strain in plane, with curvature.

condition) so that $e_x = e_y = e_T$, or, in other words, if the Poisson's ratio is the same in all directions around the axis, then it can be shown algebraically that equations (16) and (17) would apply with θ as the angle between the fibre direction and the z-axis. This conclusion also follows directly from geometry, since it is only necessary to consider what happens in the plane containing the fibre and the axis of extension.

2.6 Small-strain relation

The small strain equations can be obtained by dropping the squared terms from the earlier equations, from geometry as in figure 2, or by differentiation. We have:

$$l^2 = x^2 + y^2 \tag{18}$$

$$2l\,dl = 2x\,dx + 2y\,dy \tag{19}$$

division by $2l^2$ gives:

$$e_j = e_L \cos^2\theta + e_T \sin^2\theta \tag{20}$$

$$\rightarrow e_T \cos^2\theta, \text{ if } e_T = 0 \tag{21}$$

3. ENERGY RELATIONS

3.1 The general equation

Knowledge of fibre strain, combined with information on fibre
properties, leads to fibre stresses or strain energies. It is
now necessary to relate these to external forces.

A convenient compromise between generality and simplicity,
between breadth of application and practical relevance, is achieved
by considering a rectangular block of material, the parallelepiped
AB....H, shown in figure 6, subject to a large uniform triaxial
extension under forces P_x, P_y, P_z. However, the treatment is
easily generalised to cover shearing moments, and can also be
presented [1,2] in terms of non-uniform deformations from a collection
of forces.

The block is regarded as made up of a collection of fibre elements,
of which the element in figure 3 would be one. In many applications,
the only fibre deformation to be considered is fibre extension,
but the analysis also applies when other deformations have to be

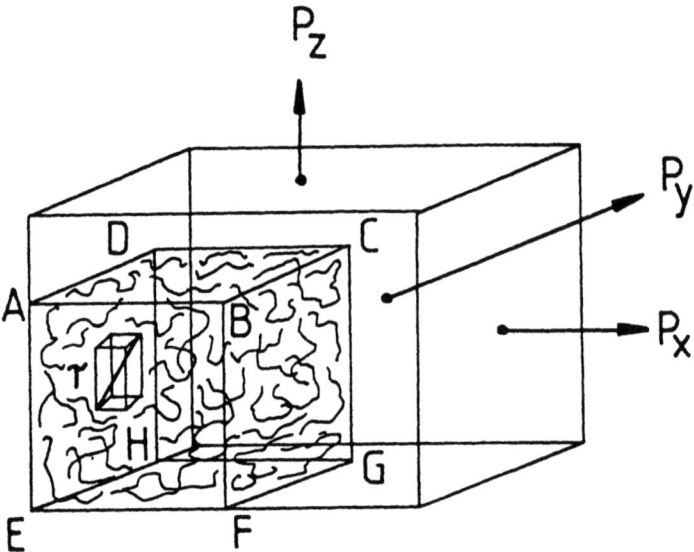

Figure 6. Element subject to extension under triaxial forces.

Table 1

Notation

Whole system

U	: energy	)
P_i	: forces	) or, for shear
X_i	: extended lengths of sides X, Y, Z	) moments and angles,) as generalised
$X_{i,o}$	: unstrained dimensions	) forces and) displacements.
f_i	: system specific stresses	)
e_i	: system strains	)
e_L, e_T	: special case of axial extension	)

Fibre elements:

m_r	: element mass
U_r	: element energy
x_j	: set of deformed fibre element dimensions*
$x_{j,o}$	: unstrained dimensions*
e_j	: fibre strains*
$f_r(e_j)$	: fibre specific strains

* suffix r is implicit

---

included. The notation used is summarised in Table 1. Note that suffix i refers to directions in the whole system; suffix r applies to individual fibre elements; suffix j refers to a mode of fibre deformation (e.g. extension, twisting, bending etc).

Consider a small variation dX_i in one of the system dimensions, with the others held constant. From conservation of energy, we have:

word done on system = work taken up in fibre elements

$$dU = P_i\, dX_i = \sum_r dU_r \qquad (22)$$

The energy associated with each element is a function of its dimensions:

$$U_r - f(x_j) \qquad (23)$$

$$P_i = \Sigma \Sigma \left(\frac{\partial U_r}{\partial x_j}\right) \left(\frac{\partial x_j}{\partial X_i}\right) \quad (24)$$
$$\quad\quad r\ j$$

This is the most general form of equation and is applicable to any set of forces P_i associated with incremental displacements dX_i. We note that it separates the fibre property effects $(\partial U_r/\partial x_j)$ from the system geometry effects $(\partial x_j/\partial X_i)$.

3.2 Application to macroscopically uniform deformation

When, as in figure 6, the system is uniformly strained, at least down to the level of the fibre elements, the relations can be converted to stress and strain - or, for our purposes, to specific stress based on mass elements.

For the whole system, we have:

$$\text{mass} = \Sigma_r m_r \quad (25)$$

$$P_i = (\Sigma_r m_r / X_{i,o})\, f_i \quad (26)$$

$$X_i = X_{i,o}\, (1 + e_i) \quad (27)$$

$$dX_i = X_{i,o}\, de_i \quad (28)$$

For the fibre element, with a stress-strain curve as given in figure 6(b), we have:

$$dU_r = \Sigma_j m_r\, f_r(e_j)\, de_j \quad (29)$$

where $f_r(e_j)$ is the specific stress in the j-mode of the r-element at the strain e_j in the deformed fibre element.

$$x_j = x_{j,o}\, (1 + e_j) \quad (30)$$

$$dx_j = x_{j,o}\, de_j \quad (31)$$

In the simplest application, x_j is the fibre length, e_j is the tensile strain, and $f_r(e_j)$ is the tensile specific stress.

Combining equations (24) to (31), we obtain:

$$f_i = \frac{\Sigma \Sigma m_r\, f_r(e_j)\, (\partial e_j/\partial e_i)}{\Sigma_r m_r} \quad (32)$$

The solution of this equation, which is in a form suitable for many applications, requires for each mass element m_r : (1) a knowledge of the necessary fibre stress-strain relations $f_r(e_j)$; (2) a knowledge of the relation between element strain e_j and system strain e_i, as given by the equations in sections 2.1 to 2.6 or other derivations. Note that e_j, and hence f_j, is generally a function of the full set of system strains e_i: it is only in the partial differentiation that the suffix becomes explicit to one direction. Similarly $f_r(e_j)$ will, in general, be a function of all the strains on the fibre, but de_j in equation (29) relates to a small variation in one strain with the others held constant.

3.3 A trivial application

It may be helpful, at this stage, to present the simplest example of the application of equation (32). If all mass elements m_r are identical, we have:

$$f_i = \overline{f_r(e_j)\,(\partial e_j/\partial e_i)} \qquad (33)$$

If we now consider a system of variously oriented fibres, subject to a small uniaxial extension e_L with no lateral contraction, taking into account only fibre extension under Hooke's Law with specific modulus E_f, we have:

$$e_j = e_L \cos^2\theta \qquad \text{[equation (21)]}$$

$$(\partial e_j/\partial e_L) = \cos^2\theta \qquad (34)$$

$$f_r(e_j) = E_f\, e_j = E_f\, e_L \cos^2\theta \qquad (35)$$

and hence:

$$\text{system stress} = f_L = \overline{E_f\, e_L \cos^4\theta} \qquad (36)$$

This is the simple relation, derived in figure 1, which gives $\cos^4\theta$ as an approximate ratio between material modulus and fibre modulus.

3.4 Some comments on applicability and validity

Equation (24) and, where the particular geometry and loading are applicable, equation (32) are rigorously correct, provided all the terms are properly included in the summation. The power — and the danger — of the energy method derives from the fact that separate effects are additions of scalar quantities and so are easily omitted: the skill lies in selecting what must be included. In contrast to this, force methods offer less choice: a mere collecting of force components as in figure 1 is necessarily a crude approximation, but any attempt at solving the equilibrium equations inevitably involves a lot of vectorial mathematical detail in a full formulation of the problem, so that solutions are difficult or impossible.

It is appropriate now to make a number of general comments on applicability and validity, which tend to be reverse sides of the same coin, of the methods described in this chapter.

Affine definition

It is implicit in the strain relations in sections 2.1 to 2.6 that the deformation is affine: in other words, the deformations of the boundaries of all fibre elements are similar to one another and to the material as a whole. For non-uniform system deformation, this statement must be valid over local zones which are large in comparison with structural elements of the material.

In order to improve on the affine approximation and obtain a more correct distribution of fibre strains, it is necessary to solve for internal equilibrium of forces or for a minimum energy conformation.

Fibre deformation and properties

For reasons which will be discussed in relation to yarns and nonwovens, it is usually reasonable to limit attention to fibre extension. We then have the implicit assumption that the value of $f_r(e_i)$ as measured in a pure tensile test on an isolated fibre is valid under a more complicated set of stresses, for example including transverse stresses in a twisted yarn. This is valid if the fibre is incompressible, since the state of strain, and hence the strain energy, must then be the same whether it is caused by tensile stress or transverse stress. Since most fibres have Poisson's ratios σ_{LT} close to 0.5, this assumption is probably justified, but it ought to be examined more closely in the light of Buckley's comments.

There is another implicit assumption, namely that there are no appreciable energy increments associated with changes of fibre cross-sectional shape. In twisted yarns, this corresponds to an assumption in Hearle's force analysis [2] that the transverse forces are the same in all directions perpendicular to the fibre axis. Huang [5] states that he has relaxed this assumption, but in fact he has imposed another restraint of geometric continuity. As indicated in figure 7(a), a set of parallel cylinders which are free to roll over one another will not be able to support

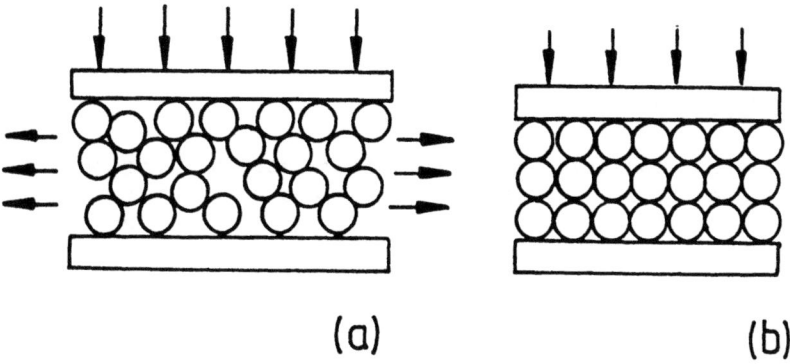

Figure 7. (a) Parallel cylinders free to roll over one another will be displaced sideways unless there is an equal balancing stress. (b) Bonded cylinders will support uniaxial stress.

asymmetric transverse pressures: it must act hydrostatically as a fluid in the transverse plane. But if the cylinders are stuck together as in figure 7(b), they will maintain a continuous structure. Reality may well lie between the extremes, but will be very difficult to analyse: in textile materials the hydrostatic analogy is usually the best assumption.

A similar effect occurs in relation to fibre shear. This is usually assumed to be relieved by slippage of fibre elements past one another, so that there is no energy contribution. If the other extreme of no slippage is adopted, then equations (8) and (9) give the shear strain and the shear mode of deformation can be included as another set in the summation.

There may be other modes of deformation, such as fibre twisting or fibre bending, which can be directly related to the system deformation, and thus easily included as additional sets in the summation if their contribution to energy change is appreciable.

The Poisson's ratio problem

Apart from the assumption of fibre incompressibility, with the fundamental implications discussed above, there is the question of compressibility by reduction of space between fibres. In practice, a full solution is usually avoided by taking yarn deformation as at constant volume ($\sigma = 0.5$) or by substituting experimental values of Poisson's ratio of nonwoven fabrics. Theoretically, there is no difficulty in formulating the necessary equations. Equations (24) and (32) apply to the transverse stresses; and, for uniaxial loading, can be solved for zero force, which is equivalent to a minimum energy formulation with $\partial U/\partial X_i = 0$.

The difficulty is to know what fibre deformation terms to substitute in the equations.

Hysteresis and viscoelasticity

In an elastic, conservative system, fibre strain energy will be a single-valued function of fibre strain, and so the equations will be soluble to give system stress as a single-valued function of system strain. But equation (22), based on conservation of energy is not limited to such systems, provided dU_r includes both stored and dissipated energy. Consequently provided $(\partial U_r/\partial x_i)$ in equation (24) and $f_r(e_i)$ in equation (32) take account not only of the current state of strain, but of the previous history, these equations for P_i and f_i are valid. For a monotonic extension, this implies following the actual fibre load-elongation curve at an appropriate rate; for recovery effects, it involves following the appropriate strain cycles.

Friction between fibres

Relative fibre movement is another source of energy dissipation. If there is a relative displacement $d\xi_p$ between a pair of fibre elements during the variation of deformation dX_i, with a frictional force F_p, the energy dissipated will be given by:

$$dU_p = F_p d\xi_p = \mu N d\xi_p \qquad (37)$$

where μ is the coefficient of friction and N is the normal load.

In principle, if $d\xi_p$ can be related to the system deformation, then it would be possible to add to equation (24) terms like:

$$\Sigma(\partial U_p/\partial \xi_p)(\partial \xi_p/\partial X_i) = \Sigma F_p (\partial \xi_p/\partial X_i) = \Sigma \mu N(\partial \xi_p/\partial X_i) \qquad (38)$$

Appropriately normalised terms would be added to equation (32).

This approach has not been followed and its use would involve a knowledge of fibre frictional properties, a treatment of the number of fibre contacts (which would involve fibre fineness, possibly compensated by changes in $d\xi_p$), a means of estimating normal stresses, and a way of handing the frictional characteristic that slip only occurs when the frictional force is exceeded*. Some more detailed analysis would be needed to

---

*This property implies that there may be a range of conditions of equilibrium stability with $\delta U/\delta x > o$ for both $\delta x \gtrless o$, instead of $\delta U/\delta x = 0$.

generate this information, though adequate approximate treatments might not be too difficult.

The usual practice is to assume, often tacitly, that frictional energy losses can be neglected *either* because slip is easy and so $F_p \to 0$ *or* because slip is difficult and so $d\xi_p \to 0$.

Multicomponent systems

Although the sum in equations (24) and (32) is strictly over all fibre elements, in practice these will be taken as groups with identical properties. It is easy to extend the treatment to fibre mixtures by taking separate sums over each fibre type. If there are other materials present, such as adhesive binder, then separate sets of terms can be added in to cover these, provided their properties are known and their deformation can be related to the system deformation.

Structural change

A dominant assumption of the method brought out in section 1.1, is that the material structure does not undergo major changes of form. However, we shall see later that the expressions can be modified to allow for some slippage at fibre ends and from bond breakage in nonwovens.

3.5 A restriction to energy methods

In order to describe the properties of a final product, it is usual to require the stress-strain curve. But, since other analyses at a higher hierarchical level use energy methods, it may be more convenient to finish with energy expressions at the intermediate level. The storage of data on properties is also most economically done by storing the energy values. For example in a solid, non-linear elastic material, one set of values of energy as a function of 6 strains (either stored numerically, with appropriate interpolation rules, or generated from theory) will correspond to 6 sets of stress values, and 36 incremental moduli. The problem is, of course, more difficult when the energy function depends on strain history and environment as well as strain but this remains the best way of trying to handle the problem.

For the dense fibre assemblies, being considered here, we have, for change dX_i:

$$dU_i = \sum_r dU_r = \sum_r \sum_j dU_{rj} = \sum_r \sum_j (\partial U_{rj}/\partial x_j)(\partial x_j/\partial X_i) \, dX_i \quad (39)$$

where dU_{rj} is the energy change associated with the j-mode of deformation of the r-element.

Hence:
$$U = \int_{X_{i,o}}^{X_i} \sum_r \sum_j (\partial U_{rj}/\partial x_j)(\partial x_j/\partial X_i) \, dX_i \qquad (40)$$

where $\int_{X_{i,o}}^{X_i}$ is the appropriately routed multiple integral (sum) over the total increments of all X_i's.

If we normalise for mass elements, we get:
$$E_i = (\Sigma\, m_r)^{-1} \int_o^{e_i} \Sigma_r \Sigma_j\, m_r\, f_r(e_j)(\partial e_j/\partial e_i)\, de_i \qquad (41)$$

We can note a link with the Hearle and Shanahan approach to the mechanics of fabrics composed of repetitive cells, referred to by Leaf. This treatment includes terms like dU_1/dl_1 where U_1 is the elongational energy and l_1 is the length of a yarn element in the unit cell of a fabric. The above formulation would give:

$$\frac{dU_1}{dl_1} = \sum_r \sum_j (\partial U_r/\partial x_j)(\partial x_j/\partial e_1)$$

$$= (l_{1,o})^{-1} \sum_r \sum_j m_r\, f_r(e_j)(\partial e_j/\partial e_1)$$

$$\xrightarrow{\text{usually}} (l_{1,o})^{-1} \Sigma\, m_r\, f(e_f)(\partial e_f/\partial e_1) \qquad (42)$$

where $l_{1,o}$ is the unstrained yarn length, e_f is the fibre strain in the yarn, as given by yarn geometry, and $f(e_f)$ is the fibre stress at the strain e_f.

4. APPLICATION TO TWISTED YARNS

4.1 The idealised twisted yarn

Early work on twisted yarn mechanics has been reviewed, with much experimental detail on structure and mechanics as well as theory, by Hearle [2], and further work is contained or referred to in two other papers [7,8].

The idealised twisted yarn, shown in figure 8 consists of a uniform circular cylinder of radius R, containing fibres uniformly packed and following concentric regular helical paths around cylinders of constant radius, with the helix pitch or period (the length h of one turn of twist) constant throughout the yarn. As a result, the helix angle takes values from zero for the straight fibre at the centre of the yarn, through θ at radius r, to a maximum at the surface twist angle α. The geometry is quasi-planar since each cylinder defining a fibre element can be opened out flat as in figure 8 and so define the geometrical relations.

Because all the fibres at a given radius will have the same strain, we can replace the sums of equation (32) by integrals over annular zones between radii r and $r + dr$. If f_y is the yarn specific stress at strain e_y, f_f is the fibre specific stress at strain e_f, and v_y is the yarn specific volume, equation (32) becomes:

$$f_y = \frac{\int_o^R (2\pi r dr/v_y) f_f (\partial e_f/\partial e_y)}{\int_o^R (2\pi r dr/v_y)} \qquad (43)$$

Usually, v_y is taken as constant and cancels out; and it is convenient to substitute the auxiliary variable $\tau = r/R$. We then get the equations of yarn extension:

$$f_y = 2 \int_o^1 f_f (\partial e_f/\partial e_y) \tau \, d\tau \qquad (44)$$

$$(1 + e_f)^2 = (1 + e_y)^2 \cos^2\theta + (1 - \sigma e_y)^2 \sin^2\theta \quad (45) \text{from}(16)$$

where Poisson's ratio = σ, is usually taken as 0.5, $\qquad (46)$

$$f_f = f(e_f) \qquad (47)$$

$$\tan \theta = \tau \tan \alpha \qquad (48) \text{ from figure 8}$$

$$\tan \alpha = 2 \pi R/h = 2 \pi v_y^{\frac{1}{2}} C^{\frac{1}{2}} T \qquad (49)$$

where C = yarn linear density and T is the yarn twist h^{-1}.

We note one important feature of the energy approach. All the geometric quantities (θ, α, r, R, h, T, C) are values in the unstrained state. It is not necessary, as it is in force methods, to deal explicitly with the geometry in the strained state.

The equations (44) to (49) would be solved by computation, with an appropriate numerical insertion of the general non-linear fibre stress-strain curve (47) and the necessary yarn geometry parameters, namely either α, or R and h, or v_y, C and T.

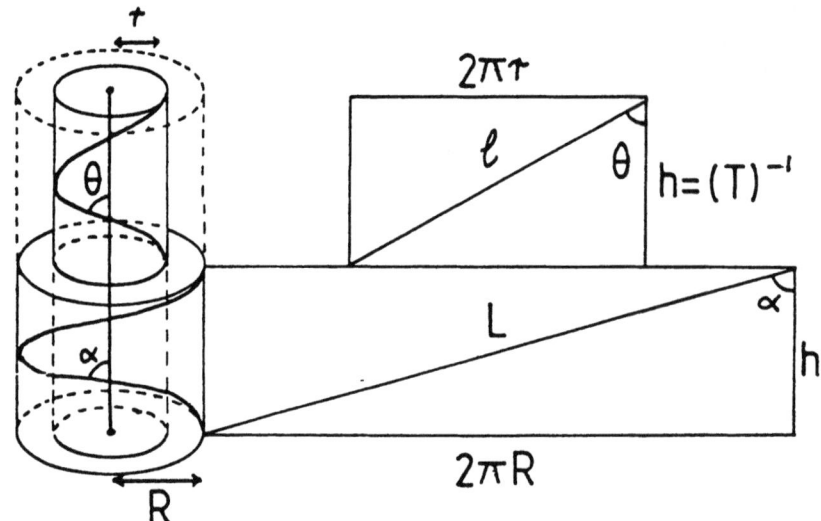

Figure 8. Geometry of idealised twisted yarn.

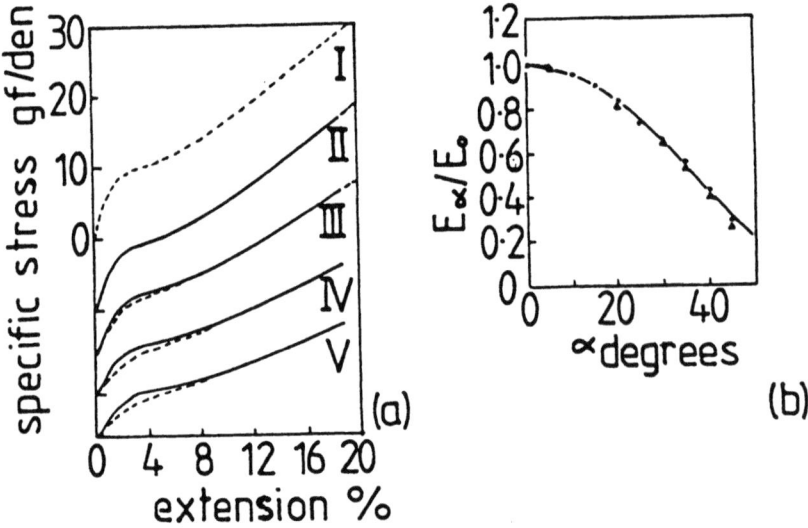

Figure 9. Theory and experiment in yarn mechanics [8].
(a) Stress-strain curves of rayon yarns (from Treloar and Riding);
------experiment, ———theory, twist angles are (I) 5° 18',
(II) 14° 55', (iii) 25° 56', (IV) 33° 49', (V) 38° 21'.
(b) Dynamic modulus (from Zorowski and Murayama); •nylon yarn,
▲ polyester yarn, — theory.

Figure 9 gives two examples of experimental results and shows the close agreement between experiment and theory for twisted continuous filament yarn.

4.2 The approximate modulus ratio

Although it is simple to compute the above equations, there are occasions when it is useful to use a rough approximation following from equation (36). It can be shown [2] that $\overline{\cos^4\theta}$ through the idealised yarn is $\cos^2\alpha$. We therefore have the approximate relation:

$$\frac{\text{yarn modulus}}{\text{fibre modulus}} = \cos^2\alpha \tag{50}$$

4.3 A force method

Hearle's force method of analysis [2] published in 1958 was based on the differential equilibrium of an element within the yarn under transverse forces. It gave rather better modulus predictions than equation (50), and had the advantage of predicting the transverse stresses within the yarn, as shown in figure 10. However, the analysis was restricted to small strains and Hooke's Law, and has not subsequently been developed to apply to large strains and non-linear stress-strain relations.

4.4 Some comments on real yarns

The major problem with the idealised yarn model is that it is not easy to obtain in a strain-free state. Apart from the bending and twisting of individual fibres, which can be neglected if (fibre radius) << (yarn radius), equal lengths are supplied to take up different path lengths in the yarn. This has to be accommodated in some combination of the following ways:

(i) buckling of the centre filaments, which gives an additional easy extension at low stresses;

(ii) permanent extension of the outer filaments which changes the fibre properties;

(ii) migration, or a change in radial position along the yarn length.

The first two effects can only be dealt with empirically - or by a full theory of yarn formation. However, an idealised yarn with a uniform complete migration can be formulated and Kilby [9] and Treloar [10] have shown that migration has little effect on the predicted yarn stress-strain curve. There is a difficulty

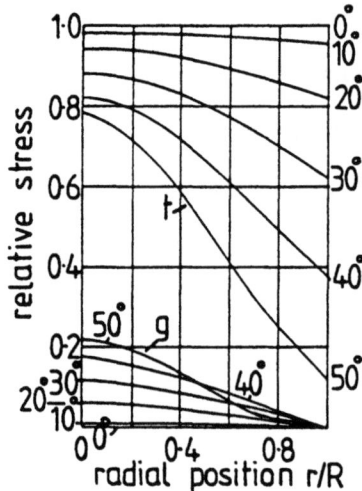

Figure 10. Values of stress in a twisted yarn, relative to fibre stress at the same strain, for the small-strain theory. The tensile stress is t and the transverse stress is g. The curves relate to different surface twist angles α in degrees.

at the centre of a yarn in combining uniform migration with uniform packing (similar to the problem at the centre of the spokes of a wheel, where the spoke thickness must go to zero) but this can be avoided in computation by leaving a small hole at the centre.

The implicit treatment of friction is interesting. The assumption is that over short lengths slip occurs and relieves the shear associated with extension of the helical structure, so that only fibre extension has to be taken into account; but that slip does not occur over long lengths, so that the fibre strain is that given by the local helix angle, and is not equalised along a migratory fibre path.

In deriving the yarn contraction on twisting, which gives rise to an increase of linear density by a factor $\frac{1}{2}(1 + \sec\alpha)$, it is found [2] that all filament lengths are equally likely. Consequently if there is a complete equalisation of strain through slippage along migratory filaments we must have:

$$e_f = \tfrac{1}{2} [(e_f \text{ at } r = 0) + (e_f \text{ at } r = R)]$$

$$= \tfrac{1}{2}\{e_y + [(1 + e_y)^2 \cos^2 \alpha + (1-\sigma e_y)^2 \sin^2 \alpha]^{\tfrac{1}{2}} - 1\} \quad (51)$$

This expression, divided by e_y, would give the factor by which the fibre stress-strain curve is reduced in this situation.

The condition in which there is no slippage but shear energy has to be taken into account has been considered by Hearle and Sparrow [11] in relation to the internal mechanics of cotton fibres.

For dense continuous filament yarns, the assumption of $\sigma = 0.5$ is adequate, but for bulky yarns Carnaby shows how greater lateral compression can be taken into account.

4.5 Developments of the treatment

Hearle and Konopasek [8] have shown that the method is easily adapted to consideration of the interrelation between the four quantities: tension, extension, torque and twist. Any two of these are independent. The method involves a generalisation of the analysis of sections 3.1 and 3.2 to cover the torque.twist contribution to energy, and a variation of section 3.3 to cover twist geometry. To the set of equations (44) to (49) we add:

specific torque = torque/linear density

$$= \mu_y = 2 \int_0^1 f_f \, (\partial e_f / \partial T) \, \tau \, d\tau \qquad (52)$$

and modify equation (45) to:

$$(1 + e_f)^2 = (1 + e_y)^2 \cos^2\theta + [1 - \sigma e_y - \sigma e_y(T/T_o) + T/T_o]^2 \sin^2\theta \qquad (53)$$

where T_o is the initial twist and T is the deformed twist.

Another development of the theory applies to yarns with other geometries, such as ply yarns. Although two directions will be required to characterise the fibre direction completely as

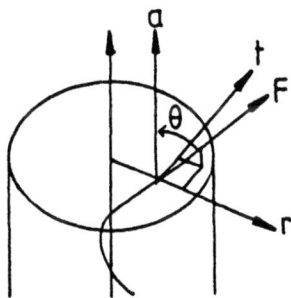

Figure 11. A general fibre direction F would require to be specified by two angles, such as two of the three direction cosines with the axial direction (a), radial direction (r), and tangential direction (t).

shown in figure 11, it follows from the comments at the end of section 3.5 that, provided the lateral contraction is the same in all directions, the uniaxial response depends only on the angle, θ, between fibre axis and yarn axis. As shown by Hearle and Konopasek (8), the specific stress equation becomes:

$$f_y = \sum_C f_f \, (\partial e_f / \partial e_y) \, \delta C / C \qquad (54)$$

$$\text{or} \quad = \sum_F f_f \, (\partial e_f / \partial e_y) \sec \theta / \sum \sec \theta \qquad (55)$$

where $\sum_C$ is taken over linear density elements, and $\sum_F$ is taken over individual fibre elements with a mass proportional to $\sec \theta$. The strain equation (45) is as before, but the distribution of the values of θ must be determined from the particular yarn geometry.

In general, the methods described here can be applied to any yarn deformation where the geometric form is maintained. It will not apply directly during compression, or untwisting, or bending when filament separation occurs and the geometrical relations change.

5. APPLICATION TO BONDED FIBRE FABRICS

5.1 Initial extension

A fuller account of experimental and theoretical studies of bonded fibre fabrics is given elsewhere [12]. Although the real distribution of binder material is rather different, it has proved reasonable to consider a bonded fibre fabric as made up of a planar assembly of elements like the one in figure 5, bonded into the network at the ends of the element. The early work used a force analysis of the fabrics as a network of straight fibre elements subject to affine deformation, as described by Seth on paper. Later workers took into account fibre curvature, and then adopted an energy method.

The fibre elements are usually assumed to be of equal mass, so that equation (32) becomes:

$$\text{fabric specific stress} = f_i = (1/N) \sum f_f \, (\partial e_r / \partial e_i) \qquad (56)$$

when the summation is taken over a representative collection of N fibres with curl c_j and orientation θ_j. The equation applies to longitudinal ($i=L$) and transverse ($i=T$) stress. The strain relation is equation (13).

For uniaxial extension, or any other case where $e_T < 0$, some values of fibre strain from equation (13) will be negative. The usual assumption made is that fibres buckle in compression without resistance, so that we put $f_f=0$ for $e_j<0$. This has the effect of making it impossible to calculate Poisson's ratio, since there would be complete lateral collapse in order to minimise energy. In principle, it is possible to include bending energy in buckling in order to enable the Poisson's ratio to be calculated. This would be valid if fibre thicknesses are small compared to the space between them, but in many instances it seems likely that lateral contraction will be hindered more by side-by-side jamming (the influence of the occupation of volume by fibres) than by line buckling. The usual practice is to substitute experimental values of Poisson's ratio. These are found to be large, often with values of 3 or 4, due to buckling out of the yarn plane with a large negative Poisson's ratio for the thickness direction.

With these assumptions a good prediction of the initial part of the stress-strain curve is made as shown in figure 12. There are many interesting features of the influence of fibre orientation and curvature in particular fabrics. It may also be noted that curl factor will be influenced by bond-to-bond distance becoming less as the bonds get closer, and short fibre elements come closer to a straight line.

As always, a rough approximation to the ratio of fabric stress to fibre stress, for straight fibres, is given by $\overline{\cos^4\theta}$, which for a planar isotropic orientation has the value 3/8.

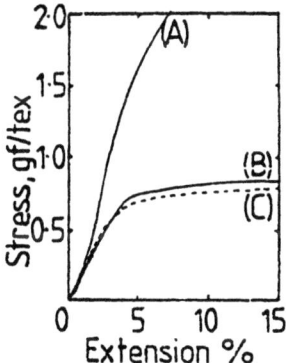
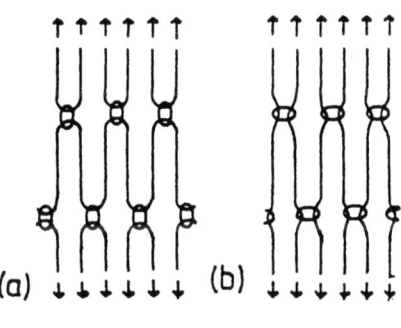

Figure 12. Comparison of theoretical (A) and experimental (C) stress-strain curves of a bonded fibre fabric. Curve (B) is the theoretical curve modified by an experimental stress limit.
Figure 13. Influence of long-range path: (a) causing stress on bonds; (b) transmitting stress along fibres.

5.2 Yield and failure

The analysis clearly breaks down at larger stresses when the fabric yields. Three different crude, and fundamentally unsound, theories can be used to predict what happens. The first is to consider the effect of slippage from fibre ends, or false ends where fibre paths curve back on themselves, as described in section 6.

The second recognises that the cause of yeilding is failure of bonds - either bond breakage or slippage of fibres through bonds. It is then proposed that this occurs whenever the fibre stress rises above some critical value f_c; and we therefore modify the computation by putting $f_f = f_c$ whenever a higher value is indicated. Figure 12 shows that this can produce a reasonable fit with experiment.

The third method [13] puts a binder element in series with each fibre element. Binder content indicates the amount of binder, and experiments on binder film give binder properties, but there remains one adjustable parameter namely the aspect ratio of the binder segment (long and thin or short and fat). The analysis includes energy terms for both fibre and binder, with the relative strains adjusted to give equal force in both segments of an element. This treatment gave good predictions for stress-strain curves in different directions in the fabrics, and the adjustable aspect ratio showed a consistent dependence on binder content. It thus seems a useful theory, but it has the defect that the binder is really between separate fibres and does not split each fibre into segments. Fibres are continuous across bonds.

Some qualitative discussion of the relevance of the geometry of fibre paths and binder distribution is given elsewhere [14]. Two fundamental points should be made here. Firstly, the assumption of affine deformation may give reasonable predictions of modulus, since these are based on average values, and only small errors will come from taking the wrong average; but failure depends on extreme values and so will be very sensitive to the distribution of strains, and the deviation from the average value. Secondly, failure is sensitive to long-range fibre paths, as well as to the orientation of short elements. For example, as long as the bond holds, the sub-systems in figure 13 will behave similarly, but bond failure is much more likely in (a) than (b). An extended long-range path will be more effective than a highly coiled one, since it will contain fewer folds which are false ends incapable of transmitting tension through the fibre material.

6. SLIPPAGE

6.1 A simple treatment

In any material composed of short fibres we must consider how tension is transmitted from one fibre to another, and what are the effects of slippage at fibre ends.

At the free end of a fibre, there can be no tension; but, provided the material is sufficiently coherent, there will then be a zone of slippage as indicated in figure 14 (a), up to a fully gripped zone, where the tension is the same as it would be in an equivalent continuous filament system without slip, evaluable by the methods already described. With the simple linear relations of figure 14 (b), with a slip length S and a fibre length L, we find:

slip factor (SF) = (mean stress with slip/stress without slip)

$$= \bar{f}/f_\infty = (L-S)/L = 1-S/L \qquad (57)$$

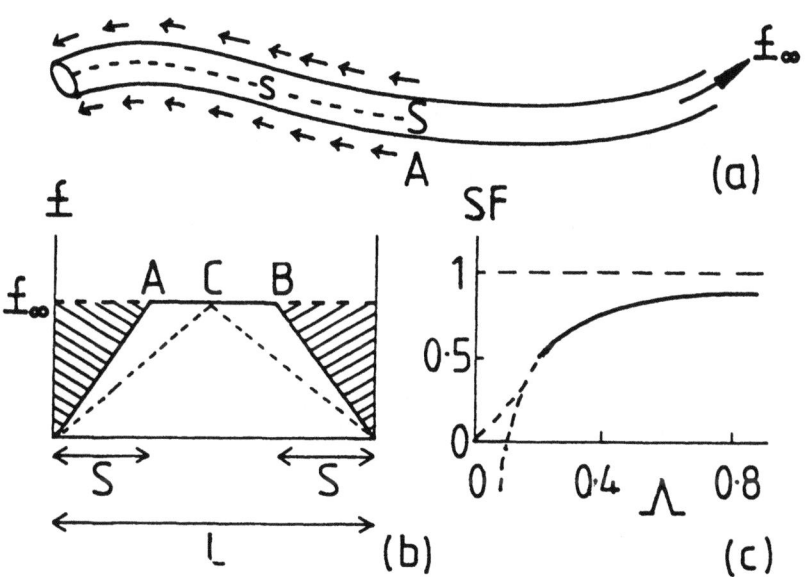

Figure 14 (a) Build-up of tension in a slipping zone from a fibre end to a point A. (b) Variation of tensile stress along fibre length L; the shaded zone represents the lost contribution to tensile stress. (c) Variation of slip factor (SF) with the factor $\Lambda = a\,f_\infty/2f_bL \; (= a/2\mu TL)$.

At A and B, the gripped tension equals the cumulative resistance to slip:

$$\pi a^2 f_\infty = 2\pi a f_b S \qquad (58)$$

where a = fibre radius and f_b is the shear stress on the surface resisting slip.

This leads to :

$$SF = \bar{f}/f_\infty = 1 - a\, f_\infty/2\, f_b L = 1 - \Lambda \qquad (59)$$

Values of the slip factor are shown in figure 14 (c). If f_b can be taken as constant, as might perhaps be the case for a bonded system, then there is no slip for small stresses, but the slip builds up as f_∞ increases. Clearly the whole analysis breaks down when A and B come together at C, S becomes greater than $L/2$, and the fibre is nowhere gripped. At this point one might expect a catastrophic drop to zero, or if there is an appreciable rearrangement of load sharing in the system a gradual fall to zero at the origin. Clearly the greater the fibre aspect ratio L/a and the stronger the bonding f_b, the less will be the effects of slippage.

In non-bonded systems, the staple fibres are held together by friction, and this usually derives from a geometry, such as twist in a yarn, which causes an externally imposed tension to be converted internally into transverse stresses between the fibre. If μ is the coefficient of friction, g is the level of transverse stress, and T is an operational factor representing the conversion of tensile stress to transverse stress, we have:

$$f_b = \mu\, g = \mu T f_\infty \qquad (60)$$

If the transverse stress is externally imposed, then f_b is constant and the behaviour described above occurs, with:

$$\bar{f}/f_\infty = 1 - a f_\infty/2\mu g L \qquad (61)$$

But in the more usual case, subsitution from equation (60) gives:

$$\bar{f}/f_\infty = 1 - a/2\mu T L \qquad (62)$$

The reduction in stress due to slippage is then independent of applied stress and is constant over the system stress-strain curve.

6.2 The exact slippage equation

The exact equation for the specific stress f acting at a distance s from the fibre end along the slip zone of the fibre in figure 14 (a) under varying transverse specific bonding stress f_b, with a system specific volume v, is:

$$f = (v/\pi a^2) \int_0^s (2a/v) f_b (1-2q/L) \, dq$$

$$= (2/a) \int_0^s f_b (1-2q/L) \, dq$$

$$\rightarrow (2/a) \int_0^s \mu g (1-2q/L) \, dq \qquad (63)$$

The parameter q is the distance from the fibre end, introduced as an auxiliary parameter for integration. The correction factor $(1-2q/L)$ is introduced because fibres which are neighbours of dq but have ends within a distance q will be slipping more severely; and so they will not merely be failing to resist slippage, but will be pulling in the direction to cause slip.

6.3 Application to twisted staple fibre yarns

Early attempts to predict the behaviour of a staple fibre yarn with idealised twist geometry ran into difficulties because the fibres on the surface were not gripped, so that $g = 0$, and hence $f = 0$, and there is no tension to grip the inner layers. An appropriate combination of twist and migration gives a self-locking structure.

Hearle [2] did carry out a full small-strain analysis of a yarn with idealised migration and a random distribution of fibre ends, taking account of the equations of fibre paths, the stress variation along fibres, and the equilibrium of forces on elements within the yarn. A set of 16 equations was obtained and Wagle [15] showed that they could be solved by numerical computation. However, apart from showing that a self-consistent treatment could be written down, the analysis was of little practical value because of its unsatisfactory combination of complication and unrealistic idealisation.

A qualitative approach based on equation (62) is more useful. The factor T will increase with yarn twist. However twist reduces the level of f_∞, the stress without slip, due to the effects of obliquity previously discussed. Consequently the yarn stress relative to fibre stress varies with yarn twist as shown in figure 15 (a). The same general arguments apply to yarn strength since, in a self-locking system, yarn breakage is triggered by fibre breakage. At high twists, the continuous filament value is approached. At lower twists, the staple fibre

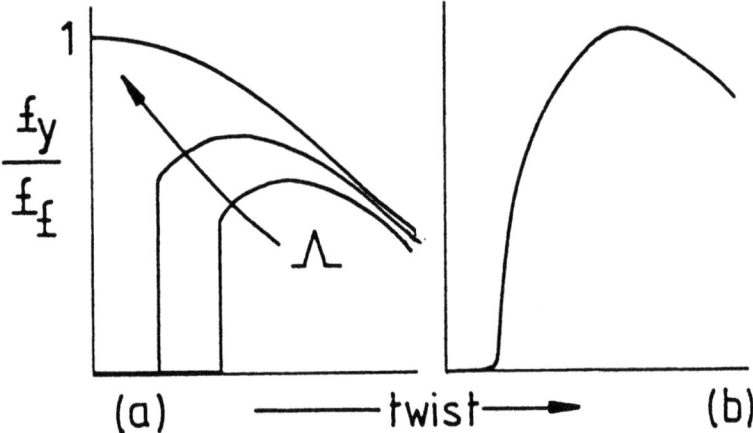

Figure 15. Variation of yarn stress f_y relative to fibre stress f_f at same strain with twist. (a) Simple theoretical prediction, with increasing Λ (decreasing slip factor). (b) Typical experimental result.

value falls increasingly lower. A critical situation is reached when half the stress has been lost, the fibres are nowhere gripped, and the stress falls to zero. At lower twists than this the material is a draftable roving in which fibres slide completely over one another. The experimental results, such as that in figure 15 (b), confirm this behaviour, except that there is some frictional resistance to drafting and some blurring of the sharp change over from a self-locking to a draftable system.

As $\mu L/a$ increases, the effects of slip will be reduced, and the optimum twist for peak strength will be reduced. Increased migration will have a similar effect, although it may be that practical yarn spinning has reached the point of diminishing returns.

An approximate off-shoot of Hearle's detailed analysis [2], considering only effects on fibres in the surface layer, led to a slippage factor $[1 - 2 \operatorname{cosec} \alpha (aQ/2\mu)^2/3L]$ where Q is the migration period. The details of the derivation are suspect, but it does suggest the empirical expression:

$$\frac{\text{yarn stress}}{\text{fibre stress at same strain}} = \cos^2\alpha \, [1 - k \operatorname{cosec} \alpha] \qquad (64)$$

where $\cos^2 \alpha$ is the approximate obliquity factor and k is a parameter dependent on L, a, μ and migration.

Such an expression does fit experimental results reasonably well.

6.4 Application to needled fabrics

Needled, spun-laced and felted fabrics are other materials dependent on frictional resistance to slip. The problem here is to decide the origins of the factor T, the transformation of tensile stress into transverse stress. Figure 16 shows that increased needling does lead from an incoherent to a self-locking structure, and a qualitative discussion of the mechanics has been given by Hearle [16].

In needled fabrics, one essential feature is the occurrence of curved fibre paths in the plane of the fabric, so that fibre tensions have normal components. However, in a simple web, there would be nothing to stop curved fibres sliding past one another and straightening. The needling induces the formation of perpendicular tufts of fibres, which act as pegs to be gripped by the fibres in the plane, thus preventing their own slippage. In spun-laced fabrics and wool felts, there will be a more complicated rearrangement with fibres passing round one another so as to give a mutual grip.

Hearle and Purdy [18] have carried out an analysis of an idealised model of a needled fabric based on the capstan equation for the increase of tension as fibres pass round the pegs, as

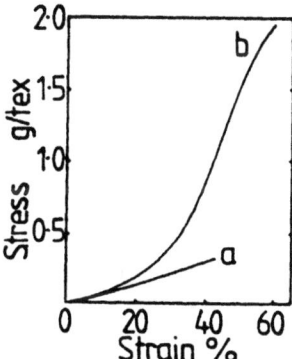

Figure 16. Stress-strain curve of an acrylic fibre needled fabric (a) at low needling density (b) at high needling density [17].

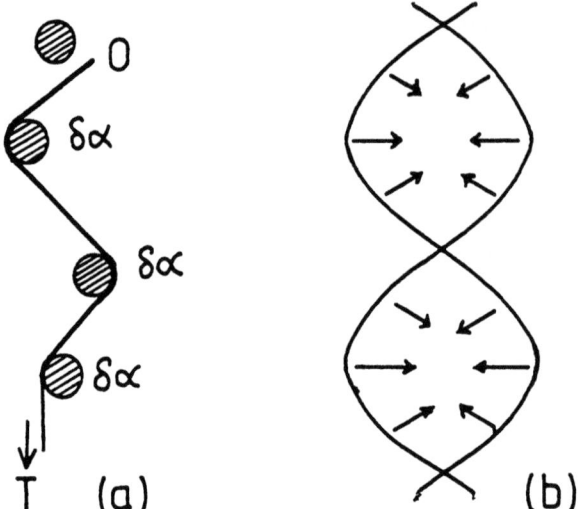

Figure 17.(a) Capstan effect in needled fabrics
(b) General pressure.

indicated in figure 17 (a). This is insufficient by itself,
since an exp ($\mu \Sigma \delta\alpha$) factor is of no value when tension at
the fibre end is zero. Consequently, a general inward pressure
derived from a tendency to area reduction within overlapping
zones, as indicated in figure 17 (b), is invoked. The capstan
equation is modified to include an external pressure. The
details of the analysis are highly dubious, but it does suggest
a dependence of the slippage factor on the curved form of the
fibre paths.

7. BREAKAGES AND STRENGTH

7.1 Continuous filament yarn strength

Extension of the idealised twisted yarn shows that fibre strain
decreases from a value equal to yarn strain for the straight
fibre at the yarn centre to a value reduced by a factor
($\cos^2\alpha - \sigma\sin^2\alpha$) at the yarn surface. Consequently in a uniform
yarn, the centre fibre will break when the yarn extension reaches
the fibre breaking extension. This always corresponds to the
maximum load, and, if there is appreciable interaction between
neighbouring fibres due to twist, leads to catastrophic breakage
of the whole yarn. The simplest rule will therefore be that yarn
breaking extension equals fibre breaking extension, and is

independent of twist; and the yarn strength falls according to the same rules as yarn stress at any strain, namely roughly as $\cos^2\alpha$.

A catastrophic break is triggered whenever there is appreciable interaction because the remote parts of the yarn are unaffected, and maintain the same load at the same overall extension. This load cannot be supported at the critical cross-section and so complete failure occurs. In a long specimen the strain reduction in the remote regions due to effects in the breaking zone is negligible. An alternative statement is that there is sufficient stored energy in the remainder of the yarn to cause the break to be completed. Different effects are found with short specimens [2]

In real yarns, as described by Hearle [2] a number of deviations from ideal structure lead to variations in breaking extension.

Fibre irregularity also plays a major role. At zero or low twists each filament breaks independently at its own weakest place, scattered along the yarn and over a range of extensions. But with a moderate degree of twist, there is local load sharing as discussed by Phoenix, weak places are not free to stretch differentially and break, but are supported by the neighbouring fibres, until eventually the stress on the weakest single cross-section leads to a break of one fibre, which in turn triggers failure over the whole cross-section. The mutual support leads to a higher peak load. Consequently continuous filament yarn strength increases with twist up to about $\alpha=10°$ and then falls due to the effects of obliquity. The support of weak places is a lesser analogue of the support of discontinuities in staple fibre yarns, and this shows that isolated breaks would not have much effect on the strength of a twisted yarn.

A number of detailed features of yarn breakage are described by Hearle [2].

7.2 Breakage of staple fibre assemblies

Similar arguments will apply to the strength of self-locking staple fibre systems, such as spun yarns and needled fabrics. Breakage must be triggered by the breakage of highly stressed fibres in the weakest cross-section. There will however be a general reduction to stress due to the reduced load carried by fibres slipping because their ends are close to the particular cross-section.

Furthermore the staple fibre systems are often bulky and contain crimped fibres. Appreciable additional extension can then occur due to fibre straightening and compacting of the material, so that breaking extension is greater.

7.3 Blended yarns

Up to the point at which fibres begin to break, the stress in a blended yarn follows a simple mixture rule, as illustrated in figure 18:

$$f_y = A f_1 + (1-A) f_2 \qquad (65)$$

where A is the mass fraction of the first fibre type, and f_y, f_1 and f_2 are specific stresses at the given strain in the blended yarn and in 100% yarns of types 1 and 2 respectively.

The problem is what happens after the first fibre type breaks. One could postulate three simple rules: (1) that a complete yarn break is triggered; (2) that the low-extension fibre ceases to contribute, but the other fibre continues to bear load up to its breaking extension; (3) that the first fibre type continues to contribute at its breaking value.

The first rule applies when there is a large fraction of type 1 and a high level of interaction, so that local load sharing throws a greater load on to the immediate neighbours than they can support.

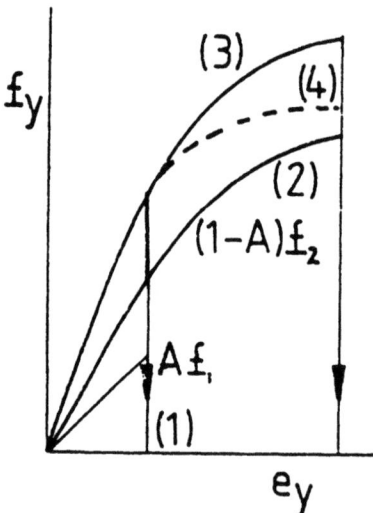

Figure 18. Possible responses of a blended yarn.

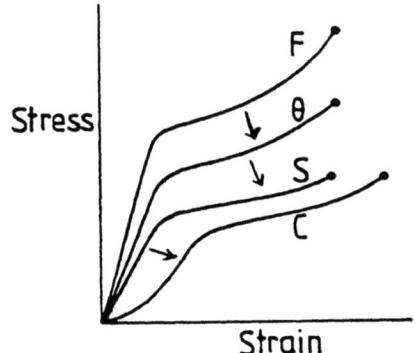

Figure 19. Modification of fibre stress-strain curve (F) by effects of orientation (θ), slippage at fibre ends (S), and fibre straightening and compacting of assembly (C).

The second rule would apply when there is very weak interaction, and independent breakage of the two types. It would apply to a low-twist continuous filament blend, but in any staple fibre system interaction is needed to give the yarn coherence and so it cannot apply.

The third rule is impossibly optimistic. However, even when fibres have broken, they can continue to carry some load, with a build-up of tension from zero at the ends, to the breaking tension in the centre. Further yarn extension leads to further breakage, and experiments have shown a progressive fibre breakage. This situation leads to the curves (4) in figure 18 (a), with a reducing contribution from the low-extension fibre. This subject has been described in more detail by Backer [2]. At some level of blend proportion and twist, the breaking extension changes from a low to a high value and the strength passes through a minimum value.

8. Summary of effects in dense assemblies.

Figure 19 summarises the behaviour of dense yarns and nonwoven fabrics in uniaxial extension, with effects which occur to varying degrees in different systems. The first approximation is the fibre stress-strain curve. This is then reduced by the effect of orientation, which can be well predicted by theoretical analysis. A further lowering comes from the effects of slippage at fibre ends, which can be treated by approximate theories. Finally there may be additional easy extension as fibres straighten and as an open, bulky structure becomes compact: treatments of this aspect are empirical. In bonded systems, there will also be a marked yielding effect when bonds fail.

REFERENCES.

1. C. Gegauff, Bull.Soc.Ind.Mulhouse, 77, 153, 1907.

2. J.W.S. Hearle in Structural Mechanics of Fibers, Yarns and Fabrics by J.W.S. Hearle, P.Grosberg and S. Backer, Volume 1. Chapters 2 to 7. Wiley-Interscience, 1969.

3. N. Jones, Int.J.Mech.Sci. 16, 679, 1974.

4. M. Konopasek and J.W.S. Hearle, J.Textile Inst., 65, 217, 1974.

5. N. C. Huang and G.E. Funk, Textile Res. J., 45, 14, 1975.

6. B. Neckar, Textile Res. J., 46, 545, 1976.

7. J.W.S. Hearle, J.Textile Inst., 60, 95, 1969.

8. J.W.S. Hearle and M. Konopasek, Appl.Polymer Symp., No. 27, 253, 1975.

9. W.F. Kilby, J. Textile Inst., 55, T589, 1964.

10. L.R.G. Treloar, J.Textile Inst., 56, T359, 1965

11. J.W.S. Hearle and J.T. Sparrow, J.Appl. Polymer Sci., in press.

12. J.W.S. Hearle in Structural Mechanics of Fibres, Yarns and Fabrics by J.W.S. Hearle, P. Grosberg and S. Backer, Volume 2, In press, Wiley-Interscience.

13. J.W.S. Hearle and V. Oszanlav, J.Textile Inst., in press.

14. J.W.S. Hearle, in Nonwovens '71. Textile Trade Press, 1971.

15. N.P. Wagle, Ph.D. thesis, University of Manchester, 1968.

16. J.W.S. Hearle in Needle-Felted Fabrics, Textile Trade Press 1972.

17. J.W.S. Hearle, M.A.I. Sultan and T.N. Choudhari, J.Textile Inst., 59, 103, 1968.

18. J.W.S. Hearle and A.T. Purdy, Fibre Sci.and Tech., 11, 127, 1978.

A CONTINUUM MODEL FOR YARN MECHANICS

J. J. Thwaites

Department of Engineering, University of Cambridge,
England

ABSTRACT. A theoretical model of yarn mechanics is presented in which the yarn is considered to be a solid rod with anisotropy corresponding to that of the filaments. Tensile and torsional properties are developed, together with the distribution of internal stress. The range of usefulness of the model is considered.

1. TWISTED YARN MECHANICS

1.1 Highly twisted yarns

Highly twisted yarns are not usually made into fabrics, but a study of their behaviour can be useful when considering the mechanics of such structures as tyre-cords, ropes and cables. Also, many yarns undergo high twist during manufacture. For example, the most important group of textured yarns is that made by the false-twist process. In this, initially untwisted yarn, generally of nylon or polyester fibre, is continuously twisted and in this state passes through a heater where the deformation is set. It is then de-twisted. Because the process is continuous and the threadline essentially linear, unsteady operation can result in large changes in tension and torque when the yarn is in the highly twisted state [1]. The properties of the polymers used depend strongly upon their thermo-mechanical history, so that these changes have considerable effect on the properties of the textured yarn produced. It is important therefore to understand the relations between these changes and the corresponding changes in yarn twist and axial strain.

In order to deal with this tensile and torsional behaviour, a theoretical model is used in which the yarn is considered to be a solid rod with anisotropy corresponding to that of the filaments. The analysis is elastic, but could readily be extended to the visco-elastic case. A similar treatment of yarn bending behaviour would be extremely difficult.

1.2 Theoretical models

Once a highly-twisted yarn is formed, the filaments are tightly packed and subject to high transverse stress. It is clear that, during any subsequent deformation of the yarn, the filaments are very much constrained. A model in which the actual constraints are included is not feasible, but one which includes a major constraint has been used by Hearle and others extensively to study tensile behaviour. Torsional behaviour has been neglected except for yarns of low (or zero) initial twist.

The theoretical model used by Hearle [2] is one in which there is, during incremental deformation, no relative movement, in the direction of their axes, between the filaments of the yarn. Transverse stresses are included, but although different analytical methods have been used, in general they involve assumptions about some stress components without relating all of these to corresponding strains. In particular, shear stresses are assumed to be negligible, on the premise that they have little effect on tensile behaviour. For the computation of torsional properties, this is hard to justify, because the transverse stresses will clearly prevent free rotation of the filaments. A model based upon the assumption that the yarn is so tightly packed as to prevent all relative movement between filaments is more reasonable.

The same filament geometry is appropriate, i.e. the filaments are assumed to lie in concentric helical paths with helix angle increasing with radius within the yarn so that the twist in any cross-section is uniform. This model is then essentially a circular cylinder possessing axial symmetry and composed of material with fibre-anisotropy, i.e. it is transversely anisotropic. At any point in the material, the axis of the 'fibre' is inclined to the cylinder axis by an angle determined by the radial position.

2. ANALYSIS OF THE MODEL [3]

The constitutive equations for elastic deformation relate to the fibre axes, so that a suitable co-ordinate system is a helical one in which the axes are obtained from the standard cylinder co-ordinate axes (r, ϕ, z) by rotation θ about the radial axis. The directions of the axes are then, radial (1) and perpendicular to the radius, inclined at $(\pi/2 - \theta)$ and θ to the cylinder axis (2 and 3 respectively). The relation between θ and radius r derives from the uniformity of twist in any cross-section. It is [2]

$$R\tan\theta = r\tan\alpha \qquad (1)$$

where R is the yarn radius and α the surface helix angle.

An element of the cylinder and the stresses on its faces are shown in Fig. 1. When the cylinder is subjected to axial tension and torque, the components of stress are, because of symmetry, independent of the cylinder co-ordinates ϕ and z and the equilibrium equations are satisfied provided

$$r\frac{\partial\sigma}{\partial r}rr + \sigma_{rr} - \sigma_{\phi\phi} = 0 = \tau_{r\phi} = \tau_{zr}$$

In the helical co-ordinate system, σ_{11} is identical with σ_{rr} and $\sigma_{\phi\phi} = \sigma_{22}\cos^2\theta + \sigma_{33}\sin^2\theta + 2\tau_{23}\sin\theta\cos\theta$. Using (1), the equilibrium equations become

$$\sin\theta\cos\theta\frac{d\sigma}{d\theta}11 + \sigma_{11} - \sigma_{22}\cos^2\theta - \sigma_{33}\sin^2\theta - 2\tau_{23}\sin\theta\cos\theta = 0 \qquad (2)$$

and $\tau_{12} = \tau_{31} = 0$.

Given an axial strain ϵ and axial twist per unit length β, because of the axial symmetry, the components of direct strain, again in cylinder co-ordinates are

$$\epsilon_{rr} = \frac{\partial u}{\partial r} \qquad \epsilon_{\phi\phi} = \frac{u}{r} \qquad \epsilon_{zz} = \epsilon$$

where u, a function of r as yet undetermined, is the radial

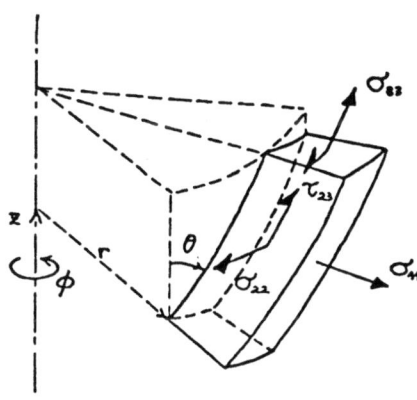

Fig. 1. Yarn element and stress components.

displacement at the point. The only non-zero component of shear strain is

$$\gamma_{\phi z} = \beta r$$

The corresponding non-zero components in 1-2-3 co-ordinates are

$$\left.\begin{aligned}
\varepsilon_{11} &= \frac{\partial u}{\partial r} \\
\varepsilon_{22} &= \varepsilon \sin^2\theta + (u/r)\cos^2\theta - \gamma \sin^2\theta \\
\varepsilon_{33} &= \varepsilon \cos^2\theta + (u/r)\sin^2\theta + \gamma \sin^2\theta \\
\gamma_{23} &= 2\{(u/r) - \varepsilon\}\sin\theta\cos\theta + \gamma\tan\theta(\cos^2\theta - \sin^2\theta)
\end{aligned}\right\} \quad (3)$$

The constitutive equations for elastic deformation of this anisotropic solid, taking account of fibre symmetry, involve the Young's modulus E_3 along the fibre axis, a transverse Young's modulus E_1, two corresponding Poisson's ratios ν_{12}, ν_{13} and a shear modulus G. They are

$$\left.\begin{aligned}
E_1\varepsilon_{11} &= \sigma_{11} - \nu_{12}\sigma_{22} - \nu_{13}\sigma_{33} \\
E_1\varepsilon_{22} &= -\nu_{12}\sigma_{11} + \sigma_{22} - \nu_{13}\sigma_{33} \\
E_1\varepsilon_{33} &= -\nu_{13}\sigma_{11} - \nu_{13}\sigma_{22} + E_1\sigma_{33}/E_3 \\
G\gamma_{23} &= \tau_{23}
\end{aligned}\right\} \quad (4)$$

In principle, the nine equations (2, 3, 4) can be solved in order to find the eight stresses and strains and the displacement u, but the mathematics involved is difficult and unrewarding. In practice, textile polymeric materials are approximately incompressible [4]. Making this further assumption about the cylinder material, a simple solution can be obtained. Thus, since the sum of the direct strain components must be zero, from (3),

$$\frac{\partial u}{\partial r} + \frac{u}{r} + \varepsilon = 0$$

and hence

$$u = \tfrac{1}{2}\varepsilon r$$

Also, in (4), the coefficients of σ_{11}, σ_{22} and σ_{33} must independently add to zero, so that there are two relations between the elastic constants:

$$\nu_{13} = 1 - \nu_{12} = E_1/(2E_3)$$

Then, directly from (3) and (4), using (1),

$$\left.\begin{array}{l} \sigma_{22} - \sigma_{11} = E\,\dfrac{1-\nu}{1+\nu}\,\sin^2\theta\,(3\varepsilon - 2\gamma) \\[1em] \sigma_{33} - \sigma_{11} = E\left\{\left(1 - \dfrac{3\nu}{1+\nu}\sin^2\theta\right)\varepsilon + \dfrac{2\nu}{1+\nu}\sin^2\theta\gamma\right\} \\[1em] \tau_{23} = G\sin\theta\cos\theta\{-3\varepsilon + (1 - \tan^2\theta)\gamma\} \end{array}\right\} \quad (5)$$

where E is written for E_3, ν for ν_{12} and $\gamma = \beta R\cot\alpha$. Substituting from (5) for σ_{22}, σ_{33} and τ_{23} the radial equilibrium equation becomes

$$\dfrac{\cot\theta}{E}\cdot\dfrac{d\sigma_{11}}{d\theta} = \left\{\dfrac{3}{1+\nu}(1+\cos^2\theta) - 2 - 3m\cos^2\theta\right\}\varepsilon$$

$$+ \left\{2 - \dfrac{2}{1+\nu}(1+\cos^2\theta) + m(2\cos^2\theta - 1)\right\}\gamma$$

where m is written for $2G/E_3$.

Since the radial stress must be zero at the cylinder surface, the appropriate boundary condition is $\sigma_{11} = 0$ when $\theta = \alpha$, so that

$$\dfrac{\sigma_{11}}{E} = \left\{\left(\dfrac{3}{1+\nu} - 2\right)\ln\dfrac{\sec\theta}{\sec\alpha} - \dfrac{3}{2}\left(\dfrac{1}{1+\nu} - m\right)\left(\cos^2\theta - \cos^2\alpha\right)\right\}\varepsilon$$

$$+ \left\{\left(\dfrac{2\nu}{1+\nu} - m\right)\ln\dfrac{\sec\theta}{\sec\alpha} + \left(\dfrac{1}{1+\nu} - m\right)\left(\cos^2\theta - \cos^2\alpha\right)\right\}\gamma \quad (6)$$

The components σ_{22} and σ_{33} follow from (5).

The required yarn tension and axial torque are, respectively,

$$T = \pi R^2\cot^2\alpha\int_0^\alpha (\sigma_{22}\sin^2\theta + \sigma_{33}\cos^2\theta - 2\tau_{23}\sin\theta\cos\theta)\,2\tan\theta\sec^2\theta\,d\theta$$

$$Q = \pi R^3\cot^3\alpha\int_0^\alpha \{(\sigma_{33} - \sigma_{22})\sin\theta\cos\theta + \tau_{23}(\cos^2\theta - \sin^2\theta)\}2\tan^2\theta\sec^2\theta\,d\theta$$

For the case considered, these reduce to

$$\left.\begin{array}{l} T = (a_{11}\varepsilon + a_{12}\gamma)\pi R^2 E \\[0.5em] Q = (a_{21}\varepsilon + a_{22}\gamma)\pi R^3 E\cot\alpha \end{array}\right\} \quad (7)$$

in which the non-dimensional stiffness coefficients a_{ij} are given by

$$a_{11} = 1 - 3f_1(\alpha) - 9f_2(\alpha)$$

$$a_{12} = a_{21} = f_1(\alpha) + 6f_2(\alpha)$$

$$a_{22} = (m/4)\tan^2\alpha - 4f_2(\alpha)$$

where

$$f_1(\alpha) = \{1 - (3m/2)\}(1 - 2\cot^2\alpha \ln \sec\alpha)$$

$$f_2(\alpha) = \left(\frac{1}{1+\nu} - m\right)\left[\tfrac{1}{2}\sin^2\alpha - 1 + 2\cot^2\alpha \ln \sec\alpha\right]$$

(8)

All the elastic moduli are functions of these coefficients. The most obvious are

tensile modulus	$\pi R^2 E a_{11}$
torque-strain ⎫ 'modulus' tension-twist ⎭	$\pi R^3 E a_{12} \cot\alpha$
torsional rigidity	$\pi R^4 E a_{22} \cot^2\alpha$

There are also torque-tension gradients,

at constant twist	$R(a_{12}/a_{11})\cot\alpha$
at constant strain	$R(a_{22}/a_{12})\cot\alpha$

Others relevant to yarns can readily be obtained, e.g.

torsional rigidity, at constant tension rather than constant strain $\quad \pi R^4 E \left(a_{22} - \dfrac{a_{12}^2}{a_{11}} \right) \cot\alpha.$

3. MODEL PREDICTIONS

The usefulness of the above analysis in predicting the dependence of the various moduli on the several parameters depends, to some extent, on experimental verification. No such verification is possible for inferences about the stress distributions, which are just as useful. Nonetheless these inferred stress distributions can be used to explain why the measured moduli are not always as predicted.

3.1 Yarn Moduli

Two of the moduli referred to above are shown in Figs. 2 and 3, as functions of the surface helix angle α, up to about 55°. For filament yarns which are heated as they are continuously twisted, as in the false-twist process, this is found to be a practical upper limit beyond which buckling occurs. For unheated yarns the corresponding limit is about 40°. The variation with twist is

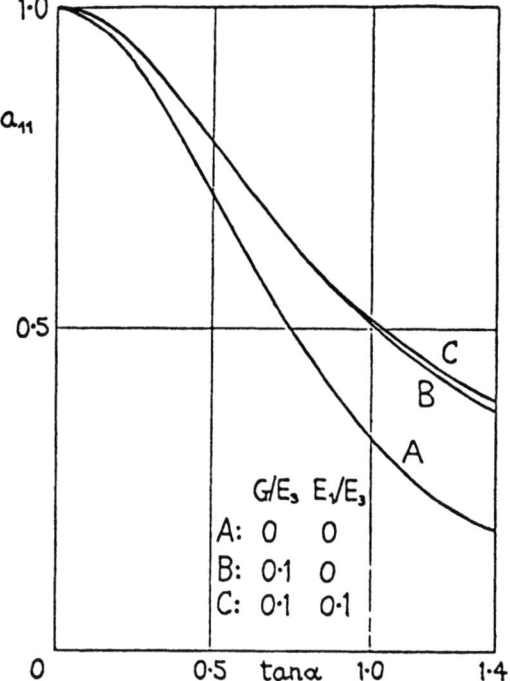

Fig. 2. Dimensionless tensile modulus.

striking. The tensile modulus reduces with increasing α, and the torsional rigidity increases, by substantial factors from the values they would have for a rod made of isotropic material. Both changes reflect the importance of the stress component σ_{33}, which at high twist completely dominates the torsional rigidity. The torque-strain 'modulus' is also strongly influenced by σ_{33}. It would, of course, be zero for an isotropic rod.

The effect of including shear stresses, or alternatively of increasing the shear modulus, is to increase both the tensile modulus and the torsional rigidity. This influence increases with increasing α for the former and decreases for the latter. The effect of shear stresses on the torque-strain 'modulus' is a reduction by a factor which does not vary greatly with twist. The greatest influence of shear stress is on the torque-tension gradients, only one of which is shown (Fig. 4). These, again, are peculiar to anisotropic structures such as yarns and are of considerable importance in textile threadline mechanics.

The choice of $G = 0.1 E_3$ for Figs. 2 to 4 represents approximately a drawn polyester fibre. The corresponding transverse Young's modulus would also be $E_1 = 0.1 E_3$. The effect of taking this value rather than $E_1 = 0$ is very small in all cases and for most practical purposes could be neglected. It is worth noting that a

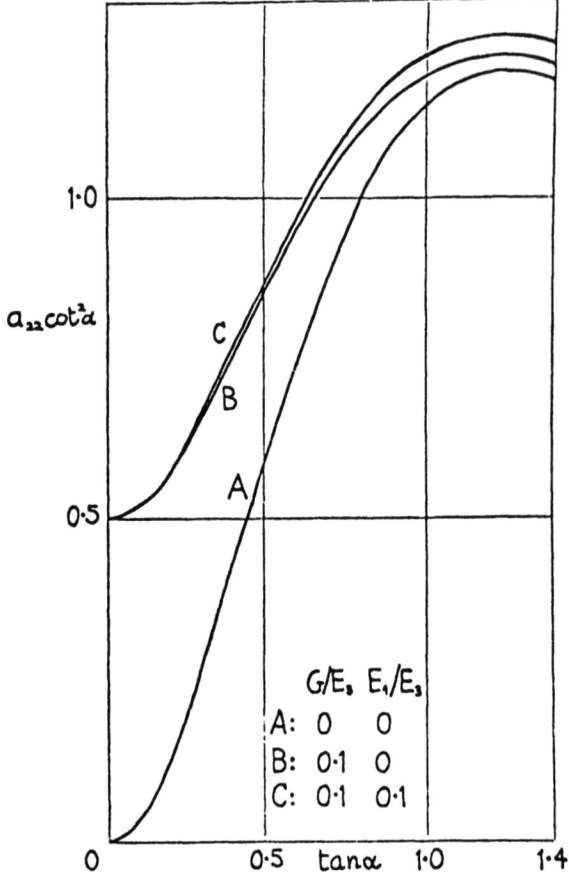

Fig. 3. Dimensionless torsional rigidity.

consequence of putting $E_1 = 0$ is to make the fibre transverse stresses, σ_{11} and σ_{22}, equal. The special case $G = E_1 = 0$ is thus the same as Hearle's constant volume deformation case.

3.2 Comparison with Experiment

One would expect that the higher the twist and/or tension, the greater would be the degree of filament constraint and, therefore, the more realistic the model. It is found that the shape of the filament cross-section is also significant.

For filament yarns twisted continuously, and heat-set, at $14° \leq \alpha \leq 52.2°$, under a tension of 2.3 cN/tex, the theoretical moduli are in good agreement with experiment. At high twist the experimental values lie close to the predictions for which shear stresses are included, i.e. those for which G is taken to be the fibre shear modulus. As twist decreases, the experimental values tend towards those predicted using $G = 0$. In all cases the experimental values

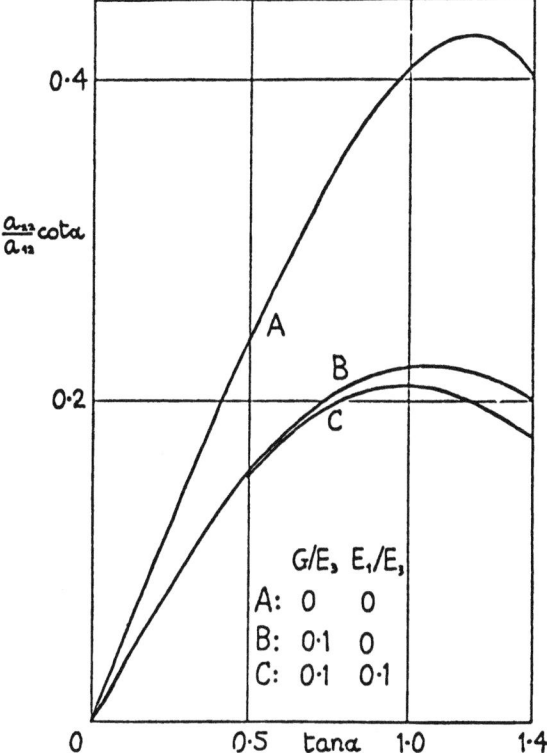

Fig. 4. Dimensionless torque-tension gradient at constant twist.

lie between the two theoretical values. Thus the yarns can be said to behave as solid rods, with appropriate material properties. Confirmation of this is provided by the measurement of hysteresis loops obtained during the various modulus tests. The ratio of energy lost per cycle to maximum energy stored for the yarn is very little different from that for the fibre material itself (in tension). This shows that there is little or no frictional energy loss due to relative motion between filaments.

For non-heat-set continuously twisted yarns, with initial tension 2.3 cN/tex, over a smaller range of twist ($12.4° \leq \alpha \leq 40.0°$) the experimental moduli are generally as predicted for $G = 0$. Substantially higher making tension is required in order to obtain values which indicate that shear stresses play a significant part. At tensions of 1 cN/tex and below the model is not useful at all. It seems clear that the difference engendered by heat-setting is due to the fact that the filament cross-sections become so much deformed that their arrangement in the yarn cross-section appears to be approximately hexagonal-closed-packed.

3.3 Filament stresses

Although there is no means of checking it experimentally, the variation of stress throughout the cross-section is of some interest. The dominance of σ_{33} when twisting has already been referred to, but for all cases the relation between τ_{23} and σ_{22} is important since it bears on the validity of the model. For this to be justifiable the shear stresses must be supported mainly by inter-filament friction, although for heat-set yarns there must be some element of interlocking, perhaps even of adhesion. The shear stress is independent of α and also of any assumption about E_1. It is therefore reasonable to consider the simpler case $E_1 = 0$ for discussion.

The stress due to an increase in twist is hydrostatic at the centre of the yarn. σ_{22} ($=\sigma_{11}$) remains compressive, σ_{33} rises to a large tensile value in the outer parts of the yarn. It is possible for inter-filament friction to support the shear stress over most of the cross-section, except at low twist (Fig. 5a). The stresses σ_{11} and σ_{22} due to an increase in extension are small and even become tensile in the outer parts of the yarn at high enough twist (Fig. 5b). If these were the total stresses then clearly no shear stress could be supported by friction. But they are incremental; a yarn in a highly twisted state is, of course, in a state of stress already. This is true for a heat-set yarn too, for example, in the cooling zone of the false-twist process, where it is held under the threadline tension and torque.

Fortunately, some assessment of this state of stress can be made. Experimental observations of the relation between torque and tension, when twisting continuously, in the steady-state, show it to be the same as the (incremental) torque-tension gradient at constant strain, for the appropriate yarn twist. This is true over a wide range of tensions, twists and different yarns. It is <u>as if</u> nearly all the twisting were achieved at very low stress and the final tension and torque were developed during the last little increment. This is clearly not how it happens, but a fair inference from the relation is that the stress distribution in the yarn is the same as it would be for a twist increment (Fig. 5a). Thus when considering the stress distribution during incremental deformation, since the incremental stresses are generally at a lower level than those in the yarn already, the total shear stress is substantially less, over most of the cross-section, than the total (compressive) transverse stress. This is true even for an increment in extension. That is, except at low twist or low tension, when the incremental stresses may be of the order of those already there, it is possible for inter-filament friction to support the shear stresses, in all cases, over most of the cross-section.

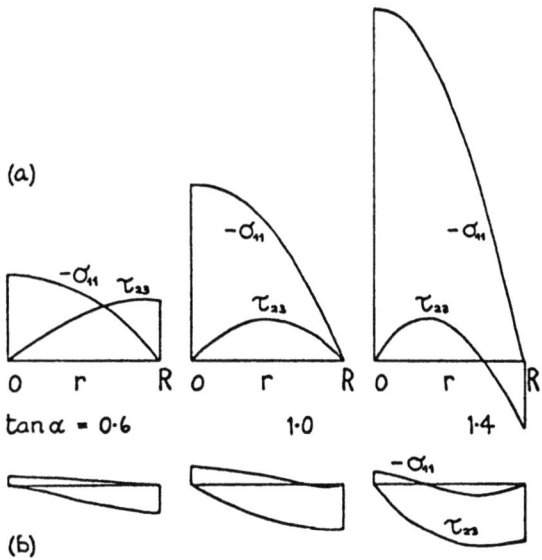

Fig. 5. Relative distribution of stress for equal increments in tension, (a) for an increase in twist, (b) for an increase in extension.

REFERENCES

1. J. J. Thwaites, J. Text. Inst., 69, 269 and 276.
2. J. W. S. Hearle, Structural mechanics of fibers, yarns and fabrics, Vol 1, Interscience, New York and London, 1969, chapter 4.
3. J. J. Thwaites, Int. J. Mech. Sci., 19, 161.
4. I. M. Ward, Mechanical properties of solid polymers, Interscience, New York and London, 1971, chapter 10.

THE COMPRESSION OF FIBROUS ASSEMBLIES, WITH APPLICATIONS TO YARN MECHANICS

G. A. Carnaby

Wool Research Organisation of New Zealand (Inc.),
Christchurch, New Zealand

ABSTRACT. Research on the mechanics of the compression of fibrous assemblies is reviewed and the results of recent studies showing that the behaviour of fibres under compression has a major influence on the structure and mechanics of spun yarns are discussed. Much of the earlier research was concerned with resilience and bulk, and the behaviour of 'random' fibre masses. However, the fibres in yarns are not usually randomly oriented but formed into an aligned and generally helical array. The compression mechanics of aligned fibres, such as those in slivers and yarns, was first studied in relation to drafting theory. The number of fibre contacts and the forces at contact points affect drafting behaviour and also relate to the resistance to bulk compression of the fibres as they are condensed into the yarn under the influence of the tensions induced by twisting. Similarly, the recovery from lateral compression of the fibrous bundle plays a crucial role during yarn relaxation, e.g., in yarn bulking. The compression properties of the fibres also affect various yarn mechanical properties, e.g., the tensile behaviour.

1. THE MECHANICS OF COMPRESSION OF A RANDOMLY ORIENTED FIBROUS MASS

The study of the compression mechanics of random fibrous masses is almost completely dominated by an early and important theoretical paper by van Wyk [1]. This is still the only fully fledged attempt at predicting the experimental results obtained from compression testing fibrous masses but in recent times, after an interval of nearly thirty years, some progress has been made in refining one of the main aspects of the van Wyk theory.

The object of this theory was to derive a relationship between the pressure applied to a wad of wool fibres and the volume of the wad. In deriving this relationship van Wyk considered the mass of fibres under compression as a 'system of bending units'. These bending units are the elements of fibre between adjacent fibre-to-fibre contacts. Twisting, extension, and frictional slippage effects were not included in his analysis. The fundamental equation governing this 'rod' bending behaviour for small strains is the following:

$$F = \frac{24 \, i \, Y}{b^3} y \quad \quad \quad \quad \quad \quad \quad \quad \quad (1)$$

where F = the vertical force applied midway along a horizontally supported rod;
i = the moment of inertia of the rod in bending ($\pi d^4/64$);
Y = Young's modulus of elasticity;
$2b$ = the length of the rod; and
y = the depression of the rod at its midpoint.

There are a number of factors to be considered in developing a pressure/volume relationship from this function. Firstly, there is the problem that in a real fibrous mass the forces at contact points will not alternate in direction at regular intervals or be of equal magnitude. This problem has not been solved either by van Wyk or by subsequent workers and at present it is overcome simply by replacing the factor 24 by a dimensionless constant k and using a calculated mean distance between contact points for the term b. As the assembly is considered to be random and the fibres have an equal chance of orientation in any direction, the effect of the angle between the fibres and the direction of compression is also simply taken into account in the rather nebulous parameter k on the assumption that the correction involves only a statistical constant of proportionality. This is only satisfactory, however, provided that the assumption of randomness applies to both the uncompressed and the compressed masses. Assuming for the moment that this is the case the fundamental equation may be written as

$$dF = \frac{k \, i \, Y}{b^3} dy \quad \quad \quad \quad \quad \quad \quad \quad \quad (2)$$

Now, all these other factors having been taken into what is essentially a dimensionless constant, k, the remaining problem is to derive a relationship for b, the distance between fibre contact points, as a function of ν, the specific volume of the assembly.

1.1 The mean distance between fibre-fibre contacts

This parameter is a fundamental property of a fibrous assembly in its own right and again van Wyk [1] was the first to study it in deriving his pressure/volume relationship. Van Wyk gave two separate approaches to the calculation of this parameter which have been termed by Duckett and Cheng [2] 'the projected-fibre-area method' and 'the projected fibre length method'; these have been shown to be equivalent [2].

Taking the first of these methods, the number of contacts is calculated from the ratio of twice the total area of the fibres projected onto the base of a cylinder containing the fibres to the area of the base itself. This ratio is actually the number of times a sphere of diameter D_f moving vertically in the container would strike one of the fibres and this is the same for a fibre held along the locus of the sphere. The mean distance b is therefore given by

$$b = 2/\pi L_v D_f \qquad \qquad (3)$$

where L_v = the total length of fibre per unit volume.

This is equivalent to

$$b = \rho_f D_f v / 2 \qquad \qquad (4)$$

where ρ_f = the density of the fibre; and
v = the specific volume of the total fibrous assembly.

It can be seen that b increases in direct proportion to the specific volume or the openness of packing.

1.2 The pressure *versus* specific volume compression curve

By taking the above result for b, inserting it in Equation (2), and using the results for a vertical cylinder with a cross-sectional area of unity

$$dp = \frac{c L_v}{b} dF$$

where p = the applied pressure; and
c = the thickness of one layer (for a large number of elements the value of c can be taken as $b/\sqrt{3}$).

If v is the volume of fibre considered

$$dv = -\frac{v}{c} dy$$

and we obtain

$$dp = -\frac{k Y m^3}{12 \rho_f^3 v^4} dv$$

where m is the mass of the fibre.

This reduces to

$$p = A\left\{\frac{1}{v^3} - \frac{1}{v_0^3}\right\} \quad \cdots \cdots \cdots \cdots \quad (5)$$

where A is a constant and v_0 is the value of v when $p = 0$.

This is the desired relationship describing the compression behaviour of the mass.

1.3 Limitations of the van Wyk theory

There are obviously a number of shortcomings in this theory, not the least of which is the necessity of introducing the term k. Since k is unknown, it is not possible to predict an explicit value for the resistance of a mass of fibres to compression from a consideration of single fibre properties. Dunlop et al. [3, 4] in fact characterised the compression behaviour of a wide range of wool fibre types by measuring A and v_0 values experimentally but found that for various reasons, especially frictional slippage effects, the inverse cubic relationship itself gave only moderate agreement with the compression behaviour of wool fibres when tested over a range of pressures and specific volumes. Another main reason for the observed discrepancies from Equation (5) has been discussed by Stearn [5], who calculated a correction factor for b, the distance between fibre contacts, to take into account the fact that as the random mass of fibres is compressed in one direction a fibre orientation effect is introduced. He showed theoretically that this preferential orientation which is introduced perpendicular to the line of compression serves to decrease b more slowly with reductions in v, the specific volume, than is predicted by Equation (4) based on a purely random orientation in both the uncompressed and compressed assemblies. This result has been confirmed by Komori and Makishima [6] who have shown that in the limit, when all the fibres lie at right angles to the applied compression force, the distance between contact points is greater than for the equivalent random case by a factor $\pi^2/8$. The significance of this in terms of the pressure/volume relationship is, however, rather more subtle than a simple correction to the term b. If the induced orientation effect is to be incorporated into b it should also be incorporated into k. This includes among other things a statistical constant of proportionality to take into account the supposedly random range of angles between the rods and their deflections in the line of action of the forces at the contact points. With a smaller component of contact-point force along the fibre direction it would appear that slippage at these points would also become less likely. In any event data taken [7-9] from log-log plots of pressure *versus* specific volume for a variety of fibres indicate a range of values for the power term for different fibres of -2.17 to -5.00. In the absence of any better theoretical model it would seem at this stage that van Wyk's inverse cubic relationship, although limited, is the most soundly based approach available.

2. THE COMPRESSION OF ALIGNED FIBRE ASSEMBLIES

While the idealised random fibre assembly is a useful mathematical abstraction it only applies to the minority of situations in which compression of fibre assemblies occurs. In nearly all the yarn structures which are to be discussed shortly, the fibres have a preferential orientation along the yarn axis and it is necessary to consider the effect of this on the nature of the loading/compression curve.

Clearly, the main parameter in van Wyk's theory which should be reconsidered is that of b, the mean distance between contact points. Three different authors working on the drafting of slivers [10-12] have all provided formulae for calculating the mean distance between contact points in slivers based on the statistical probability distribution of the angle of contact between the fibres in the sliver. In practice, the fibres are never perfectly aligned and even in slivers made from perfectly straight man-made fibres a range of fibre angles to the sliver axis is always found. For combed tops made from straight fibres an average angle of deviation from the sliver axis might be in the region of $6\frac{1}{2}°$ [11] while for crimped fibres such as wool the deviation is usually in the region of $20-25°$ [12]. However, in all these papers it is shown that the formula for b involves simply a correction factor $f(\theta)$ to account for the scatter of fibre orientations. The equations therefore generally take the form

$$b = \frac{\rho_f D_f \nu}{2} f(\theta) \quad \cdots \cdots \cdots \cdots \quad (6)$$

Very recently a further refinement to the calculation of b in fibre assemblies has been introduced by Komori and Makishima [6, 13]. These authors have shown that by measuring the number of fibres crossing secant planes cutting the assembly it is possible to calculate theoretically a density function for the fibre orientation. To specify orientation in three-dimensional space requires a set of two angles. If the distribution function of their orientation found in this way is given by $\Omega(\theta, \phi)$ then

$$b = \frac{\rho_f D_f \nu \pi}{2 I} \quad \cdots \cdots \cdots \cdots \quad (7)$$

where I is the correction factor for orientation effects and is given by

$$I = \int_0^\pi d\theta \int_0^\pi d\phi \int_0^\pi d\theta' \int_0^\pi d\phi' \left|1 - \{\cos\theta\cos\theta' + \sin\theta\sin\theta'\cos(\phi - \phi')\}^2\right|^{\frac{1}{2}}$$
$$\times \Omega(\theta,\phi)\Omega(\theta',\phi')\sin\theta\sin\theta' \quad . \quad (8)$$

From the point of view of the loading/compression curve,

however, the most significant aspect of all these theories is the predicted linear relationship between b and ν. This is important because it leads to the simple result that orientation effects do not affect the basic inverse cubic nature of the relationship between applied pressure and specific volume. They simply provide correction factors for the A term.

If we now go one step further and consider the density function of orientation for fibres in a yarn it is clear that the structure is not homogeneous. In order to work out a precise relationship for the resistance of fibres to packing within the yarn it would be necessary to calculate $\Omega(\theta,\phi)$ as a function of both radius and axial position. It is clear that the mean fibre-fibre contact distance and hence the resistance to close packing will vary with the radius but this analysis has not been carried out at present.

3. THE EFFECT OF THE COMPRESSION PROPERTIES OF FIBRES ON YARN STRUCTURE

The resistance of roughly parallel fibres to condensation under the transverse pressures in twisted structures was considered by Smith [14] who showed in an early rather elementary model that the fibres were not close-packed as is often assumed in theories of yarn structure, but that the specific volume of the structure decreased with increases in twist, and hence transverse pressure, in a manner that might be explained by a relationship of the van Wyk type.

Several workers [15-17] have made experimental studies of the compaction of yarns under the influence of twist and in the most detailed of these studies [16] fibre crimp, the main fibre characteristic affecting the compression behaviour of loose wool [18], was also found to be the main fibre property affecting the bulk of the yarn after spinning. It was shown that the resultant yarn packing density was determined principally by the interaction of the opposing effects of fibre crimp (and hence resistance to compaction) and yarn twist.

Mention should also be made of some recent Japanese [19] work relating to the packing together of helices with different phase angles. By treating the results for three coils statistically, the authors derive a structure and diameter for the yarn. Although very much more satisfactory than the simple models of close packing often employed, the approach is still purely geometrical as the fibres in the relaxed yarn are taken as being in their undeformed helical shapes. It would be interesting to see this model developed to take into account mechanical interaction between the helices under the influence of an applied yarn twist.

This incorporation of mechanical interactions is important

because, in order to describe theoretically many aspects of yarn
behaviour and particularly yarn structural modification, it is
imperative that the structure of a yarn which is not subject to any
external forces be still regarded as an equilibrium state arising
from a balance of internal forces. Grosberg [12] has shown that
even the fibres in an untwisted sliver contain significant elastic
strain energy which is locked in by the friction at contact points.
In a yarn, which is a relatively dense structure, the locked-in
strains are very much larger and include as a major factor the
deflections due to the steric interference of the fibres during the
close packing itself. Unless this is recognised it is quite impossible to model the behaviour of yarns in a variety of important
commercial applications. Some examples of these applications in the
wool carpet industry include bulking during yarn steaming or dyeing,
tuft burst in carpet finishing, twist balance and snarling effects,
and yarn twist setting.

Some progress has recently been made towards adequately modelling this behaviour. It has been shown [20] that the energy associated with the resistance of fibres to close packing is in fact one
of the two main terms in the energy minimisation. When a wool yarn
is relaxed by steam, or a wetting and drying sequence, the attempts
of the fibres to regain their natural crimped form are a major cause
of the yarn bulking that normally occurs. The approach to this problem was theoretical [20] and the resistance of the crimped fibres
to close packing in a parallelised helical array was incorporated
through an energy term based on a van Wyk relationship between
pressure and the inverse cube of specific volume. Residual fibre
tensile strains from processing were also included and it was shown
that bulking continued until these two driving mechanisms became
balanced by an internal build-up in tensile and axial strains that
varied with the radius. It was indeed shown theoretically how all
unstrained yarn structures are equilibrium states in which the
axial fibres are subject to axial compression and that this may
lead to buckling of fibres along the yarn axis depending on the
geometry of the yarn and the distance between fibre-to-fibre contact points in the central regions.

4. THE EFFECT OF THE COMPRESSION PROPERTIES OF FIBRES ON YARN MECHANICS - TENSILE BEHAVIOUR

In 1976 Carnaby and Grosberg [21] showed by testing staple-fibre
yarns at short gauge lengths, where slippage near the fibre ends is
virtually eliminated, that the relative weakness of staple-fibre
yarns could not be explained solely by correcting successful continuous filament yarn theories to allow for fibre slippage. They showed
in fact that slippage was a minor effect especially at low strains
and that consolidation of the yarn structure had a far greater
effect than slippage on the strain distribution and the resulting

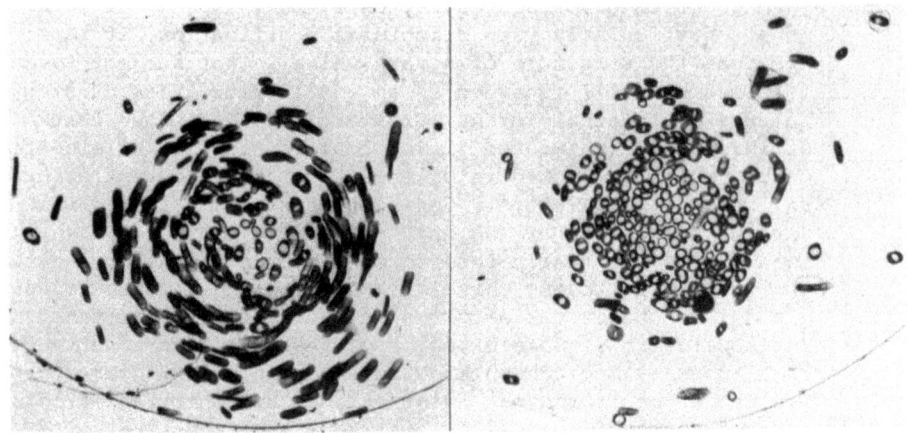

Fig. 1. Lateral fibre movements in a tensioned semi-worsted wool carpet yarn. A: before extension; B: after extension

tensile properties.

Staple-fibre yarns, especially those made from crimped fibres, are very open structures; this is illustrated in Figure 1A, which shows the cross-section of an unstrained wool carpet yarn. Figure 1B shows the same yarn under the influence of tensile loading; there has been a considerable radial displacement of the fibres from their initial positions and clearly the movements will have had a major influence in reducing the overall strains in the fibres.

In most previous theories of yarn tensile behaviour it had been conventional [22] to make a number of assumptions which artificially determined these sideways movements. In particular it was usual to assume that the yarn could be considered to have a circular cross-section both before and after straining and that the density of fibre packing in the yarn should be uniform in both states. This is the equivalent of assuming that the fractional change in radial position of fibres as a result of yarn straining should be independent of radius. In addition the fractional radial contraction (or yarn Poisson's ratio) was generally obtained empirically or set at some arbitrary level. There have, however, been a number of analyses in which some attempt has been made to provide more realistic experimental information on contraction for the calculations [23-25].

4.1 The theoretical model

The main development in the author's analysis [21] was to provide a means of theoretically predicting the actual lateral displacements of the fibres in deformed yarns. It was shown not only that these were large but that they varied with radial position.

The assumption of uniform packing densities was discarded as these had been shown to be quite inappropriate for staple-fibre yarns [26, 27]. The equilibrium state of the deformed yarn was predicted from its minimum energy state [28] and the stress-strain behaviour of the yarn was derived from the first differential of the minimum energy values with respect to yarn strain in a manner similar to the energy method of Treloar and Riding [28, 29].

In order to obtain detailed variations of the lateral contraction ratio of the yarn with radius it was necessary to divide the cross-section of the yarn into a number of concentric annular rings. This is in principle [30] a 'finite element approach' but the normal variational approach to the solution was not used. Instead the solution was found by the simpler and mathematically equivalent approach [31] of minimising the stored elastic strain energy. The differences between this energy method and a conventional finite element method have been investigated recently in New Zealand [32] and have been found to give almost identical results.

The two main energy terms for each ring come from the tensile strain energy in the fibres and their bending energy due to steric interference. These two terms are calculated separately to obtain the total strain energy in the annular ring. This is obviously only a first-order approach to the problem and one might have a number of misgivings about the independence of these two terms even at small strains. The problem is not so difficult for uncrimped fibres since most of the steric interference effects are those associated with adjacent radial layers of fibres and the perturbations to the general helical array caused by migration and other effects. A van Wyk-type relationship, modified if necessary by an orientation function to account for the local distribution of fibre orientations, is probably the most practical means of estimating the bending energy. The tensile energy can be calculated on the basis of an idealised helical geometry using an average helix angle for the radial position of each annular ring. When the fibres are crimped, however, the problem is a little more complex since the lateral compression properties of the aligned assembly are no longer independent of the tensile loading of the fibres. This effect has been ignored in developing a simple theory and the fibre input data are considered to be the average single-fibre stress-strain curve and the pressure/volume behaviour of a sliver of the fibres compressed at right angles to the fibre direction in a channel device [33].

In an independent approach to this problem of the lateral compression resistance of the fibres in the yarn Neckář [34] has put forward the concept of 'effective elastic bonds', to describe the resistance of fibres to close packing. He assumes that the fibres are effectively joined by non-linear springs and oppose each other's approach according to a relationship of the form

$$dF = \alpha (1 + \eta)^\beta d\eta$$

where F is the force between any two neighbouring fibres, η is the strain in the effective elastic bond, and α and β are constants. Neckář's approach [34] is based on a continuum mechanics model [35] and might be expected to give similar results to the one under discussion [21], but Neckář gives no detailed derivation of this power function describing the stiffness of the 'effective elastic bonds'.

In the analysis that follows, other energy terms including those associated with bending and torsion of the fibres to accommodate the generally helical configuration have been ignored. Various recent more vigorous stress analyses [36-38] have incorporated these terms and shown them to be important in particular situations. However, in predictions of the stress-strain behaviour of wool yarns using energy methods to take these terms into account [21] the author has found them to be negligible and for the sake of this simplified analysis they have been ignored.

4.2 Mathematical development

If the yarn is divided into n concentric annular zones each of the same width, then the main energy terms follow.

The tensile strain ε_j in the fibres is given by

$$\varepsilon_j = \frac{l_j}{l_{j_0}} - 1 \quad (j = 1, 2, \ldots \ldots n, n+1) \quad \ldots \ldots (9)$$

where j is a hypothetical typical fibre at the boundary of one of the n annular cylinders (the fibre $j = 1$ lies along the axis of the yarn);
l_j is the strained length of fibre j; and
l_{j_0} is the original length of fibre j.

$$l_j = \sqrt{4\pi^2 \gamma_j^2 r_{j_0}^2 + \beta^2 h_0^2} \quad \ldots \ldots \ldots (10)$$

$$l_{j_0} = \sqrt{4\pi^2 r_{j_0}^2 + h_0^2} \quad \ldots \ldots \ldots (11)$$

All helices have the same initial pitch h_0 and fibre j has an initial radial position r_{j_0}. The terms γ_j and β refer to the lateral and axial extension ratios.

The stored elastic tensile energy in fibre j, U_{T_j}, can then be found from the experimental stress-strain curve, shown for a typical fibre in Figure 2. The curve is far from linear for wool fibres.

$$U_{T_j} = f_T(\varepsilon_j) \quad \ldots \ldots \ldots \ldots (12)$$

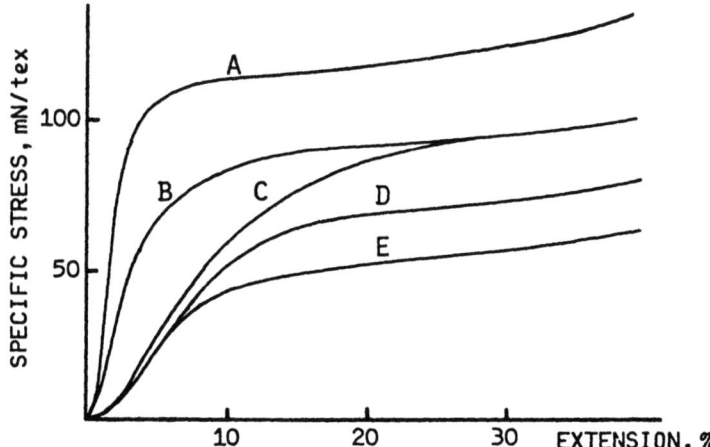

Fig. 2. Actual fibre and yarn tensile behaviour compared with theoretical predictions of yarn behaviour based on interpolation of the fibre curve. A: fibre behaviour; B: predicted yarn behaviour by Treloar and Riding's theory with a Poisson's ratio of 0.5; C: predicted yarn behaviour by the analysis in this paper; D: measured yarn behaviour (short gauge); E: measured yarn behaviour (long gauge)

The total stored tensile energy for zone i, W_{T_i}, should be calculated by an area-weighted average of the energy in its two peripheral fibres in the zone, but just taking a simple average

$$W_{T_i} = \tfrac{1}{2}(U_{T_i} + U_{T_{i+1}}) N_i \quad (i = 1, 2, \ldots n) \quad \ldots \quad (13)$$

where N_i is the number of fibres in zone i.

The resistance of the oriented fibres to close packing may be characterised as suggested earlier by a van Wyk-type expression. It is assumed that this term contains only energy contributions from fibre bending. To ensure that the fibres are not compressed beyond their jamming point it is expedient to incorporate an extra term v_f (the fibre specific volume) into the pressure/specific volume relationship, i.e.,

$$p = A\left\{\frac{1}{(v-v_f)^3} - \frac{1}{(v_0-v_f)^3}\right\} \quad \ldots \ldots \quad (14)$$

This expression can be suitably integrated to give an expression for the total stored bending energy in the fibres as a function of the zone specific volume. This is done simply by integrating expression (14) for compression of the zone from a specific volume v_0 to a specific volume v.

The resulting stored energy W_{B_i} for the ith zone is

$$W_{B_i} = f_B(v_i)$$

where $v_i = \pi(\gamma_{i+1}^2 r_{(i+1)_0}^2 - \gamma_i^2 r_{i_0}^2) h_0 \beta / mass_i$

and $mass_i$, the mass of fibres in zone i, can be calculated from N_i, the fibre radius, and the fibre density.

The total stored elastic energy in the yarn, W, is then given by

$$W = \sum_{i=1}^{i=n} (W_{T_i} + W_{B_i})$$

Apart from measured fibre characteristics (radius, specific volume, tensile stress-strain curve, sliver compression terms A, v_0) and variables describing the unstrained yarn structure (helix pitch, number density of fibre packing in each of n equal radial zones N_i) this expression gives W as a function of γ_j ($j = 1, 2, \ldots n, n+1$) and β. These latter variables are those describing the structural modification in the strained yarn. As these are kinematically indeterminate [21] they may be considered independent and the equilibrium state and stored elastic strain energy can be found for any imposed β by minimising W with respect to the $n+1$ terms γ_j. The yarn stress-strain curve is then found by differentiating (numerically) the minimum energy versus extension curve.

4.3 Results

The results of the new theory are shown in Figure 2 along with the predictions of earlier theories and experimental fibre and yarn results. It can be seen that the new theory, which takes into account the compressible nature of the initial structure, gives much better agreement over the early part of the stress-strain curve.

It has been shown [21] that while this improvement in the prediction of modulus and yield stress is due in part to the inclusion of fibre-crimp effects [25, 39], a large 'toe-in' region is still found for the yarn curve if Hookean fibre behaviour is assumed. This is illustrated in Figure 3.

The implication of this is that the reduction in modulus and yield stress is largely due to greater unevenness in the strain distribution so that the central fibres actually yield before the outer fibres are expected to take up significant strains. The theory [21] has been used to show that in bulky yarns the central fibres might even be expected to break before the crimp is removed from the outer fibres.

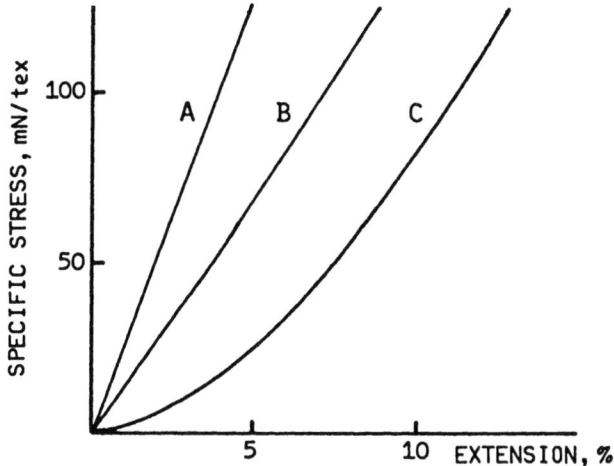

Fig. 3. Theoretical yarn tensile behaviour on the assumption of Hookean fibres. A: fibre behaviour; B: predicted yarn behaviour by Treloar and Riding's theory with a Poisson's ratio of 0.5; C: predicted yarn behaviour by the analysis in this paper.

5. CONCLUSIONS

In this general review of the mechanics of compression of fibrous assemblies the author has attempted to show how the results from studies of the mechanics of random and oriented fibrous masses have been introduced as a simple means of dealing theoretically with lateral compression effects in bulky yarns. Special attention has centred on how this approach has been used in explaining the dimensional changes in yarns during relaxation. There is clear evidence that the same mechanisms apply to fabric relaxation [40] and yarn diameter changes as high as three times have been reported [41] in woven cloths. However, a full treatment of this problem is still awaited.

A more detailed discussion has been given in relation to the tensile behaviour of yarns where it is shown that the same approach to lateral compression effects has led to a theoretical method of predicting the variation of lateral contraction with radius in extended yarns. This in turn has led to better predictions of the strain distribution, particularly in the more open staple-fibre yarns, and has provided an explanation for the typically lower initial moduli and yield stresses in these yarns.

*   *   *

The three Figures are reproduced from the *Journal of the Textile Institute* (Ref. 21) by kind permission of the editor.

REFERENCES

1. C. M. van Wyk, *J. Text. Inst.*, 1946, **37**, T285-92.
2. K. E. Duckett and C. C. Cheng, *J. Text. Inst.*, 1978, **69**, 55-9.
3. J. I. Dunlop, *J. Text. Inst.*, 1974, **65**, 532-6.
4. J. I. Dunlop, G. A. Carnaby, and D. A. Ross, *WRONZ Commun.* No. 28, 1974.
5. A. E. Stearn, *J. Text. Inst.*, 1971, **62**, 353-60.
6. T. Komori and K. Makishima, *Text. Res. J.*, 1977, **47**, 13-7.
7. E. Sebestyen and T. S. Hickie, *J. Text. Inst.*, 1971, **62**, 545-60.
8. D. S. Varma and R. Meredith, *Text. Res. J.*, 1973, **43**, 627-33.
9. A. K. Basu and U. Mukhopadhyay, *Text. Res. J.*, 1976, **46**, 841-4.
10. S. L. Anderson, D. R. Cox, and L. D. Hardy, *J. Text. Inst.*, 1952, **43**, T362-79.
11. D. G. Medley, J. E. Stell, and P. A. McCormick, *J. Text. Inst.*, 1962, **53**, T105-43.
12. P. Grosberg, *J. Text. Inst.*, 1963, **54**, T223-33.
13. T. Komori and K. Makishima, *Text. Res. J.*, 1978, **48**, 309-14.
14. P. A. Smith, *J. Text. Inst.*, 1962, **53**, T511-28.
15. J. W. S. Hearle and V. B. Merchant, *Text. Res. J.*, 1963, **33**, 417-24.
16. W. J. Onions, E. Oxtoby, and P. P. Townend, *J. Text. Inst.*, 1967, **58**, 293-315.
17. B. E. van Issum and N. H. Chamberlain, *J. Text. Inst.*, 1959, **50**, T599-623.
18. M. A. Chaudri and K. J. Whiteley, *Text. Res. J.*, 1968, **38**, 897-906.
19. S. Yamaguchi and S. Kawabata, *J. Text. Mach. Soc. Japan (Engl. ed.)*, 1976, **22**, 38-44; 62-6.
20. G. A. Carnaby and P. Grosberg, *J. Text. Inst.*, 1977, **68**, 24-36.
21. G. A. Carnaby and P. Grosberg, *J. Text. Inst.*, 1976, **67**, 299-308.
22. J. W. S. Hearle, P. Grosberg, and S. Backer, 'Structural Mechanics of Fibers, Yarns and Fabrics', vol. 1, Wiley-Interscience, New York, 1969, p. 180.
23. M. Konopasek, Ph.D. Thesis, UMIST, 1970, pp. 179-97.
24. C. J. Monego, Ph.D. Thesis, UMIST, 1973.
25. S. Norota, S. Kawabata, and H. Kawai, *J. Text. Mach. Soc. Japan (Engl. ed.)*, 1970, **16**, 41-52.
26. T. S. Hickie and M. Chaikin, *J. Text. Inst.*, 1974, **65**, 433-7.
27. G. A. Carnaby, Ph.D. Thesis, Leeds, 1976, pp. 96-100.
28. J. W. S. Hearle, *J. Text. Inst.*, 1969, **60**, 95-101.
29. L. R. G. Treloar and G. Riding, *J. Text. Inst.*, 1963, **54**, T156-70.
30. H. C. Martin and G. F. Carey, 'Introduction to Finite Element Analysis', McGraw-Hill, New York, 1973.
31. G. A. Carnaby, Ph.D. Thesis, Leeds, 1976, pp. 280-4.
32. C. J. van Luijk, A. J. Carr, G. A. Carnaby, and P. J. Moss, *Proc. Third Int. Conf. Finite Element Methods, Australia*, July 1979.
33. J. M. D. Pitts, *P.D. Report* No. 123, I W S, Ilkley, 1971.
34. B. Neckář, *Text. Res. J.*, 1976, **46**, 545-62.
35. C. C. Cheng, J. L. White, and K. E. Duckett, *Text. Res. J.*, 1974, **44**, 798-803.
36. M. Konopasek and J. W. S. Hearle, *J. Text. Inst.*, 1974, **65**, 217-21.
37. S. K. Batra, *J. Text. Inst.*, 1973, **64**, 209-222.
38. S. K. Batra, A. Tayebi, and S. Backer, *J. Text. Inst.*, 1973, **64**, 363-73.
39. S. Kawabata and T. Sasai, *J. Text. Mach. Soc. Japan (Engl. ed.)*, 1978, **24**, 13-8.
40. S. de Jong and R. Postle, *J. Text. Inst.*, 1977, **68**, 316-23.
41. P. Ellis and D. L. Munden, *J. Text. Inst.*, 1973, **64**, 279-94.

STATISTICAL MODELS FOR THE TENSILE STRENGTH OF YARNS AND CABLES*

S. L. Phoenix

Sibley School of Mechanical and Aerospace Engineering
Cornell University, Ithaca, New York 14853 U.S.A.

ABSTRACT. This paper discusses recent theoretical results for the chain-of-bundles model of the strength of yarns, ropes and cables. Within each bundle the fiber strengths vary randomly, and the nonfailed fibers share the applied load according to a load sharing rule. The structure fails when the weakest bundle fails and typically the number of fibers in each bundle and the number of bundles in the chain will be large. Thus, we seek asymptotic results which allow us to compare the statistical strength of the structure with that of a single fiber with respect to size effect, strength efficiency, variability and load sharing. Important conclusions and unanswered questions are discussed.

1. INTRODUCTION

The flexible fiber assemblies of interest in this paper are yarns, cords, ropes and cables. All take the form of a bundle of fibers in parallel, and transverse integrity is maintained to varying degrees by the existence of twist or a flexible matrix to bind the fibers together. Our intent is not to model all obvious geometric and mechanical features of such structures, but rather to focus on the primary factors which govern the statistical aspects of the strength of such structures. Initially, we will mention some empirical facts. We will then describe the model

---

*This work was supported in part by the United States Department of Energy under Contract EY-76-S-02-4027.

and develop some simple examples to elucidate several key ideas. Finally we will consider some recent advanced results, and discuss the possibilities for further developments.

1.1 Motivation

In most applications, yarns, cords, ropes and cables are very long and narrow. When an individual fiber in these structures breaks, it carries little or no load over a relatively <u>short</u> region along the length of a few fiber diameters in magnitude; farther away from the break the load returns to normal. This phenomenon is due to the existence of shear tractions on the fiber surface near the break. These shear tractions result from friction in the case of a twisted structure, or from inter-fiber bonding when a flexible matrix is present. The length along the structure over which the load carrying capacity of a fiber is significantly reduced is called the <u>ineffective length</u>. From a probabilistic point of view, the significant aspect is that the total fiber assembly has a length which is <u>several orders of magnitude</u> greater than the ineffective length.

At the site of a broken fiber, the load of that fiber is transferred in the lateral direction to neighboring fibers. Intuitively, one might expect the mildest situation to occur when this load is redistributed <u>uniformly</u> on all lateral survivors. The most severe situation would be perhaps when this load is redistributed <u>locally</u> only onto two or three immediate neighbors. From a statistical point of view, the nature of the load sharing among nonfailed fibers will be very important, and we will introduce two well defined <u>load sharing rules</u> into our model.

Individual fibers exhibit considerable statistical variation in their strength. This variation is the result of flaws which occur randomly in both position and severity along the fiber. A fiber loaded in simple tension will fail at the site of the <u>weakest</u> flaw. What is required is a simple but realistic model for fiber strength which is based on a model for random flaw occurrence. Moreover, this model must reflect the rather severe <u>size effect</u> typically exhibited by high strength fibers wherein the mean strength of such fibers diminishes quite rapidly with increasing length. It happens that such a model is easily posed, and a useful form of the model will yield the often used <u>Weibull distribution</u> for fiber strength.

1.2 Key questions about strength

We are motivated to consider the following important questions about how the probability distribution of the strength of a fiber assembly (yarn, rope, or cable) compares with that of the individual fibers:

1. What changes occur in the mean strength and in the variability in strength in passing from single fiber to fibrous assembly?

2. What are the statistical and mechanical factors that influence most the strength translation efficiency in passing from individual fibers to fibrous assemblies?

3. How does the size effect (weakening with increasing length) of a fiber assembly compare with that for individual fibers?

4. What is the effect of the type of load sharing among fibers on the probabilistic character of the strength of a fiber assembly?

Empirical evidence indicates that simple averaging rules are inadequate for providing answers for these questions. In what follows we develop a straightforward model for failure which is believed to capture the essential features of the failure process.

2. THE MODEL FOR FAILURE

The flexible fiber assembly (yarn, rope, cable) is viewed as a chain-of-bundles as shown in Figure 2.1. Each bundle contains n fiber elements, and there are m such bundles arranged in series. The length of each bundle is δ, the ineffective length referred to earlier, so that the length of the fiber assembly is mδ. Typically, m and n will be large, and the strength of the chain-of-bundles is equal to the strength of the weakest bundle.

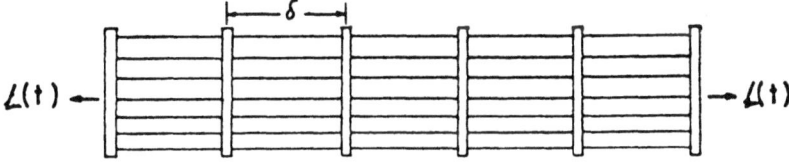

m – bundles in the chain
n – elements per bundle
$\mathcal{L}(t)$ – applied load

Fig 2.1 The chain-of-bundles model for tne strength of yarns ropes and cables.

2.1 Load sharing among fiber elements

We begin with the total load $\mathcal{L}$ which is applied to the chain-of-bundles. This load must be supported in turn by <u>each</u> bundle. To compare results for bundles of different sizes n, we speak in terms of the <u>applied load</u> $L = \mathcal{L}/n$ and consequently L is the <u>nominal load per fiber</u> in each cross-section of the chain-of-bundles. (Of course, L is proportional to the applied stress in common textile units.)

Now the <u>actual</u> load supported by a surviving fiber element in a bundle is typically greater than L since the element may be supporting part of the load shed by failed fiber elements in the same bundle. (However, this actual fiber element load may also be lower than L if the element is geometrically slack relative to other fiber elements in the same bundle.) Thus, we introduce the concept of the load sharing rule: Let the set $N = \{1, 2, \ldots, n\}$ be the numbers assigned (beforehand) to the fibers in a given bundle. For each proper subset S of N, we assume that a set of constants $\{K_i(S), i \in N \backslash S\}$ is given satisfying

$$\sum_{i \in N \backslash S} K_i(S) = n \quad \text{and} \quad K_i(S) > 0. \tag{2.1}$$

These constants are called <u>load sharing constants</u> and define a <u>load sharing rule</u>. They have the interpretation that if the elements in S are the failed fiber elements of the bundle when the applied load is L, then the load is redistributed on the elements in $N \backslash S$ (which is the set of <u>nonfailed</u> elements) such that fiber element i is subjected to the load $K_i(S) \cdot L$. In typical applications the values of the $K_i(S)$ will be independent of the applied load L, though in some applications involving random fiber slack, some dependence will occur.

<u>The equal load sharing rule</u>. In this simplest case, all surviving fiber elements in a given bundle share the total load equally. Thus we have

$$K_i(S) = n/|N \backslash S| \quad \text{for} \quad i \in N \backslash S \tag{2.2}$$

where $|N \backslash S|$ is the number of fiber elements in $N \backslash S$, the set of survivors. In other words, if j of the n elements have failed, then $K_i(S) = n/(n-j)$ which exceeds unity. Notice that $K_i(S)$ does not depend on the geometric arrangement of failed and surviving fiber elements in the bundle. For this reason, the equal load sharing rule may be unsatisfactory when interfiber friction or bonding is severe.

The local load sharing rule. For geometric simplicity, we envision the fiber elements in each bundle to be arranged in a circular array, so that the fiber assembly is viewed as a long tube. Let S be the set of failed fiber elements, and for a surviving fiber i let s(i) be the total number of failed elements adjacent to i or connected to i by a sequence of failed elements. Then

$$K_i(S) = 1 + s(i)/2 \quad \text{for} \quad i \in N \setminus S. \tag{2.3}$$

Thus, if a surviving fiber element i has only one immediate failed neighbor, then $K_i(S) = 3/2$, and this surviving element supports its own original load plus half of that of the failed element. Two immediate failed neighbors result in $K_i(S) = 2$ for the surviving fiber element.

Obviously the two load sharing rules described above represent highly idealized situations, and the reader can appreciate the existence of other more realistic possibilities. However, for analytical purposes, these two rules have advantages: First, they are easily described and envisioned. Second, the mathematical and numerical analysis required to obtain key statistical quantities is tractable. Third, the two rules lie at the extremes of the actual load sharing to be expected. If the fiber assembly has low twist and no inter-fiber bonding, some variation on equal load sharing would be likely, the length δ of the bundles would tend to be large, and a failed bundle would appear "frayed" in the plane of failure. On the other hand, if the twist is high, or strong bonding between fibers exists, severe local effects around broken fibers would be expected, the bundle length δ would be small, and failure of a bundle would proceed by "tearing" or "catastrophic crack growth." The local load sharing rule is designed to capture the essence of this phenomenon. The actual load sharing rule in typical structures would lie somewhere in between these two extremes. Thus, results for these two rules will give us "bounds" on the statistical strength behavior.

Example. To illustrate the previous ideas on load sharing, we consider a bundle with 8 fiber elements as shown in Figure 2.2. At the time instant under consideration, elements 4, 7 and 8 are failed while elements 1, 2, 3, 5 and 6 are surviving. We have $S = \{4, 7, 8\}$ and $N \setminus S = \{1, 2, 3, 5, 6\}$. For the equal load sharing rule, we have $|N \setminus S| = 5$ and $K_1(S) = K_2(S) = K_3(S) = K_5(S) = K_6(S) = 8/5$. For the local load sharing rule we find $K_1(S) = 2$, $K_2(S) = 1$, $K_3(S) = 3/2$, $K_5(S) = 3/2$ and $K_6(S) = 2$. Fiber elements 1 and 6 are more highly loaded under local load sharing than under equal load sharing while the reverse is true for the remaining surviving elements. Typically,

Equal load sharing rule

$S = \{4,7,8\}$

$K_1(S) = K_2(S) = K_3(S)$
$= K_5(S) = K_6(S) = 8/5$

Local load sharing rule

$S = \{4,7,8\}$

$K_1(S) = 2 \qquad K_5(S) = 3/2$
$K_2(S) = 1 \qquad K_6(S) = 2$
$K_3(S) = 3/2$

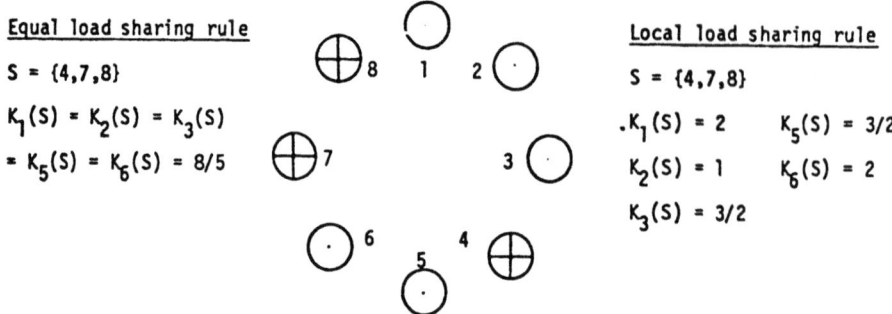

Fig 2.2 Bundle of 8 fiber elements with elements 4, 7 and 8 being failed. Load sharing constants for both rules are shown.

uneven distribution of load on surviving elements increases the likelihood of further failures as time progresses; our problem is to quantify these effects with respect to the statistical strength of the chain-of-bundles.

2.2 The Weibull model for the static strength of single fibers

We assume that the strengths of the fiber elements in a given bundle of the chain are independent and identically distributed (i.i.d.) random variables with common cumulative distribution function (c.d.f.) $F(x)$. In principle, we can obtain results for quite general $F(x)$; however, to gain insight into the behavior of the individual fibers and their fiber assemblies we must be more specific about the form of $F(x)$. Phoenix [1, 2, 3] discusses the relationship between fiber strength and the random aspects of flaw occurrence. While many forms for $F(x)$ are plausible, the form which has the most advantageous properties is the Weibull distribution

$$F(x) = 1 - \exp\{-(x/x_0)^\rho\}, \quad x \geq 0 \qquad (2.4)$$

where x_0 is the scale parameter, and ρ is the shape parameter. (Actually, the form which is familiar for tension testing of single fibers is $F(x) = 1 - \exp\{-\delta(x/\tilde{x}_0)^\rho\}$, $x \geq 0$ where δ, the fiber length appears explicitly and may be varied with $\tilde{x}_0$ and ρ remaining constant. Since the bundle length δ is essentially a fixed quantity in our investigation, we are taking $x_0 = \tilde{x}_0 \delta^{-1/\rho}$.) The mean fiber strength under (2.4) is

$$\bar{x} = x_0 \Gamma(1 + 1/\rho), \qquad (2.5)$$

and the coefficient of variation is

$$\overline{cv} = \left[\frac{\Gamma(1+2/\rho)}{\Gamma(1+1/\rho)^2} - 1 \right]_0^{1/2} \quad (2.6)$$

Typically the shape parameter ρ takes a value $5 \leq \rho \leq 25$ with the lower values being characteristic of brittle fibers. Consequently, the scale parameter x_0 is numerically close to the mean fiber strength. Notice that the coefficient of variation $\overline{cv}$ depends only on ρ; its range is typically 0.04 to 0.25, and its value diminishes as ρ increases.

Lower tail behavior of the Weibull distribution. Later we will require an expression for the lower tail behavior of the Weibull distribution (2.4), that is, the behavior of $F(x)$ for small loads x. This is because failure of the chain-of-bundles has weakest link aspects wherein the failure process is initiated at relatively rare isolated breaks. In fact, the chain-of-bundles will have a median strength which is only a small fraction of x_0. The lower tail property of interest is obtained from (2.4) using a Taylor series expansion, and is

$$F(x) = (x/x_0)^\rho + o(x^\rho), \quad x \geq 0 \quad (2.7)$$

where $o(z)/z \to 0$ as $z \to 0$. Essentially $o(x)$ represents terms of order $x^{2\rho}$ and higher.) The key feature in (2.7) is that for small probabilities of failure (or small loads x) the probability of failure of an individual fiber element grows as the load x raised to the power ρ. From an empirical point of view, such behavior is certainly plausible.

2.3 The size effect for single fiber strength

It is well known that longer fibers are weaker on the average than shorter fibers. To understand this we imagine that a long fiber is say m short fiber elements arranged in series, and that failure occurs with the failure of the weakest element. Under load x, the probability of survival of a single fiber element is $1 - F(x)$, and that of the long fiber is $[1 - F(x)]^m$. Consequently the c.d.f. for the strength of the long fiber (of length m) is

$$F_m(x) = 1 - [1 - F(x)]^m, \quad x \geq 0 \quad (2.8)$$

which is recognized to be of the form of the weakest link rule. Taking $F(x)$ to be the Weibull distribution (2.4) we find

$$F_m(x) = 1 - \exp\{- m(x/x_0)^\rho\}, \quad x \geq 0. \quad (2.9)$$

Clearly (2.9) is also a Weibull distribution with shape parameter

ρ and scale parameter $x_0 m^{-1/\rho}$. The mean strength of the long fiber is now

$$\bar{x}_m = x_0 m^{-1/\rho} \Gamma(1+1/\rho), \qquad (2.10)$$

but the coefficient of variation remains as $\overline{cv}$ of (2.6).

We have uncovered important properties of the Weibull model for fiber strength. These are as follows:

1. The strength of a fiber follows a Weibull distribution regardless of the length.

2. The mean strength $\bar{x}_m$ diminishes in proportion to $m^{-1/\rho}$ as the length $m\delta$ increases. On log-log coordinates, the mean strength is linear in the length m with negative slope $-1/\rho$; empirically, such a feature is observed quite frequently over changes of length of one to three orders of magnitude.

3. The variability in strength, as reflected by the coefficient of variation, is unchanged with increasing length.

It is not being suggested that all types of fibers exhibit precisely the above behavior. However, significant deviations are believed to be uncommon, and it is believed that the Weibull model captures the essence of the statistical character of fiber strength for the purposes of our models.

2.4 Statistical quantities of interest

Earlier we let $F(x)$, $x \geq 0$ be the c.d.f. for the strength of a single fiber element of length δ. We now let $G_n(x)$, $x \geq 0$ be the c.d.f. for the strength of a single bundle under the load L, and let $H_{m,n}(x)$, $x \geq 0$ be the c.d.f. for the strength of the chain-of-bundles, that is, of the fibrous assembly. Our main interest is in comparing various measures of the behavior of $F(x)$ for single fibers with those of $H_{m,n}(x)$ for the fibrous assembly.

Since the strengths of the individual fiber elements are independent and identically distributed random variables, so are the strengths of the individual bundles. Failure of the chain-of-bundles occurs with the failure of the weakest bundle and thus, we have the straightforward result

$$H_{m,n}(x) = 1 - [1 - G_n(x)]^m, \quad x \geq 0 \qquad (2.11)$$

by the weakest link rule. This is the easy part of the analysis:

the difficult task is to obtain $G_n(x)$ for the strength of a single bundle in terms of the load sharing rule and $F(x)$ for single fibers.

As mentioned earlier, m the chain length and n the number of fibers in each bundle may be large ($1 \leq m \leq 10^8$, $1 \leq n \leq 10^6$) since yarns, ropes and cables may be very long, and may have many fibers in a cross-section. When m is large, we will require extreme numerical precision in the lower tail of $G_n(x)$. Unfortunately when n is large it is virtually impossible to obtain exact expressions for $G_n(x)$ which are manageable. Thus, we must consider various approximations for $G_n(x)$ as n grows large for use in (2.11) when m is any value. Since our aim is to gain insight into the strength behavior of fibrous assemblies, we will build on some simple examples which follow. In so doing we will keep in mind several key questions:

1. What are the general shapes of $G_n(x)$ and $H_{m,n}(x)$ as compared with $F(x)$? When $F(x)$ is Weibull, to what extent do $G_n(x)$ and $H_{m,n}(x)$ have Weibull features?

2. What size effect in strength results when m the chain length or n the number of fibers is increased in the fibrous assembly? How does this size effect compare with that for single fibers as _their_ length is increased?

3. How does the median strength for $H_{m,n}(x)$ compare with that for $F(x)$? How do corresponding measures of the variability compare?

4. How does the load sharing rule affect the behavior of $G_n(x)$ and $H_{m,n}(x)$?

5. To what extent does "weakest link" character prevail for the chain-of-bundles as length m or bundle size n increase (possibly together)?

3. ANALYSIS FOR FIBROUS ASSEMBLIES WITH SMALL BUNDLES

To gain insight we begin by studying the chain-of-bundles model when n is small ($1 \leq n \leq 9$) under both load sharing rules. In this case, we will be able to compute some exact results.

3.1 Expressions for the strength distribution of small bundles

Consider first a bundle of _two_ fibers. Let X_1 and X_2 be the strengths of the two fibers and let Q_2^* be the strength of the bundle. Now for some fixed applied load x, we have $Q_2^* \leq x$ if

$X_1 \leq x$ and $X_2 \leq x$, or $X_1 \leq x$ and $x < X_2 \leq 2x$, or $X_2 \leq x$ and $x < X_1 \leq 2x$. (Recall that $K_i(S) = 2$ for both load sharing rules when one fiber has failed.) These events are disjoint and exhaust all possible ways that the bundle can fail. The probability for the first event is $F(x)^2$ while the probability for each of the latter two events is $F(x)[F(2x) - F(x)]$. Summing probabilities we arrive at

$$G_2(x) = F(x)^2 + 2F(x)[F(2x) - F(x)]$$
$$= 2F(x)F(2x) - F(x)^2, \quad x \geq 0. \qquad (3.1)$$

A similar analysis may be carried out for a bundle of <u>three</u> fibers, and here again the results coincide for the two load sharing rules. Letting X_1, X_2 and X_3 be the strengths of the fibers and Q_3^* be the strength of the bundle, we have $Q_3^* \leq x$ if $X_2 \leq x$ and $x < X_1 \leq 3x/2$ and $3x/2 < X_3 \leq 3x$ as one possible event where $3/2$ and 3 are the appropriate values for the $K_i(S)$. There are several other possible events, all of which are disjoint. It turns out that

$$G_3(x) = 6F(x)F(\tfrac{3x}{2})F(\tfrac{6x}{2}) - 3F(\tfrac{3x}{2})^2 F(x)$$
$$- 3F(x)^2 F(3x) + F(x)^3, \quad x \geq 0. \qquad (3.2)$$

For a bundle of <u>four</u> fibers, results differ for the two load sharing rules for the first time (as we increase n). For the <u>local load sharing rule</u> we show in Figure 3.1 all sixteen states of failed and surviving fibers; the components of the vectors are the associated load sharing constants with zeros meaning that the particular fiber has failed. For the <u>equal load sharing rule</u>, the vectors are the same except for b,c,d, and e, where the nonzero constants are all $4/3$. (Of course, differences in the constants increase as n increases.) To compute $G_4(x)$, we again must enumerate all disjoint events or sequences for the way the bundle can fail. For the <u>equal load sharing rule</u>, we have the following example sequences:

$a \to b \to n \to p$ w.p. $F(x)[F(4x/3) - F(x)]^2[F(4x) - F(4x/3)]$

$a \to f \to n \to p$ w.p. $F(x)^2[F(2x) - F(x)][F(4x) - F(2x)]$

$a \to 0 \to p$ w.p. $F(x)^3[F(4x) - F(x)]$

and

$a \to p$ w.p. $F(x)^4$

where w.p. stands for "with probability." For the <u>local load sharing rule</u>, the probabilities for these four sequences are the same except for the first which is

a	(1, 1, 1, 1)	i	(2, 0, 0, 2)
b	(0, 3/2, 1, 3/2)	j	(2, 0, 2, 0)
c	(3/2, 0, 3/2, 1)	k	(2, 2, 0, 0)
d	(1, 3/2, 0, 3/2)	l	(4, 0, 0, 0)
e	(3/2, 1, 3/2, 0)	m	(0, 4, 0, 0)
f	(0, 0, 2, 2)	n	(0, 0, 4, 0)
g	(0, 2, 0, 2)	o	(0, 0, 0, 4)
h	(0, 2, 2, 0)	p	(-, -, -, -) failure

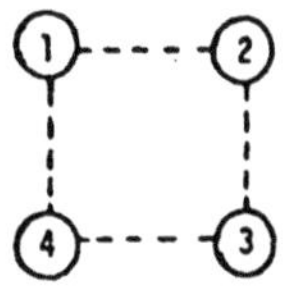

Fig 3.1 All states of failed and surviving fibers for a bundle of four fiber elements under the local load sharing rule. The components of the vectors are the associated load sharing constants.

$$a \to b \to n \to p \quad w.p. \quad F(x)[F(3x/2) - F(x)]^2[F(4x) - F(x)]$$

The rule for generating these probabilities is evident on inspection; they are the probabilities that the fiber strengths lie in the appropriate range for such a failure sequence to occur. Summing all such probabilities, we obtain $G_4(x)$. For the <u>equal load sharing rule</u> we obtain

$$\begin{aligned}G_4(x) = &\ 24F(x)F(4x/3)F(2x)F(4x) \\ &- 12F(x)F(4x/3)F(2x)^2 \\ &- 12F(x)F(4x/3)^2F(4x) \\ &+ 4F(x)F(4x/3)^3 \\ &- 12F(x)^2F(2x)F(4x) + 6F(x)^2F(2x)^2 \\ &+ 4F(x)^3F(4x) - F(x)^4, \quad x \geq 0. \end{aligned} \quad (3.3)$$

For the <u>local load sharing rule</u> we obtain

$$\begin{aligned}G_4(x) = &\ 16F(4x)F(2x)F(3x/2)F(x) - 4F(4x)F(2x)F(x)^2 \\ &- 4F(4x)F(3x/2)^2F(x) + 4F(4x)F(x)^3 \\ &- 8F(2x)^2F(3x/2)F(x) + 2F(2x)^2F(x)^2 \\ &- 8F(4x)F(3x/2)F(x)^2 \\ &+ 4F(3x/2)^2F(x)^2 - F(x)^4, \quad x \geq 0 \end{aligned} \quad (3.4)$$

In comparing $G_4(x)$, $G_3(x)$, $G_2(x)$ and $G_1(x)$, we notice that the complexity and number of terms increases rapidly in n; indeed little intuition into strength behavior is afforded by these results. For larger n, the written expressions for $G_n(x)$ quickly become unmanageable and desired numerical results must be obtained by suitable computer algorithms. (Results for $G_5(x)$ are given by Harlow and Phoenix [4].)

The c.d.f. $H_{m,n}(x)$ for the strength of the fibrous assembly is obtained from (2.11) by simple substitution of the appropriate version of (3.1) to (3.4) with $F(x)$ for single fibers taken as the Weibull distribution (2.4).

3.2 Lower tail behavior of the strength distribution for single bundles

Recall that the Weibull form of $F(x)$ for single fibers has the lower tail behavior $F(x) = (x/x_0)^\rho + o(x^\rho)$, $x \geq 0$. We now consider the corresponding result for $G_n(x)$ and we begin with $G_2(x)$ for a bundle of <u>two</u> fibers. Upon substituting (2.7) into (3.1), we obtain

$$G_2(x) = 2(x/x_0)^\rho (2x/x_0)^\rho - (x/x_0)^{2\rho} + o(x^{2\rho})$$

$$= (2^{\rho+1} - 1)(x/x_0)^{2\rho} + o(x^{2\rho}) \tag{3.5}$$

or

$$G_2(x) = (x/x^*_{0,2})^{2\rho} + o(x^{2\rho}), \quad x \geq 0 \tag{3.6}$$

where

$$x^*_{0,2} = x_0 \{2^{\rho+1} - 1\}^{-1/(2\rho)}. \tag{3.7}$$

Comparing (3.6) and (3.7) with (2.7), we notice that the exponent has <u>doubled</u> to 2ρ and the scale parameter has <u>diminished</u> somewhat. Using a similar analysis, we determine

$$G_4(x) = (x/x^*_{0,4})^{4\rho} + o(x^{4\rho}), \quad x \geq 0 \tag{3.8}$$

where for the <u>equal load sharing rule</u>

$$x^*_{0,4} = x_0 \{24(32/3)^\rho - 12(16/3)^\rho$$
$$- 12(64/9)^\rho + 4(64/27)^\rho$$
$$- 12(8)^\rho + 10(4)^\rho - 1\}^{-1/(4\rho)}, \tag{3.9}$$

and for the local load sharing rule

$$x_{0,4}^* = x_0\{16(12)^\rho - 4(8)^\rho - 4(9)^\rho$$
$$+ 6(4)^\rho - 16(6)^\rho + 4(9/4)^\rho - 1\}^{-1/(4\rho)} \quad (3.10)$$

Harlow, Smith, and Taylor [5] have shown that in general

$$G_n(x) = (x/x_{0,n}^*)^{n\rho} + o(x^{n\rho}), \quad x \geq 0 \quad (3.11)$$

for some scale parameter $x_{0,n}^*$ and all n. The scale parameter $x_{0,n}^*$ depends on the load sharing rule, and becomes difficult to compute as n increases. In Table 3.1 below, we give the lower tail scale parameters for $1 \leq n \leq 9$, three values of ρ and both load sharing rules. (Computer costs limit us to $n \leq 9$.)

Table 3.1 Lower tail scale parameters for bundles under local load sharing (LLS) and equal load sharing (ELS).

	$x_{0,n}^*/x_0$					
	$\rho=5$		$\rho=10$		$\rho=15$	
n	LLS	ELS	LLS	ELS	LLS	ELS
1	1.0000	1.0000	1.0000	1.0000	1.0000	1.0000
2	.6608	.6608	.6830	.6830	.6910	.6910
3	.5406	.5406	.5708	.5708	.5821	.5821
4	.4708	.4773	.5015	.5116	.5131	.5249
5	.4215	.4375	.4504	.4744	.4612	.4890
6	.3833	.4100	.4095	.4485	.4198	.4641
7	.3523	.3896	.3762	.4294	.3857	.4457
8	.3265	.3738	.3485	.4145	.3568	.4314
9	.3044	.3612	.3245	.4026	.3323	.4200

Notice that the values for equal load sharing are somewhat higher than and diverge from those for local load sharing, as n increases.

3.3 Asymptotic Weibull distribution for strength of fibrous assembly

We now determine the asymptotic behavior of the probability distribution for the strength of a long chain-of-bundles. We recall (2.11) and (3.11), and consider taking a limit for large n. Replacing x by $ym^{-1/(n\rho)}$ we have

$$\lim_{m\to\infty} H_{m,n}(ym^{-1/(n\rho)})$$

$$= 1 - \lim_{m\to\infty} [1 - (y/x^*_{0,n})^{n\rho}/m + o(1/m)]^m$$

$$= 1 - \exp\{-(y/x^*_{0,n})^{n\rho}\}, \quad y \geq 0, \qquad (3.12)$$

since $\lim_{m\to\infty} [1 - z/m + o(1/m)]^m = e^{-z}$ which is a well known analytical fact. (Notice that as m was allowed to grow, we diminished $x = ym^{-1/(n\rho)}$ at the correct rate to get a nontrivial Weibull limit. Such steps are standard procedure in extreme value theory [5].) The consequence is that as m grows large, we have the approximation

$$H_{m,n}(x) \cong 1 - \exp\{-(x/x^*_{m,n})^{n\rho}\}, \quad x \geq 0 \qquad (3.13)$$

where

$$x^*_{m,n} = x^*_{0,n} m^{-1/(n\rho)}. \qquad (3.14)$$

Thus the probability distribution for the strength of a long chain-of-bundles is asymptotically Weibull with shape parameter $n\rho$ and scale parameter $x^*_{m,n}$.

Before investigating the accuracy of this result, we list some implications:

1. The Weibull shape parameter for the fibrous assembly is $n\rho$ which is the shape parameter ρ for a single fiber times n the number of fibers in a cross-section. Recall that as the shape parameter increases, the coefficient of variation of strength decreases. Thus, a fibrous assembly will have much less variability in strength than a single fiber.

2. The (asymptotic) mean strength of the fibrous assembly is

$$\bar{x}_{m,n} = x^*_{0,n} m^{-1/(n\rho)} \Gamma(1 + 1/(n\rho)). \qquad (3.15)$$

Typically $\bar{x}_{m,n}$ is only a fraction of $\bar{x} = x_0 \Gamma(1+1/\rho)$ which is the mean strength of a <u>single</u> fiber element since $x^*_{0,n} < x_0$ according to Table 3.1, and $m^{-1/(n\rho)} < 1$. (Recall that strength is in terms of force <u>per</u> <u>fiber</u> in a cross-section of the assembly). But if we compare the mean strength of the fibrous assembly with $\bar{x}_m = x_0 m^{-1/\rho} (1+1/\rho)$ which is the strength of an equally long fiber, we find that eventually $\bar{x}_{m,n}$ exceeds $\bar{x}_m$ as the length m grows large; it is easy to see that $m^{-1/\rho}$ diminishes far more rapidly than $m^{-1/(n\rho)}$ when $n \geq 2$.

3. The size effect for the fibrous assembly is typically far less than the size effect for the constituent fibers as discussed in Section 2.3. The mean strength asymptotically diminishes in proportion to $m^{-1/(n\rho)}$ as m increases. If $n = 5$ and $\rho = 5$ then $n\rho = 25$ and more than a 10^7 fold increase in length is required for a reduction of the mean strength by <u>one half</u>; for a <u>single</u> fiber, only a 32 fold increase in length is required! The structural benefit of the bundle arrangement becomes manifest.

3.4 Examples comparing exact and asymptotic results

We consider numerical results for two cases to illustrate important features. The graphical scaling which is most convenient for plotting $F(x)$, $G_n(x)$ and $H_{m,n}(x)$ is the scaling of Weibull probability paper. On such a scaling, the Weibull c.d.f. $F(x) = 1 - \exp\{-(x/x_0)^\rho\}$ versus load x plots as a straight line; the slope is proportional to the scale parameter ρ and the horizontal position depends on the scale parameter x_0. Furthermore, upon fixing n, results for various values of the chain length m over a wide range are obtained by a constant shift (simple translation) of the graph on the vertical scale. To see this we recall from (2.11) that

$$1 - H_{m,n}(x) = [1 - G_n(x)]^m \qquad (3.16)$$

whence

$$\log(-\log\{1 - H_{m,n}(x)\}) = \log m + \log(-\log\{1 - G_n(x)\}) \qquad (3.17)$$

where the log is the natural logarithm. Since Weibull probability paper has a vertical scale which is linear in $\log(-\log(1-p))$ where p is probability, then $H_{m,n}(x)$ for the chain-of-bundles is simply $G_n(x)$ for a single bundle shifted vertically the amount $\log m$ <u>without any change in shape</u>. Of course neither $G_n(x)$ and $H_{m,n}(x)$ will plot as a straight line.

On Figure 3.2 we have plotted $F(x)$ for a single fiber segment, $F_4(x)$ for a <u>chain</u> of four fiber segments, and $G_4(x)$ for a <u>bundle</u> of four fiber segments. We have taken $\rho = 5$ as a typical value of the Weibull shape parameter, and both load sharing rules have been assumed. Also we have plotted $(x/x_0^*,_4)^{4\rho}$ of the approximation for $G_4(x)$ given by (3.8). We make the following observations:

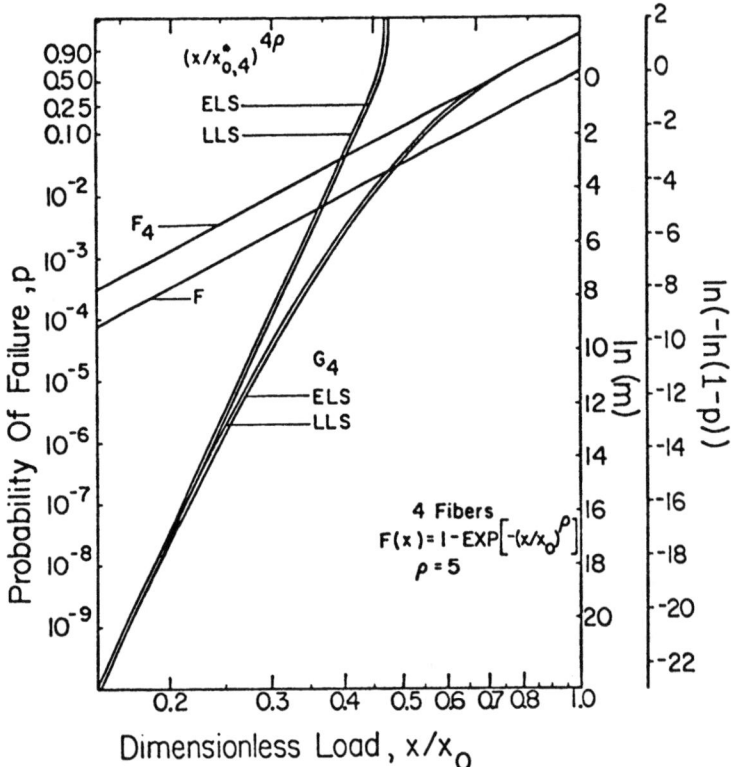

Fig 3.2 Probability distribution for the strength of a four fiber bundle under equal (ELS) and local (LLS) load sharing.

1. At <u>high</u> loads, where failure is highly likely, the bundle c.d.f. $G_4(x)$ is asymptotic to $F_4(x)$ for a chain (the weakest) of 4 fibers. This means that the <u>first</u> fiber to fail triggers the failure of the bundle, and the bundle is essentially a weakest link structure.

2. For a bundle of only four fibers (n=4), the equal load-sharing rule and the local load sharing rule yield almost identical results. (Recall from Figure 3.1 that the load state vectors differed in only four cases.) This situation will <u>not</u> prevail as the bundle size n increases.

3. The lower tail approximation $(x/x_{0,4}^*)^{4\rho}$ for $G_4(x)$ is conservative but accurate only at probabilities below 10^{-6}. Notice that the lower tail of $G_n(x)$ has an effective shape parameter of $4\rho = 20$.

4. At <u>lower</u> loads x, the probability of failure for the single bundle is orders of magnitude below that for a single fiber segment. This is by far the most important feature since it is at the heart of the explanation as to why the chain-of-bundles has excellent statistical strength.

5. $H_{m,n}(x)$, the c.d.f. for the strength of the chain-of-bundles, is simply $G_n(x)$ shifted vertically the amount log m for which a scale is provided. (Notice that $F_4(x)$ for the chain-of-fibers is just $F(x)$ shifted vertically log 4 = 1.39 units on the log m scale.)

6. The accuracy of the Weibull approximation given by (3.13) and (3.14) is easily seen: For m = 1 the right hand side of (3.13) is a straight line tangent to $G_4(x)$ in the lower tail, or is $(x/x_{0,4}^*)^{4\rho}$ without the curved section at the top of the graph. Since $G_4(x)$ and this tangent diverge at log m = 14 or m $\sim 10^6$. The Weibull approximation will be conservative but increasingly inaccurate as the chain length m is reduced below 10^6. Nevertheless, for this small value of n = 4, the Weibull approximation is useful.

7. By rotating Figure 3.2 counterclockwise 90°, we obtain a plot of the median strength of the chain-of-bundles (median of $H_{m,n}(x)$) versus chain length m; the vertical scale is the x/x_0 scale, the horizontal scale is the log m scale, and the graph is that of $G_4(x)$. In comparing the median strength of the fibrous assembly with that of a single fiber of length m (using the $F(x)$ graph), we see that the medians are equal when m $\sim$ 30. When m exceeds 30, which is the typical situation, <u>the fibrous assembly is far stronger per fiber than a single fiber</u>; indeed the size effect for the fibrous assembly is quite mild.

On Figure 3.3, we have plotted the same quantities as in Figure 3.2 but with <u>eight</u> fibers per bundle (n = 8). Most of the basic features are preserved, but we note the following changes:

1. The power approximation $(x/x_{0,8}^*)^{8\rho}$ to $G_8(x)$ is conservative and inaccurate except at <u>extremely</u> low probability levels, say below 10^{-12}. This means that the Weibull approximation (3.13) for $H_{m,n}(x)$ will be conservative except when m takes on astronomical values exceeding 10^{12}. This is unfortunate and the situation worsens as n is increased or as ρ increases. Thus, the Weibull approximation for $H_{m,n}(x)$ diminishes in value as n is increased above about ten, and Table 3.1 thus extends only to values which are of practical interest.

2. The results for the two load sharing rules have diverged

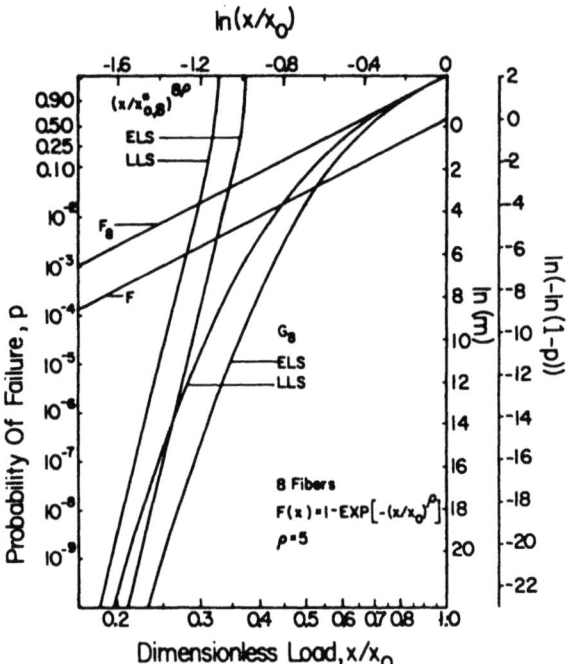

Fig 3.3 Probability distribution for the strength of an eight fiber bundle under equal (ELS) and local (LLS) load sharing.

considerably; equal load sharing yields stronger bundles and fibrous assemblies than local load sharing. This divergence continues as bundle size n is increased, and later we discuss the limiting behavior in the two cases as n is increased indefinitely.

3. The effective Weibull shape parameter is now $8\rho = 40$; the limiting size effect for the fibrous assembly becomes very mild indeed.

4. ANALYSIS FOR FIBROUS ASSEMBLIES WITH LARGE BUNDLES UNDER LOCAL LOAD SHARING

In comparing Figures 3.2 and 3.3, we notice that $G_n(x)$, under the local load sharing rule, is shifting as n increases. We ask

"What asymptotic behavior occurs for $G_n(x)$ or $H_{m,n}(x)$ as $n\to\infty$?"

4.1 Asymptotic weakest link structure for the fibrous assembly ($n\to\infty$)

Harlow and Phoenix [6] considered the transformation

$$W_n(x) = 1 - [1 - G_n(x)]^{1/n}, \quad x \geq 0 \qquad (4.1)$$

which is a weakest link scaling in reverse of $G_n(x)$ back to single fiber size. Indeed notice that

$$G_n(x) = 1 - [1 - W_n(x)]^n, \quad x \geq 0. \qquad (4.2)$$

They performed a numerical study which led to the <u>conjecture</u> that

$$\lim_{n\to\infty} W_n(x) = W(x), \quad x \geq 0 \qquad (4.3)$$

for some limiting c.d.f. $W(x)$ <u>which depends on $F(x)$ and the actual load sharing rule.</u> The important extension of this conjecture was that

$$G_n(x) \equiv 1 - [1-W(x)]^n, \quad x \geq 0 \qquad (4.4)$$

and

$$H_{m,n}(x) \equiv 1 - [1-W(x)]^{mn}, \quad x \geq 0 \qquad (4.5)$$

differ negligibly from $G_n(x)$ and $H_{m,n}(x)$ respectively for applications, even for quite small values of m and n. Since $H_{m,n}(x)$ is the c.d.f. for the strength of the chain-of-bundles, the consequence of (4.5) is that <u>the fibrous assembly would behave effectively in a weakest link manner in terms of the limit c.d.f. $W(x)$ and the total number of fiber elements mn in the assembly.</u>

The numerical evidence presented in Harlow and Phoenix [6] is very compelling. In Figure 4.1 we present the numerical results for $\rho = 10$ and $1 \leq n \leq 9$; the c.d.f. $G_n(x)$ was computed using an algorithm which computed probabilities for all failure sequences (as described in Figure 3.1 for example). Computer costs limited the results to a maximum bundle size of <u>nine</u> fibers. Notice that even at probability levels of 10^{-10} the convergence is virtually complete for $n \geq 5$. The graph for $W_9(x)$ is believed to be identical to the conjectured limit $W(x)$ for all practical purposes. Thus an accurate approximation

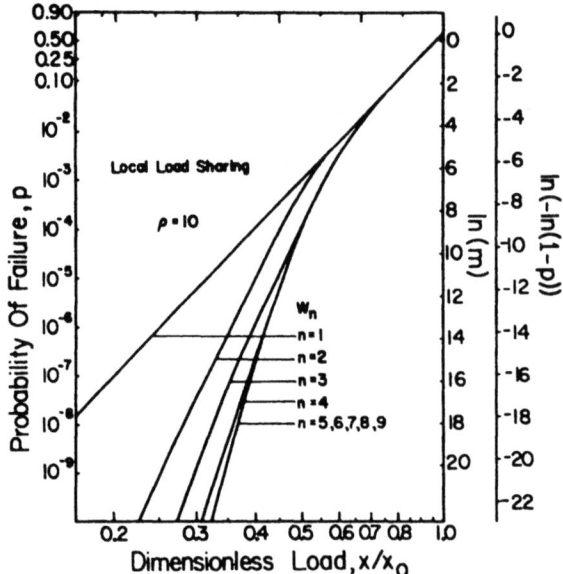

Fig 4.1 Convergence of the transformed distribution $W_n(x)$ to the conjectured limit W as bundle size n increases. Local load sharing is assumed.

to $H_{m,n}(x)$ for $n \geq 5$ and $1 \leq m \leq 10^9$ is believed to be $W_9(x)$ shifted vertically the amount $\log(mn)$.

4.2 Consequences of the weakest link feature

We interpret the differences between $W(x)$ and the c.d.f. $F(x) = W_1(x)$ for the strength of <u>single</u> fiber elements. At loads x above 0.7 x_0 (which are associated with probabilities above 10^{-2} on Figure 4.1) $F(x)$ and $W(x)$ are virtually identical. This means that at such loads, failure of the fibrous assembly will be triggered with the failure of the <u>first</u> fiber element. However, <u>at lower loads</u> x, <u>the probabilities for</u> $W(x)$ <u>are orders of magnitude below those for</u> $F(x)$, and it is here that the redundancy associated with load sharing among fibers yields key benefits. While the probabilities for $W(x)$ may seem small, we observe that (4.5) gives $H_{m,n}(x) \simeq mnW(x)$: for typical values of mn (say $10^4 \leq mn \leq 10^9$), $W(x)$ may be extremely small yet the probability $H_{m,n}(x)$ for the failure of the fibrous assembly may be substantial (say > 0.1). We point out that if the fibrous assembly <u>always</u> failed when the first fiber failed, then $H_{m,n}(x) \simeq mnF(x)$; obviously very low loads x would still yield almost certain failure as the assembly would behave no better than a single fiber of length $mn\delta$.

Fortunately, this isn't the case.

The speed of convergence of $W_n(x)$ to the conjectured limit $W(x)$ appears to slow somewhat when the Weibull shape parameter ρ is decreased, although it is still very rapid even for $\rho = 5$. (See Harlow and Phoenix [6].) It is also believed that the speed of convergence will decrease as the local load sharing becomes more diffuse, that is, as redistributed loads are shared by a larger number of neighboring survivors.

In Figure 4.2 we have plotted the conjectured limit $W(x)$ (actually $W_9(x)$) for various values of the Weibull shape parameter ρ. It is interesting to note that the lower tails are quite parallel in spite of the fact that the associated lines for $F(x)$ would have a variation in slope of more than an order of magnitude. Evidently variability in the strength of the individual fibers seems to affect the <u>location</u> far more than the <u>slope</u> of the conjectured limit $W(x)$.

4.3 Practical features of the strength distribution for a fibrous assembly

Engineers frequently use the Weibull distribution to represent the strength of yarns, ropes and cables. The scale and shape

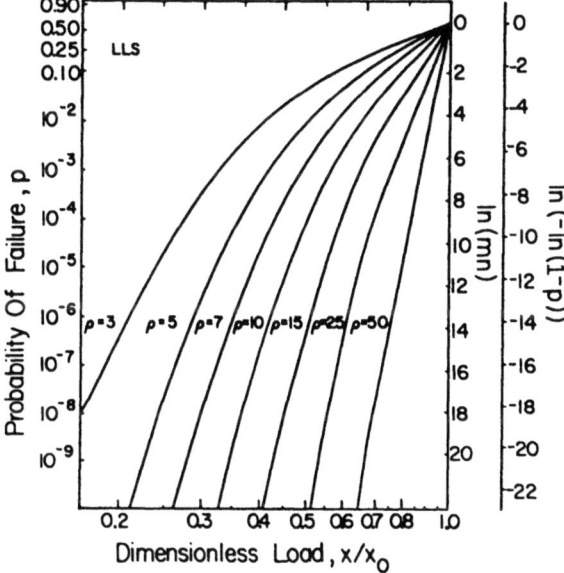

Fig 4.2 Conjectured limiting distribution $W(x)$ plotted for various values of the Weibull shape parameter ρ for single fibers.

parameters are estimated from experimental data, and thus, the distribution of strength for specimens of a variety of lengths is generated easily. (Recall the discussion of Section 2.3.) The results of this section will support this practice; the theoretical predictions of the Weibull shape and scale parameters will agree with those values that are often obtained experimentally.

In Figure 4.3 We give the (conjectured) probability distributions of strength for fibrous assemblies with $mn = 10^6$ fiber elements under a variety of Weibull shape parameters ρ for the individual fibers. (If $n = 9$, the results are exact.) These graphs for $H_{m,n}(x)$ were constructed from those of Figure 4.2 by a simple vertical shift on Figure 4.2 the amount $\log(mn) = \log(10^6) = 13 \cdot 82$. We make the following important observations.

1. When mn is large, $H_{m,n}(x)$ for the fibrous assembly is approximately Weibull, that is, the graphs on Figure 4.3 are nearly straight over a wide probability range.

2. The _effective_ Weibull shape parameter for the fibrous assembly, as measured from the slope of the graphs, varies from about 25 to 50 even though ρ for the fibers varies between 5 and 50. An effective value of 30 is typical for yarns, ropes and cables. Notice that the reduction in _variability_ in strength, in passing from fiber to fibrous assembly, diminishes as ρ increases.

3. The median strength of the fibrous assembly diminishes rapidly as ρ decreases, that is, as the variability in fiber strength increases. For the typical value $\rho = 7$, the median strength of the fibrous assembly is only about 30% of that of a single fiber element (but may be a much higher fraction of that of a longer fiber).

5. ANALYSIS FOR FIBROUS ASSEMBLIES WITH LARGE BUNDLES UNDER EQUAL LOAD SHARING

In comparing Figures 3.2 and 3.3, we notice that $G_n(x)$ under the equal load sharing rule is shifting as n increases. The question naturally arises "What asymptotic behavior occurs for $G_n(x)$ or for $H_{m,n}(x)$ as bundle size n is further increased?

5.1 Asymptotic distribution for the strength of a single bundle ($n \to \infty$).

For the present discussion, we let the random variables $X_1, \ldots, X_n$ represent the strengths of the n fiber elements in a

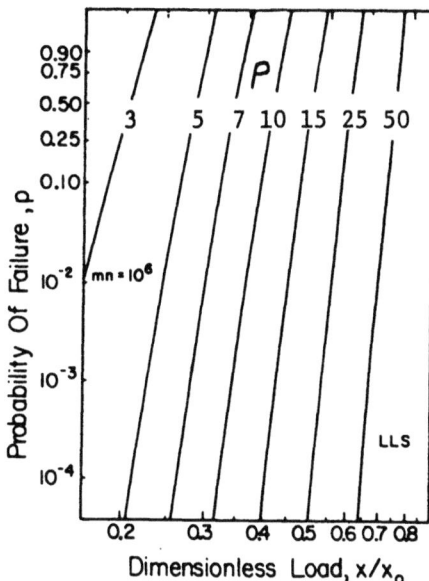

Fig 4.3 Distribution of strength $H_{m,n}(x)$ for a fibrous assembly under local load sharing and with 10^6 fiber elements. Graphs are plotted for several values of the Weibull shape parameter ρ.

single bundle, and note that these strengths are independent with common c.d.f. $F(x)$, $x \geq 0$. We let $X_{1,n} \leq \ldots \leq X_{n,n}$ be these strengths <u>arranged in increasing order of magnitude</u>. Under the equal load sharing rule it is not difficult to see that the <u>bundle strength</u> Q_n^* is

$$Q_n^* = \max\{X_{1,n}, \frac{n-1}{n} X_{2,n}, \ldots, \frac{1}{n} X_{n,n}\}, \qquad (5.1)$$

that is, Q_n^* is the maximum force divided by the number of fibers n that the bundle will sustain. Still another representation results from considering <u>the empirical c.d.f.</u>

$$\tilde{F}_n(x) = \begin{cases} 0 & \text{for } 0 \leq x < X_{1,n} \\ i/n & \text{for } X_{i-1,n} \leq x < X_{i,n} \text{ and } i=2,\ldots,n \\ 1 & \text{for } X_{n,n} \leq x. \end{cases} \qquad (5.2)$$

Then (5.1) is identical to

$$Q_n^* = \sup_{x \geq 0} x[1 - \tilde{F}_n(x)], \qquad (5.3)$$

that is, Q_n^* is the maximum value that the function $x[1 - \tilde{F}_n(x)]$ achieves as x is increased from zero.

We let

$$\mu(x) = x[1 - F(x)], \quad x \geq 0, \qquad (5.4)$$

and

$$\Gamma(x_1, x_2) = x_1 x_2 F(x_1)[1 - F(x_2)], \quad x_2 \geq x_1 \geq 0, \qquad (5.5)$$

and let

$$\mu_{max} = \sup_{x \geq 0} x[1 - F(x)], \qquad (5.6)$$

that is, μ_{max} is the maximum value that the function $x[1 - F(x)]$ achieves as x is increased from zero. Also, let x^* be the point x (assumed unique) where $\mu(x)$ achieves its maximum μ_{max}, that is,

$$x^* = \{x; \mu(x) = \mu_{max}\}. \qquad (5.7)$$

H. E. Daniels [7] was the first to show that <u>the bundle strength Q_n^* is asymptotically normally distributed with mean</u> μ_{max} <u>and variance</u>

$$\phi_n = \Gamma(x^*, x^*)/n. \qquad (5.8)$$

(See Phoenix and Taylor [8] for a proof under broader assumptions on load sharing.) When $F(x)$ for single fibers is the Weibull distribution (2.4), we compute (after Coleman [9])

$$\mu_{max} = x_0 \, \rho^{-1/\rho} \exp\{-1/\rho\} \qquad (5.9)$$

and

$$\phi_n = n^{-1} x_0^2 \, \rho^{-2/\rho} \exp\{-1/\rho\}(1 - \exp\{-1/\rho\}). \qquad (5.10)$$

The result given above is a classic analytical result and we list important aspects as follows:

1. The mean strength of a single fiber element is $\bar{x} = x_0 \Gamma(1 + 1/\rho)$, and we have the <u>strength efficiency ratio</u>

$$\mu_{max}/\bar{x} = \frac{\rho^{-1/\rho} \exp\{-1/\rho\}}{\Gamma(1 + 1/\rho)}. \qquad (5.11)$$

As ρ decreases from ∞ (and the variability in fiber strength increases) $\mu_{max}/\bar{x}$ decreases from unity; however, for typical values of ρ the strength efficiency ratio exceeds 1/2. Thus a large bundle retains a major portion of the mean strength of a single fiber element. (See Coleman [9] for a graph of $\bar{\mu}_{max}/\bar{x}$ versus coefficient of variation of fiber strength.)

2. The variability in the bundle strength Q_n^* diminishes as bundle size n increases. Indeed, the asymptotic standard deviation $\sqrt{\phi_n}$ diminishes in proportion to $1/\sqrt{n}$ as $n \to \infty$. The consequence is that the bundle strength Q_n^* approaches μ_{max} with certainty as $n \to \infty$. Recall that such a result did not prevail under the local load sharing rule.

The question naturally arises, "How close is the asymptotic normal distribution to the exact distribution $G_n(x)$ for the bundle strength?" To answer this question we require an efficient means for computing $G_n(x)$ under the equal load sharing rule. The algorithm used in the case of local load sharing is not suitable for $n > 10$. However, Daniels [7] obtained a recursive formula wherein $G_n(x)$ for a given bundle size n could be expressed in terms of $G_{n-1}(x), G_{n-2}(x), \ldots, G_1(x) = F(x)$ for all smaller bundles:

$$G_n(x) = \sum_{k=1}^{n} (-1)^{k+1} \binom{n}{k} F(x)^k G_{n-k}\left(\frac{nx}{n-k}\right), \quad x \geq 0. \quad (5.12)$$

(See Suh et al. [10] for a straightforward proof.) Even with this recursion, accuracy is rapidly lost as n increases. In Figure 5.1 we plot exact results for n up to thirty and $\rho = 5$; the horizontal scale is chosen to be linear in the standardized quantity $(Q_n^* - \mu_{max})/\sqrt{\phi_n}$ so that the asymptotic normal distribution for bundle strength plots as the same straight line for all n. We note that the convergence is somewhat slow and, as might be expected, the normal approximation is inaccurate in the lower tail. Nevertheless, in this typical case, the normal approximation is conservative and thus, is quite useful in applications where n is of much higher order.

5.2 Asymptotic distribution for the strength of the fibrous assembly as $m \to \infty$ and $n \to \infty$

Earlier we obtained an asymptotic Weibull distribution for the strength of a chain-of-bundles when the bundle size n was held fixed and the chain length m grew large. On the other hand, we have just seen that the strength of a single bundle is asymptotically normal as $n \to \infty$. Now in classical extreme value theory, an important problem is to let $Y_1, \ldots, Y_m$ be independent random variables which follow a common normal distribution, and to examine the distribution of the smallest of these

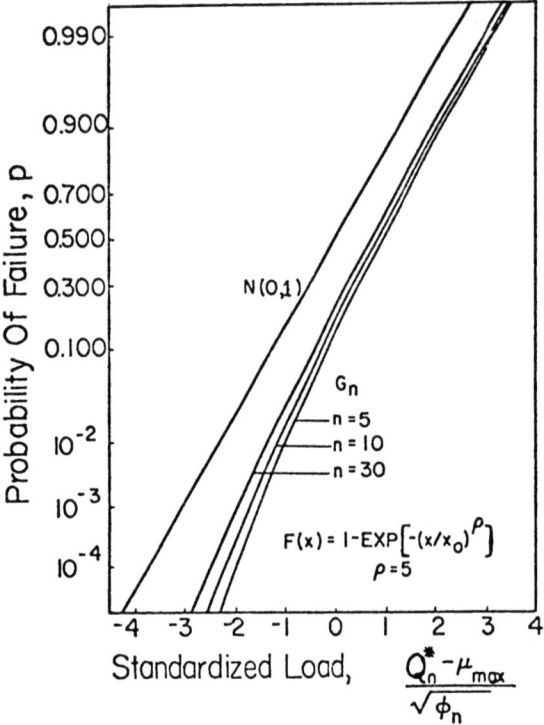

Fig 5.1 Convergence of the exact distribution G_n for bundle strength to the limiting normal distribution.

$Y_{1,m} = \min\{Y_1,\ldots,Y_m\}$ when m grows large. The classic result is that the random variable $(Y_{1,m}-b_m)/a_m$ has the asymptotic distribution

$$H(x) = 1 - \exp\{-\exp(x)\}, \quad -\infty<x<\infty \tag{5.13}$$

as $m \to \infty$ where the a_m and b_m are a sequence of normalizing constants which we need not give here. Thus $Y_{1,m}$ is asymptotically distributed as $H([y-b_m]/a_m)$ for $-\infty<y<\infty$. The question here is: "If the individual bundle strengths are <u>asymptotically</u> normal (but not truly normal) as n becomes large, do we obtain the same limiting distribution $H(x)$ of (5.13) for the strength of the chain-of-bundles as $m \to \infty$, and how fast must the bundle size n grow with respect to the chain length m?" The answers are "yes" and "very slowly" respectively:

Let $m = m(n)$ for some function $m(\cdot)$ such that

$$n^{-1/3} \log m(n) \to 0 \quad \text{as} \quad n \to \infty. \tag{5.14}$$

For example, we could let $m(n) = n^\gamma$ where the exponent γ is a positive constant so that the chain length m could grow as any power of the bundle size n. Define

$$\alpha_n = \sqrt{\phi_n} \, (2 \log m(n))^{-1/2} \qquad (5.15)$$

and

$$\beta_n = \mu_{max} + \sqrt{\phi_n} \, [\log(4\pi) + \log \log m(n)$$
$$- 4 \log m(n)] \cdot [8 \log m(n)]^{-1/2} \qquad (5.16)$$

where ϕ_n is given by (5.8). Then the c.d.f. $H_{m,n}(x)$ for the strength of the chain-of-bundles satisfies

$$H_{m,n}(\alpha_n x + \beta_n) \to H(x) \quad \text{as} \quad n \to \infty$$

for $-\infty < x < \infty$ and $H(x)$ given by (5.13). In other words, the strength of the chain-of-bundles or fibrous assembly is asymptotically distributed as $H([x-\beta_n]/\alpha_n)$ for $-\infty < x < \infty$. This result was proven recently by R. L. Smith [11].

Although this result has not been studied numerically as yet, it does indicate that central measures of the strength of the fibrous assembly diminish extremely slowly as the length m is increased when n is appreciable in magnitude. If we increase the bundle size n at the chain length rate m raised to any small power, the strength of the fibrous assembly approaches the asymptotic mean μ_{max} of (5.6) for a single bundle and the variability shrinks to zero. Again, we have an extremely mild "size effect."

6. FURTHER ANALYSIS AND CONCLUDING COMMENTS

Various extensions are possible for the models of this paper and many have been carried out. Phoenix [12] has studied the strength of twisted bundles under variations on equal load sharing which allow random fiber slack and fiber strain variations resulting from twist. The analysis is based on key results of Phoenix and Taylor [8]. Harlow and Phoenix [13] and Harlow [14,15] have made considerable progress in developing efficient bounds for $H_{m,n}(x)$ in the case of local load sharing, and in proving the conjecture that $W_n(x) \to W(x)$. We also draw attention to a wealth of material on the time dependent fatigue of fibrous assemblies. See Phoenix [16] and Smith [11] for an extensive discussion. The author knows of work in progress under local load sharing rules which are more general (and realistic) than the one considered here. This work is to be reported on later.

REFERENCES

1. S. L. Phoenix and R. G. Sexsmith, J. Composite Materials, 6 (1972) 322-337.

2. S. L. Phoenix, Composite Materials: Testing and Design (Third Conference), ASTM STP 546, American Society for Testing and Materials, (1974) 130-151.

3. S. L. Phoenix, Composite Reliability, ASTM STP 580, American Society for Testing and Materials, (1975) 77-89.

4. D. G. Harlow and S. L. Phoenix, J. Composite Materials, 12 (1978) 195-214.

5. D. G. Harlow, R. L. Smith and H. M. Taylor, The Chain-of-Bundles Probability Model for the Strength of Fibrous Materials III: Asymptotic Distributions for Long Structures, (1979) submitted.

6. D. G. Harlow and S. L. Phoenix, J. Composite Material, 12 (1978) 314-334.

7. H. E. Daniels, Proceedings of the Royal Society, A183 (1945) 405-435.

8. S. L. Phoenix and H. M. Taylor, Advances in Applied Probability, 5 (1973) 200-216.

9. B. D. Coleman, J. Mechanics and Physics of Solids, 7 (1958) 60-70.

10. M. W. Suh, B. B. Bhattacharyya and A. Grandage, J. Applied Probability, 7 (1970) 712-720.

11. R. L. Smith, Limit Theorems for the Reliability of Series-Parallel Load-Sharing Systems, Ph.D. Thesis, Cornell University, Ithaca, NY (1979).

12. S. L. Phoenix, Statistical Theory for the Strength of Twisted Fiber Bundles with Applications to Yarns and Cables to appear in Textile Research Journal (1979).

13. D. G. Harlow and S. L. Phoenix, Bounds on the Probability of Failure of Composite Materials II: Two-Level Failure Analysis, (1979) submitted.

14. D. G. Harlow, Two-Level Failure Analysis for the Strength Distribution of a Fiber Bundle under Local Load Sharing, (1979) submitted.

15. D. G. Harlow, The K-level Failure Analysis for the Strength Distribution of a Fiber Bundle Under Local Load Sharing, (1979) submitted.

16. S. L. Phoenix, International J. Fracture, 14 (1978) 327-344.

WOVEN FABRIC TENSILE MECHANICS

G.A.V. Leaf

Department of Textile Industries, University of
Leeds, England.

ABSTRACT

Three approaches to the analysis of the tensile behaviour
of plain woven fabrics are described. The first, based on
Castigliano's theorem, is for small deformations only. A force
equilibrium method and an energy approach are used for the
analysis of large deformations.

1. INTRODUCTION

The tensile properties of woven fabrics, together with other
mechanical properties such as their behaviour in bending and
shear, are of considerably importance in determining how the
fabric will perform in use. The applications of woven fabrics
vary widely. In apparel uses the tensile strains likely to be
encountered in normal wear will usually be relatively small, but
in some industrial applications the strains involved may be quite
large.

In this paper three analyses of the tensile behaviour of
plain woven fabrics are presented. One of these is applicable
only to small strains but is useful because it gives a relatively
simple closed form solution to the problem. The other analyses
examine the tensile behaviour at all levels of strain. The
methods of attack in all cases are quite different, and the
analyses are of interest for that reason also.

A sketch of a typical load-extension curve of a plain woven
fabric is given in Figure 1a, and three distinct regions can be
discerned. The initial region OA exhibits a high modulus which

is probably caused by the high initial resistance to bending of yarns, produced by friction between the fibres. Once this resistance has been overcome, the fabric becomes easier to extend and the region AB is likely to be governed primarily by the ease with which the yarns can bend. In the final section BC the yarns have straightened as much as possible and the fabric modulus is affected mainly by the extension properties of the yarns.

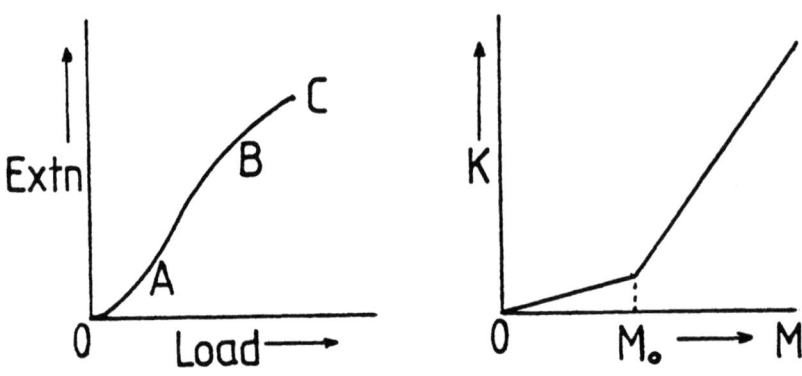

Fig 1a.   Fig. 1b

The tensile properties of woven fabrics are thus determined by, among other things, the tensile and bending behaviour of the yarns composing the fabric. We shall assume that the yarns obey Hooke's law in tension and, following Huang's[1] modification of Grosberg's[2] idealization of yarn bending behaviour, that the yarn bends according to the equations

$$K = M/B, \qquad (M \leq M_o), \qquad (1)$$

$$= M_o/B + (M-M_o)/B^*, \qquad (M > M_o), \qquad (2)$$

where K is the change in curvature produced by the applied couple M, and B and B^* are initial and final flexural rigidities ($B > B^*$). This behaviour is illustrated in Figure 1b.

Any theoretical analysis of the tensile behaviour of woven fabrics requires a starting point, i.e. a model of the unstrained fabric. The models usually used are those provided by Peirce[5] in his early, pioneering paper. Both models suggested by Peirce assumed that the cross-sections of the yarns in the fabric are circular. The first model also assumed that the yarns are flexible, i.e. their bending rigidity is zero. This leads to the model sketched in Figure 2a in which the yarn path is composed of circular arcs and straight lines. The relations between the dimensions

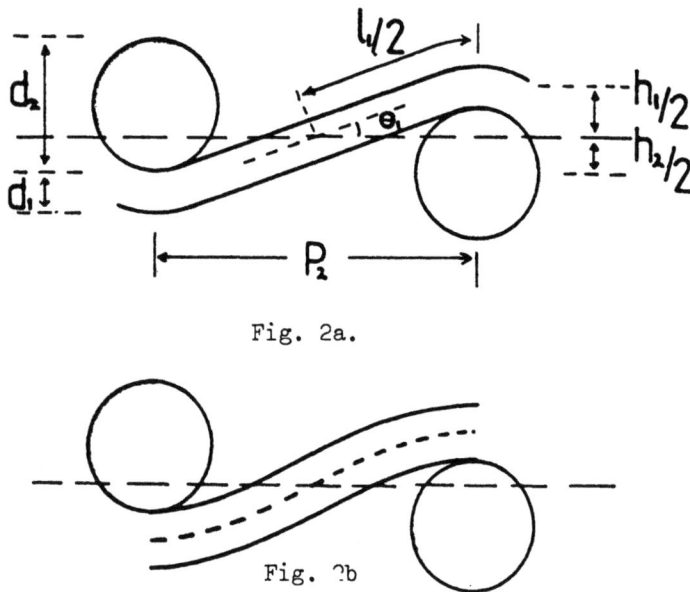

Fig. 2a.

Fig. 2b

shown on the diagram are then given by equations (23) - (27) of this paper. In the other model the yarn is assumed to have a non-zero bending rigidity. The central line of the yarn path then becomes an elastica and the parameters are given by equations (8) to (14).

The assumption of circular yarn cross-sections in the fabric is, of course, quite unrealistic since yarns are easily compressed by the lateral forces imposed on them at the points of yarn intersection. Several authors have attempted to modify Peirce's geometry to take account of this, notably Kemp[10], who suggested a "race-track" cross-section, and Shanahan and Hearle[8] who have considered a "lenticular" or "rugby football" cross-section.

2. A SMALL STRAIN ANALYSIS

The analysis presented here is a modification by Leaf and Kandil[3] of an approach first described by Grosberg and Kedia[4], the main difference being in the model used to represent the undeformed woven fabric. Grosberg and Kedia used Peirce's rigid thread model[5] but, in order to simplify the analysis, we shall adopt the "saw tooth" model sketched in Figure 3, which shows a single intersection of warp and weft. The threads are modelled as straight lines rigidly jointed at H_1 and H_2 and it is assumed that the yarns are set in the shape shown. We shall use the following notation, suffices 1 and 2 being used to distinguish the warp and weft directions.

ℓ_1 = warp modular length = $A_1H_1B_1$ in Figure 3

p_1 = warp thread spacing = A_2B_2

h_1 = warp amplitude = $2CH_1$

θ_1 = warp weave angle = $C\hat{A}_1H_1$

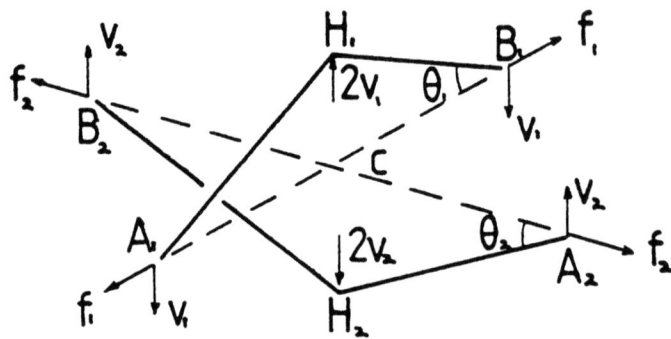

Fig. 3

The yarns are assumed to be inextensible and incompressible. Because of the latter assumption we have

$$h_1 + h_2 = \text{constant} = 2H_1H_2 \tag{3}$$

Suppose the fabric is deformed by forces F_1 and F_2 per unit width along the warp and weft directions respectively. Then the forces f_1 and f_2 acting on individual warps and wefts are

$$f_1 = F_1p_1, \quad f_2 = F_2p_2 \tag{4}$$

We shall suppose that f_1 and f_2 are small enough to ensure that equation (1) describes the bending behaviour of the yarns. When deformation takes place forces $2v_1$ and $2v_2$ will be generated along the line H_1H_2 between the threads. Of course, $v_1 = v_2$, but for the moment we retain the separate notation. The tension and shear in the yarn at a point like A_1 can be resolved into components f_1 and v_1 as shown in Figure 3 and, by symmetry, there will be a point of inflexion at A_1 so that the bending moment there is zero. Assuming small deformations the bending moment at any point on A_1H_1 distant s from A_1 is approximately

$$(f_1\sin\theta_1 - v_1\cos\theta_1)s,$$

and the strain-energy in A_1H_1 is therefore

$$U_1 = (1/2B_1)\int_0^{l_1/2} (f_1\sin\theta_1 - v_1\cos\theta_1)^2 s^2 ds$$

$$= l_1^3(f_1\sin\theta_1 - v_1\cos\theta_1)^2/48B_1.$$

The strain-energy U_2 in A_2H_2 is given by a similar expression so that the total strain energy in the unit is

$$U = 2U_1 + 2U_2 = l_1^3(f_1\sin\theta_1 - v_1\cos\theta_1)^2/24B_1 +$$

$$+ l_2^3(f_2\sin\theta_2 - v_2\cos\theta_2)^2/24B_2.$$

Using Castigliano's theorem[6], the change in h_1 corresponding with the force v_1 is

$$\Delta h_1 = \partial U/\partial v_1 = -l_1^3 \cos\theta_1 (f_1\sin\theta_1 - v_1\cos\theta_1)/12B_1,$$

with a similar expression for Δh_2. But, because of equation (3), $\Delta h_1 + \Delta h_2 = 0$, and putting $v_1 = v_2 = v$ we get

$$v = \frac{f_1 B_2 l_1^3 \sin\theta_1\cos\theta_1 + f_2 B_1 l_2^3 \sin\theta_2\cos\theta_2}{B_2 l_1^3 \cos^2\theta_1 + B_1 l_2^3 \cos^2\theta_2}.$$

Also, the change in p_2 corresponds with f_1, so that

$$\Delta p_2 = \partial U/\partial f_1 = l_1^3(f_1\sin\theta_1 - v\cos\theta_1)\sin\theta_1/12B_1$$

The strain ε_1 in the warp direction is therefore

$$\varepsilon_1 = \frac{\Delta p_2}{p_2} = \frac{l_1^3 l_2^3 \sin\theta_1\cos\theta_2(F_1 p_1 \sin\theta_1\cos\theta_2 - F_2 p_2 \sin\theta_2\cos\theta_1)}{12 p_2 (B_2 l_1^3 \cos^2\theta_1 + B_1 l_2^3 \cos^2\theta_2)}$$

(5)

where we have used equations (4) and the above expression for v. A similar equation for the strain ε_2 in the weft direction can be found by interchange of suffices.

Equation (5) is an expression for the strain produced by biaxial loading, which Grosberg and Kedia did not consider. In the special case of uniaxial loading, say when $F_2 = 0$, equation (5) leads to

$$F_1 = \frac{12B_1 p_2}{p_1 l_1^3 \sin^2\theta_1}\left(1 + \frac{B_2}{B_1}\frac{l_1^3 \cos^2\theta_1}{l_2^3 \cos^2\theta_2}\right)\varepsilon_1 = E_1 \varepsilon_1, \qquad (6)$$

say, where E_1 is the initial Young's modulus of the fabric. We also find that in this case

$$\varepsilon_2 = -F_1 l_1^3 l_2^3 \sin\theta_2 \cos\theta_1 \sin\theta_1 \cos\theta_2 / 12(B_2 l_1^3 \cos^2\theta_1 + B_1 l_2^3 \cos^2\theta_2)$$

so that the initial Poisson's ratio of the fabric is

$$\sigma_1 = -\varepsilon_2/\varepsilon_1 = p_2 \tan\theta_2 / p_1 \tan\theta_1.$$

The above theory has been tested experimentally on a range of fabrics made from cotton, vyncel and cotton/vyncel yarns. In order to estimate the values of θ to be used in equation (6), values of l and p were measured. Then from Figure 3 it can be seen that $\cos\theta_1 = A_1C/A_1H = p_2/l_1$ but when values of θ calculated from this equation were used in equation (6) the estimated values of E were considerably greater than those found experimentally. However, it may be shown that, for Peirce's rigid thread model[5], values of

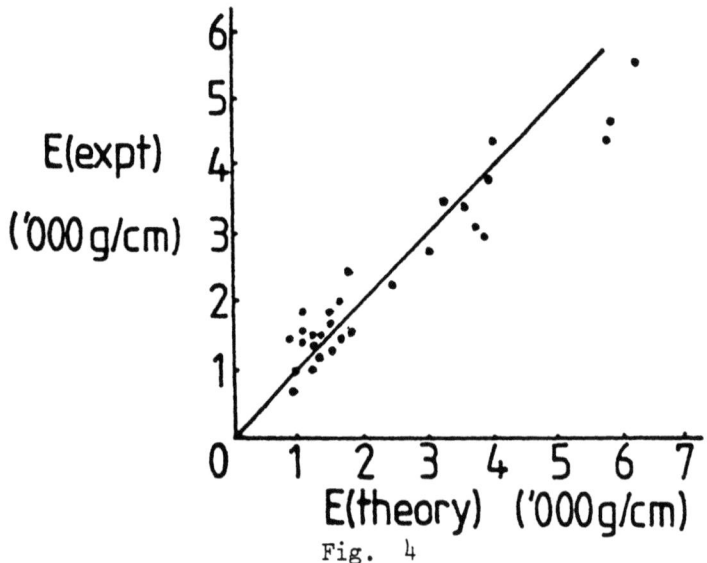

Fig. 4

θ(in degrees) can be estimated from

$$\theta_1 = 106(l_1/p_2 - 1)^{\frac{1}{2}}, \quad \theta_2 = 106(l_2/p_1 - 1)^{\frac{1}{2}}, \tag{7}$$

and values of θ calculated from these equations gave the results shown in Figure 4. Considering the difficult series of measurements involved the agreement between theory and experiment is reasonably good. There is, of course, an element of inconsistency in using equations (7) in conjunction with an analysis derived from a different fabric model but the experimental results suggest that the procedure is justified in practice.

3. LARGE STRAIN ANALYSIS I

The large strain analysis, to be described next, is essentially that given recently by Huang[1]. It assumes that the fabric is completely set before deformation and uses Peirce's rigid thread model as a starting point. Figure 5(a) shows the central axis A_1H_1 of the undeformed warp thread referred to rectangular axes A_1xy through the point of inflexion A_1. The notation is the same as in the previous section but we use an additional suffix 0 to

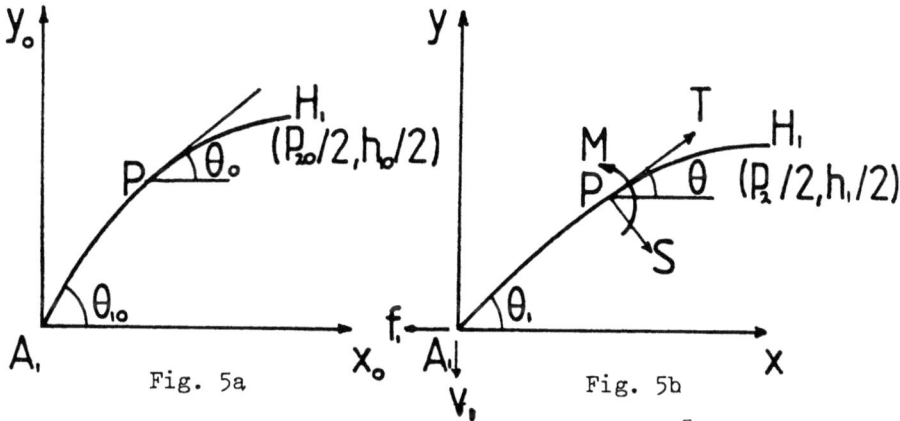

Fig. 5a   Fig. 5b

denote quantities in the undeformed state. Peirce[5] showed that at any point P on the undeformed yarn distant s_0 from A_1,

$$d\theta_0/ds_0 = -4e\{F(e,\pi/2) - F(e,\emptyset_H)\}\cos\emptyset/l_1, \tag{8}$$

$$\cos\theta_0 = 2e\sin\emptyset(1 - e^2\sin^2\emptyset)^{\frac{1}{2}}, \tag{9}$$

and that

$$h_{10} = l_{10}\left\{1-2\cdot\frac{E(e,\pi/2)-E(e,\phi_H)}{F(e,\pi/2)-F(e,\phi_H)}\right\}, \quad (10)$$

$$p_{20} = 2el_{10}\cos\phi_H/\left\{F(e,\pi/2)-F(e,\phi_H)\right\}, \quad (11)$$

where $F(e,\phi)$ and $E(e,\phi)$ are elliptic integrals of the first and second kind respectively, and

$$e = \sin(\pi/4+\theta_{10}/2), \quad (12)$$

$$e\sin\phi = \sin(\pi/4+\theta_0/2), \quad (13)$$

$$\phi_H = \sin^{-1}(e^{-1}2^{-\frac{1}{2}}) \quad (14)$$

When the fabric is deformed by forces f_1 and f_2 per thread in the warp and weft directions respectively the situation is as shown in Figure 5(b). A_1 remains a point of inflexion and forces 2v are generated at the intersection of warp and weft. The tension and bending moment at P are then given by

$$T = f_1\cos\theta + v\sin\theta, \quad (15)$$

$$M = f_1 y - vx. \quad (16)$$

If the Young's modulus of the warp thread is K_1 the axial strain in the yarn is

$$\vartheta_1 = T/K_1 = (f_1\cos\theta + v\sin\theta)/K_1 \quad (17)$$

and, if s is the arc length A_1P in the deformed state,

$$ds = (1+\vartheta_1)ds_0 \quad (18)$$

Further, the change in curvature at P is

$$\kappa = d\theta/ds - d\theta_0/ds_0$$

from which

$$d\theta/dx_0 = \frac{1+\vartheta_1}{\cos\theta_0}\cdot d\theta/ds = \frac{1+\vartheta_1}{\cos\theta_0}(d\theta_0/ds_0+\kappa) \quad (19)$$

after using equation (18) and the fact that $dx_0/ds_0 = \cos\theta_0$.

Since $dx/ds = \cos\theta$ and $dy/ds = \sin\theta$ we also get

$$dx/dx_0 = (1+\nu_1)\cos\theta/\cos\theta_0 , \qquad (20)$$

and

$$dy/dx_0 = (1+\nu_1)\sin\theta/\cos\theta_0 , \qquad (21)$$

where, from equations (9) and (17),

$$\frac{1+\nu_1}{\cos\theta_0} = \frac{1+(f_1\cos\theta+v\sin\theta)K_1^{-1}}{2e\sin\phi(1-e^2\sin^2\phi)^{\frac{1}{2}}} . \qquad (22)$$

By regarding x_0 as an independent variable and θ, x, y as dependent variables, equations (19) to (22) enable the shape of the deformed warp thread to be calculated, provided that κ, v and $d\theta_0/ds_0$ are known. The latter is given by equation (8) and κ by equations (1) and (2) via equation (16). It remains to determine the value of v. This is found from the fact that the warp and weft threads remain in contact at H_1 throughout the deformation. Figure 6(a) shows that the height of the actual contact point I_1 above the fabric plane before deformation is $\frac{1}{2}h_{10}-R_{10}$, where R_{10} is the initial radius of the yarn. After deformation, this height is $\frac{1}{2}h_1-R_1$, where R_1 is the compressed radius of the yarn. The change in height of the contact point during the deformation is therefore

$$\tfrac{1}{2}h_{10}-R_{10}-\tfrac{1}{2}h_1+R_1 .$$

h_{10} is found from equation (10) and h_1 from the integration of equation (21). Huang assumes that the decrease in yarn radius

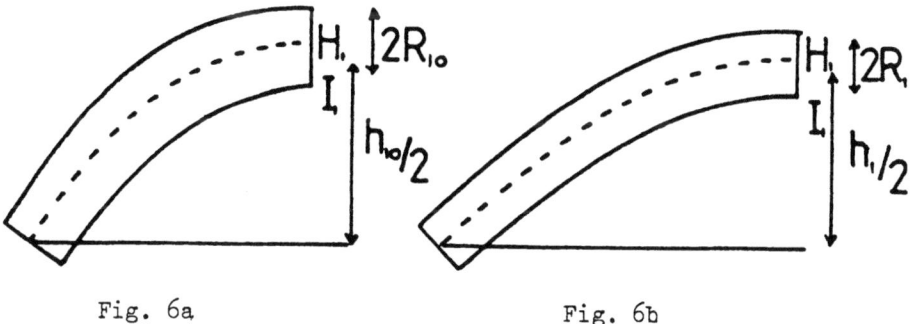

Fig. 6a   Fig. 6b

from R_{10} to R_1 is caused by two factors. These are

(a) a yarn Poisson effect, for which

$$(R_1-R_{10})/R_{10} = -\sigma_1 \nu_{1H},$$

where σ_1 is the yarn Poisson ratio and ν_{1H} is the value of ν_1 at H_1, i.e. $\nu_{1H} = f_1/K_1$ from equation (17) since $\theta = 0$ at H_1;

(b) a yarn compression effect for which it is assumed that

$$(R_{10}-R_1)/R_{10} = \lambda_1 v,$$

where λ_1 is a constant. Superposing these effects we get

$$(R_1-R_{10})/R_{10} = -(\sigma_1 \nu_{1H} + \lambda_1 v)$$

and the change in height of the contact point is therefore

$$\tfrac{1}{2}(h_{10}-h_1) + R_{10}(\sigma_1 \nu_{1H} + \lambda_1 v).$$

A similar expression for the change in height of the weft contact point can be written down, and if the yarns are to remain in contact the sum of these changes must be zero, i.e.

$$\tfrac{1}{2}(h_{10}-h_1) + R_{10}(\sigma_1 \nu_{1H} + \lambda_1 v) + \tfrac{1}{2}(h_{20}-h_2) + R_{20}(\sigma_2 \nu_{2H} + \lambda_2 v) = 0, \quad (22a)$$

which determines v.

The boundary conditions for the integration of equations (19) to (22a) are $x = y = 0$ at A_1, where $x_0 = 0$, and $\theta = 0$ at H_1, where $x_0 = p_{20}/2$. The integration uses a standard iterative procedure to satisfy these boundary conditions and in this way the shape of the deformed yarn path is found. Consequently, h_1 and p_2 are known from which the fabric strains can be calculated from

$$\varepsilon_1 = (p_2-p_{20})/p_{20}, \quad \varepsilon_2 = (p_1-p_{10})/p_{10}.$$

Huang provides some numerical results for the cases when $f_2 = 0$ and when $f_1 = f_2$. Some typical results for the former case are shown in Figure 7, which show a marked resemblance to

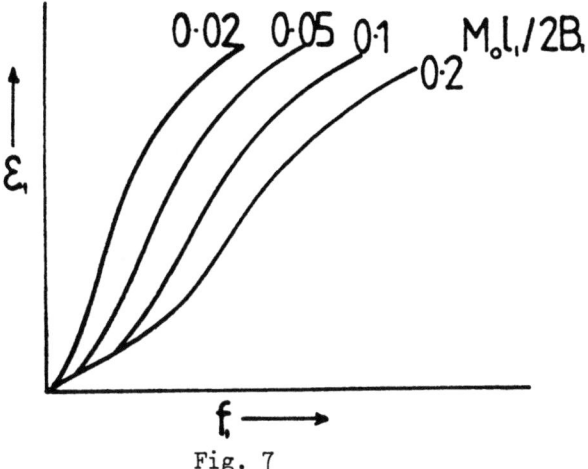

Fig. 7

the stress-strain curves of real fabrics, sketched in Figure 1. Huang gives no experimental data to compare with the theory and more work on these lines is obviously necessary.

4. LARGE STRAIN ANALYSIS II

Recently, an energy approach to the solution of problems in fabric mechanics has been described by Hearle and Shanahan[7,8]. In order to simplify the numerical calculations involved it is assumed that the form of the geometry of the structure is known both before and during the deformation. For example, so far as woven fabrics are concerned, a relatively simple model to consider is Peirce's flexible thread model, in which

$$p_2 = (\ell_1 - D\theta_1)\cos\theta_1 + D\sin\theta_1 \,, \tag{23}$$

$$p_1 = (\ell_2 - D\theta_2)\cos\theta_2 + D\sin\theta_2 \,, \tag{24}$$

$$h_1 = (\ell_1 - D\theta_1)\sin\theta_1 + D(1-\cos\theta_1) \,, \tag{25}$$

$$h_2 = (\ell_2 - D\theta_2)\sin\theta_2 + D(1-\cos\theta_2) \,, \tag{26}$$

$$h_1 + h_2 = D \tag{27}$$

Suppose that, in general, the geometry can be expressed by one or more equations, such as (23)-(27), which are equivalent to

$$f(x_1,x_2,x_3,\text{---},x_\ell,\ell_1,\ell_2,\ell_3,\text{---},\ell_n) = 0, \tag{28}$$

where the x_i ($i = 1,2,\text{---},\ell$) are generalized dimensions or displacements associated with the external deformation (e.g. p_1 and p_2 in the above equations) and the ℓ_j ($j = 1,2,\text{---},n$) are a set of independent geometrical parameters (e.g. ℓ_1 and ℓ_2). Suppose also that α_k ($k = 1,2,\text{---},m$) are a set of dependent geometric parameters (such as the weave angles θ_1,θ_2) and that F_i are generalized external forces associated with the generalized displacements x_i. It is assumed that the total energy of the deformed system is given by

$$U = -\sum_{i=1}^{\ell} F_i x_i + E(\ell_1,\ell_2,\text{---},\ell_n,\alpha_1,\alpha_2,\text{---},\alpha_m), \tag{29}$$

where E is the strain-energy of the system. The x_i are not all independent. For example, when a woven fabric is extended to a given value of p_2 the corresponding value of p_1 will be determined. Hence, one of the x_i (say x_1) must be chosen as a dependent variable. In the position of equilibrium the energy U will have a minimum value, i.e. we must have

$$\frac{\partial U}{\partial x_i} = -F_1 \frac{\partial x_1}{\partial x_i} - F_i + \sum_{k=1}^{m} \frac{\partial E}{\partial \alpha_k} \cdot \frac{\partial \alpha_k}{\partial x_i} = 0, \; (i = 2,\text{---},\ell) \tag{30}$$

and

$$\frac{\partial U}{\partial \ell_j} = -F_1 \frac{\partial x_1}{\partial \ell_j} + \frac{\partial E}{\partial \ell_j} + \sum_{k=1}^{m} \frac{\partial E}{\partial \alpha_k} \cdot \frac{\partial \alpha_k}{\partial \ell_j} = 0, \; (j = 1,\text{---},n). \tag{31}$$

These provide $(\ell+n-1)$ equations; the geometrical relation (28) provides a further equation so that there is a system of $(\ell+n)$ equations that describes the behaviour of the structure. The α_k may, in principle, be eliminated from the equations so that there are $(2\ell+n)$ unknowns, the F_i, x_i, ℓ_j ($i = 1,\text{---},\ell$; $j = 1,\text{---},n$). Thus the system is completely determined if ℓ of the latter, say the F_i, are given.

As an example of the application of this method consider the fabric, defined by the equations (23)-(27) deformed biaxially by forces f_1 and f_2 per thread in the warp and weft directions respectively. It is assumed that the fabric is unset so that the strain-energy of bending in the contact regions is

$$E_B = 2(B_1\theta_1 + B_2\theta_2)/D,$$

assuming a simple bending law for the yarns given by equation (1). We suppose also that the yarns can extend during the deformation, keeping their volume constant. The latter condition leads to

$$D = d_1(\ell_{10}/\ell_1)^{\frac{1}{2}} + d_2(\ell_{20}/\ell_2)^{\frac{1}{2}}, \qquad (32)$$

where d_1 and d_2 are the original warp and weft diameters, and ℓ_{10} and ℓ_{20} are the original modular lengths. The strain-energy of yarn extension is

$$E_e = \tfrac{1}{2}K_1(\ell_1 - \ell_{10})^2/\ell_{10} + \tfrac{1}{2}K_2(\ell_2 - \ell_{20})^2/\ell_{20},$$

where K_1 and K_2 are the yarn moduli. Hence, in this case,

$$E = 2(B_1\theta_1 + B_2\theta_2)/D + \tfrac{1}{2}K_1(\ell_1 - \ell_{10})^2/\ell_{10} + \tfrac{1}{2}K_2(\ell_2 - \ell_{20})^2/\ell_{20}. \qquad (33)$$

In the notation of the general equations we have

$$x_1 = p_2, x_2 = p_1, F_1 = f_1, F_2 = f_2, \ell_1 = \ell_1, \ell_2 = \ell_2, \alpha_1 = \theta_1, \alpha_2 = \theta_2, \alpha_3 = D$$

and the minimum energy conditions thus give

$$f_1 \partial p_2/\partial p_1 + f_2 = \frac{\partial E}{\partial \theta_1}\frac{\partial \theta_1}{\partial p_1} + \frac{\partial E}{\partial \theta_2}\frac{\partial \theta_2}{\partial p_1},$$

$$f_1 \frac{\partial p_2}{\partial \ell_1} = \frac{\partial E}{\partial \ell_1} + \frac{\partial E}{\partial \theta_1}\frac{\partial \theta_1}{\partial \ell_1} + \frac{\partial E}{\partial \theta_2}\frac{\partial \theta_2}{\partial \ell_1} + \frac{\partial E}{\partial D}\frac{\partial D}{\partial \ell_1},$$

$$f_1 \frac{\partial p_2}{\partial \ell_2} = \frac{\partial E}{\partial \ell_2} + \frac{\partial E}{\partial \theta_1}\frac{\partial \theta_1}{\partial \ell_2} + \frac{\partial E}{\partial \theta_2}\frac{\partial \theta_2}{\partial \ell_2} + \frac{\partial E}{\partial D}\frac{\partial D}{\partial \ell_2},$$

where, from equation (33),

$$\partial E/\partial \ell_1 = K_1(\ell_1 - \ell_{10})/\ell_{10}, \; \partial E/\partial \ell_2 = K_2(\ell_2 - \ell_{20})/\ell_{20},$$

$$\partial E/\partial \theta_1 = 2B_1/D, \; \partial E/\partial \theta_2 = 2B_2/D, \; \partial E/\partial D = -2(B_1\theta_1 + B_2\theta_2)/D^2,$$

while, from equation (32),

$$\partial D/\partial \ell_1 = -\tfrac{1}{2}d_1 \ell_{10}^{\tfrac{1}{2}} \ell_1^{-3/2}, \quad \partial D/\partial \ell_2 = -\tfrac{1}{2}d_2 \ell_{20}^{\tfrac{1}{2}} \ell_2^{-3/2}.$$

The other partial derivatives, such as $\partial \theta_1/\partial p_2$, can be found from the geometrical relations (23)-(27).

We now have a complete set of equations that describe the behaviour of the fabric under specified loading conditions, i.e. when f_1 and f_2 are given. Hearle and Shanahan solved these equations using the interactive computing system QAS developed at UMIST by Konopasek[9], which is particularly convenient for solving non-linear simultaneous equations of the type involved in the analysis. They provide some comparisons of the results obtained by this method and a more conventional (exact) solution using Peirce's rigid thread model. The comparison is reproduced in Figure 8 from which it can be seen that the agreement is good in the cases considered.

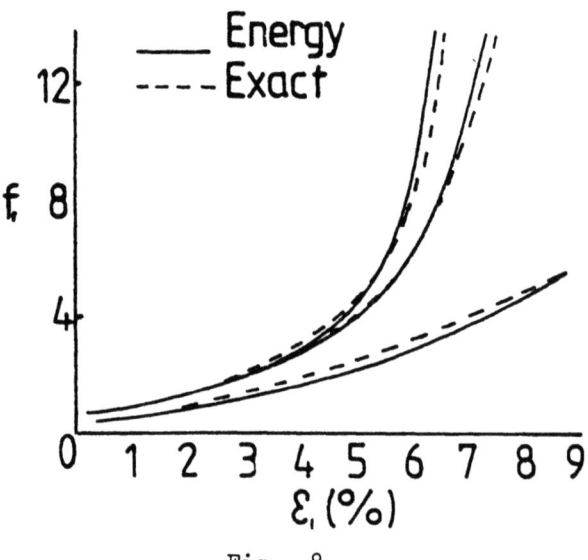

Fig. 8

REFERENCES

1. N.C. Huang. Technical Report SM7801 Sept. 1978. Solid Mechanics Group, Dept. of Aerospace & Mechanical Eng., U. of Notre Dame.
2. P. Grosberg. Text.Res.J., 36, 1966, 205-211.
3. G.A.V. Leaf and K.H. Kandil. To be published.
4. P. Grosberg and S. Kedia. Text.Res.J., 36, 1966, 72-79.
5. F.T. Peirce. J.Text.Inst., 28, 1937, T45-T96.
6. R.V. Southwell. "An Introduction to the Theory of Elasticity" Oxford Univ. Press, 2nd edn., 1941, 15.

7. J.W.S. Hearle and W.J. Shanahan, J.Text.Inst., 69, 1978, 81.
8. J.W.S. Hearle and W.J. Shanahan, J.Text.Inst., 69, 1978, 92.
9. M. Konopasek. Proc.European Computing Congress, London, 1975
10. A. Kemp, J.Text. Inst., 49, 1958, T44 - T.48.

FROM FIBERS TO WOVEN FABRICS*

S.R. Moghe

BFGoodrich Research and Development Center,
Brecksville, Ohio

ABSTRACT. A mathematical model for predicting the load deformation behavior of woven fabrics based upon the weave and geometric parameters is developed. The analysis takes into account the measured stress-strain behaviors of fiber yarns, crimps and take-ups and jamming conditions. Excellent agreement between the analysis and experimental results for fabrics is obtained.

INTRODUCTION. A typical reinforced rubber product is a laminated composite structure consisting of various fabric and rubber components. Therefore, its properties depend strongly upon the properties of individial components and the interaction between them. Since fabric is the main load bearing member in the composite construction, its design and manufacture is crucial for satisfactory product performance. For example, load deformation, strength, ultimate elongation and splice strength depend upon the fabric properties and composite construction. Deliberate efforts are made by the designer to choose a suitable weave, fibers and other construction parameters so as to manufacture the reinforced product to suit particular applications. Accurate choices in routine everyday applications are based upon experience and technical expertise without much difficulties.

---

*Invited presentation at the NATO Advanced Study Institute on Mechanics of Flexible Fiber Assemblies, to be held in Kilini, Greece on August 19, 1979.

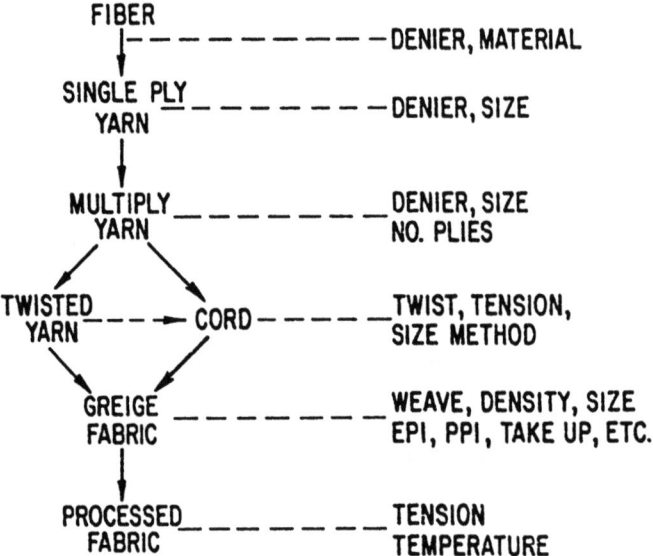

FIGURE 1.

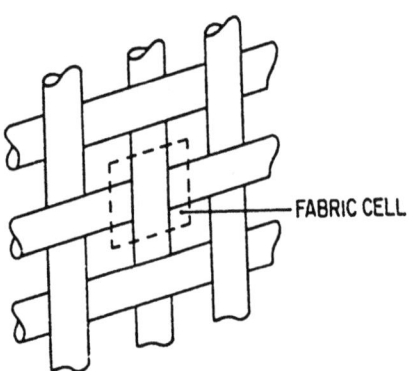

PLAIN WEAVE FABRIC

FIGURE 2.

Analytical approaches to fabrics and product designs are necessary when evaluating new fibers and constructions where prior experience may be non-existent. It is also useful in reducing and identifying perhaps better choices amongst the myriad of possibilities. Of course, an analytical approach becomes valuable only if actual material properties, weave parameters and manufacturing variables are included explicitly in the analysis. We, at BFGoodrich, have developed an analysis and computer program which allow predictions of load deformation properties of fabrics and composites under different load conditions from fiber properties and various processing operations involved (Figure 1). The analysis and computer program have shown excellent agreement with the experimental results. Historically, various attempts to analyze woven fabric structures have been made in the past.[1,4] There must be many more investigations reported in various publications to which I cannot refer here. I would like to mention, however, two investigations which are somewhat related to my work. Hearle, et al, have approached the fabric structure from the fundamental mechanics approach, namely the energy considerations. Kawabata, et al, have systematically investigated the plain weave fabrics under complex loading conditions and large deformations. Their approach is based upon force equilibrium for the fabric cell. Theirs was the first comprehensive attempt to analyze the fabric structure.

This paper is based upon a mathematical model which utilizes actual fiber properties in the analysis. The analysis also takes into account every possible fabric parameter such as crimp or take-up, weave parameters, jamming conditions and nonlinear fiber properties. The analysis predicts complete nonlinear stress-strain behavior up to strength for woven fabrics. The analysis utilizes energy principles, simplified slender curved rod analysis and many observations on fiber and fabric behaviors.

MATHEMATICAL MODEL. A woven fabric consists of a large number of repeated units referred hereafter as 'fabric cells'. The cell pattern repeats throughout the fabric by translation or reflection (Figures 2, 3). It is tacitly assumed that the cords and cell pattern are homogenously distributed throughout the entire fabric. For example, the actual curved lengths of typical cord elements are assumed to be identical from cell to cell. The fabric behavior depends upon the load deformation response of a representative cell unit. Therefore, we need to consider in detail the load deformation response of a typical cell unit only (Figure 4).

The cell may consist of two or more cords with a number of crossover points amongst them. Typically, the plain weave fabric cell consists of only one warp and fill yarn which have one common contact area. The complicated fabric cells generally have multiple

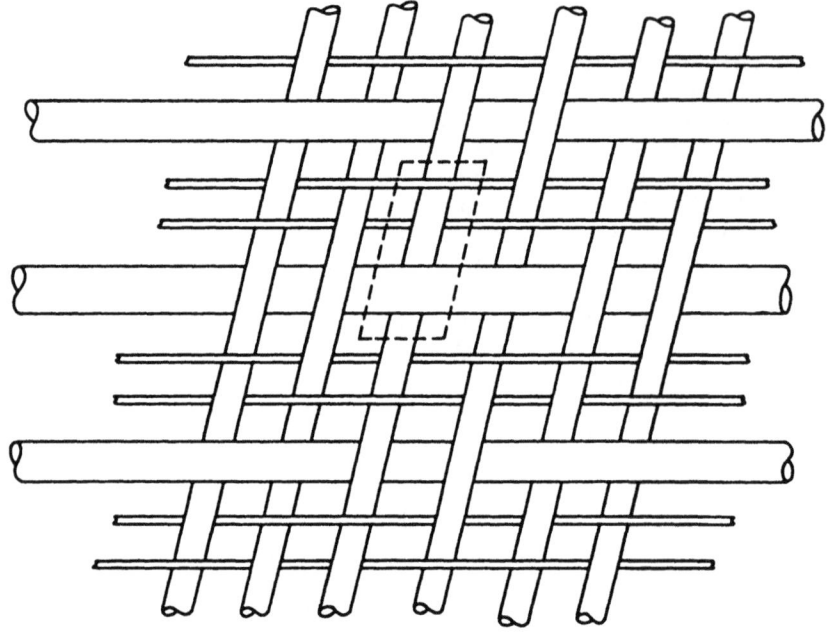

Fig. 3   **COMPLEX FABRIC and MODEL CELL**

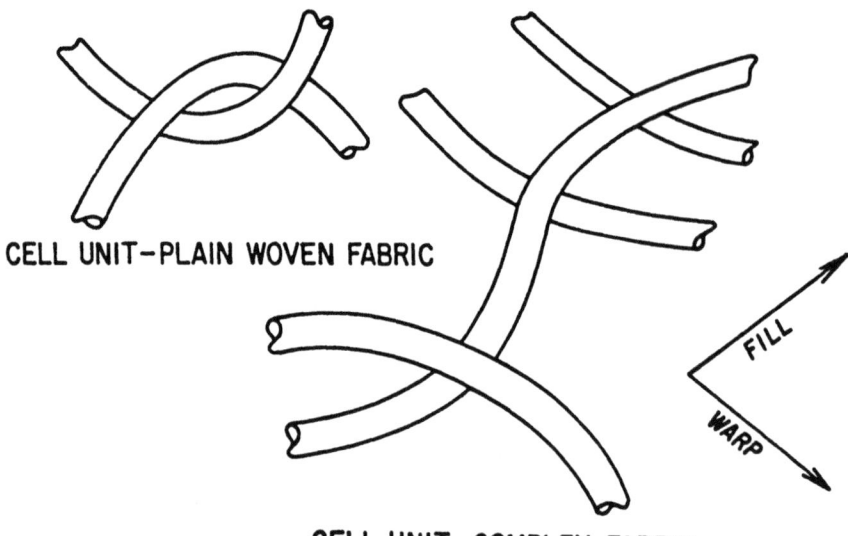

Fig. 4.

yarns in one or more directions with many contacts amonst them (Figure 4). In most common woven fabrics, warps are located in parallel planes and so are the fills. The weave pattern is always symmetrical about the fabric middle plane. The cell shape, size and the curved lengths of each cord are precisely known from fabric parameter specifications. For example, ends per inch and the fiber crimp or take-up define uniquely the goemtrical coordinates for the cell model unit.

In order to calculate the load deformation response of a cell, I will assume that the cords behave as slender curve rods and have nonlinear stress-strain properties (Figure 5). It is clear that the cords have axial, bending and torsional stiffnesses, with the axial stiffness being considerably higher than the other two. The actual measured stress-strain behavior is used in the analysis so that the analysis is realistic and practical. The influence of yarn sizes and bulk are considered in the analysis by imposing appropriate jamming conditions. It should be noted that the assumption of slender curved rods representing cords automatically disregards the cord size or bulk.

Under external load, the cell structure will respond in such a way that the cords in contact always remain in contact. As a result, the displacement and stress distribution follow usual laws of continuity in these regions. Furthermore, the boundary conditions in the form of external load or displacements are also satisfied. The analysis then finds a compatible state of deformation on the basis of the principle of virtual work.

Let us now consider deformation response of a typical cord in the fabric cell (Figure 4) and apply slender rod analysis and follow the notations and terminology adopted by Huang[5]. Let the origin be at one end of a typical cord in a given cell (Figure 6) and X, axis in the fabric middle plane passing through the other end. Choose X_3 axis vertically up in the cord plane. The X_2 axis is chosen such that $0-X_1X_2X_3$ forms a conventional right handed rectangular coordinate systen. If 's' is the distance measured along the cord center line then the unit tangential, principal normal and binormal vectors λ_i, μ_i, ν_i are defined as

$$\lambda_i = X_i' \tag{1}$$

$$\mu_i = \frac{1}{\kappa} \lambda_i' \tag{2}$$

$$\nu_i = \frac{1}{\tau} (\mu_i' + \kappa\lambda_i) \tag{3}$$

$$\nu_i' = -\tau\mu_i \tag{4}$$

and $\quad ds^2 = dX_i \cdot dX_i$, $( )' = \frac{d}{ds}$ \qquad (5)

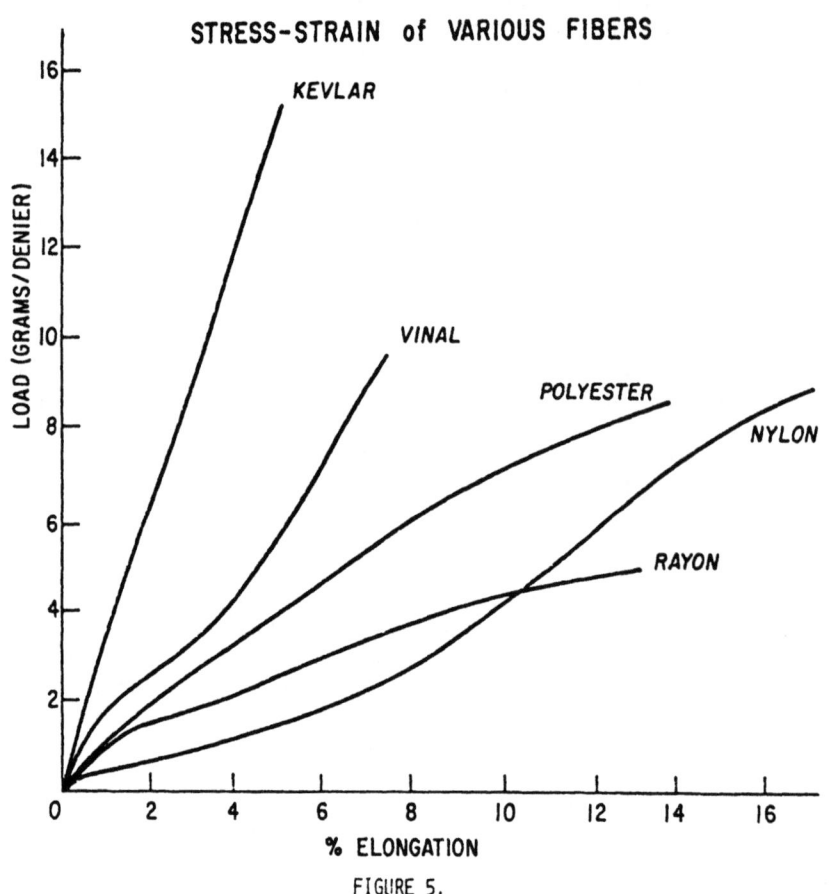

FIGURE 5.

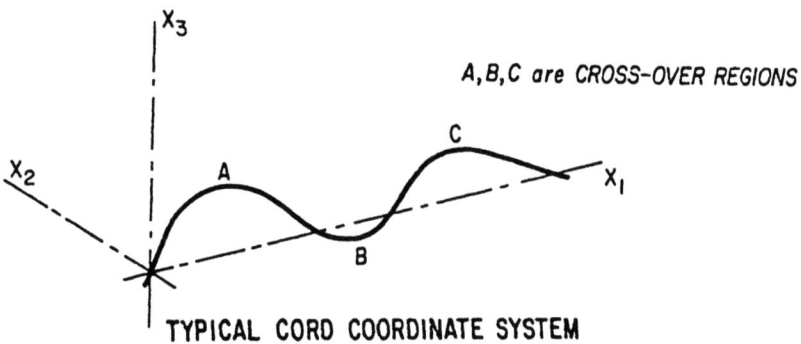

FIGURE 6.

where prime denotes differentiation with respect to s, κ and τ are curvature and torsion respectively. We can define similar quantities and relationships in the deformed state by barred quantities and differentiation by an overhead dot.

Let us further define components of displacement U_i, external forces p_i and m_i in the two states as:

$$U_i = U_\lambda \lambda_i + U_\mu \mu_i + U_\nu \nu_i$$
$$= \overline{U}_\lambda \overline{\lambda}_i + \overline{U}_\mu \overline{\mu}_i + \overline{U}_\nu \overline{\nu}_i \tag{6}$$

$$P_i = \overline{P}_\lambda \overline{\lambda}_i + \overline{P}_\mu \overline{\mu}_i + \overline{P}_\nu \overline{\nu}_i \tag{7}$$

and

$$m_i = \overline{m}_\lambda \overline{\lambda}_i + \overline{m}_\mu \overline{\mu}_i + \overline{m}_\nu \overline{\nu}_i \tag{8}$$

The stress resultants F_i and moments M_i at any cross-section are defined as

$$F_i = \overline{F}_\lambda \overline{\lambda}_i + \overline{F}_\mu \overline{\mu}_i + \overline{F}_\nu \overline{\nu}_i \tag{9}$$

and

$$M_i = \overline{M}_\lambda \overline{\lambda}_i + \overline{M}_\mu \overline{\mu}_i + \overline{M}_\nu \overline{\nu}_i \tag{10}$$

If 'e' is defined as the engineering strain then

$$\overline{s}' = 1 + e \tag{11}$$

and

$$\overline{\lambda}_i = \frac{1}{1+e} \left[(1 + U_\lambda' - \kappa U_\mu) \lambda_i \right.$$
$$+ (U_\mu' + \kappa U_\lambda - \tau U_\nu) \mu_i \tag{12}$$
$$\left. + (U_\nu' + \tau U_\mu) \nu_i \right]$$

Now if $\delta \overline{U}_\lambda$, $\delta \overline{U}_\mu$ and $\delta \overline{U}_\nu$ are the components of virtual displacement at any point on the center line and $\delta \overline{\phi}_\lambda$, $\delta \overline{\phi}_\mu$ and $\delta \overline{\phi}_\nu$ the virtual rotations then virtual work principle gives[5]

$$I_E + E - I_I = 0 \tag{13}$$

where

$$I_E = \int_0^L \left[\bar{P}_\lambda \delta\bar{U}_\lambda + \bar{P}_\mu \delta\bar{U}_\mu + \bar{P}_\nu \delta\bar{U}_\nu \right. $$
$$\left. + \bar{m}_\lambda \delta\bar{\phi}_\lambda + \bar{m}_\mu \delta\bar{\phi}_\mu + \bar{m}_\nu \delta\bar{\phi}_\nu \right] d\bar{S} \tag{14}$$

$$E = \left[\bar{F}_\lambda \delta\bar{U}_\nu + \bar{F}_\mu \delta\bar{U}_\mu + \bar{F}_\nu \delta\bar{U}_\nu \right.$$
$$\left. + \bar{M}_\lambda \delta\bar{\phi}_\lambda + \bar{M}_\mu \delta\bar{\phi}_\mu + \bar{M}_\nu \delta\bar{\phi}_\nu \right]_{\bar{S}=0}^{\bar{S}=L} \tag{15}$$

$$I_I = \int_0^L \left[\bar{F}_\lambda (\delta \dot{\bar{U}}_\lambda - \bar{\kappa}\, \delta\bar{U}_\mu) \right.$$
$$+ \bar{F}_\mu (\delta \dot{\bar{U}}_\mu + \bar{\kappa}\, \delta\bar{U}_\lambda - \bar{\tau}\, \delta\bar{U}_\nu - \delta\bar{\phi}_\nu)$$
$$+ \bar{F}_\nu (\delta\dot{\bar{U}}_\nu + \bar{\tau}\, \delta\bar{U}_\mu + \delta\bar{\phi}_\mu)$$
$$+ \bar{M}_\lambda (\delta\dot{\bar{\phi}}_\lambda - \bar{\kappa}\, \delta\bar{\phi}_\mu)$$
$$+ \bar{M}_\mu (\delta\dot{\bar{\phi}}_\mu - \bar{\tau}\, \delta\bar{\phi}_\nu + \bar{\kappa}\, \delta\bar{\phi}_\lambda)$$
$$\left. + \bar{M}_\nu (\delta\dot{\bar{\phi}}_\nu + \bar{\tau}\, \delta\bar{\phi}_\mu) \right] d\bar{S} \tag{16}$$

I_E is the work done by distributed external loading, E is the work done by end loads whereas I_I is the work done by internal stresses and moments when virtual displacements and rotations are applied to the system. Since the distributed external loads on the cords in a fabric result from the interaction with other cords in contact we need only evaluate the integral I_E over the contact regions. Therefore, the integral I_E is nothing but the work done by the interaction forces and moments in the contact regions.

The second term E is considerably simplified because of the symmetry of the fabric cell about the middle plane. The middle plane symmetry implies

$$\delta\bar{\phi}_\lambda = \delta\bar{\phi}_\mu = \delta\bar{\phi}_\nu \equiv 0 \tag{17}$$

at the cord ends $\bar{S} = 0, L$.

Moreover, symmetry requires that the permissible virtual displacements must be prescribed in a such way that the cord ends move in the $X_1 X_2$ plane only.

The internal energy integral I_I is simplified in our application because of the considerably large axial stiffness of cords compared to the stiffness in other directions and symmetry of the cell geometry. Recall that we have not yet stipulated any particular set of boundary conditions on the fabric cell. No doubt each class of boundary conditions will further simplify equations (13) through (16). For example, absence of shear in the fabric plane will make displacement is X_2 direction and rotations about normal axis zero. As a result, each cord will deform only in the $X_1 - X_3$ plane.

It can then be shown that for each cord we obtain simplified versions of equations (14) through (16) as

$$I_E = \sum_{\text{no. of contacts}} \oint_{\text{contact area}} [\overline{P}_\lambda \delta \overline{U}_\lambda + \overline{P}_\mu \delta \overline{U}_\mu + \overline{P}_\nu \delta \overline{U}_\nu + \overline{m}_\mu \delta \overline{\phi}_\mu + \overline{m}_\nu \delta \overline{\phi}_\nu] \, d\overline{s} \quad (18)$$

$$E = [\overline{F}_\lambda \delta \overline{U}_\lambda + \overline{F}_\mu \delta \overline{U}_\mu + \overline{F}_\nu \delta \overline{U}_\nu] \Big|_{\overline{S}=0}^{\overline{S}=L} \quad (19)$$

and

$$I_I = \int_0^L \Big[\overline{F}_\lambda (\delta \dot{\overline{U}}_\lambda - \overline{\kappa} \, \delta \overline{U}_\mu) + \overline{M}_\mu (\delta \dot{\overline{U}}_\mu - \overline{\tau} \, \delta \overline{\phi}_\nu + \overline{\kappa} \, \delta \overline{\phi}_\lambda) + \overline{M}_\nu (\delta \dot{\overline{\phi}}_\nu + \overline{\tau} \, \delta \overline{\phi}_\mu) \Big] d\overline{s} \quad (20)$$

The principle of virtual work for each cord on the fabric cell then is described by equation (13) with these new definitions for I_E, E and I_I. We can write similar equations for each cord in the fabric cell. Knowing the cell geometry, shape and size, we must solve the resulting set of equations for each fabric cell within the restrictions of the compatibility which are as follows:

1. The cords, which are parallel to each other, must undergo the same end displacements.

2. The stresses and displacements for either cords in contact regions must be continuous so that contact is always maintained.

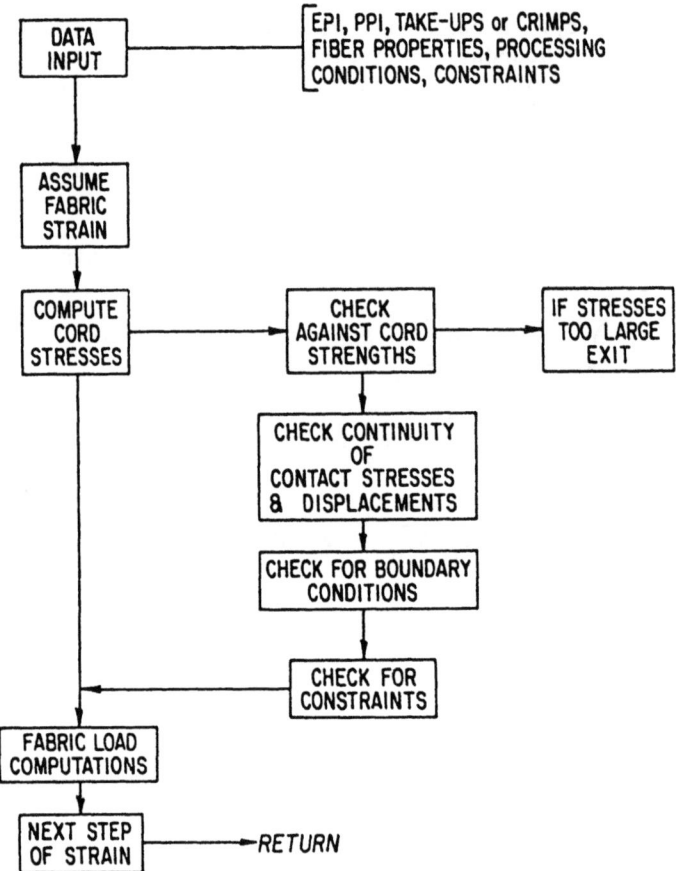

SCHEMATIC COMPUTER LOGIC

FIGURE 7

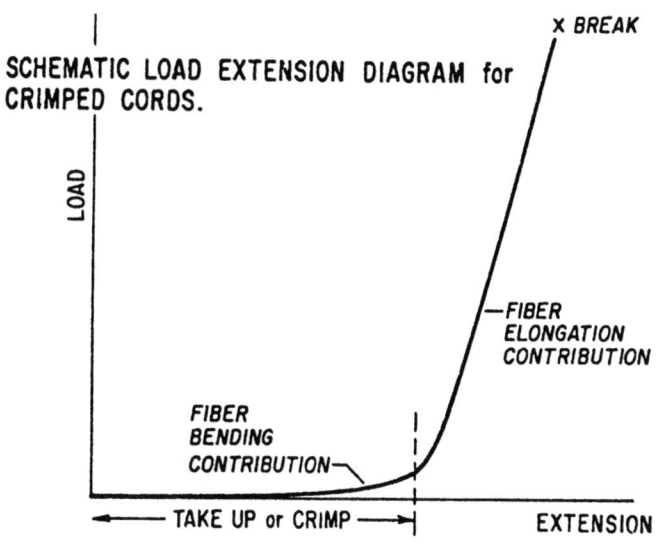

FIGURE 8.

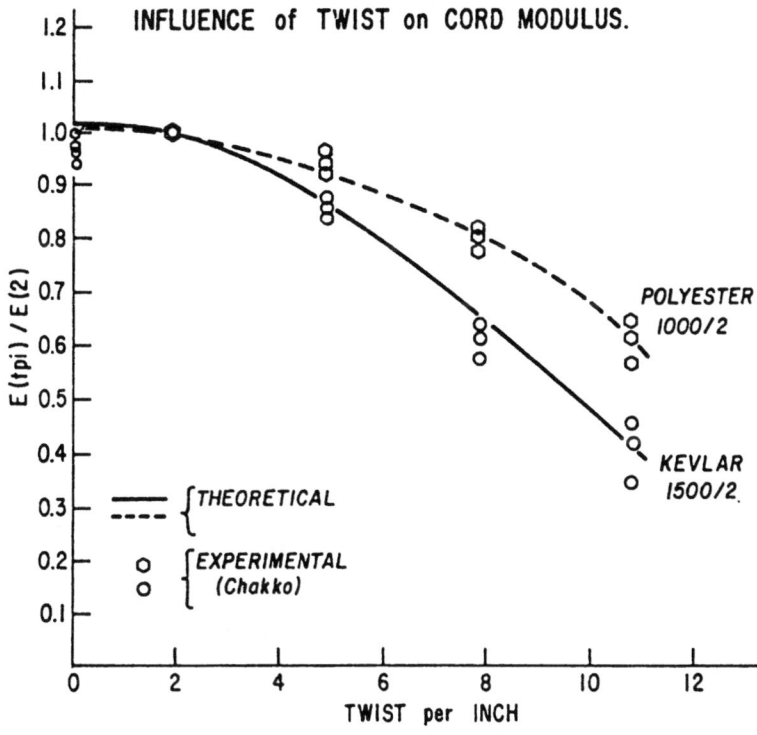

FIGURE 9.

3. The prescribed boundary conditions for the cell must be satisfied.

I have developed a method and computer program to solve these equations successfully. The computer program allows us to predict complete stress-strain behaviors of many types of woven fabrics. I will show later an excellent agreement we have seen between this analysis and actual experimental data on many woven fabrics. It is needless to emphasize that the complexity of the analysis and computer calculations depends upon the fabric weave and external boundary conditions.

COMPUTER LOGIC AND DATA INPUT. The computer logic is quite simple and straightforward. The program solves a set of nonlinear equations and computes loads for prescribed fabric displacements. The solutions are sought within the constraints on the cord loads not to exceed breaking stresses and satisfy boundary conditions and compatibility at contacts. The inputs to the program are the geometrical fabric weave parameters, cord properties, constraints such as jamming conditions. The cord properties were measured on crimped and uncrimped fiber cords. The elongation and bending behavior of cords is calculated from these data (Figure 8). The computer program also accepts basic fiber properties and cord/yarn construction parameters if the experimental cord properties are not available. In this case cord/yarn behavior is computed separately. A typical result is shown in Figure 9[6]. The program also has capabilities of predicting the behavior of processed fabrics if processing conditions are specified as input data.

NUMERICAL RESULTS AND DISCUSSIONS. Since the reinforced composite products are used in many applications, the fabrics in them are made from different fibers and weave patterns and different geometric parameters. To establish the validity we computed the results of the computer analysis to experimental data on many fabrics. Following few figures show the versatility of the computer program.

Figure 10 shows typical data for a plain fabric with both warps and fills made from twisted cotton yarns. The yarn properties were measured on crimped yarns taken out from the fabric. Excellent agreement can be seen from this comparison.

Figure 11 shows similar comparison for both greige and processed plain woven fabrics. The computer predictions for processed fabric were made first on the basis of yarn stress-strain measured on processed yarn and then on the basis of known fabric processing conditions. The computer results in both cases were close to each other and compared very well with the experimental results. The

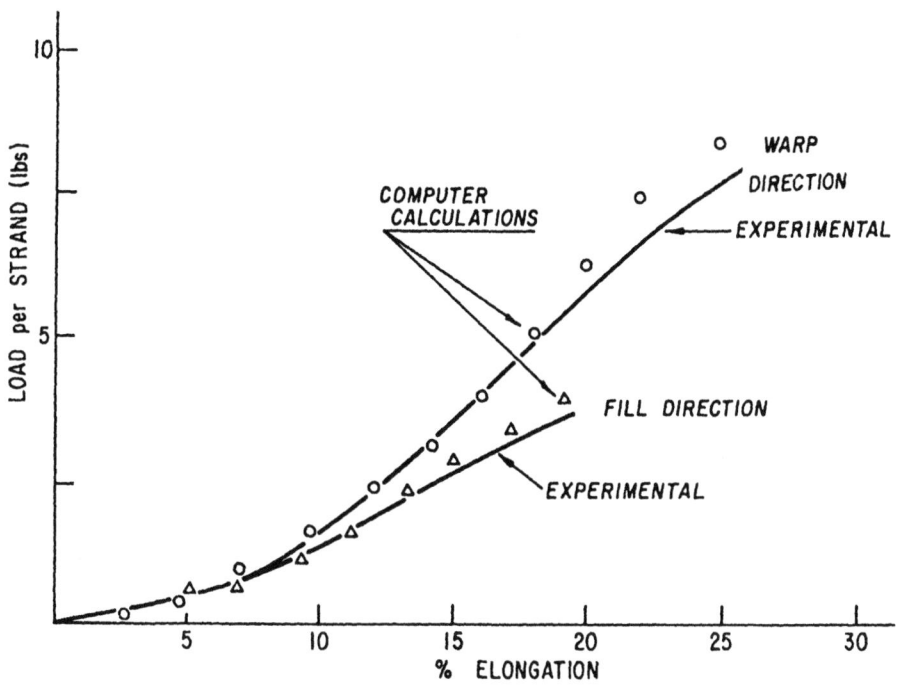

FIGURE 10.

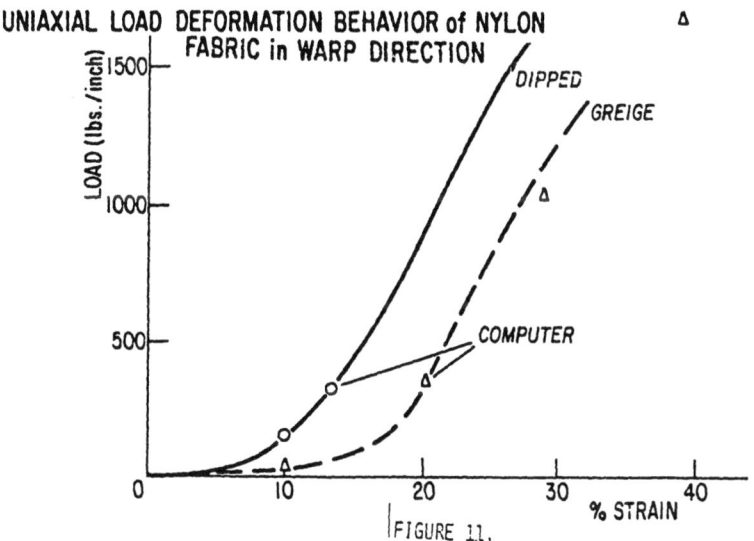

FIGURE 11.

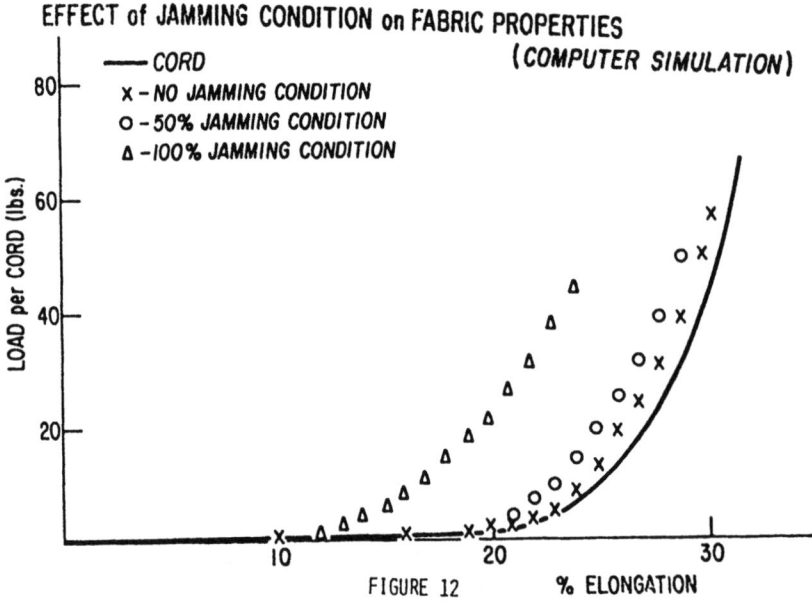

FIGURE 12

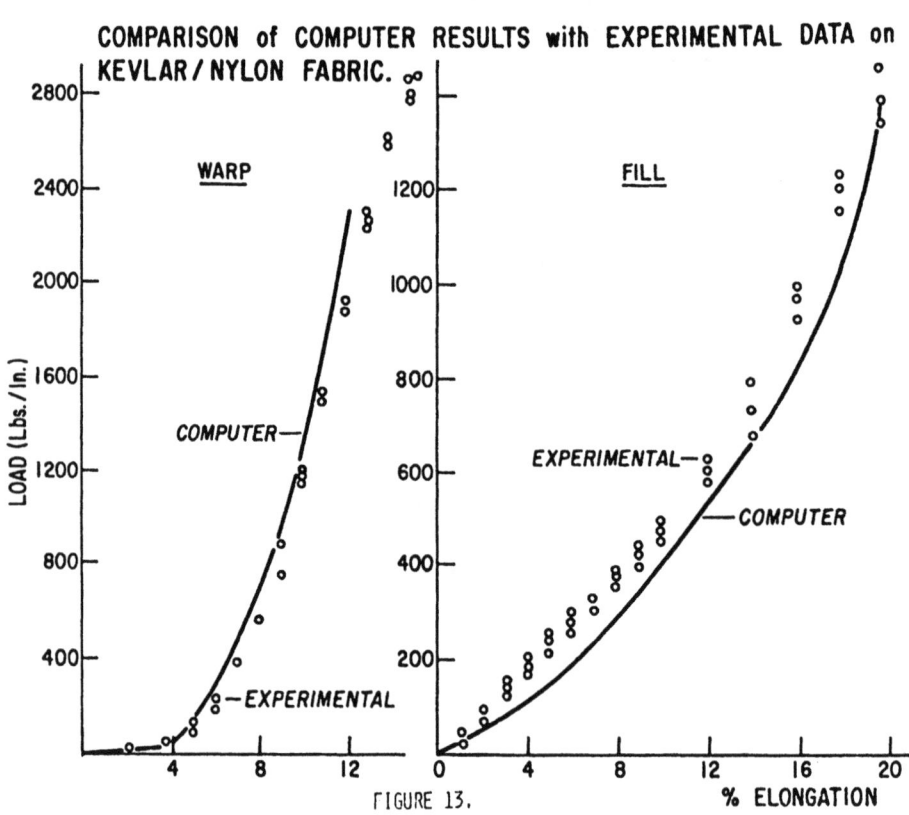

FIGURE 13.

fabric processing stiffens the fabric in the warp direction as can be seen in the figure.

The computer program was used to determine an influence of cord size on fabric behavior. Small cord diameters allow better crimp exchange whereas larger cord sizes tend to jam the fabric and prevent mobility. This constraint on mobility results in higher stiffness, lower strength and ultimate elongations. This trend is very clearly indicated by the results of a computer simulation on a plain fabric in Figure 12. I might mention that the computer program can be useful in evaluating the influence of various parameters economically compared with the routine experimental evaluation.

The good agreement between analysis and the experimental data is demonstrated in Figure 13 for a typical complex fabric made from Kevlar and Nylon. In this construction the warps were of two different Kevlar cord constructions and the fills were made from Nylon as shown in Figure 3.

I have shown, with the help of these few illustrations, the versatility of the computer analysis in predicting load deformation of complex woven fabrics. There are many possible applications, yet to be explored, of this computer analysis.

Finally, I would like to emphasize that the computer predictions can be valuable if realistic modelling and analysis are based on true fiber behavior and fabric parameters.

REFERENCES

1. Peirce, F.T., J. Textile Institure, 28, T45 (1937).
2. J.W.S. Hearle and W. Shanahan, J. Textile Institute (1978).
3. J.W.S. Hearle et al, "Structural Mechanics of Fibers, Yarns and Fabrics" Vol. 1 John Wiley.
4. S. Kawabata et al., J. Textile Institute (1973).
5. Huang, N.C., J. Applied Math Phys. (ZAMP), Vol. 25, p. 1, (1973).
6. S.R. Moghe, "Textiles and Textile Mechanics" presented at Engineering Mechanics Seminar, University of Michigan, Fabruary 26, 1979.

SOME ASPECTS OF THE MECHANICS OF A MODEL OF PLAIN WEFT-KNITTING

R.B. Hepworth

Department of Textile Industries, University of
Leeds, England.

ABSTRACT. The dimensions and mechanical properties of a knitted
fabric are determined by the interactions between adjacent loops
in the fabric and by the physical properties of the yarn. An
analysis of the interactions leads to six simultaneous
differential equations for the shape of the yarn constituting a
loop. The numerical solution of these equations determines the
loop shape and the fabric dimensions. Since the known boundary
values of the equations are shared between end points of the loop
their solution involves the use of an optimization procedure.
Further information to emerge from the solution includes the
magnitude and direction of inter-yarn forces, the load-extension
properties and bending behaviour of the fabric.

1. INTRODUCTION

The structure of plain weft-knitting is illustrated
topographically in Fig. 1. A continuous length of yarn forms
the loops in each course, and loops in adjacent courses interlace
with each other in such a way that, in general, the yarn is bent
into a three dimensional shape. The length of yarn in each loop,
typified by $A_0B_0A_1B_1A_2$, contributes to the length, width and
thickness of the fabric. Its contribution to the length and
width is represented by a unit cell of fabric defined by joining
together corresponding points of loops in adjacent wales and
courses. Thus wale-spacing is represented here by A_0A_2, which can
be seen to depend entirely on loop shape, while course-spacing

Fig. 1. The basic elements of plain weft-knitting

represented by $A_2A_2^1$ depends also on the relative position of interlacing courses. However, if the way in which interlacing loops interact is defined, their relative position is related to their shape. Thus the study of fabric dimensions can be reduced to a study of loop shape. But the loop, initially a straight piece of yarn, is held in shape by inter-yarn forces acting in the interlacing region, and possibly also at other points if loops in adjacent wales touch as at J_w (Fig. 2) or if loops in alternate courses touch as at J_c, a condition known as jamming.

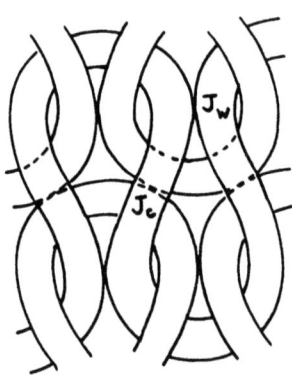

Fig. 2. Jamming between loops

There are thus two particular areas of interest. One is a
study of the system of forces acting on each loop and the way in
which they determine its shape; the other is a study of the
geometrical constraints imposed by inter-loop contact. Both
require a knowledge of the physical properties of the yarn. For
the first, relationships must be established linking the applied
forces and couples with the resulting shape of the yarn. For
the second, the behaviour of the yarn in the interlacing region
must be clearly defined.

Because the physical properties of yarn are in general complex,
simplifying assumptions must be made about its behaviour. Other
assumptions must also be made about the fabric as a whole and
about the loops within the fabric. In other words a model of the
structure must be defined. This will retain those characteristics
of the actual fabric that are of particular interest while
rejecting others, either because they are considered relatively
unimportant or because their inclusion would make the solution,
or even the definition of the problem, difficult or impossible.
By setting up a clearly defined model of the structure an exact
solution can be obtained to a precisely stated problem which
throws light on at least some of the characteristics of the
actual fabric.

2. THE IDEALIZED STRUCTURE

The method of solution to be described was developed for,
and will be applied here to, a structure lying in a plane, but
it will be shown later that it can be modified to apply to
certain non-planar configurations.

The idealized yarn is naturally straight, inextensible,
incompressible and frictionless. It behaves as a long thin
cylinder, having circular cross-sections that remain circular
and perpendicular to the yarn central axis after bending. It
follows from this that, at a point where two yarns touch (Fig. 3)

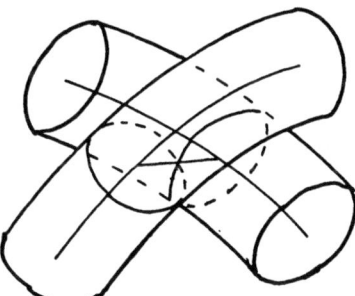

Fig. 3. Contacting yarns

their central axes are separated by a distance equal to the yarn diameter, and the common normal to their surfaces at this point (which is the line of action of the reaction force) passes through the central axis of each yarn. Thus, in the absence of friction, no twisting couples about the yarn axes are introduced at contact points. The yarn behaves in bending and twisting according to the "ordinary approximate theory" described, for example, by Love (1).

The fabric is held in equilibrium by uniform tensions of T_c per course parallel to the courses and T_w per wale parallel to the wales.

It is assumed that the fabric is loaded in such a way that all the loops remain equal and symmetrical, implying that the wales are perpendicular to the courses. It follows that it is only necessary to consider the shape of, and the forces acting upon, a quarter-loop A_1B_1 which interlaces with A_2B_2 (Fig. 4) and may or may not jam with other loops in neighbouring wales and courses. Its coordinates will be referred to rectangular axes, with OX parallel to the wales, OZ parallel to the courses, and OY perpendicular to the plane of the fabric. Fabric dimensions

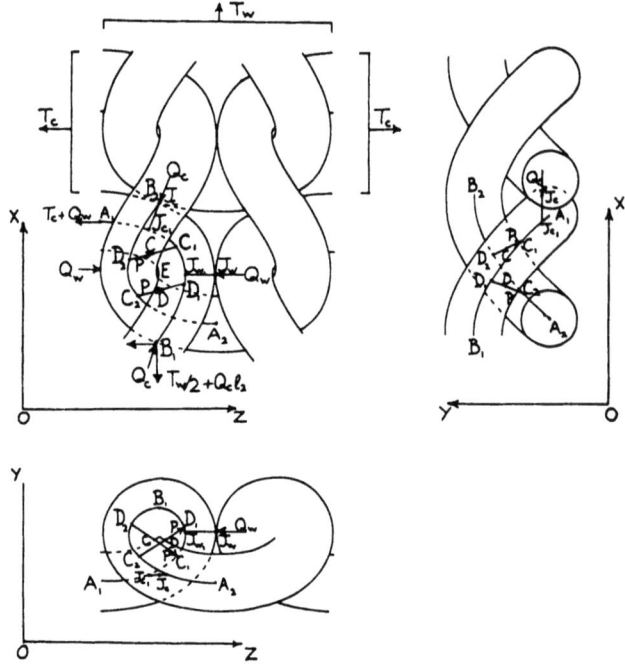

Fig. 4. Forces on the fabric and on a quarter-loop

can then be derived from

$$\text{wale-spacing} = 4(Z_{B1}-Z_{A1}) ,$$

$$\text{course-spacing} = X_{B2}-X_{B1} .$$

3. THE SHAPE OF A QUARTER-LOOP

The object now is to derive as much information as possible about the forces that give the quarter-loop A_1B_1 its shape, and also about the geometrical constraints imposed on it by its contact with neighbouring loops. This information will be drawn from a study of the implications of two assumptions embodied in the model, namely the assumption that all loops are equal and symmetrical and the assumption that the fabric rests in equilibrium under the applied tensions.

3.1 From the symmetry of the loops

The fact that the loops are symmetrical leads to the following conclusions. At A_1 there is no twisting couple about the yarn axis (which is here parallel to OZ) since, if there were a positive couple at this point acting on A_1B_1, then this would be balanced by a negative couple on the adjoining quarter-loop, which would be contrary to the assumption of symmetry. Similarly there is no shear stress at this point. The only possible forces at A_1 are a tension parallel to OZ and a bending moment about an axis perpendicular to OZ.

By similar reasoning, at B_1, where the tangent to the central axis is parallel to the ZOX plane, there is no bending moment about an axis parallel to OY and no shear stress parallel to OY.

In the interlacing region, because the yarn is incompressible, contact occurs at two points, C and D, at which common normals to the yarn surfaces cut the central axes of the quarter loops at points C_1, D_1, C_2 and D_2 as in Fig. 4. In the absence of friction, these normals are the lines of action of the inter-yarn forces at these points which, by symmetry, have equal magnitude P, say and if C_2D_1 has direction cosines (ℓ_1, m_1, n_1) then D_2C_1 has direction cosines $(\ell_1, -m_1, n_1)$.

The fact that the yarn central axes are at C and D separated by a distance equal to the yarn diameter, d, leads to two geometrical relationships which will later be used in calculating loop shape. First, from symmetry, the points D_1 and D_2 have the same Y coordinate i.e. $Y_{D1} = Y_{D2}$.

But,
$$Y_{D2} - Y_{C1} = m_1 d ,$$

hence
$$Y_{D1} - Y_{C1} - m_1 d = 0 . \qquad (3.1)$$

Now if E is the mid-point of the line joining C and D, then by symmetry
$$Z_E = Z_{B1} = Z_{B2} .$$

But
$$Z_E = (Z_C + Z_D)/2$$
$$= (Z_{C1} - n_1 d/2 + Z_{D1} - n_1 d/2)/2$$

giving
$$2Z_{B1} - Z_{C1} - Z_{D1} + n_1 d = 0 \qquad (3.2)$$

When the courses jam $A_1 B_1$ touches a loop in an alternate course J_C which, by symmetry has the same X and Z coordinates at B_2,

i.e. $X_{B2} = X_{Jc}$ and $Z_{B2} = Z_{Jc}$.

The line of action of the inter-yarn force Q_c is, by symmetry parallel to the ZOX plane and meets the central axis of $A_1 B_1$ at J_{c1}. If this has direction cosines $(-\ell_2, 0, -n_2)$ then the fact that the length of $J_c J_{c1}$ is d/2 leads to the following relationships between the coordinate of J_{c1} and other points on the yarn central axis:

$$Z_{Jc1} = Z_{Jc} - n_2 d/2 ,$$

i.e. $Z_{Jc1} - Z_{B1} + n_2 d/2 = 0,$ \qquad (3.3)

and since
$$X_{B2} = 2(X_E - X_{B1})$$
$$= 2 (X_C + X_D)/2 - X_{B1}$$

then

$$X_{Jc1} - X_{c1} - X_{D1} + X_{B1} + \ell_1 d + \ell_2 d/2 = 0 \qquad (3.4)$$

These two relationships hold only when the courses jam.

When the wales jam A_1B_1 touches a loop in an adjacent wale at J_w, the widest point of the loop. The line of action of the inter-yarn force Q_w is, by symmetry parallel to OZ and meets the yarn central axis at J_{w1}. Since the yarn is incompressible

$$Z_{Jw1} = Z_{J2} - d/2 .$$

But by symmetry

$$Z_{Jw} - Z_{A1} = 2(Z_{B1} - Z_{A1})$$

Thus

$$Z_{Jw1} - 2Z_{B1} + Z_{A1} + d/2 = 0 \qquad (3.5)$$

This relationship holds only when the wales jam.

3.2 From the state of equilibrium of the fabric

Further information about the forces on A_1B_1 can be deduced from the fact that the fabric lies in a plane acted on by uniform tension T_c per course parallel to the courses and T_w per wale parallel to the wales.

As has already been shown, the only force at A_1 is a tension parallel to OZ. The magnitude of the tension can be deduced in the following way. Consider the fabric to be cut parallel to the XOY plane through A_1 and held in its equilibrium position by forces at points such as A_1 equal to the yarn tension there, and by forces equal to Q_w at jamming points.

These forces must balance the external load of T_c per course so that, by resolving parallel to OZ, the tension at A_1 is equal to $(Q_w + T_c)$. Similarly, considering the fabric to be cut parallel to the courses through points such as B_1 shows that the resultant parallel to OX of the tension and shear at B_1 is $(T_w/2 + Q_c \ell_2)$. The OZ resultant of tension and shear at B_1 can be obtained by resolving all the forces on the quarter loop in this direction. In order to maintain equilibrium this must be $(T_c + 2Q_w + Q_c n_2 - 2Pn_1)$. A relationship between inter-yarn forces can also be obtained in the following way. At any point of the quarter loop between C_1

and D_1 the OX resultant of tension and shear is $T_W/4$, and thus, resolving the forces acting on the part of the loop between A_1 and C_1 parallel to OX gives

$$P_1 - Q_c \ell_2 = T_W/4 .$$

4. THE DIFFERENTIAL EQUATIONS OF THE YARN

After obtaining all the available information about the forces acting on the quarter-loop the next step is to set up equations relating these forces to the resulting shape of the yarn. The yarn in the loop can be treated as separate sections, $A_1 J_{c1}$, $J_{c1} C_1$, etc., each assumed to behave as an elastic rod acted on by forces and couples at its ends, care being taken to maintain geometrical and physical continuity between sections. The theory of the behaviour of such rods, of which a complete account is given in Love (1), will be briefly summarised here.

Consider a rod, initially straight and parallel to the Z axis, now bent by a force R in direction (ℓ,m,n) and by a couple with components M_X, M_Y, M_Z, all acting at the end of the rod, E (Fig. 5). At a section through any point P on the central axis of the rod, moving axes Px,Py,Pz are defined such that Pz is along the tangent to the central axis and Px and Py are along elements of the rod initially parallel to the fixed OX,OY axes. If the point P were considered to move with uniform velocity along the strained

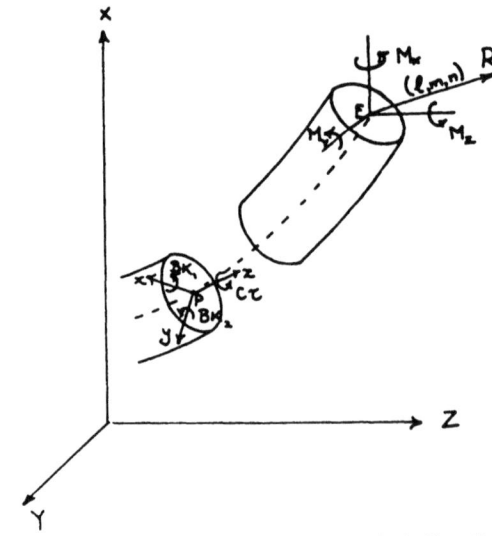

Fig. 5. Stress couples at a point P of a rod bent by forces and couples at the end

central axis of the rod, the angular velocity with which the moving axes rotated about their instantaneous positions would have components K_1, K_2 and τ, where K_1 and K_2 are components of curvature of the central axis and τ is the twist of the rod about its central axis.

The "ordinary approximate theory of bending" used by Love and assumed to apply for this model states that, if the stress couples at P have components (G_1, G_2, G_3) about the moving axes then

$$G_1 = BK_1, \quad G_2 = BK_2, \quad G_3 = C\tau, \quad (4.1)$$

where B and C are the flexural and torsional rigidities of the rod.

Basically, these are the equations that link the external forces acting on the rod with its resulting shape, since the couples on one side of the equations can be expressed in terms of the forces and couples at the end of the rod, while the curvatures and twist on the other side of the equations can be expressed in terms of coordinates of the central axis. The coordinates used in this study were rectangular coordinates (X,Y,Z) which define the position of point P, together with Euler angles (θ, ϕ, ψ) which define the inclination of axes Px, Py, Pz relative to the fixed axes as shown in Fig. 6.

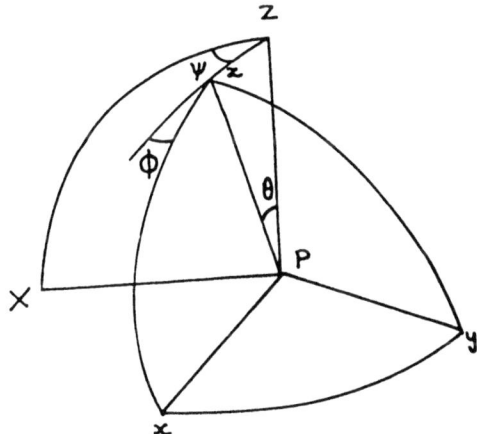

Fig. 6. Euler angles

The equations relating curvature and twist to the Euler angles are

$$K_1 = \frac{d\theta}{ds} \sin\phi - \frac{d\psi}{ds} \sin\theta\cos\phi$$

$$K_2 = \frac{d\theta}{ds} \cos\phi + \frac{d\psi}{ds} \sin\theta\sin\phi \qquad (4.2)$$

$$\tau = \frac{d\phi}{ds} + \frac{d\psi}{ds} \cos\theta$$

where s is the length measured along the central axis of the rod.

By considering the equilibrium of a small element ds of the rod, Love shows that, when the cross-section is circular

$$\frac{Cd\tau}{ds} = 0$$

i.e. the twist τ is constant throughout its length. It follows that, for the quarter loop, since all the inter-yarn forces act through the yarn central axis and no twisting couples are introduced, τ is constant between the two end points A_1 and B_1, But it has been shown that, by symmetry, there is no twisting couple at A. Thus, at all points of the loop there is no twisting couple and $\tau = 0$. The third of equations (4.2) thus becomes

$$\frac{d\phi}{ds} + \frac{d\psi}{ds} \cos\theta = 0 \qquad (4.3)$$

Substituting the expressions for K_1, K_2 and τ in equations (4.1) and expressing the stress couples (G_1, G_2, G_3) in terms of the external force and couple and the coordinates of P and E leads to the following equations

$$B\frac{d\theta}{ds} = M_y \cos\psi - M_x \sin\psi + R\cos\psi \left\{(Z_E - Z) - n(X_E - X)\right\}$$

$$-R\sin\psi \left\{n(Y_E - Y) - m(Z_E - Z)\right\} \qquad (4.4)$$

$$B\sin^2\psi \frac{d\psi}{ds} = M_z + mR(X_E - X) - \ell R(Y_E - Y) \qquad (4.5)$$

which together with equation (4.3) and the relations

$$\frac{dX}{ds} = \sin\theta\cos\psi$$

$$\frac{dY}{ds} = \sin\theta\sin\psi \qquad (4.6)$$

$$\frac{dZ}{ds} = \cos\theta$$

provide six simultaneous differential equations which can be integrated numerically to give the coordinates and the Euler angles at any point of the central axis.

5. THE METHOD OF SOLUTION

Returning to the quarter loop A_1B_1, a set of six simultaneous differential equations can now be obtained for each of its sections. These can be integrated numerically and the method chosen was a fourth-order Runge-Kutta method due to Merson (2). Before the equations can be integrated, however, initial values must be given to all the variables and all the coefficients of the equations must be known; the latter require the magnitudes and directions of the forces and couples at the ends of each section of loop. From the information set down in section 3 it is apparent that there is no point of the loop at which all the boundary values are known. Furthermore, such information as is available about the forces and couples relates to their direction only and at no point is their magnitude known, so that the coefficients of equations (4.4) and (4.5) are unspecified. However, some boundary values are known at A_1 and some at B_1 and equations (3.1) to (3.5) link coordinates of other points on the yarn axis. This situation is typical of a boundary value problem and the solution described here follows one of the standard methods available.

One end of the loop is chosen as the starting point for integration; in this case it was A_1. This point can be taken as origin of coordinates so that X=Y=Z=0 and it is also possible to set $\emptyset=0$ at A_1. It is known that $\theta_A = \overline{\pi}$ (the angle between the tangent to the yarn axis and the Z axis). The force and couple at A_1 are a tension $-(T_c+Q_w)$ parallel to OZ and a bending moment $B(K_1)_A$ about an axis perpendicular to OZ, inclined at $-\psi_A$ to OY. Since the values of $Q_w, (K_1)_A$ and ψ_A are not known they are estimated in order that numerical integration can begin at A_1 and proceed as far as J_{c1}. Since there is no information about the position of J_{c1} this must be defined by means of an estimated parameter. The boundary values emerging from the integration at J_{c1} can then be used as starting values for the next section $J_{c1}C_1$ but before integration can proceed they must be supplemented by one further estimated value since the magnitude of the force Q_c at J_{c1} is not known. (Its direction is determined by the requirement that its line of action is perpendicular to the tangent at J_{c1}.) Integration proceeds in this way from one section to another of the loop, estimated values being supplied wherever necessary to specify the positions of inter-yarn forces or their magnitudes or directions. For the case where both courses and wales jam seven quantities in all must be estimated.

Having integrated as far as B_1 using these estimated values the loop shape that has been calculated is only approximate and the known boundary conditions at B_1

$$\left(\frac{d\theta}{ds}\right)_B = 0 \quad \text{and} \quad \psi_B = 0$$

are not satisfied, say instead that the values obtained are

$$\left(\frac{d\theta}{ds}\right)_B = R_1 \quad \text{and} \quad \psi_B = R_2$$

where R_1 and R_2 are known as "residues". Similarly the boundary conditions of equations (3.1) to (3.5) are not satisfied but have residues R_3 to R_7 associated with them. (Note that there are seven boundary conditions just as there are seven estimated values.)

A function F is now defined such that

$$F = \sum_{i=1}^{7} R_i^2$$

and the estimated values are adjusted systematically until F=0 (or what is judged to be a sufficiently small quantity for the accuracy required) in which case it is known that all the boundary conditions are satisfied, an exact solution has been obtained, and the estimated values must now be the true ones. Thus, not only has the shape of the loop been calculated, but the magnitudes, directions and points of application of all the forces and couples have also been determined.

The method by which the estimated values are adjusted is known either as a "minimization procedure" since a minimum value is sought for the function F (in this case known to be zero) or otherwise as an "optimization procedure". The procedure used in this study was one due to Powell (3) which sets up conjugate directions of search for a minimum in a space defined by the variables, which in this case are the estimated values.

Although minimization procedures are usually applied to functions of several variables their method of working is more easily visualised when applied to a function of two variables $F(x_1,x_2)$, say. It is usually found in physical problems that, when sufficiently close to a minimum, the function behaves approximately like a quadratic function. These procedures are therefore designed to work well with quadratic functions. Near the minimum of a quadratic function equal values of F are represented by a set of elliptical contours in the x_1x_2 plane (Fig. 7) with the minimum point at the centre of the system of ellipses. Starting from any point A a possible method would be to search for a minimum of F first along a line parallel to the x_1 axis, then parallel to the x_2 axis (Fig. 7a), the minimum along a line being the point where that line touches an ellipse of the system. This method can take a large number of iterations unless the major axes of the ellipses happen to be parallel to a coordinate axis. A search along conjugate directions however (Fig. 7b), would locate the minimum at the first iteration, following from the property that any diameter of an ellipse is conjugate to the tangent at its ends. A similar property applies to a quadratic function of n variable and it is this property that Powell's method seeks to exploit in setting up conjugate directions of search.

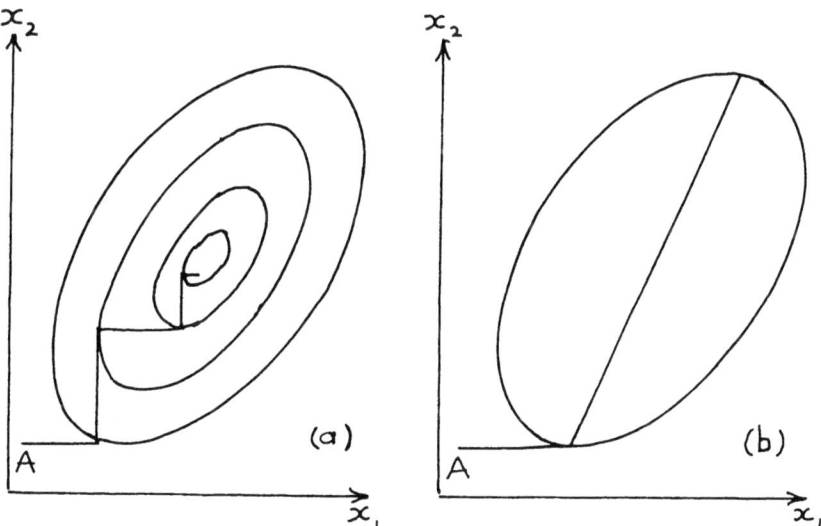

Fig. 7. A search for the minimum of a function

6. RESULTS OF CALCULATIONS

Before proceeding with the solution all equations were written in dimensionless form, lengths being expressed relative to loop length (e.g. $X/\ell, Y/\ell, Z/\ell$) and forces being represented by dimensionless groups such as $(P\ell^2/B)$. Equations (3.1) to (3.5), which express the fact that at contact points the yarn axes are separated by a distance equal to yarn diameter, now contain the ratio (d/ℓ). Since these equations are used as boundary conditions in the solution it follows that, for any specified system of fabric loading, a different loop shape (and hence fabric dimensions) would be obtained for each value given to d/ℓ. This fact was of particular interest in studying the special case of relaxed fabrics, i.e. $T_c=T_w=0$, and results showing the variation with d/ℓ of loop shape and fabric dimensions for that case are given elsewhere (4).

By varying the external tensions on the fabric, load-extension curves were obtained for a range of values of d/ℓ. The results, which are discussed at greater length elsewhere (5) will be briefly summarised here.

6.1 Fabrics loaded parallel to the wales only (Fig. 8)

Most of the curves show marked discontinuities in slope which in every case coincide with changes in jamming conditions in the fabric. For all values of d/ℓ represented in the graph the courses jam at zero load, which means that although the fabric as a whole is not loaded, there is a pressure Q_c between courses resulting in a tension in the yarn in the region of B_1. As the load on the fabric increases this pressure decreases to zero and the courses separate, but the tighter fabrics in particular require a considerable load before this separation occurs and the extension process really begins. Thus jamming in the unloaded fabrics causes them to behave like pre-stressed structures. After this separation the slacker fabrics ($d/\ell < 0.06$) have a period of easy extension corresponding to a state of no jamming of either kind, but ultimately all fabrics are in a state of single jamming (between the wales) for which the elastic modulus approaches a constant value.

6.2 Fabrics loaded parallel to the courses only (Fig. 9)

For all values of d/ℓ represented here the courses jam in the unloaded state and remain in contact under load. For values of $d/\ell > 0.06$ the wales also jam at zero load and these again behave like pre-stressed structures before the wales separate.

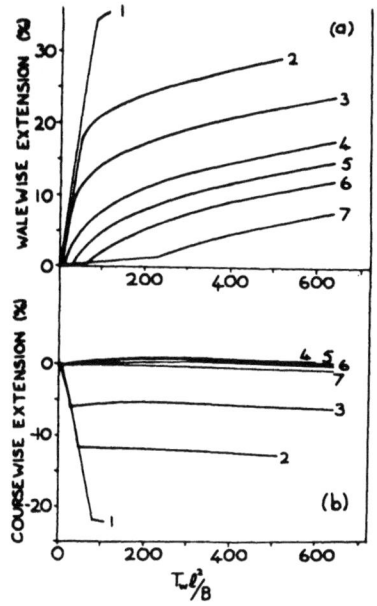

Key for Figs. 8 and 9.

d/ℓ
(1) 0.0500
(2) 0.0550
(3) 0.0575
(4) 0.0600
(5) 0.0625
(6) 0.0650
(7) 0.0700

Fig. 8. Load-extension curves for fabrics loaded parallel to the wales only

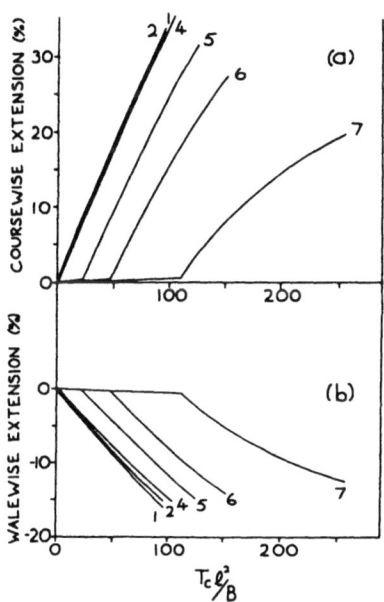

Fig. 9. Load-extension curves for fabrics loaded parallel to the courses only

As in the previous section the major part of the graph illustrates a situation of single jamming, this time between courses, for which the near constant and equal slopes suggest constant elastic moduli for the fabrics which are proportional to B/ℓ^2 for yarns of given diameter.

6.3 Fabrics loaded parallel to both wales and courses (Fig. 10)

For biaxial loading in which the load per course is equal to the load per wale the load-extension curves again illustrate behaviour which is complicated by the onset or removal of jamming. The tighter fabric shown $(d/\ell = 0.065)$ in particular passes through four separate phases. As always the period of easiest extension corresponds with the removal of all jamming. For this roughly equal loading in both directions the coursewise extension is always positive and the walewise extension almost always negative indicating that the structure is more extensible in the coursewise direction.

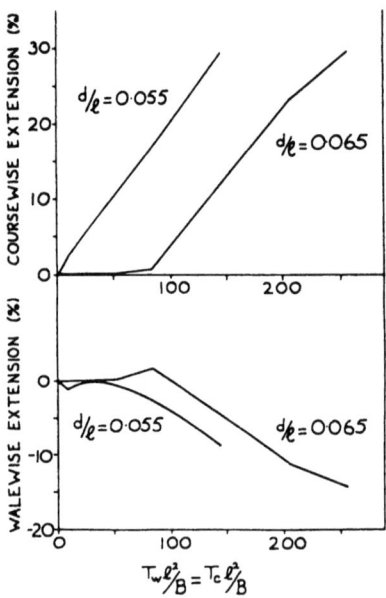

Fig. 10. Load-extension curves for fabrics loaded parallel to both wales and courses

7. BENDING OF THE FABRIC

The solution giving the dimensions of the fabric lying in a plane and subjected to external tensions has been obtained by the study of the mechanics of one repeating cell of the fabric consisting of two interlaced quarter loops (Fig. 11). From that solution there emerges also the values of the forces between these quarter loops, internal to the cell, and the forces between adjacent such elements - tensions, shears and bending moments - which are external to the cell. These are transferred to adjacent courses and wales, and so on, until they appear at the edges of the fabric where the sum of the tensions make up the applied tension, the sum of the shears balance to give zero resultant, but the couples remain. In other words the configuration which has been studied can only exist if external couples are applied along the edges of the fabric; without them the yarns in the end courses or wales tend to "unbend" giving rise to the curling well known in plain-knitted fabrics. In practice this curling often extends only a little way into the fabric because coursewise and walewise curls

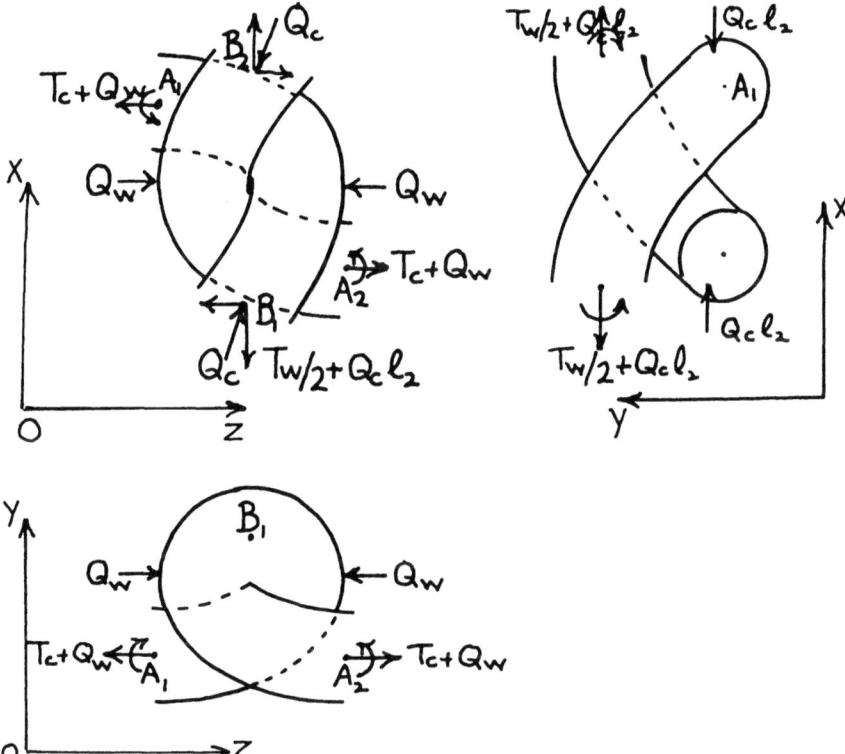

Fig. 11. Forces and couples external to two interlaced quarter-loops

interfere giving rise to a complex configuration in which all the cells of the fabric are not alike, which is those outside the scope of this solution. If this interference did not occur curling would continue throughout the fabric.

Thus the plane configuration of the fabric is one special case, and the couples on the edges necessary to maintain this state can be calculated from bending moments emerging from the solution described in earlier sections, together with a further couple to account for the effect of jamming when it occurs. For the case of a relaxed fabric, expressions for the couples required to hold it in a plane were given in an earlier paper (4), together with graphs showing their variation with d/ℓ.

Two other special cases which can be studied in the same way are those for which the fabric describes the surface of a circular cylinder whose generators are parallel to either the courses or the wales, and by obtaining solutions for different values of curvature, the variation of curvature with applied couples can be calculated. Although it is possible to combine with these couples a uniform tension on the fabric in the direction of the generators the method will be illustrated here by the case where this tension is zero and the generators are parallel to the courses.

It is again assumed that the loops remain symmetrical so that two interlacing loops A_1B_1 and A_2B_2 will be studied. In the interlacing region these touch at two points C and D and fixed rectangular axes are chosen in such a way as to exploit the symmetry of the system with OZ parallel to the courses and OY perpendicular to the line joining C and D.

As in section 3, a great deal of information can be deduced from the symmetry and this will now be summarised. In the projection on the XOY plane shown in Fig. 12 the perpendiculars to the tangent planes at B_1 and B_2 meet at L, a point on the axis of the cylinder on which the fabric lies. The angle between LB_1 and LB_2 is 2γ.

As before, at A_1, where the tangent to the yarn axis is parallel to OZ, the only force is a tension parallel to OZ, and a bending moment about an axis perpendicular to OZ.

At B_1, where the tangent to the yarn axis is perpendicular to LB_1 (and hence $\psi_B = \gamma$) there is no component of bending moment about LB_1 so that $\left(\frac{d\theta}{ds}\right)_B = 0$.

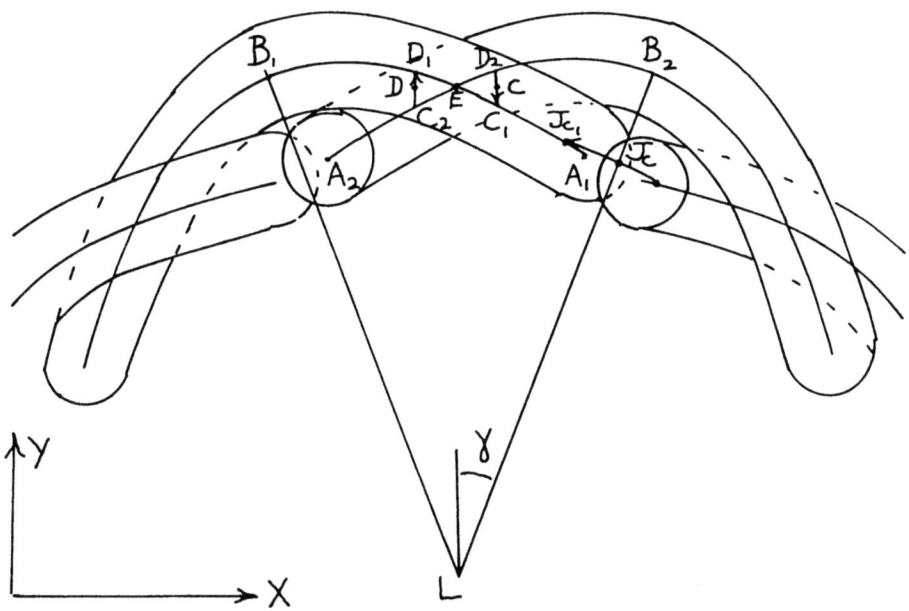

Fig. 12. Forces on a quarter-loop for a bent fabric

At points C and D equal inter-yarn forces P have direction cosines $(\ell_1, -m_1, n_1)$ and (ℓ_1, m_1, n_1) respectively. Because the symmetry of the inter-lacing region relative to the coordinate axes has been preserved, the relationships between coordinates of C_1 and D_1 established in section 3 namely equations (3.1) and (3.2) still hold, i.e.

$$Y_{D1} - Y_{C1} - m_1 d = 0 \qquad (7.1)$$

and

$$2Z_{B1} - Z_{C1} - Z_{D1} + n_1 d = 0 \qquad (7.2)$$

When the courses jam they touch at J_C which lies on LB_2, and the force Q_C, which is perpendicular to LB_2 has direction cosines $(-\ell_2 \cos\gamma, \ell_2 \sin\gamma, -n_2)$. Since the Z coordinate of J_C is the same as that of B_2 equation (3.3) still applies i.e.

$$Z_{Jc1} = Z_{B1} - n_2 d/2 . \qquad (7.3)$$

Now

$$X_{B2}-X_{B1} = 2(X_E-X_{B1})$$

where E is the mid-point of CD.

But

$$X_E = (X_C+X_D)/2 = (X_{C1}+X_{D1}-\ell_1 d)/2$$

and

$$X_{B2} = X_{Jc1}+d\ell_2/(2\cos\gamma)+(Y_{B1}-Y_{JC1})\tan\gamma$$

giving

$$X_{Jc1}+d\ell_2/(2\cos\gamma)+(Y_{B1}-Y_{Jc1})\tan\gamma-(X_{C1}+X_{D1}-\ell_1 d)+X_{B1}= 0 \quad (7.4)$$

When the wales jam at J_W there is a force Q_W parallel to OZ and as in section 3 if the line of action of its force meets the yarn axis at J_{w1}

$$Z_{Jw1}-2Z_{B1}+Z_{A1}+d/2 = 0 \tag{7.5}$$

The relationship between the forces P and Q_C now becomes

$$P\ell_1 = Q_c\ell_2\cos\gamma .$$

The equations (7.1) to (7.5) together with

$$\left(\frac{d\theta}{ds}\right)_B = 0$$

and $\psi_B-\gamma = 0$

provide seven boundary conditions for the solution of the differential equations of the yarn.

7.1 The couples on the fabric

The couples applied to the edge of the fabric parallel to the OZ axis must be equivalent to the internal couple in this direction at a section through the fabric at B_2. Suppose the fabric to be cut at this point by a plane through B_2L perpendicular to the XY plane and held by a couple equal to the bending moment in the yarn at B_2 and forces equal to the tension and shear in the yarn at B_2 and the jamming force at J_c. Let the bending moment have a component M_z about an OZ axis. The jamming force Q_c at J_c has a component $Q_{c\,2}$ perpendicular to the section which, together with the resultant of tension and shear in the yarn at B_1 (which is also equal to $Q_c\ell_2$) is equivalent to a couple of $Q_c\ell_2 x B_2 J_c$.

Thus the couple on the edges of the fabric needed to hold it in this configuration = $M_z + Q_c\ell_2 x B_2 J_c$.

The value of this couple can be calculated for different values of the angle γ and the results for the case where $d/\ell = 0.050$ are shown in Fig. 13.

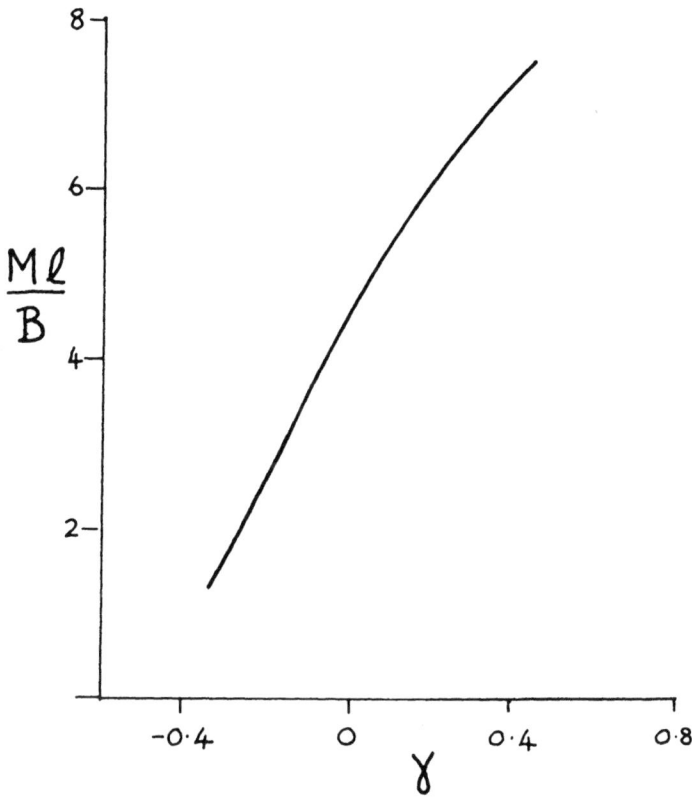

Fig. 13. Variation of couple on edge with angle for a bent fabric

8. CONCLUSIONS

It has been shown that by defining an idealized model the dimensions of a fabric under different systems of loading can be calculated and also the magnitude and directions of internal forces and couples. Obviously care must be taken in interpreting these results in the context of actual fabrics but a particularly interesting result to emerge from the solution is the effect of jamming conditions on the modulus of the fabric.

REFERENCES

1. A.E.H. Love, "A Treatise on the Mathematical Theory of Elasticity", Dover Publications, New York, 1944, p.381.
2. R.H. Merson, "An Operational Method for the Study of Integration Processes" Proceedings of Symposium on Data Processing, Weapons Research Establishment, Salisbury, S.Australia, 1957.
3. M.J.D. Powell, Computer Journal, 1964, 7, 155.
4. R.B. Hepworth and G.A.V. Leaf, J.Text.Inst., 1976, 67, 241.
5. R.B. Hepworth, J.Text.Inst., 1978, 69, 101.

THE BENDING OF YARNS AND PLAIN WOVEN FABRICS

P. Grosberg

Department of Textile Industries, The University of Leeds, England.

ABSTRACT

The bending of yarns is analysed in terms of the independent bending of a set of helices. Reasons for deviations from this model are considered and the effect of friction on the bending behaviour of yarns is considered. It is shown that for the bending of fabrics the effect of steric hindrance to the bending of the yarn must also be considered.

1. INTRODUCTION

The bending of fibre structures differs markedly from the behaviour of solid structures in that by and large the fibres in the structure bend independently of each other so that the resistance to bending of the structure becomes the sum of the bending resistances of the fibres and is not proportional to the moment of inertia of the cross section of the structure. Thus to a first order of accuracy the bending resistance of a yarn is proportional to the number of fibres in its cross section which is proportional to the square of the yarn radius, R^2, while the moment of inertia of the yarn cross section is proportional to R^4. Consequently it is possible to use relatively thick yarns without producing exceedingly rigid yarns. The bending of a yarn or fabric is usually described in terms of its flexural rigidity, B, defined by the product of the bending couple, M, and the resulting radius of curvature ρ. If the structures bending behaviour is linear M is directly proportional to the curvature, $\varkappa$, i.e. $1/\rho$, and B is a constant. The analysis of the bending of yarns has therefore usually proceeded by the determination of the bending resistance of the fibres in the yarn, these fibres being assumed to lie in simple helices.

2. THE BENDING OF A YARN

The simplest analysis of the bending of a fibre helix is that due to Livesey and Owen, 1964. They used what is in essence a Costiliagno treatment in which it is assumed that the deformed geometry is practically identical with the initial geometry.

If therefore a helix of radius, r, and helix angle, θ, is deformed by a couple, M, it is relatively easy to show that if ϕ is the angle made by the helix radius to a point on the helix and the plane of bending, then bending and torsional couples are applied at this point equal to $M(1-\sin^2\phi\sin^2\theta)^{\frac{1}{2}}$ and $M\sin\phi\sin\theta$ respectively, since the geometry is assumed unchanged. The strain energy per unit length produced by these couples is by first principles equal to

$$M^2(1-\sin^2\phi\sin^2\theta)/B_f + (M^2\sin^2\phi\sin^2\theta)/C_f$$

where B_f is the fibre flexural rigidity and C_f its torsional rigidity. Integrating over one turn of complete helix the strain energy becomes

$$(M^2 B_f r/\sin\theta)(1+\tfrac{1}{2}(N-1)\sin^2\theta)$$

where $N = B_f/C_f$.

If this couple results in a radius of curvature, ρ, the work done on unit axial length of helix becomes $M/2\rho$, and hence for one turn of helix, axial length $2\pi r/\tan\theta$ the work done is $\pi r M/\rho\tan\theta$.

By equating the work done to the strain energy $M\rho$, i.e. the bending rigidity of the helix, B_H can be found to be

$$B_H = B_F\cos\theta/(1+\tfrac{1}{2}(N-1)\sin^2\theta)$$

As each helix in a simple model of a yarn consisting of idealized helices has a different value of θ depending on its radius the value of the flexural rigidity of the yarn B_y can only be obtained by integrating B_H times the number of helices passing through a section dr from zero to the outer radius of the yarn R. By this means Livesey and Owen obtain the result

$$(B_y/B_f) = (t_y/t_f)(2/(1+N\tan^2\alpha))\ln(1+\tfrac{1}{2}(1+N)\tan^2\alpha)$$

$$\approx (t_y/t_f)(1-\tfrac{1}{4}(1+N)\tan^2\alpha)$$

where α is the value of θ when r = R, t_y is the linear density of the yarn and t_f the linear density of the fibre.

This analysis clearly only applies for small deflections. An earlier analysis by Platt, Klein and Hamburger, 1959, avoids this difficulty by assuming the geometry of the bent helix. It was assumed, following Backer, 1952, that the helix became a torus in which the angle ∅ as previously defined is directly proportional to the angle made by the radius of curvature of the now bent helix axis to the plane in which ∅ lies.

Platt et al. then proceed to calculate the moment required to hold a fibre in this position. They deal with three cases. (1) Where the fibre is originally straight, (2) where it has the curvature of the fibre in a helix before bending into a torus and (3) where it has the torsion of the fibre in a helix before bending into a torus. These three conditions are difficult to relate to any real yarn bending situation, but as all three provide approximately the same answer they were assumed to provide a measure of the bending of a yarn. The two solutions are very different in form, but at low twists (small values of α), the solution by Platt et al. tends to

$$(B_y/B_f) \quad (t_y/t_f)(1 - \tfrac{1}{2}\alpha^2)$$

The two solutions by Platt et al. and by Livesey and Owen are clearly similar when N=1 and for this value the solutions are also very nearly similar for large values of α.

In a recent publication Leaf, 1979, examined the geometry of a helix after a couple had been applied to it. Using the standard methods for analysing the deformation of a long thin rod originally developed by Euler, and discussed in detail by Konopasek in another section of this volume, he was able to show that the equation developed by Livesey and Owen for the bending of a single helix was accurate to within 2% for even the largest deflections considered. It would appear that in Livesey and Owen's simple analysis we do have an accurate idea of the factors which govern the bending of yarns. Comparison with actual yarns, however, show very marked divergences. A simple example will suffice. The bending rigidity of yarns made with increasingly larger values of α, that is with more and more twist, should result in yarns whose flexural rigidity decreases by some 30% as α varies from 0° to 45°. The variation of real yarns over such a range in twist angle is not uniform but it is quite commonplace for the flexural rigidity of yarns to increase by 200% to 300% over this range in twist angle. Clearly the analysis is not at fault but the model of the yarn structure assumed differs from a real yarn in some vital way.

This difference almost certainly lies in the assumption that the fibres bend independently. There are two fundamental reasons for doubting this assumption as being universally valid. The first arises from Leaf's analysis of the position of the neutral axis in a bent helix. He has shown that the neutral axis no longer lies symmetrically placed inside a curved cylindrical envelope round which the helix is wrapped. The neutral axis moves further and further towards the inner part of the helix facing the centre of curvature as the curvature of the helix is increased. The shift in the neutral axis depends on the radius of the helix. As a result a structure consisting of two helices in which one lies within the other and which are in contact cannot bend independently.

Backer also showed in his initial analysis of the bent yarn geometry, and this has been confirmed by Leaf's analysis, that when a helix is bent the length of the "helix" below the neutral axis plane decreases and that above increases as the curvature increases. The size of these changes becomes larger as the radius of the helix increases. As a result two helices in which one lies within the other will experience relative motions. If there is any pressure between the two helices this relative motion will result in frictional work.

Platt et al. initially considered this possibility by considering the bending of a yarn in which, due to friction, groups of p helices are prevented by friction from moving relatively to each other. As a result they showed that the flexural rigidity of the yarn is increased by a factor approximately equal to p. This analysis is rough and ready but it shows the large effect that can be produced by fibres being prevented from bending independently.

To understand the bending of yarns under conditions where friction prevents or tries to prevent the relative motion of the fibre helices, it is heuristically useful to analyse the bending of a set of plates which are pressed together. In this analysis it will be assumed that:

(i)   along each plate there is a uniform pressure per unit length, v,

(ii) the number of plates, n, is large,

(iii) each plate is identical and can be considered to be long and thin so that shear effects during bending can be neglected.

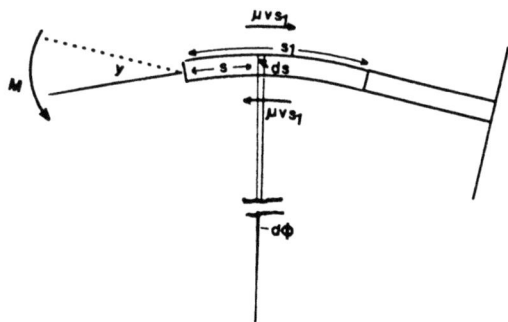

Fig. 1 shows the forces acting on a typical plate when a couple, M, is exerted on all the plates. If up to a point s_1 from the free end friction is limiting so that relative motion occurs at any point a distance s from the free end, the couple available to cause bending is given by $(M/n) - \mu v s \Delta$ where Δ is the thickness of a plate and provided $0 < s < s_1$.

If $d\phi$ is the angular deflection of an element of length ds the curvature is given by $d\phi/ds$ which must equal the couple divided by the flexural rigidity of a plate B. Hence for $0 < s < s_1$,

$$\frac{d\phi}{ds} = \frac{M - \mu v s n \Delta}{nB} = \frac{M - \mu v s d}{nB}$$

where d is the sum of the thickness of all the plates.

For $s > s_1$ friction has not become limiting and no relative motion occurs. This section can therefore be treated as a solid beam and from elementary analysis it can be shown that $(d\phi/ds) = M/Bn^2$. Since n was assumed to be large we can write that for $s > s_1$, $(d\phi/ds) = 0$.

Hence the total angular deflection of the beam is given by

$$\int d\phi = \int_0^{s_1} \frac{(M - \mu v s d) ds}{Bn} = \frac{M s_1 - \frac{1}{2}\mu v s_1^2 d}{Bn}$$

where s_1 equals $(M/\mu v d)$ when friction is only partially overcome and equals s, the total length of the plates when friction is completely overcome.

By definition the average curvature, K, of the set of plates is given by the angular deflection of the beam divided by s, hence

$$nBK = \frac{Ms_1}{S} - \frac{\mu v s_1^2 d}{2S} \quad \text{for} \quad 0 < s_1 < S \tag{1}$$

$$\text{and} \quad nBK = M - \tfrac{1}{2}\mu vSd \quad \text{for} \quad s_1 = S$$

$\tfrac{1}{2}\mu vSd$ is a constant and the second equation shows that for large values of M when $s_1 = S$ there is a linear relationship between the curvature and the applied couple, but the bending moment M has an intercept on the M axis when the curvature is zero given by $\tfrac{1}{2}\mu vSd$. This is a common phenomenon in the bending of fibrous assemblies and this intercept has been called the coercive couple M_o. Denoting μvSd by M_o and noting that $s_1/S = 2M/M_o$ we can rewrite equation (1) in the form

$$nBK = M - M_o \qquad M \geqslant 2M_o$$

$$nBK = M^2/4M_o \qquad M < 2M_o$$

Popper, 1966, analysed in detail the bending of multilayer beams each layer being of finite thickness, and his analysis has shown that for thin beams the analysis given above is reasonably accurate for length to width ratios such as would apply to fibres in a yarn.

There are two points to be noted from this analysis. The slope of the graph of M against K becomes constant for M $\geqslant 2M_o$. If therefore a yarn were to behave in the same way there should be a linear portion to the graph after an initial non-linear region. This is in fact the case. In addition the slope of the graph in this region is equal to nB, that is the slope is equal to the sum of the flexural rigidities of the component plates. Hence once more if the yarns behaved as a series of plates, the slope of the graph should be equal to the solution given by Livesey and Owen. There is little experimental information available on whether this is true for real yarns, but it is usually assumed to be true. If one can therefore determine the value of M_o for a yarn, the basic behaviour of yarns would be understood. From the above argument it is only necessary to determine the work done against the internal friction by calculating the slippage between fibres which would occur between contiguous fibres when no friction is present to enable us to find the value of the coercive couple.

Such an analysis has been carried out by Abbott et al. 1971. The necessary elements in the model used in this analysis are as follows.

(i) To calculate the number of contact points where friction was acting, close hexagonal packing of the fibres was assumed.

(ii) The pressure throughout the cross section was assumed to vary linearly across the cross section, being zero at the upper face of the yarn.

(iii) The relative movement of the fibres was calculated from Backer's toroidal geometry of the bent helices in the yarn. From this geometry it can be shown that no relative fibre movement takes place at the top and bottom of the helix most distant from the neutral axis. At the neutral axis the movement is a maximum and is given approximately by

$$KrR\cos\theta/\tan\alpha = KrR/\tan\alpha \sqrt{1+(r\tan\alpha/R)^2}$$

This clearly varies with r and the relative motion of two fibres at different radii can be calculated. By this means it was found that

$$M_o = (4\sqrt{2}/\pi)\mu v R^2(1-\cos\alpha)/\tan^3\alpha \approx (2\sqrt{2}/\pi)\mu v R^2/\tan\alpha$$

Of the assumptions made above (i) assumes an excessively close packing while the pressure almost certainly decreases as the radius decreases. These assumptions however, become more reasonable when considering the bending of yarns inside a fabric when in addition the value of v can be estimated very accurately.

Dhingra and Postle, 1976, have on the other hand calculated the pressure v by considering that a yarn has a lateral force between fibres due to an axial tension T on the yarn applied during the spinning of the yarn. This was assumed to decay to 40% of its value when the tension was removed leaving a calculable pressure between the fibres. In their analysis v is no longer constant across the cross section. Dhingra and Postle define a value N which is the force pressing two fibres together/unit length of fibre and they find

$$M_o = 0.08 \ \mu d_f n_f NR/\tan\alpha$$

where d_f is the fibre diameter and n_f the number of fibres in the cross section.

A similar analysis was carried out by Popper 1966, who calculated the value of M_o when pressure P is acting on the yarn. He obtains

$$M_o = (2/\pi)\mu v(d_f n_i)R/\tan\alpha$$

where d_f is the fibre diameter and n_i is the number of fibre interfaces in a yarn. For hexagonal packing this agrees with the result given by Abbott et al. 1971. He also calculated M_o when the yarn was not under a pressure v but experiences a tension T and obtained

$$M_o = (2/\pi)\mu T \emptyset R/\tan\alpha$$

where $\emptyset$ is the yarn packing factor.

It can therefore be seen that satisfactory analyses of the bending of yarns are available. These analyses have assumed linear fibre elasticity, modifications to include non-linear relaxation behaviour are discussed later in another section. It is the purpose of the next half of this chapter to consider the bending of plain fabrics, again restricting ourselves to linear elasticity.

3. THE BENDING OF PLAIN FABRICS

Introduction

The simplest possible case of fabric bending is the case where the fabric contains yarns which have been bent into their shape, and there is only point contact between the yarns in their cross over regions. As a result, bending in a plane containing one of the yarn axis consists only of the bending of the yarn. The yarn, however, is not straight but is permanently shaped into its crimped form.

If a length ds of yarn is considered it will have an initial curvature K_o and a final curvature K. The couple applied to the width of a fabric divided by the number of yarns in that width, M say, will therefore be equal to

$$M = B_y(K-K_o) = B_y\left(\frac{d\emptyset}{ds} - \frac{d\emptyset_o}{ds}\right)$$

or $\quad M_s = B_y\left[\int d(\emptyset - \emptyset_o)\right]$

The bending of the fabric, however, will be determined by the fabric curvature $\int d(\phi - \phi_o) / s'$ where s' is the length of the fabric centre link. Hence

$$Ms = B_F \int (\phi - \phi_o)$$ and it is clear that

$B_F/B_y = 1/(1+C)$ where C is the crimp ratio, s/s', as usually defined, and B_F is the fabric rigidity modulus per yarn.

Abbott, Coplan and Platt, 1960, were the first to point out that in the majority of fabrics the yarns are prevented from bending in the cross over regions by being in contact with the cross over yarns for a fraction p of their length. If one were to use the analysis given above it can be seen that when integrating $d\phi - d\phi_o$ this should only be done for the lengths of yarn which can be bent and hence

$$B_F/B_y = 1/(1+C)(1-p)$$

Owen, 1968, showed that there was a correlation between experimental and calculated values of B_F if rather arbitrary values were assigned to p. It is clearly necessary to look more closely at what is meant by the section of yarns "which cannot bend". Abbott, Grosberg and Leaf, 1973, have carried out a more basic analysis of the bending of plain-woven fabrics based on the following model.

The yarns were assumed to be circular and incompressible and at the contact points there are distributed forces between the contact yarns. The resultant vertical component of these forces being V_o in the initial condition. The methods used originally by Peirce, 1937, were used to obtain the energy of the yarn in the fabric in its configuration and the vertical pressure V_o at the contact point before bending. These are straight forward and are based on the equation

$$V_o B_y = \int (\text{Bending moment})^2 ds$$

where V_o is the elastic strain energy in the yarn. If $2\beta_o$ is the angle subtended by the contact at the yarn centre and the yarn radius is R, then $V_o = 2B_y/(Rp - 2R^2 \sin\theta_o)$ and

$$V_o = B_y (\beta_o/R) + 2V_o^{\frac{1}{2}} [E(k, \pi/2) - E(k, X) - (1-k^2) F(k, \pi/2) - F(k, X)]$$

where p is the distance between successive yarns, θ is the weave angle and $k = \sin(\pi/4 + \frac{1}{2}\theta_o)$ and $k \sin X = \sin(\pi/4 + \frac{1}{2}\beta_o)$.

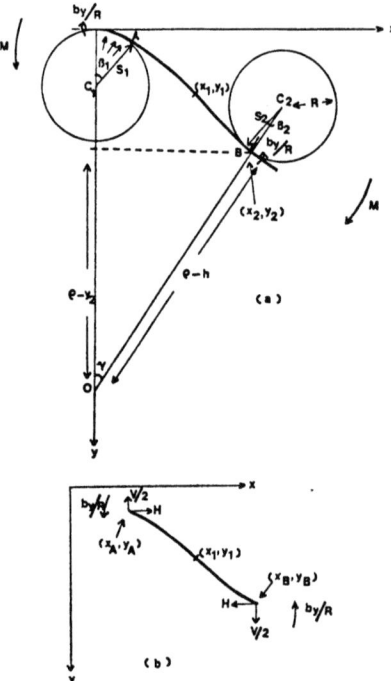

Fig. 2 shows the configuration of the yarn after the fabric has been bent. There are three alterations to the geometry which can be readily seen.

(i) The angle β_1 is larger than β_0 because extra yarn has been "wrapped around" the yarn contact lying above the neutral axis.

(ii) Similarly the angle β_2 is less than before because the yarn has been "unwound" from the lower contact region.

(iii) The region between A and B no longer has the same length as before but it can be found knowing the values of the forces V and H. These can be found remembering that the component of the forces along C_1O and C_2O must be equal.

There is a fourth change in the geometry which cannot be seen so readily but is due to the fact that x_1, y_1 have moved relative to the radius of curvature and consequently the crimp height of the crossing yarn has changed. It was assumed that as the change was small the change in V was proportional to the change in the crimp height.

In this way the strain-energy change in both yarns could be found and hence the couple required to bend the yarn to unit

curvature. It should be noted that it has been assumed that the
yarns in the fabric have a natural curvature of zero before the
fabric is bent, i.e. if they were removed from the fabric. This
is usually known as the case for the bending of unset fabrics.
For set fabrics, i.e. for fabrics in which the yarn retains its
shape when removed from the unbent fabric, a similar analysis
has been carried out. Two points of detail should be noted for
the set fabric one must write

$$M = B_y(K-K_o)$$

and the value of β_2 is always zero when the fabric is bent.

Closed form solutions were not possible for this case, but numerical solutions are readily obtained.

The following conclusions resulted from this analysis.

(i) For unset fabrics the ratio B_F/B_y is approximately equal to the ratio of the total length of yarn in the intersection to the free length, i.e. $\ell/(\ell-2\beta R)$ see Fig. 2.

(ii) For set fabrics the ratio B_F/B_y is approximately equal to the ratio of the total length of yarn to the free length plus half the length wrapped around the lower contact, i.e. $/(-\beta R)$ see Fig. 2.

(iii) The increase, during bending, of V, the contact pressure, is negligible for unset fabrics but is quite marked for set fabrics.

Shanahan and Hearle, 1978, have shown that one of these cases can be solved relatively simply by using a simplified geometry for their bent fabric and using their standard QAS computer system to obtain the energy functions from the geometry. In their simplified geometry they have assumed that $\beta_1 = \beta_2+y$ which applies only to the unset case and have also assumed that AB is always a straight line. Whether this second assumption produces results which are in reasonable agreement with the complete force solution has not been answered, as Shanahan and Hearle have not calculated any values using their method which would allow such a comparison to be made.

de Jong and Postle, 1977, have also used their generalised energy method to analyse the bending of plain woven fabrics. Their method allows for the inclusion of two other factors. (i) Partially set fabrics can also be analysed and (ii) it is no longer necessary to assume that the yarn is incompressible but it is possible to allow for a change in shape of the cross section of the yarn as it is bent. However, in the calculations reported by these authors the second possibility does not appear to have been used. The definition of curvature constraint used by these authors is difficult to relate simply to any of the other models of a woven fabric but the ratio of B_F/B_y does appear to be larger than the values obtained by other methods where similar conditions apply. For example, for unset fabrics de Jong and Postle obtain values of B_F/B_y greater than unity when there is point contact, while Abbott et al. obtain values less than unity for this case. As the models analysed in both cases would appear to be identical it is difficult to understand the reason for this discrepancy.

In conclusion it should be pointed out that all the above analyses provide a knowledge of the force per contact point V which presses the two yarns together at the contact region. If the length of yarn between one intersection and the next is then it is clear that the average pressure per unit length of yarn is V/ and the coercive couple or frictional work done during the bending of a fabric can be carried out by using the analysis for the frictional work done during the bending of a yarn which has been given above. It should be noted that the length of a twist repeat in a yarn is much longer than the distance between the intersections of two yarns so that the work done against friction in the yarn becomes almost independent of the fabric structure (see Abbott, Grosberg and Leaf, 1971).

Some of the interesting results of these analyses can be summarised as follows.

(i) The bending rigidity of a yarn is approximately equal to the sum of the bending rigidities of the constituent fibres.

(ii) The initial bending of yarns is, however, nonlinear due to frictional restraint.

(iii) The bending rigidity of a fabric is usually greater than the bending rigidity of its constituent yarns. The ratio becomes greater when (a) there is a larger constraint on yarn bending due to the interference of the cross yarns and (b) the fabric is less well set i.e. $K_o \rightarrow 0$.

(iv) The frictional resistance to bending only affects the slope of the initial bending curve.

(v) The initial frictional resistance to bending decreases as the degree of set is increased, becoming zero when the fabric is completely set.

(vi) The frictional resistance remains constant with radius of curvature for unset fabrics, but increases with curvature for partially set fabrics and changes from zero to a noticeable value for completely set fabrics.

BIBLIOGRAPHY

Peirce F.T., 1937, J.Text.Inst., $\underline{28}$, 45.
Backer S., 1952, Text.Res.J., $\underline{22}$, 668.
Platt M.M., Klein W.G. and Hamburger W.J., 1959, Text.Res.J. $\underline{29}$, 611.
Livesey R.G. and Owen J.D., 1964, J.Text.Inst., $\underline{55}$, T516.
Abbott N.J., Coplan M.J. and Platt M.M., 1960, J.Text.Inst., $\underline{51}$, 1384.
Popper P., 1966, D.Sc. thesis, Massachusetts Institute of Technology.
Abbott G.M., Grosberg P. and Leaf G.A.V., 1971, Text.Res. J., $\underline{41}$, 345.
Abbott G.M., Grosberg P. and Leaf G.A.V., 1973, J.Text.Inst., $\underline{64}$, 346.
de Jong S. and Postle R., 1977, J.Text.Inst., $\underline{68}$, 362.
Shanahan W.J. and Hearle J.W.S., 1978, J.Text.Inst., $\underline{69}$,92.
Dhingra R.C. and Postle R., 1976, J.Text.Inst., $\underline{67}$, 426.
Leaf G.A.V., 1979, J.Text.Inst., $\underline{70}$ to appear shortly.

SHEAR OF WOVEN FABRICS

J. Skelton

FRL, An Albany International Company
Dedham, MA 02026

INTRODUCTION

   The ability of a woven fabric to accept shear deformation is a necessary condition for a conformable fitting to a general three-dimensional surface, and is the basis for the success of woven textiles as clothing materials and in a large number of industrial applications involving forming or molding operations. This paper presents a general survey of the theoretical and experimental progress that has been made in understanding both the geometrical and the physical aspects of fabric shear, and explores the relationships between the fabric performance and its structural and mechanical characteristics for this mode of deformation.

GEOMETRICAL ASPECTS OF CONFORMAL FITTING

   This subject, which is central to an understanding of the practicalities of fabric shear behavior, was first explored by Mack and Taylor [1] and some of their results are repeated here. They consider that a woven fabric is composed of two sets of inextensible yarn elements, pin-jointed at their intersection points so as to permit unlimited rotational freedom. Such a structure can be fitted to a general macroscopic three-dimensional surface only by local shearing or skewing of the elements, and the fact that the elements are unaltered in length by the deformation makes the description of the fitting analytically tractable. Simple fitting to a surface of revolution is particularly easy to visualize and helps to clarify some of the issues. Consider a small square of fabric in the flat state with the sides of the square parallel to the yarn elements (Fig. 1a): when this is fitted to a surface the square will be deformed into a rhombus, and if the element is

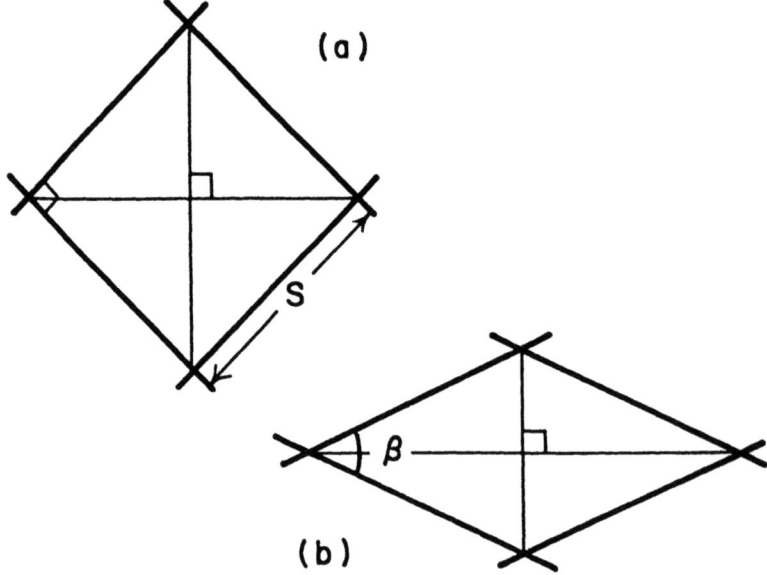

Fig. 1. Fabric element in: a) undeformed; b) deformed configuration.

sufficiently small compared with the curvature of the surface the rhombus can be considered to be plane. The acute angle included between the yarn elements is β in the fitted state (Fig. 1b). Now the diagonals of a rhombus always intersect at 90°, so any bias lines drawn at 45° to the thread direction in the plane state will still cut at 90° in the sheared form when the fabric is fitted to the surface. If the surface to be fitted is a surface of revolution, the cylindrical polar coorindates of a point on the surface are R, θ, Z (Fig. 2), and it is possible to fit a fabric to this surface so that the lines that were originally at +45 to the warp became lines of latitude with Z constant, and lines at -45 to the warp became meridians with θ constant. These latitudinal and meridianal lines intersect at 90°, which is exactly the condition that is fulfilled by the intersecting diagonals of the deformed rhombus, and a conformal fitting is assured.

The diagonals of the rhombus, which in the plane state have length $\sqrt{2}s$, in the sheared condition have lengths of

$$2s \cos(\beta/2); \quad 2s \sin(\beta/2) \tag{1}$$

and the surface of revolution can be fitted completely by a rectangle of unsheared fabric of length L and width W provided that:

$$W \cos(\beta/2) = \pi R \sqrt{2} \tag{2}$$

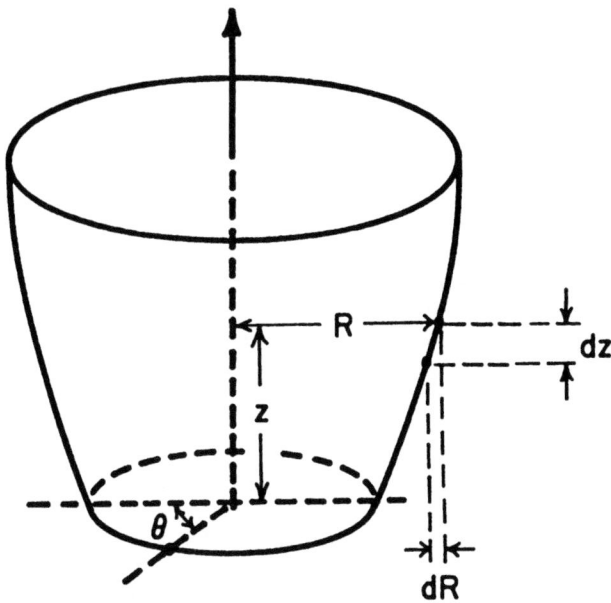

Fig. 2. Geometry of surface of revolution - nomenclature.

$$\text{and } L = \int_{Z_1}^{Z_2} \left\{ \frac{1 + (dR/dZ)}{2 - (4\pi^2 R^2/W^2)} \right\}^{1/2} dZ \qquad (3)$$

Equation 2 shows the variation of the shear angle β with the radius of the surface of revolution, and also indicates that for complete coverage the minimum width of the undeformed rectangle must be:

$$W \geq \pi R \sqrt{2} \qquad (4)$$

since cos (β/2) cannot exceed unity.

The simple fitting described above involves a rectangle of fabric which in the plane condition is aligned so that the boundaries are at ±45° to the warp direction. A more general fitting is possible in which the alignment of the rectangle is not so restricted. This fitting is characterized by the fact that lines:

x + ky = constant

in the plane condition becomes lines of latitude (R = constant) on the surface, and the simple fitting can be seen to be the special case where k = 1. In the general case the meridianal lines represented by θ = constant do not correspond to straight lines in the plane state.

In the general state the angle β between the sheared yarn elements is given by:

$$\cos \beta = (k^2 + 1 - R^2/a^2)/2k \qquad (5)$$

where k determines the orientation of the plane fabric and a is a normalizing parameter for the linear dimensions of the unsheared fabric. Again, the theoretical lower limit for β is zero, and this places some restrictions on the range of values that are possible for k and a. The practical consequence of this is that complete coverage of a particular surface may not be possible for all sizes and orientations of the fabric, even for a fabric which has a zero resistance to the shear deformation. This point has also been recently discussed by Shanahan, Lloyd and Hearle [2], who point out that for the condition β = 0 the fabric elements that originally were finite in extent degenerate into regions of zero area. In practice, the limiting shear deformation is reached well before β is reduced to zero, and the limit is set by the interference between the finitely sized fabric elements. The geometrical consequences of this interference are discussed in detail below.

The limiting configurations of sheared fabrics have been described by Skelton [3]. For a symmetrical model fabric consisting of interwoven incompressible cylinders the limiting condition can be estimated by consideration of the sequence of cross sections shown in Figure 3. These sections are taken through the fabric perpendicular to the line bisecting the acute angle β = 2α between the sheared elements. At the center of the contact region, cross-sections through the element appear as vertically-stacked, contacting ellipses with major axes (d/cos α) and minor axes d. At the point midway between contact points the sections are ellipses on the same vertical level with major axes (d/cos α) as before, but with minor axes which are slightly greater than d as a consequence of the inclination of the yarn axes at this point. Interference is inevitable and unavoidable if the projected sections overlap at the midpoint. It can be deduced from the figure that this condition is expressible as:

$$\sin \alpha = a/(1/2n) = 2n(d/2 \cos \alpha), \qquad (6)$$

leading to:

$$\sin 2\alpha = 2nd. \qquad (7)$$

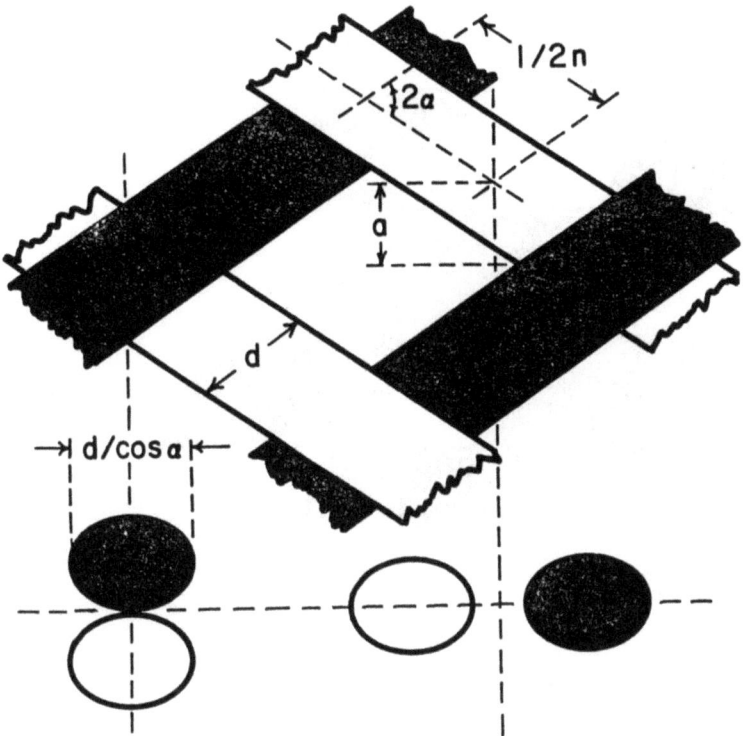

Fig. 3. Shearing of interwoven incompressible circular cylinders.

Side-by-Side Contact

A limiting condition of more direct practical significance can occur when adjacent elements of a structure come into side-by-side contact. This type of limit configuration is not confined to "square" structures, and it is of particular interest where there is a crimp imbalance between the two sets of elements, as is often the case. This limit configuration is shown in Fig. 4 and it is defined by the condition:

$$\cos \theta' = d(1/n) = nd, \qquad (8)$$

where θ is related to the angle α through the relationship:

$$\theta' = (\pi/2) - 2\alpha. \qquad (9)$$

The limiting conditions defined by Equations 7, 8, and 9 are shown graphically in Fig. 5. The most stringent condition, allowing the least amount of shear distortion at a given fractional cover, defined by the parameter nd, is found for the interwoven cylinder model. The theoretical limit for nd for a jammed, square, plain weave fabric is also shown. Note that this limit occurs at

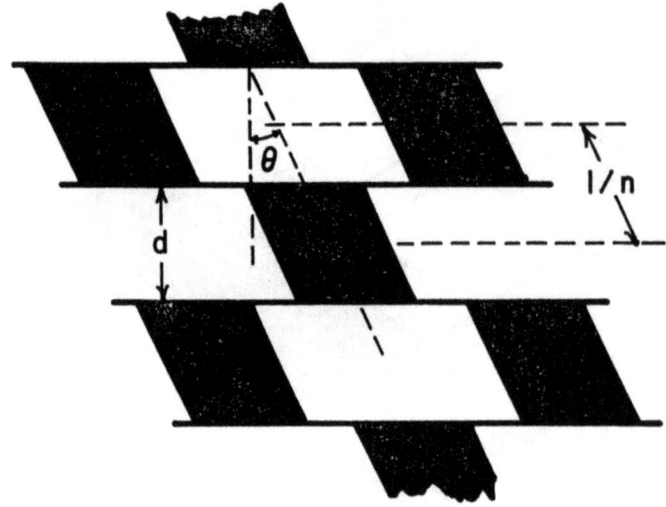

Fig. 4. Shearing limit in side-by-side configuration.

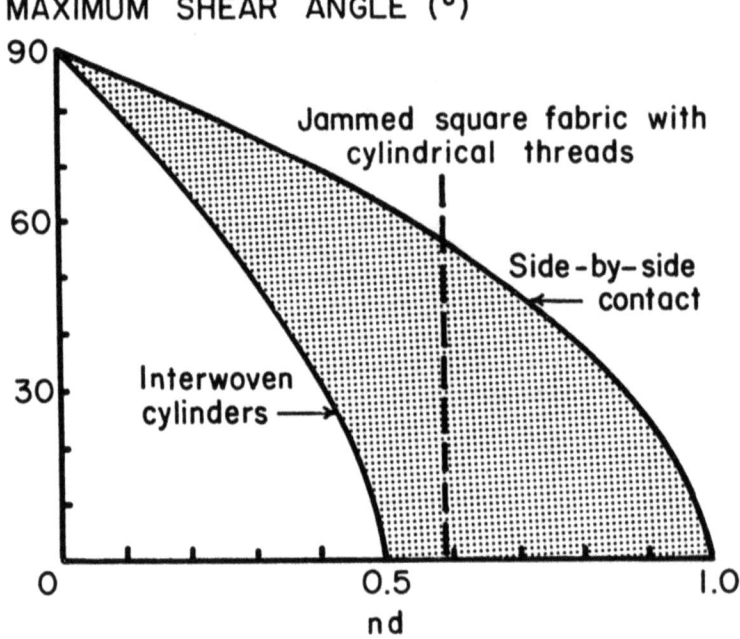

Fig. 5. Shearing limits for various geometrical configurations.

a higher value of nd than that for which the shearing potential is zero, leading to the expectation that square fabrics can be woven sufficiently tightly to prevent the occurrence of shear by the rotation of the interyarn contact regions. This is found to be the case in practice: tightly woven suiting fabrics are very resistant to shear, and such deformation as exists is mainly as a consequence of yarn distortion; tightly woven monofilament screen fabrics, in which this distortion mechanism is not available, do not shear significantly and behave and feel like laminar sheet materials. Most practical fabrics fall within the shaded area of Fig. 5 with the exact location of a fabric depending on the closeness of fit with a particular limit model. A maximum shear angle of about 45° seems to represent a reasonable compromise between good conformability and an adequate degree of cover in the unsheared regions of the fabric.

SHEAR STIFFNESS OF WOVEN FABRICS

In view of the complexity of observed shearing behavior it is perhaps unreasonable to speak of shear stiffness as a single entity. We are concerned rather with the overall stress-strain characteristic of the sheared fabric, and with an understanding of the mechanisms that are operative when a fabric undergoes shear deformation. The general stress-strain curve in shear shows several distinct characteristics that are illustrated in Fig. 6.

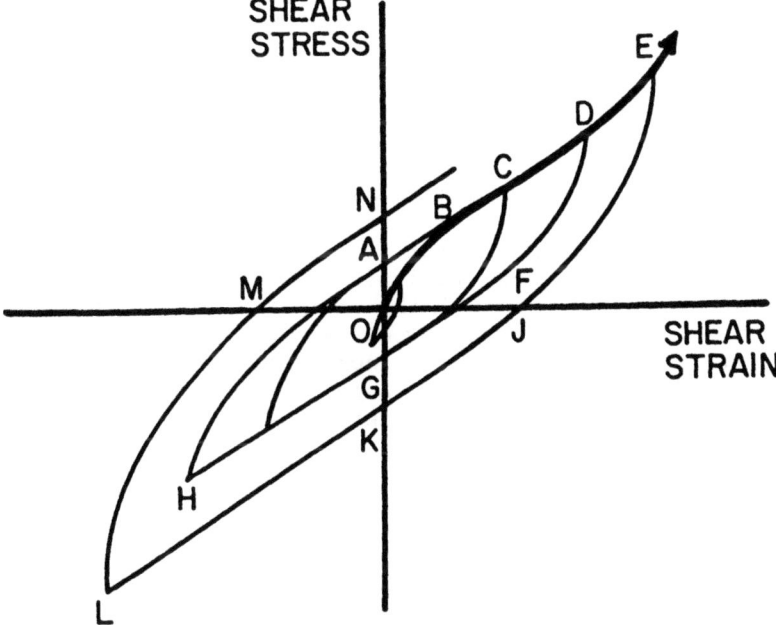

Fig. 6. General stress-strain curve for shear deformation.

If a fabric cycled between low levels of stress and strain, the shear stiffness is initially large, and decreases with increasing strain. Closed curves are generated if the cycling is continuous, and the hysteresis loss, which is very small for small deformations, increases with increasing amplitude of stress and strain. In this region (OBG) of the characteristic the behavior is dominated by frictional mechanisms and the generally accepted model concept is that the decreasing incremental stiffness is attributable to the sequential movement of frictional elements. At a particular amplitude of stress or strain (point B) the incremental stiffness reaches a minimum level, and remains constant over a range of amplitudes (region BCD). The closed curves generated by cycling within this region of the characteristic are bounded by two parallel lines, with slopes that are controlled by the deformation of purely elastic elements in the fabric. At amplitudes greater than D the incremental stiffness again begins to rise, and the closed curves increase in width with increasing amplitude. The increase in incremental stiffness can be attributed to the onset of the jamming phenomenon described in the previous section, and it signals a marked change in the nature of the deformation that is occurring at the yarn crossover points.

Woven fabrics encompass an enormous range of structural densities, material properties, and states of stress relaxation, and it is not surprising that the observed shear characteristics are equally varied within the overall outline described above. Relatively open loomstate fabrics at low shear angles typically show the parallel sided curves within the OBCDFGHA envelope, with the width of the curve and the slope of the straight portions being set by the magnitude of the frictional restraints and the stiffness of the elastic elements respectively. At shear angles greater than 5° to 10° all but the most open fabrics show a transition to the EJKLMN envelope behavior, and tightly woven fabrics can make this transition at shear strains so low that the region of constant slope is completely surpressed, and only an inflection in the curve indicates the changeover from frictional to incipient jamming behavior. Well-relaxed fabrics, in which the frictional effects are minimized, and in which the crimp form in the threads is fairly set, will often show narrow curves with a very marked and continuing increase in slope with increasing shear strain. Some of the theoretical and practical aspects of these various characteristics are discussed in more detail below.

<u>The Decreasing Stiffness Region</u>

This region of the curve is of interest since it is here that the hysteresis loss in the cyclic deformation is determined. At extremely low levels of shearing stress there is not sufficient moment available to move any of the elements of the fabric at the interaction regions, and they behave as if they were welded junctions. Under these circumstances the total deformation in the fabric elements takes place in the free length between interactions and the deformation is essentially that of a collection of parallel beams. The deformation in a unit cell of a square fabric under these conditions is shown in Fig. 7. The free length of the cantilever is $(\ell - d)/2$ and the maximum deflection y is:

$$y = \frac{f(\ell - d)^3}{24G} \tag{10}$$

where f is the force/thread and G is the flexural rigidity of the yarn.

The angular deflection of each arm is $2y/\ell$ approximately and this is one half of the total angular shear deflection of the fabric if it shared equally between the two sets of yarn elements. If the total shearing force on a sample of fabric of length L is F, then the angular deformation θ of the fabric can be written:

$$\theta = \frac{4F\ell}{L} \cdot \frac{(\ell - d)^3}{24G\ell} = \frac{F}{6GL\ell} \cdot (\ell - d)^3 \tag{11}$$

and the effective shear stiffness is:

$$\theta/F = \frac{(\ell - d)^3}{6GL} \tag{12}$$

Grosberg and Park [4] have investigated the initial shear stiffness of woven fabrics in this region, and have found reasonable agreement between the theoretical and observed values. In particular it appears that the effective area of contact at the intersections in the fabrics that they tested was very similar to the apparent area of contact based on projected yarn diameters, and does not appear to be particularly sensitive to the tensions in the threads.

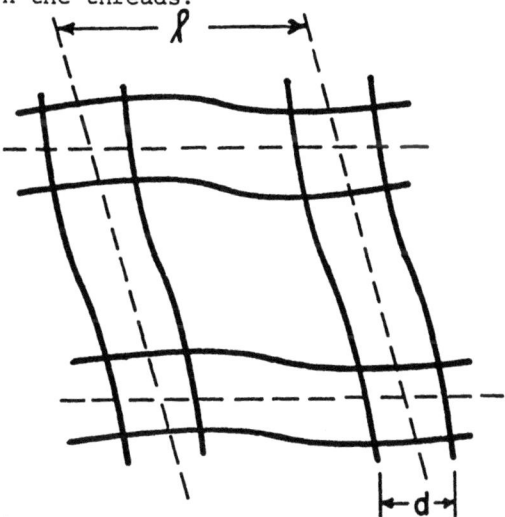

Fig. 7. Deformation of unit cell of fabric in shear.

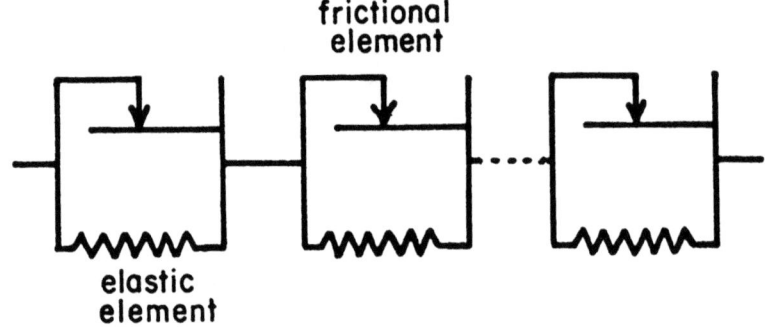

Fig. 8. Assembly of frictional and elastic elements.

As soon as the stresses are large enough to overcome the smallest of the frictional restraints that are acting at the intersection regions, the system starts to slip, and the incremental stiffness falls. The detailed mechanisms that are operating are extremely complex and it is difficult to devise a convincing model that is adequate to explain the behavior. However, in a general sense the behavior of an array of elastic and frictional components has been studied by Oloffson [5], and by Skelton [6], and Skelton and Schoppee [7]. It can be shown that the stress-strain behavior of a series assembly of nearly similar frictional/elastic units such as those illustrated in Fig. 8 can be reasonably represented in the initial, non-linear region by an expression of the form:

$$\sigma = K\varepsilon^{1/2} \tag{13}$$

where σ and ε are the stress and strain respectively. This type of stress-strain behavior is found in a wide range of friction controlled assemblies, including fabric bending, which is also a non-linear phenomenon under cyclic loading: we probably have a right to be suspicious of any detailed models for the slippage process that do not give this type of characteristic. Grosberg and Park have made the best analytical attack on this problem and their results can be simplified with appropriate approximations to give the square law variation at low strains. Perhaps the worst feature of their analysis is the assumption that there is line contact between the yarns at the intersection regions, a supposition that is some distance removed from reality.

There are two interrelated parameters that control the extent of the non-linear region. These are the slope of the stress-strain

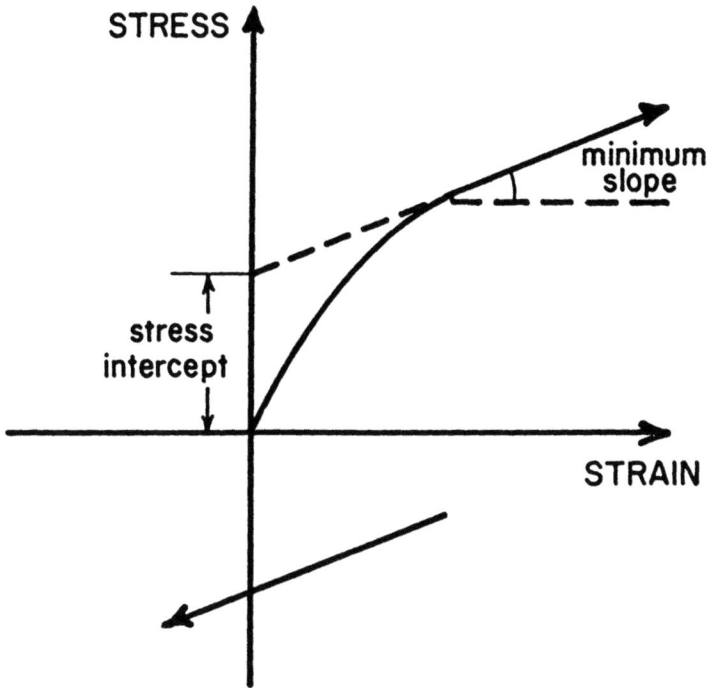

Fig. 9. Important parameters in stress-strain behavior of frictional/elastic elements.

curve where it attains its minimum value, and the intercept of projection of this line on the stress axis (Fig. 9). The minimum slope of the curve represents the contribution of the purely elastic elements of the assembly, and when this value is achieved it is assumed that all the frictional contact points are in motion. This slope is clearly finite in all woven fabrics, and this consideration makes any unmodified slipping clutch type of model invalid, since such a model would have a zero incremental stiffness once all the slip mechanisms are activated. The intercept represents the magnitude of the stress required to activate the frictional elements if there were no elastic elements present - that is, the idealized slipping clutch described above. It is only when the elastic and friction elements are considered simultaneously that the model can be made to fit at both boundary conditions, and this is another of the problems with the Grosberg-Park model.

The Constant Stiffness Region

In many woven fabrics there is a substantial region of the shear stress-strain curve over which the slope is essentially constant, and affected to only a very small degree by the external loading conditions. This type of behavior is characteristic of a simple, linearly elastic system, and it is a little disappointing that all the models that have been proposed so far to explain the behavior in this region are so complex. The nature of the torque-twist relationship for a single, isolated model yarn intersection has been studied by Kawabata, Niwa and Kawai [8], and they confirmed that the behavior was essentially linear over the experimental range. Their graphs of torque against angle of rotation show intercepts on the torque axis, but all their experimental lines project to a common intercept on the twist axis, and seems likely that the problem is associated with the fact that two intersecting twisted yarns are not necessarily orthogonal in the zero torque configuration because of the effect of yarn twist interactions.

It is possible that we have been ignoring important clues for some time in trying to understand this linear behavior. Many reports have drawn attention to the close relationship that exists between the shearing behavior and the bending behavior of woven fabrics, particularly with respect to recovery properties. For example, several authors [9,10,11], have suggested that the shear energy loss is a good guide to the kind of behavior that can be expected in bending or creasing. A little thought will show that the shear and bending are not merely related - they are essentially identical. If we go back to the sheared junction discussion of Fig. 7, in a condition just after slippage, then the picture that we see is of a yarn with a region that is restrained in some way by the crossing yarn, together with a free length of bent yarn. The shear angle associated with the repeated form of this deformation is θ which is divided equally between the two sets of yarns and the effective curvature of the yarns is:

$$\text{curvature} = \theta/\ell. \tag{14}$$

For a shear angle of a few degrees (5°), and a fabric with 20 threads/cm, the effective curvature in the yarns is:

$$\frac{5 \times 20}{54} \sim 2 \text{ cm}^{-1},$$

which is typical of the curvature levels that have been investigated in studies of fabric bending behavior. Thus the magnitude and distribution of the curvature in the yarns in the bent state are almost identical with those in the sheared state - the only difference is that in the bent condition the increments of angular deformation along any yarns are additive, while in the sheared state the average deformation remains constant. Notice that the onset of high deformation non-linearity takes place in both modes of deformation at about the same strain levels (5°+; 2 cm^{-1}+).

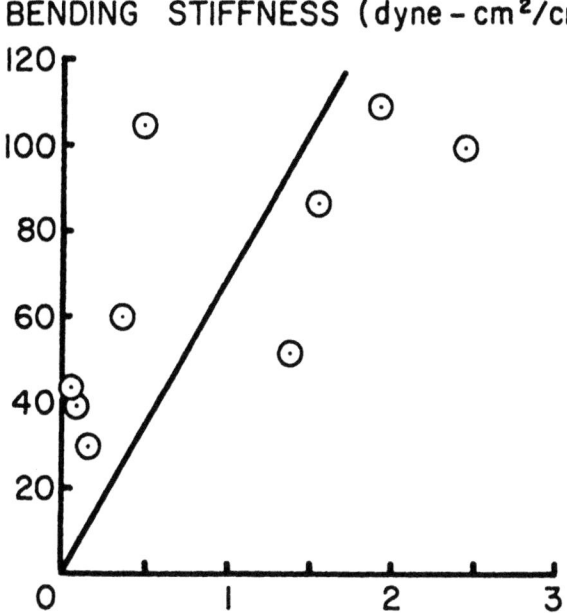

Fig. 10. Relationship between stiffness in bending and in shear for a range of fabrics.

There are other minor differences, of course, that modify the relationship between the two modes of deformation. If there is any yarn flattening, the yarn will have curvature about an axis parallel to its major axis in the bent state, and about an axis parallel to its minor axis in the sheared state, with a concomitant change in yarn stiffness. In addition, the nature of the steric hinderances, which have been shown to affect the yarn bending behavior, will be different in the two cases. Nevertheless, it is to be anticipated that there will be a good correlation between the elastic stiffness in shear and bending for plain weave woven fabrics.

In spite of the large amount of experimental work that has been reported in those areas separately, very few workers have made measurements in the same fabrics in both modes of deformation. The only unambiguous data that I know of is in a paper by Dawes & Owen [12], and their measurements are plotted in Fig. 10. The correlation is quite good and the constant of proportionately can be related to the fabric structural parameters and the specimen size. An additional piece of information that adds credence to

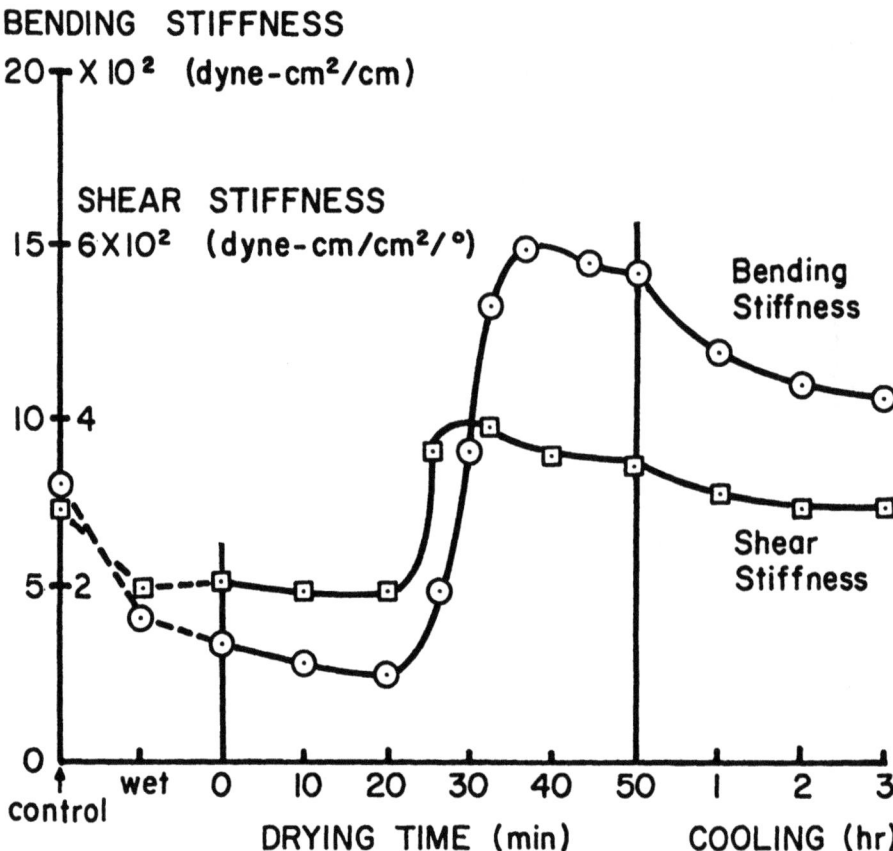

Fig. 11. Variation of fabric bending and shear stiffness through a range of environmental conditions.

this concept is shown in Fig. 11. This gives the measured variations of bending and shearing stiffness in the same fabric throughout a cycle of wetting and drying; the overall changes in mechanical properties as the environmental conditions change are almost identical in the two modes of deformation and it is hard to resist the idea that they share a common origin.

The Increasing Stiffness Region

It is a commonly observed fact that above a relatively low level of shear strain (5° → 10° is typical) the shear stiffness increases with increasing strain. It seems almost certain that this is due to steric hinderance between the two bent intersecting yarns, leading to transverse distortion of the yarns, or riding up of the intersection, or both. Some preliminary theoretical work

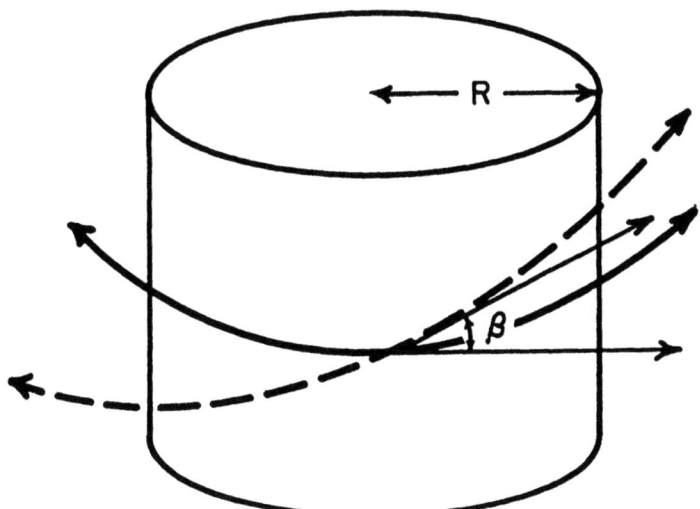

Fig. 12. Yarn bent at various angles around a fixed cylinder.

has been carried out on these aspects of shear, but a reasonable idea of what to expect can be had from a consideration of the simplest possible case of a yarn bent around a cylinder (Fig. 12).

The curvature in the yarn when the yarn and cylinder axes are perpendicular is $1/R$ and it changes to $(1/R)\cos^2\beta$ when the angle changes by β. The change in curvature ΔC is

$$\Delta C = (1/R) - (1/R)\cos^2\beta = (1/R)\sin^2\beta \qquad (15)$$

and the moment M associated with this change is:

$$M = G\Delta G = G/R \sin^2\beta \sim GR\beta^2. \qquad (16)$$

The effective stiffness for the rotation is:

$$dM/d\beta = 2GR\beta \qquad (17)$$

which represents a stiffness that increases linearity with increasing angle of deformation. If the interaction of <u>two</u> bent cylinders is considered, the mathematics is obviously more complicated, but the form of the expression remains the same. If there is any free space between the yarns in the orthogonal configuration, or a region of easy yarn transverse compaction at low strains then the onset of this region will be delayed, but in general terms the explanation seems valid.

CONCLUSIONS

The characteristics of the various regions of the stress-strain curves of a woven fabric in shear can be explained in fairly simple general terms which give the correct analytical forms for the relationships. The limiting shear configurations of fabrics, which determine the extent to which they can be molded and formed, can be deduced from the fabric geometrical parameters. All the phases of shear behavior are sufficiently complex that they can bear an almost unlimited amount of analytical probing, but it is hoped that the simple ideas described in this account will help in developing some physical insight into this fascinating phenomena.

REFERENCES

[1] C. Mack and H. M. Taylor, J. Text. Inst. 1956, 47, T477.

[2] W. J. Shanahan, D. W. Lloyd and J. W. S. Hearle, Text. Res. J. 1978, L18, 495.

[3] J. Skelton, Text. Res. J. 1976, 46, 862.

[4] P. Grosberg and B. J. Park, Text. Res. J. 1966, 36, 420.

[5] B. Olofsson, J. Text. Inst. 1967, 58, 221.

[6] J. Skelton, Text. Res. J. 1974, 44, 746.

[7] J. Skelton and M. M. Schoppee, Text. Res. J. 1976, 46, 661.

[8] S. Kawabata, M. Niwu and H. Kawai, J. Text. Inst., 1973, 64, 62.

[9] R. M. Butler, G. E. R. Lamb and D. C. Prevorsek, Text. Res. J. 1975, 45, 426.

[10] F. C. Brenner and C. S. Chen, Text. Res. J., 1964, 34, 505.

[11] J. Lindberg, B. Behre and B. Dahlberg, Text. Res. J., 1961, 31, 99.

[12] V. Dawes and J. D. Owen, J. Text. Inst. 1971, 62, 233.

ENERGY OPTIMISATION METHODS IN FABRIC MECHANICS*

S. de Jong and R. Postle

School of Textile Technology, University of New South
Wales, Kensington, 2033, Australia.

ABSTRACT. A general energy approach to the analysis of fabric
mechanics is presented and applied to a number of different structures and deformations. The analysis is chiefly concerned with the
recoverable mechanisms of fabric deformation, although yarn set has
been included in the analysis.
   The methods presented here are based on the fundamental principle that elastic structures always assume a configuration of minimum strain energy, regardless of the deformation applied. The
resulting minimisation problem is solved with the use of a specific
optimisation technique, i.e. it is treated as an optimal control
problem.
   The present general analysis of fabric mechanics using energy
optimisation techniques makes no inherent assumptions about the
shape of the yarns in the fabric structure, although restrictions
on the yarn shape can be conveniently introduced without limiting
the information gained from the solution. The analysis will be
presented in its general form and then applied to several specific
structures by way of illustration.

1. ENERGY METHODS OF ANALYSIS FOR FABRIC STRUCTURES

Energy methods are used widely in complex mechanics problems[1],
where geometric intuition is replaced by algebraic relations deduced
from the energy principle. The starting point of an energy method
of analysis for elastic structures consists of identifying and formulating each of the individual contributions to the energy of a
structure. This energy needs to be carefully defined and may be

---
* This work has been supported by the Australian Wool Corporation.

chosen to represent[1] (1) the total potential energy, (2) the complementary energy, or (3) the strain energy. The choice depends both on convenience and on the nature of the applied or boundary forces and couples.

In the presently used method, the total strain energy of a structure is formulated (consisting of the sum of the individual yarn bending, torsion, lateral compression and longitudinal extension strain energies). This total strain energy is minimised subject to certain constraints which give rise to internal forces and couples within the structure itself. These constraints lead to the necessity of minimising an augmented energy expression which includes the work done by the forces of constraint.

The necessary conditions of force and moment equilibrium can then be established by a suitable condition of minimum energy. Provided the problem is correctly formulated, the greater simplicity of the energy method naturally leads to a more rigorous approach and often a number of unnecessary assumptions and approximations can be eliminated.

Energy methods of analysis have been applied to fabric structures in a number of different ways, usually in conjunction with a descriptive geometric model of the shape of the yarns in the structure[3,4]. However, when assumptions are made about the geometry of the structure this may be done at the expense of gaining information about the interyarn forces in the fabric structure[3]. These interyarn forces are very important in determining the fabric dimensional and mechanical properties.

2. ANALYSIS

2.1 General Outline of Method

The statement of the problem, and its solution is summarised in Table I, to which reference should be made. The total strain energy of a fabric structure is represented by an energy functional, U in equation (1) where s is the arc parameter of the yarn axis, m is the control variable and z defines the shape of the yarn axis in the fabric. It is desired to find the minimum of U, overall admissable variations of the control variable m. The admissable variations in m are restricted by a differential equation constraint (2), where g is some function of the yarn shape z, and the control m.

To arrive at some very general necessary conditions of the optimal control[5] m which minimises U in expression (1) subject to the differential equation constraint (2), it may be possible to substitute expression (2) into (1) and then to solve the minimisation of U according to the ordinary rules of the calculus of variations. Alternatively, the constraints (2) may be appended to the energy functional U by means of Lagrange multipliers, λ_i, which enable the control variable variations to be treated quite independently from (2).

The latter method implies forming a new energy functional, U_a, as in (4). Clearly, when constraint (2) holds, minimising U_a in (3) is equivalent to the minimisation problem of (1) subject to (2).

The necessary conditions for a minimum can be very simply stated by defining a new function, H (the Hamiltonian), as in (3). It can be shown[5] that a necessary condition for a minimum is $\partial H/\partial m = 0$, which in the present problem represents the equation of equilibrium (7) together with the <u>state</u> and <u>costate</u> equations shown in Table I. The <u>state</u> equation (5) is just a restatement of the constraint equations (1). The <u>costate</u> equation is an additional equation (6) which is used in conjunction with (5) and (7) to solve for the variables m, z, and λ.

2.2 Assumptions

Since energy is a scalar quantity, the total strain energy E of the unit cell may be represented by the sum of the energy terms of each individual loop in the repeat, i.e.:

$$E = \sum_{i=1}^{n} \int_0^{\ell_i} (E_b + E_\tau + E_c + E_t)_i \, ds* \qquad (8)$$

where E_b, E_τ, E_c, E_t are the strain energy terms per unit length of yarn for bending, torsion, lateral compression and longitudinal tension, respectively; ℓ_i is the length of the ith loop in the repeat, n is the number of differently shaped loops in the unit cell, and s* the arc length.

It will be assumed that the yarn bending and torsional energies per unit length can be written as $E_b = B\chi^2/2$ and $E_\tau = G\theta^2/2$ where B and G are the yarn bending and torsional rigidities and χ and θ the yarn total curvature and twist. The yarn compression energy will be represented as $E_c = Cg(r)$, where C is the yarn compression rigidity and g(r) is a function of the interyarn distances, r. In the general three-dimensional analysis, the energy term due to yarn extension is neglected. It will need to be introduced when the method is applied to the plain-weave fabric structure.

2.3 Minimum Strain Energy

To arrive at the condition of equilibrium for the fabric structure represented by (1), it is necessary to find the absolute minimum of E (8) over all the admissable variations, i.e. those variations of yarn shape, torsion, extension and lateral compression which

TABLE 1

OPTIMAL CONTROL FORMULATION

Problem Minimise: $U(m) = \int_s f(z,m) \, ds$ \hfill (1)

Subject to: $dz/ds = g(z,m)$ \hfill (2)

Solution Define: $H(z,m,\lambda) = f(z,m) + \lambda g(z,m)$ \hfill (3)

Minimise without constraint:
$$U_a = \int \left[f(z,m) + \lambda\left(g(z,m) - dz/ds\right) \right] ds \quad (4)$$
$$= \int \left(H(z,m,\lambda) - \lambda \, dz/ds \right) ds$$

CONDITIONS FOR A MINIMUM

State equations	Costate equations
$\partial H/\partial \lambda = dz/ds$ \quad (5)	$\partial H/\partial z = -d\lambda/ds$ \quad (6)

$\partial H/\partial m = 0$ EQUATION OF EQUILIBRIUM   (7)

---

conform to the boundary conditions and any other restrictions inherent in the problem. Because the theorem of minimum strain energy holds regardless of the boundary conditions applied to the fabric, the general method of solution must be independent of the boundary conditions. There are restrictions on the variations of yarn shape, however, because the cartesian co-ordinates $z=(z_1,z_2,z_3)$ are not all independent. In fact, the following condition must hold: $\dot{z}_1^2 + \dot{z}_2^2 + \dot{z}_3^2 = L^2$, where L is the curvilinear yarn length, and (˙) denotes differentiation with respect to the normalised arc parameter s which is related to the real arc length parameter s^* by $Ls = \bar{s}^*$. This differential constraint is analogous to (2) and can be rewritten in an alternative form, by defining new variables m, as in the state equations in Table II. It should be noted that the dependence on the yarn loop length L is being explicitly stated here, and is slightly different from the treatment elsewhere.[6]

Following the scheme of Table I, it is possible to write down the relevant equations for the minimisation of U in (9) subject to constraints (10) or (12). This is depicted in Table II. Because H does not contain s explicitly, it can be proved[5] that H is constant along the loop.

TABLE II

GENERAL THREE-DIMENSIONAL SOLUTION WITHOUT YARN EXTENSION

Problem Minimise: $U(m) = \int \left[\frac{1}{2}(m_1^2 + m_2^2 \sin^2 z_4) + \frac{G}{2B}(m_2 \cos z_4 + m_3)^2 \right.$
$\left. + \frac{CL^3}{B} g(r) \right] ds$ (9)

Subject to: arc parameter constraints, $\dot{z}_1^2 + \dot{z}_2^2 + \dot{z}_3^2 = L^2$ (10)

Solution Hamiltonian: $H = \frac{1}{2}(m_1^2 + m_2^2 \sin^2 z_4) + \frac{G}{2B}(m_2 \cos z_4 + m_3)^2 + \frac{CL^3}{B} g(r)$
$+ L(\lambda_1 \cos z_4 + \lambda_2 \sin z_4 \cos z_5 + \lambda_3 \sin z_4 \sin z_5)$
$+ \lambda_4 m_1 + \lambda_5 m_2 + \lambda_6 m_3$ (11)

NECESSARY CONDITIONS FOR MINIMUM:

State equations:
$\partial H / \partial \lambda_i = \dot{z}_i$ (12)

$\dot{z}_1 = L \cos z_4$

$\dot{z}_2 = L \sin z_4 \cos z_5$

$\dot{z}_3 = L \sin z_4 \sin z_5$

$\dot{z}_4 = m_1$

$\dot{z}_5 = m_2$

$\dot{z}_6 = m_3$

Costate equations:
$\partial H / \partial z_i = -\dot{\lambda}_i$ (13)

$\dot{\lambda}_1 = (CL^3/B) g'(r) (\bar{z}_1 - z_1)/r$

$\dot{\lambda}_2 = (CL^3/B) g'(r) (\bar{z}_2 - z_2)/r$

$\dot{\lambda}_3 = (CL^3/B) g'(r) (\bar{z}_3 - z_3)/r$

$\dot{\lambda}_4 = -m_2 \sin z_4 ((\frac{G}{B} - 1) m_2 \cos z_4 + \frac{G}{B} m_3) + \lambda_1 L \sin z_4$
$\quad - \lambda_2 L \cos z_4 \cos z_5 - \lambda_3 L \cos z_4 \sin z_5$

$\dot{\lambda}_5 = \lambda_2 L \sin z_4 \sin z_5 - \lambda_3 L \sin z_4 \cos z_5$

$\dot{\lambda}_6 = 0$

Equilibrium Equations:

$$m_1 + \lambda_4 = 0$$
$$m_2 \sin^2 z_4 + \frac{G}{B}(m_2 \cos z_4 + m_3) \cos z_4 + \lambda_5 = 0$$
$$\frac{G}{B}(m_2 \cos z_5 + m_3) + \lambda_6 = 0 \quad (14)$$

2.4 Physical Interpretation

If $C g(r)$ represents the compression energy per unit length, then $\partial/\partial z_i (C g(r)) = C g'(r) (\bar{z}_i - z_i)/r$ are the forces per unit length acting along the yarn z in the 1,2,3 directions due to yarn $\bar{z}$. Consequently the λ_i, which were introduced to treat the differential equation constraints (12), have an actual physical significance. The negative λ_i, i = 1,2,3 in Table II are the dimensionless distributed forces along the length of the yarn z (proportional to L^3/B). Upon integrating with respect to s, $L\lambda_i$

$i = 1,3$ depict the shear and axial forces (i.e. $L\lambda_i$, $i = 1,2,3$ are proportional to L^2/B). From the costate equations in Table II, λ_5 is the dimensionless incremental bending moment in the yarn z, (acting always in the 1-direction) due to shear forces $L\lambda_2, L\lambda_3$ The last three terms in the λ_4 equation represent that part of the incremental bending moment in the m_1 direction due to shear forces $L\lambda_i$, $i = 1,2,3$. The first term in this equation represents the contribution in the m_1 direction, due to the rate of change in direction of the incremental bending moment, for λ_4 rotates around the yarn in the same way as m_1 [6].

It follows that the equilibrium equations of Table II can be viewed as the condition of moment equilibrium. λ_i, $i = 4,5,6$ represent the negative of the couples or bending moments (divided by B/L) acting on the yarn z. In a planar curve, e.g. if $m_2 = 0$ everywhere, the first of these equilibrium equations in Table II states that the curvature is proportional to the bending moment. The third of the equilibrium equations in Table II implies that the twisting couple is proportional to the twist inserted. The relationship between the couples λ_5, λ_6 and m_2 and m_3 in non-planar curves is more complex. If both λ_5, λ_6 are constant, the second two of the equilibrium equations are equivalent to those derived by Love. No analogy can be found for the first of these equations, although it follows from a consideration of moment equilibrium.

H can be identified as the negative total energy of the structure [6] $H = -(T \mp V)$, where T represents the bending and torsion energy of the yarns and V the potential of the internal (positive) or externally applied (negative) forces.

2.5 Computing Algorithm

The following algorithm was used to solve the state, costate, and moment equilibrium equations of Table II on a digital computer.
  (a) Feed in a guess for m such that, after step (b), a loop shape with the correct interlocking characteristics is obtained.
  (b) Integrate the state equations to find the loop shape.
  (c) Obtain the shapes of the surrounding yarns (including all types of possible jamming or contact) either from symmetry considerations or from steps (a) and (b) in the case of differently shaped yarns.
  (d) Calculate the distances, r between contacting yarns.
  (e) For the k'th loop shape, calculate from r the forces per unit length due to all yarn contacts (which automatically includes all types of jamming).
  (f) Integrate the costate equations.
  (g) Check whether the energy gradients $\frac{\partial H}{\partial m_j} \to 0$, $j = 1,2,3$

everywhere; if so, the moment equilibrium and the
energy minimum have been reached and computation ceases.
If this condition is not satisfied, update m according
to:

$$m_j^{k+1} = m_j^k + \alpha \left[\frac{\partial^2 H}{\partial m_j \partial m_\ell}\right]^{-1} \frac{\partial H}{\partial m_\ell} \quad j = 1, 2, 3 \quad (19)$$

where α is the stepsize, and go to (b). The step
size α is usually best determined by a line search
along the direction of descent [6].

An alternative method of computing the optimum solution for
the control variables is to consider the moment equilibrium
equations of Table II as a set of residuals (together with
residuals on any boundary conditions). Fast, efficient algorithms
exist at most computing libraries to systematically search for an
optimum m among the residuals.

3. SPECIFIC EXAMPLES

Integration of the state and the costate equations in Table II is
usually done simply by Simpson's rule. What needs to be supplied
are the boundary conditions, i.e. the integration constants, for
each of the state and costate equations. These boundary conditions
have been determined so far for the plain-knitted, 1*1 rib knitted and
plain-weave fabrics under general biaxial tension and for the plain
weave fabric also in pure bending. [6] Other structures and
deformations are being treated currently.

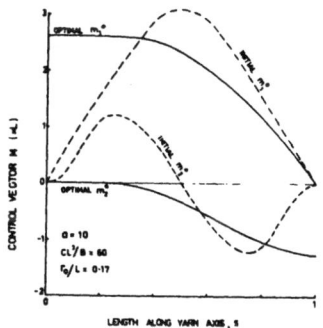

Fig. 1. Plain knitted guess for m_1, m_2 (broken line) and optimal
solution (full line) shown for quarter loop.

3.1 The plain-knitted fabric solution

As an example of the preceding analysis, the guess for the control
vectors $m = (m_1, m_2)$, as shown in fig.1, was fed into the program

(ignoring yarn twist) together with the correct boundary
conditions and appropriate yarn constants. The guess for m
was such that the loop shape defined by the state equations
(Table II) had the correct interlacing characteristics. At
the solution, i.e. when the minimum energy conditions was
determined to within a specified tolerance, the optimal control
vector m was found as shown in fig. 1. Typical distributed
forces between the interlocking yarns at the equilibrium point
were found to be as in Fig. 2, where the 1 and 2 directions
correspond to the course-and wale-wise directions and the 3
direction is perpendicular to the fabric plane. The distributed
forces that existed at the start of the iteration procedure, i.e.
as calculated from the initial guesses or the control m, were
roughly of a similar shape but about four magnitudes greater,
due to the fact that the guessed loop shape was too flat for the
yarn diameter to be incorporated in the three dimensional loop.
Jamming forces between courses and wales can be found simultaneously.[6]
Analogous results have been obtained for the 1*1 rib fabric.[6]

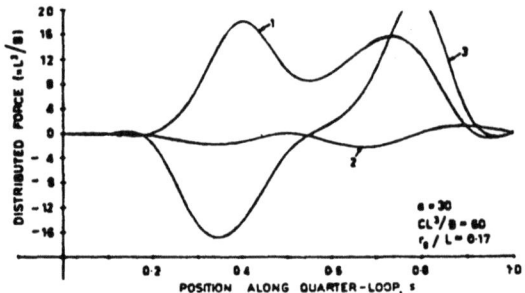

Fig. 2. The distributed interyarn forces acting in the 1,2,3
directions along quarter loop of plain-knit fabric.

It is important to note that no assumptions have been made
in this analysis about the nature of yarn contact, or the geometry
of the loop shape. The only assumption made concerning the inter
yarn forces is that, as a result of the assumed compression
potential function depending only on the interyarn distances, the
interyarn forces must pass through the yarn axis and apply therefore
no twisting couple to the yarns.[6] A result of the analysis is that
the contact between interlocking yarns is dynamic, depending on the
yarn compression characteristics, and on the applied fabric tension.
The effect of different yarn compression properties is not so much
that the possibility of yarn compression gives the fabric more
extensibility (the reduction interyarn distance being small on
extension). The major mechanism that governs the ability of a
fabric to extend is the freedom of yarn movement within the structure.
The more compressible yarns need to be knitted more tightly and

therefore have greater effective contact along the length of yarn in the repeat, thereby reducing fabric extensibility.

3.2 Plain-weave fabric in pure bending

A non-square plain-weave fabric under biaxial tension or pure bending represents an example where more than one yarn is considered in the structural repeat unit of the fabric. Yarn extension and set are included in this example. Yarn extension energy is represented by $Y(m_h-1)^2/2$, where Y is Young's modules of the yarn and m_h is the ratio of extended local length to unextended local length of the yarn element. Set is introduced by the set control vector $m_1{}^0$. The continuously distributed forces acting between the yarns within the fabric are replaced in this example by the point forces Q. The woven fabric structural repeat unit in bending, as represented by its yarn axes, is shown in Fig. 3, with the forces P, Q acting at ends A and B of the yarn.

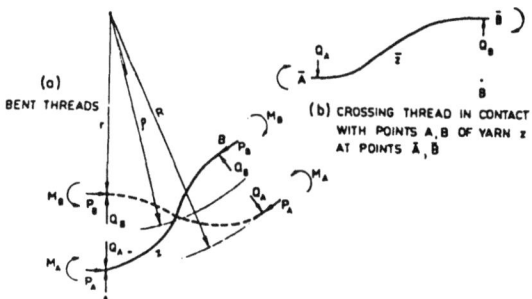

Fig. 3. Forces and couples acting at end-points A,B of half-wave repeat of yarn axes in a plain-weave fabric during bending.

The energy integral, U can be written simply as in Table III, where h represents the work done by the point forces and couples at point s=1. The simplicity and symmetry of the method applied to the two yarn structure is aptly shown in Table III, where the computational scheme is identical to the preceding Tables I, II. There are two sets of state and costate equations, one for each yarn in the repeat. Because of the planar nature of the yarns within a woven fabric, which is not subjected to shear, the state and costate equations are very much simplified, in that torsion (m_3, z_6) and the z_3, m_2 components of yarn bending and torsion can be neglected.

TABLE III

GENERAL EQUATIONS FOR PLAIN WEAVE UNDER TENSION OR BENDING

Minimise: $U = \dfrac{EL}{B} \sum_{j=1,2} \int \tfrac{1}{2}\{(m_{1j}-m^0_{1j})^2 + \dfrac{YL^2}{B}j(m_{hj}-1)^2\}\,ds + h(z(1))$

subject to state equations

Hamiltonian: $H = dU + \sum_{j=1,2}\sum_{i=1,2,4} \lambda_{ij}\dot{z}_{ij}$

State equations: Costate equations:
$\partial H/\partial \lambda_{ij} = \dot{z}_{ij}$ $\partial H/\partial \dot{z}_{ij} = -\lambda_{ij}$

$\dot{z}_{1j} = L_j m_{hj}\cos z_{4j}$ $\dot{\lambda}_{1j} = 0$

$\dot{z}_{2j} = L_j m_{hj}\sin z_{4j}$ $\dot{\lambda}_{2j} = 0$

$\dot{z}_{4j} = m_{1j}$ $\dot{\lambda}_{4j} = L_j m_{hj}\{\lambda_{1j}\sin z_{4j} - \lambda_{2j}\cos z_{4j}\}$

At equilibrium:

$\partial H/\partial m_{ij} = 0, \quad i=1,h; \quad j=1,2$

$m_{1j} - m^0_{1j} + \lambda_{4j} = 0 \quad \begin{Bmatrix}\text{only where constraint not} \\ \text{violated else } m_{1j}=\text{constraint}\end{Bmatrix}$

$\dfrac{YL^2}{B}j(m_{hj}-1) + \lambda_{1j}L_j\cos z_{4j} + \lambda_{2j}L_j\sin z_{4j} = 0$

Boundary condition:

$\dfrac{\partial}{\partial z(1)}\{h(z(1))\} = \lambda(1)$

In figure 3, P and Q are the axial end transverse yarn forces. During bending, Q may be considered to remain constant.[6] If point A (s=0) of the bent yarn shown in full is fixed, any variation in yarn shape from its equilibrium position will cause forces P,Q and couple M_B at point B (s=1) to move and do work. The reason the end point function h(z(1)) is included in the energy expression for U (Table III) but was not included in Table II, is that in this case the boundary conditions for the shear forces in the yarn in the 1,2 directions and the couple at point B depend on the Q force, and on its direction at point B. Furthermore, it is necessary to restrict the possible variations to those in which the forces P,Q are as shown in Fig.3. The Q force at point B changes direction every iteration, so that the $\lambda_1, \lambda_2, \lambda_4$ would change also. Specifying a function which depends on the Q force and on its direction allows the program to calculate the bending characteristics of one particular fabric.

(The size of the Q force determines the fabric construction).
If on the other hand the yarn shear forces λ_1, λ_2 and the
couple λ_4 at point (s=1) were specified independently, the
program would certainly iterate towards an equilibrium solution,
but not necessarily for the fabric construction requested, nor
necessarily for the case of a fabric in pure bending. Consequently
specifying an appropriate function h restricts the variations in
yarn shape to purely admissable variations. The minimum of the
energy functional U is then found only over those variations of
the yarn shape which give rise to pure fabric bending.

It is now required to investigate whether the approximation
provided by the point forces is realistic in fabric bending.
For this, constraints are imposed on the yarn curvature which
have the effect of preventing the control or curvature, m to
increase beyond a certain limit. This limit may be due to the
continuous contact between the crossing yarns in a repeat, and
the curvature limit might be some function of the interyarn
distance. Optimal control theory shows that when constraints are
placed on the control variable, the state, costate and equilibrium
equations still define the equilibrium solution, except where the
constraint would be violated, in which case the control variable
equals the constraint[5]. Consequently geometric constraints have
now been introduced which do not affect the basic minimisation
process so that the internal forces are still part of the
solution.

The results show that the interyarn forces, and dimensions
of the structural repeat are only slightly affected when the crimp
of the yarns is low (less than 10% crimp)[6]. The constraints
increase the bending rigidity of the fabric. Typical computed
yarn shapes are as shown in Fig. 4.

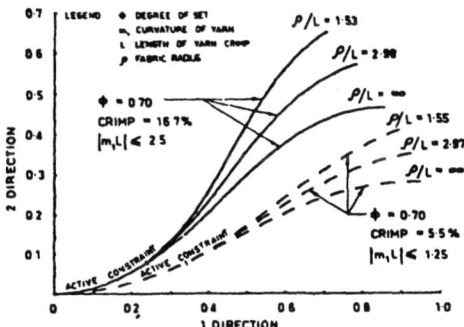

Fig. 4. Typical shapes of the yarn axis within a bent plain-weave
fabric for two different levels of weave crimp (5.5 and 16.6%).

3.3 Introduction of yarn set ϕ for the relaxed and tensioned plain weave

When the possibility of yarn compression is included, then the function $h(z(1))$ represents the point compression energy of the yarn at point B. The Q forces can be calculated from the compression energy $C'g(d_0/d)$ where C' is the point compression rigidity and g is a function of the ratio of a fixed yarn diameter d_0 to the actual yarn diameter (or interyarn distance). Thus the boundary condition in Table III shows that $Q = \partial/\partial z_2(1)\{C'g(d_0/d)\}$

The force P on the other hand, is predetermined when the structure is extended to a fixed P, and is simply introduced as the boundary condition $\lambda_1(1) = -P$, because[6] λ_1 represents the shear force acting in the opposite direction.

When it is recognised that yarns have an initial curvature, so that upon unravelling from the plain-weave fabric they do not return to a straight configuration then it is possible to write the bending energy of the yarns as $E_b = B(m_1 - m_1^0)^2/2$, where the superscript ^{0} refers to the set yarn shape. The moment equilibrium condition for each yarn is simply $m_1 - m_1^0 + \lambda_4 = 0$ (see Table III), which can be interpreted in terms of physical reasoning, i.e. the couples are proportional to the difference in real and set curvatures.

In general, for fabrics under applied deformations, the set curvature control m^0 is not related to the actual control m. For fabrics in their relaxed state (no applied deformation), it can be shown that[6], to a good approximation, the set curvature is proportional to the real curvature of the yarn, i.e. $m_1^0 = \phi m_1$ where ϕ is the degree of set of the yarn. This relationship can be substituted, into the energy integral, U (Table III). However simply applying the relationship, $\partial H/\partial m_1 = 0$, which is to hold for an equilibrium solution would yield that the couples λ_4 are proportional to $(1-\phi)^2$. This contradicts intuitive reasoning about the expected proportional factor of $(1-\phi)$ between the couples and curvature.

This represents another example where the minimisation procedure cannot be simply applied without considering possible restrictions on the allowed variations. Just as restrictions were placed on the possible variations in yarn shape due to the differential constraints $\dot{z}_1^2 + \dot{z}_2^2 + \dot{z}_3^2 = L^2$, an assumption implicit in the approximating relationship $m_1^0 = \phi m_1$ must be introduced. The moment equilibrium equation in Table III shows that the couples are proportional to $(1-\phi)$, and hence the forces that produce the couples, i.e. the Q forces, must also be proportional $(1-\phi)$. This can be achieved by writing the point compression energy for this special case as $(1-\phi)C'g(d_0/d)$. Substituting this for the function h in the energy functional U in Table I, and relationship $m_1^0 = \phi m_1$ yields the correct relationship between the curvature and couple, i.e. $(1-\phi)m_1 + \lambda_4 = 0$.

The effect of yarn curvature constraints, similar to those applied previously in the bending of a plain-weave fabric, is to reduce fabric extensibility significantly[7]. Yarn extension is a very important mechanism in the extension of woven fabrics[6].

3.4 The relationship between geometric and mechanistic models of the plain weave.

To illustrate how the energy analysis can be used to derive a model that is partly mechanistic and partly geometric[3] in nature, the values of weave-crimp angle θ versus h/L (the ratio of crimp height to curvilinear yarn length) are plotted in Fig.5, due to Knoll[8], using the current energy equations in conjunction with the control constraints. Two curves were derived from the energy equations, one for the balanced square-weave construction ($L_2/L_1=1$) and one for a non-square weave where $L_2/L_1=1.5$). Both curves are compared with the corresponding curves derived for Peirce's mechanistic and geometrical models[9].

In order to facilitate comparison with Peirce's models, the curves shown in Fig. 5 derived from the energy equations were based on the assumption of incompressible yarns of circular cross-section, but the analysis can be readily extended to include compressed yarn with flattened lenticular cross-sectional shapes.

Considering the curves for the square-weave structure shown in Fig.5, it is clear that for values of $\theta<35°$ (corresponding to crimp values less than 11% approximately), the Peirce mechanistic model[9] based on single-point contact between warp and weft is applicable. As θ increases towards $35°$, i.e. as the weave-crimp approaches 11%, the warp and weft yarn systems make contact over a larger area and the curve derived from the energy equations approaches the limiting case of Peirce's geometrical model (as structural jamming is approached). Complete jamming of the structure occurs at the maximum or limiting value of h/L=0.478; the reaction forces Q between warp and weft threads (not shown) rises steeply as jamming is approached.

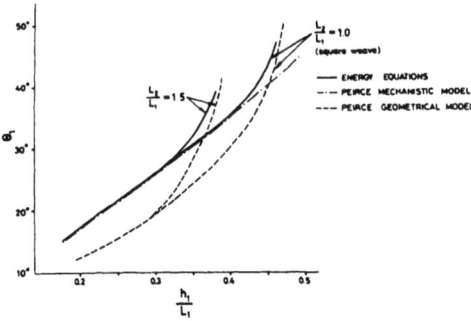

Fig. 5. The weave-crimp parameters θ and h/L derived from energy equations, Peirce's mechanistic and geometric models of the plain weave[8]

For the non-square weave (where $L_2/L_1=1.5$), the condition of continuous yarn contact is reached at lower values of crimp and h/L than for the square-weave.

Similar curves may be obtained for other crimp ratios (different values of L_2/L_1) and for pairs of fabric parameters other than θ and h/L.

It may be concluded that the basic energy/optimal-control formulation of woven fabric mechanics can be used to derive the elastic mechanistic model of the relaxed plain-weave structure for low to moderate values of the weave crimp. However, the introduction of an inequality constraint, limiting the maximum value of the yarn curvature to that governed by the shape of the crossing thread, means that the energy equations produce results corresponding to Peirce's purely geometrical model[9] as the weave crimp increases and the structure becomes very tight and approaches the jamming condition. The actual value of the weave crimp at which jamming occurs is very dependent on the ratio of curvilinear yarn lengths in warp and weft, L_2/L_1 and on the actual yarn cross-sectional shape.

4. CONCLUSION

The energy method presented here, based on the optimal control formulation[3], is an extremely powerful tool in the study of fabric mechanics. It is completely general in its application, but the method of solution of each particular problem is in each case similar. The first step consists of formulation of each of the strain energy contributions of the structure. The second step is to recognise the restrictions on the possible variations from the equilibrium solution i.e. to identify the admissable variations. Subsequently, routine application of the method presented here yields all the necessary equations, and a framework for a computational scheme.

In the general energy analysis no reference was made to the boundary conditions placed on the fabric. When it is applied to the plain knit fabric, these boundary conditions merely specify the tension under which the fabric is held, and do not add other constraints. For the plain-weave fabric under pure bending conditions, the boundary conditions need to be included in the minimisation process because they place restrictions on the admissable variations considered by the computing program. For the plain weave in tension, with compressible yarns, the compression strain energy is computed at a simple point (i.e. point forces are considered) and hence this strain energy can not be included in the energy integral but gives rise to an additional boundary condition.

A good way of checking the computed results is to verify that the area under the computed stress-strain curve equals the increase in total strain energy of the structure.

The use of inequality constraints on the control variable allows the introduction of geometric restraints on the yarn shape. It is shown that in this way the mechanistic (or force determined) model and the geometric model of the plain-weave, for example, are not mutually exclusive, but are both applicable at different levels of weave-crimp. The current energy method (optimal control formulation) has thus provided a unified basis for the detailed study of the mechanical properties of fabrics. When geometric restraints are applied to the yarns, the solution still yields all the necessary information about internal forces.

REFERENCES

(1) Goldstein, H., "Classical Mechanics", Addison Wesley, 1968.

(2) Sokolnikoff, I.S., "Mathematical Theory of Elasticity", McGraw-Hill, 1956, Chapter 7.

(3) Hearle, J.W.S. and Shanahan, W.J., J.T.I., 69, 81-99, 1978.

(4) Grosberg, P. in 'Structural Mechanics of Fibres, Yarns and Fabrics', Vol.1, by J.W.S. Hearle, P. Grosberg, and S. Backer, Wiley, New York, 1969, P343.

(5) Kirk, D.E., "Optimal Control Theory", Prentice Hall, 1970.

(6) De Jong, S., "A Study of Fabric Mechanics using Energy Methods of Analysis", Ph.D. Thesis, 1976, UNSW.

(7) De Jong, S. and Postle, R. Letter to the editor, J.T.I., in publication.

(8) Knoll, A.L., J.T.I., in publication.

(9) Peirce, F.T., J.T.I., 1937, 28, T45.

TEARING BEHAVIOR OF WOVEN FABRICS

J. Skelton

FRL, An Albany International Company
Dedham, MA 02026

INTRODUCTION

Tearing, which may be described as the sequential breakage of yarns or groups of yarns along a line through a fabric, is one of the most common types of failure in textile materials and in many cases serves to terminate their useful life. Consequently the tearing strength, which is usually measured as the force required to propagate a tear, may often be used to give a reasonably direct assessment of serviceability, and a fabric with a low tearing strength is generally an inferior product. In contrast to the tensile strength, which involves the force required to break a large number of yarns simultaneously, and which is relatively insensitive to yarn and fabric structural parameters, the tearing strength is considerably affected by changes in yarn and fabric geometry, the state of relaxation of the fibers, and their frictional characteristics. This paper deals with some of the more important factors influencing fabric tearing strength, and attempts to outline some of the theoretical considerations underlying the observed effects.

THEORETICAL ANALYSIS OF TONGUE TEAR TEST

Although many techniques have been proposed for the measurement of tearing strength, the most common method is probably the one known in the USA as a tongue tear, which employs the configuration shown in Fig. 1. In this test an initial longitudinal cut is made part way down the center of a strip of fabric and then the two "tails" thus formed are pulled apart, so that a tear proceeds through the uncut portion of the fabric. This test imitates quite closely the type of failure that occurs when one tears a piece of

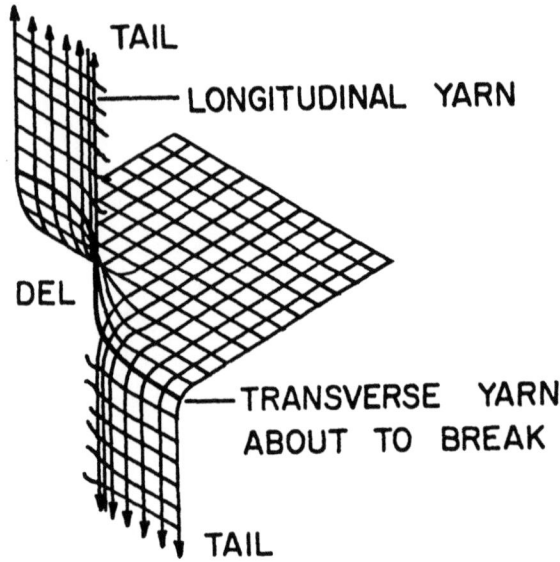

Fig. 1. Tear test configuration.

fabric or paper by hand. The mechanism of this type of tear has been analyzed from photographs made at various stages of a test [1,2], and it has been demonstrated that the force applied to the specimen first causes the longitudinal threads held in the grips of the tester to stretch and lose crimp, and then to start slipping along the transverse threads to the point from which the tear is about to extend to form a triangular opening in which there are no longitudinal threads. The increasing load causes the longitudinal threads to continue to slip and crowd together, so that the size of the "del" is increased, and the density of yarn crossover points increases in the small areas adjacent to the del. In these regions yarn slippage becomes increasingly more difficult and the load is progressively transferred to the transverse threads. As the load is increased further, some of the transverse threads eventually fail, and longitudinal threads again slip across each other to reform the del; the entire cycle is then repeated until the specimen is torn completely.

This qualitative description explains the characteristic occurrence of maxima and minima on a tearing test diagram, Fig. 2, but it was left to Taylor, in a remarkably comprehensive paper [3], to place the mechanics of tearing tests on a quantitative basis: the treatment that is presented below follows his pioneering analysis. We consider a situation in which the failing threads break singly, though as a result of the fabric distortion, several are under tension at any instant. Accordingly, the tearing strength of the cloth is usually a small multiple of the single-thread

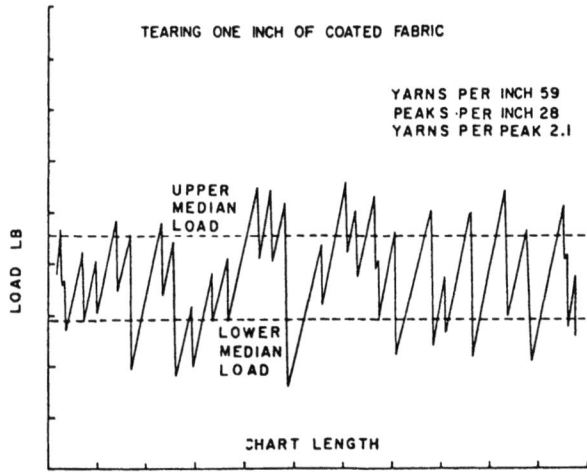

Fig. 2. Typical tearing test load trace.

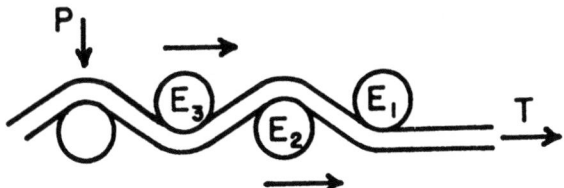

Fig. 3. Section along stressed yarn in a woven fabric.

strength [4]. In the region of the advancing tear the fabric is violently distorted in shear and in bending, and there is considerable interchange of crimp between the two sets of threads, but these factors are neglected in this treatment. Moreover, only changes in the longitudinal thread spacing are taken into account. Fig. 3 represents a cross section of a woven fabric. If a force T is applied to a thread as shown, and, at the same time, the front crossing thread E_1 is prevented from moving, the tensioned thread will, if T be great enough, slip over E_1 and draw up the fabric so that E_2 approaches E_1; slipping over two threads will now occur until E_3 approaches E_2. As the number of threads slipped over increases, so does the tension in the yarn until, eventually, when T = f, the yarn breaking strength, it will break. If the slipping thread were perfectly flexible and the crossing threads fixed cylinders, then the tension in the slipping thread would be multiplied by a constant factor β at each crossing thread. Because of their rigidity, however, the yarns exercise a mutual pressure P at each intersection and, on this account, there is an additional frictional force, αP, at each intersection. Thus in a space between two crossing threads where slipping is about to occur, the tension will be αP, in the next space, $\alpha \beta P + \alpha P$, and, in succeeding spaces $\alpha \beta^2 P + \alpha \beta P + \alpha P$ and so on. If slipping has occurred at N crossing places, the tension at the thread end is:

$$\alpha P(1 + \beta + \beta^2 \ldots \beta^N) = \frac{\alpha P(\beta^{N+1} - 1)}{\beta - 1} \tag{1}$$

and, if the breaking strength of the yarn is f, we can write:

$$f = \frac{\alpha P(\beta^{N+1} - 1)}{\beta - 1} \tag{2}$$

to establish the greatest possible value of N. The tension f will pull out of the fabric a length of thread NS where S is the average distance by which the space between the threads may be reduced (measured along the slipping thread); the movement of the crossing threads towards each other is assumed to be uniform, and determined principally by the structural geometry of the fabric.

In the "del" the thread that is about to break has slipped out a distance NS from each side, and the maximum number of parallel threads in the ladder of the del will be NSt + 1. The tension in the outermost thread is f and the tension on the first thread which does no slip is zero. Assuming a linear progression of tension along the threads in the ladder, the total force exerted on each of the tails will fluctuate as the threads break one by one, and the maximum force is (NSt + 2)f/2; this is the force which must be supplied to tear the fabric in a tear test.

It is necessary now to consider the values of α and β. This may be done by measuring the force f_s necessary to withdraw a known length of thread from the woven fabric, which is given by:

$$f_s = \frac{\alpha P(\beta^{n+1} - 1)}{\beta - 1} \tag{3}$$

where f is the measured force to cause slippage of a thread over n crossing threads.

Thus we have:

$$\frac{\beta^{N+1} - 1}{\beta^{n+1} - 1} = f/f_s. \tag{4}$$

If it can be assumed that both β^{N+1} and β^{n+1} are much larger than unity, then:

$$N = \frac{\log f/f_s}{\log \beta} + n \tag{5}$$

and the tear strength T is given by:

$$T = \left\{ \left[\frac{\log f/f_s}{\log \beta} + n \right] St + 2 \right\} f/2 \tag{6}$$

a relation that, for convenience, may be put in the form:

$$\frac{R-1}{t} = S/2 \left\{ \frac{\log f/f_s}{\log \beta} + n \right\} \tag{7}$$

where $R = T/f$.

The foregoing analysis applies to the plain weave only. To modify for any other weave, it is assumed that, in place of:

$$f = \alpha P \frac{(\beta^{N+1} - 1)}{\beta - 1}, \tag{8}$$

it is permissible to write:

$$f = \alpha P \frac{(\beta^{N/w+1} - 1)}{\beta - 1}, \tag{9}$$

where w is the number of crossing threads in the weave repeat divided by the number of times the longitudinal threads change from face to face in this distance. This leads to:

$$\frac{R-1}{t} = S/2 \left\{ w \frac{\log f/f_s}{\log \beta} + n \right\}. \tag{10}$$

Taylor showed that there was a reasonable correlation between $(R - 1)/t$ and $\log f/f_s$ for a range of woven cotton fabrics; he also showed that the parameter s was of the order of the free space between the yarns of the fabric, and that the parameter β was of the order of $e^{\mu\theta}$ where μ is the coefficient of friction between the yarns and θ the angle of contact at crossing places.

Taylor's theoretical analysis has obvious shortcomings, but it forms an extremely useful basis for many of the explanations of the observed variations in tear strength. Some of these variations are discussed in detail below, at least in a qualitative way. A useful survey of this topic also appears in reference [5].

FACTORS AFFECTING TEARING STRENGTH

Yarn Strength

It is explicit in the analysis, and has been amply confirmed experimentally, that the tearing strength is at least roughly proportional to the single yarn strength, at least in fabrics in which the damage propagates by the rupture of single yarns. If the fabric structure is such as to encourage the breaking of small groups of yarns, then the breaking strength of the group becomes the controlling factor.

Yarn Breaking Elongation

This parameter does not appear explicitly in the analytical expression, but it has obvious influence on the interyarn spacing at the point of break, and hence on the size of the del and the total number of yarns involved in the break. Accordingly, a high elongation at break normally produces a high tearing strength.

Yarn Twist

Increasing yarn twist affects the tensile strength directly, producing a maximum in the strength of staple yarns and continuous filament yarns made from materials with very low breaking elongation, and a continuous decrease in other continuous filament yarns; these changes are often reflected in tearing strength variations. However, increasing twist produces a more compact yarn with reduced diameter, and hence offers the possibility of a greater amount of free space for structural readjustment of the fabric near the del (increased S in Taylor's analysis). The problem is complicated if finishing treatments or coatings are used, since the degree of penetration and adhesion are also affected by the twist.

Fabric Density

The tightness of the fabric structure has a large effect on the tearing strength. In a densely woven structure there is less free space, less ultimate deformation potential, and hence a lower tearing strength. Additionally, a tight construction favors high interyarn frictional forces, and a further expected reduction in tearing strength.

Weave Pattern

The weave pattern influences tearing strength by control of the relative frequency of yarn crossing points in the fabric. This has a direct effect both on the ease of deformation of the fabric, and on the number of yarns breaking together. Generally multiple basket weaves, which have a low density of crossing points and a number of identical yarns formally grouped to give the equivalent of single, larger yarns, have been found to perform well.

Finish

The influence of finishing treatments, exclusive of any major geometrical effects is normally through modification if the level of frictional interaction (Taylor's β factor). Increasing the level of frictional restraint, via scouring, for example, decreases the tearing strength, and decreasing the level, either by reduction of the coefficient of friction or the normal force usually leads to an increase in tearing strength. Application of a coating material can be considered as a rather special case of finishing treatment, in which the yarns are essentially completely immobilized, and thus inevitably produces a large reduction in tearing strength.

CRACK PROPAGATION

The discussion up to this point has centered on situations in which the fabric responds to a steadily increasing load by tearing when the load reaches a critical level. When the load is again reduced below this critical value, the tearing does not continue. In contrast, another type of tear failure, usually called crack propagation, has also been observed in which a small region of localized damage introduced into a fabric while it is under load produces a tear that propagates rapidly across the fabric and leads to almost immediate failure [6]. Some experimental characteristics of this type of failure are described below, and a preliminary theoretical investigation of the phenomenon is outlined.

Fig. 4 shows the results of a series of measurement on two coated fabrics of different weights and constructions. There is a critical tension, T_c, below which crack will not propagate from a deliberately induced failure site. At tensions above the critical tension the crack propagates at high speed (several hundred feet per second is typical) and the propagation velocity appears to vary linearly with increments of tension up to the breaking load of the fabric. The critical tension is a smaller fraction of the breaking load for lightweight fabrics and the rate of change of propagation velocity with tension appears to be inversely related to fabric areal density, as might be expected in a kinetic system of this type, in which inertial forces will play a part in determining the energy balance around the propagating crack.

The factors controlling the phenomenon can be discussed in a general way. When a yarn in a fabric is strained to the point where it breaks, energy is suddenly released due to the rapid recovery of the yarn to a condition of stress equilibrium. This released energy, E_R, is distributed in a number of ways:
   As kinetic energy, due to the resulting movement of elements of the structure, E_K;
   As shear energy, due to the shearing deformation of the structure which results from the sudden failure of one of its load-bearing elements, E_S;
   As frictional heat, E_H;
   As an increase in strain energy, due to the stretching of one or more yarns, $E_\varepsilon = \Delta E_1 + \Delta E_2 + \ldots$ ahead of the apex of the crack as a result of the fabric shearing.

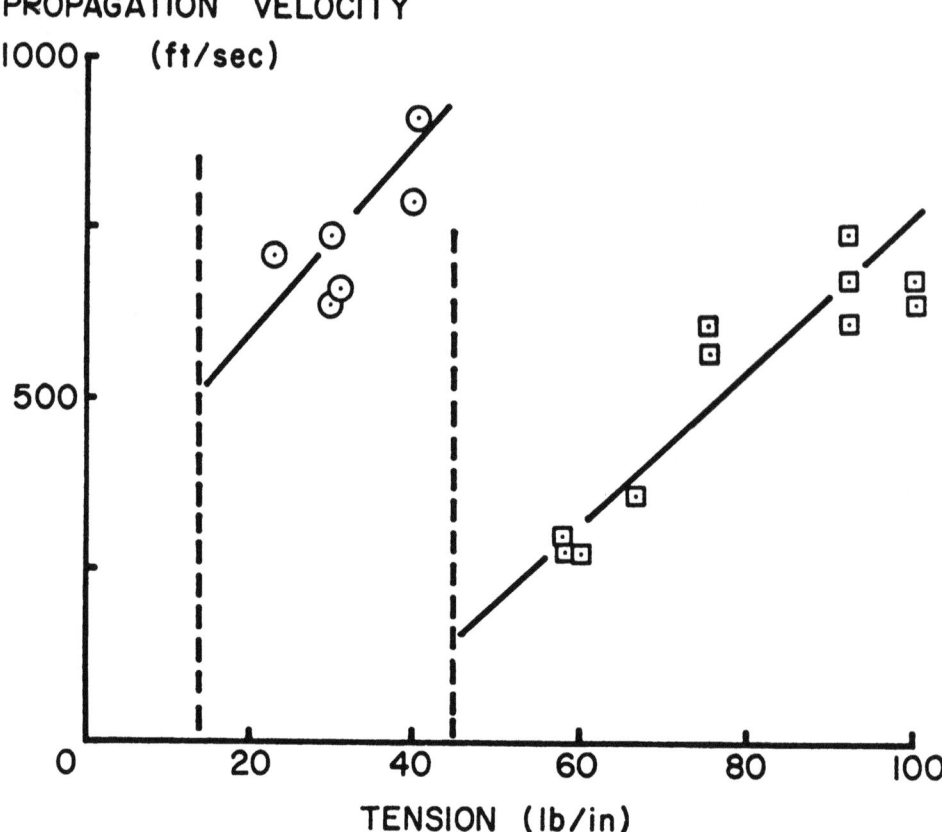

Fig. 4. Crack propagation velocities.

At any given stress level, σ, we can expect that the terms E_K, E_S, and E_H will be essentially constant with a sum that is also constant and is denoted by K_σ, and can also anticipate that K_σ will probably be dominated by the kinetic energy term. Accordingly we can write for the condition in which a crack will just fail to propagate:

$$E_R \leqslant K_\sigma + E_\varepsilon, \qquad (11)$$

since each yarn, as it moves in a stepwise manner closer to the apex of the crack, absorbs a total amount of energy equal to E_ε.

The calculation of the recovery energy requires care in order to account for the effects of the geometrical restraints in the force. When a yarn in a fabric is broken, it recovers differently from a free yarn, since it is restrained along its length by the crossing yarns. Because of this, it is neither strained uniformly

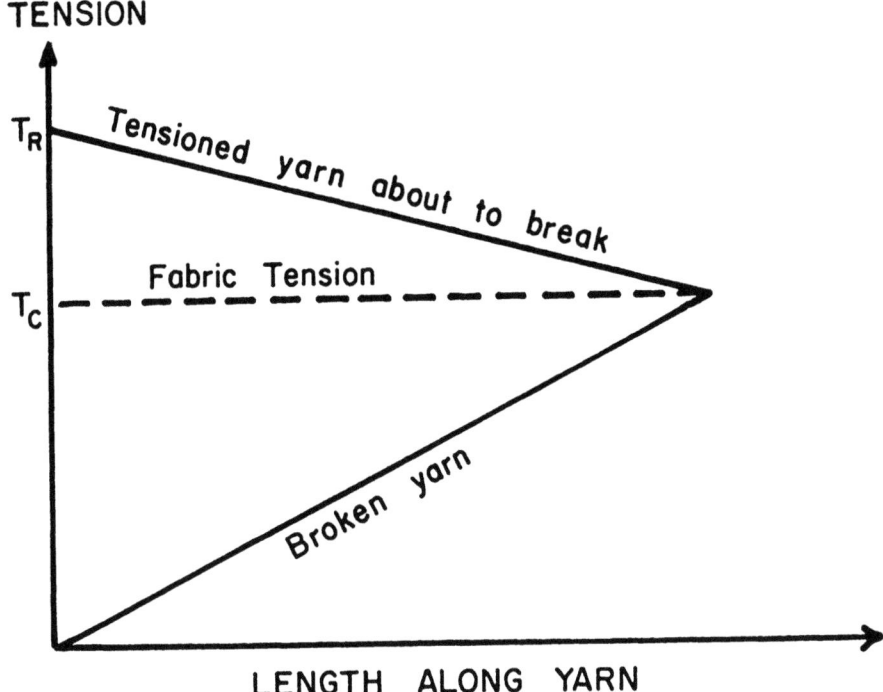

Fig. 5. Distribution of tensions in loaded fabrics.

to rupture, nor is the strain uniform after rupture. This situation is depicted in Fig. 5 in which the assumption is made that tensions are distributed linearly along the yarn away from the point of rupture. The top line represents the tension distribution in a yarn about to break. This falls from the rupture tension T_R, to the tension resulting from the steady loading of the fabric, T_C, over the length, S, which depends upon the shear stiffness of the fabric. Immediately after rupture, the tension distribution is represented by the bottom line, rising from zero at the point of rupture to T_C over the same length S of yarn. Thus, tension changes vary linearly from T_R to zero and, assuming a linear stress-strain curve, the strain changes from ε_R to zero over length S.

The work done in extending a linearly elastic yarn from an untensioned state to a total stretched length S with a final distribution of tension extension along the stretched length given by $T_s = f_1(S)$, $\varepsilon_s = mf_1(S)$ can be shown to be:

$$E = \int_0^S \frac{T_s \varepsilon_s}{2(1 + \varepsilon_s)} \cdot ds \qquad (12)$$

where m is the tensile modulus of the material, assumed to be constant, and $ds = ds_0(1 + E_s)$ for the unstretched elemental length ds_0.

For a uniform tension T along the yarn this expression reduces to:

$$E_U = mT^2 s/2(1 + mT), \tag{13}$$

while for a distribution such that the tension is zero at one end of the specimen and increases linearly with distance along the yarn to a value T at the other end, the work done reduces to:

$$E_D = \frac{mT^2}{2S} \int_0^S \frac{s^2 ds}{S + msT}. \tag{14}$$

Carrying out the integration and simplifying the ratio E_D/E_U can be shown to be approximately:

$$E_D/E_U = 1/3(1 + \varepsilon_T/4) \tag{15}$$

where ε_T is the elongation produced by uniform tension T. Thus the strain energy of a yarn with a linear distribution of tension up to tension T is approximately one third of the strain energy of the same stretched length of yarn with uniform tension T, for all values of ε_T appropriate for normal textile materials ($\varepsilon_T/4 \ll 1$).

The work needed to extend a yarn to unit stretched length having a uniform tension T_R is given by Equation (13) as:

$$E_U = 1/2 \frac{mT_R^2}{1 + mT_R}. \tag{16}$$

Therefore, the work needed to extend the same yarn to a tension and extension linearly distributed between the limits $0 \rightarrow T_R$, $0 \rightarrow \varepsilon_R$ is given approximately by

$$E_D = 1/6 \frac{mT_R^2}{1 + mT_R}. \tag{17}$$

This quantity is independent of the fabric tension T_c, and is, therefore, a constant which is independent of stress, and depends only upon the stress-strain properties of the yarns involved. If this yarn breaks, recovery energy is released which is some fraction, f, of the stored strain energy. Thus,

$$E_R = f/6 \frac{mT_R^2}{1 + mT_R}. \tag{18}$$

Part of the recovery energy, $(E_R - K_\sigma)$ of Equation (11), is used in straining yarns ahead of the apex of the crack. This is represented by E_ε in Equation (11). As was shown, this is the total energy required to change the tension in the first yarn from a uniform T_c to the distribution represented by the top line in Fig. 5. The magnitude of this tension change varies linearly over a length S from zero to $(T_R - T_C)$, and over this same length, the strain change varies from zero to $(\varepsilon_R - \varepsilon_C)$.

The case of a uniformly tensioned linear elastic yarn whose tension is varied from T_C to T_R is represented in Fig. 6. The work required to accomplish this is given by the area of the shaded trapezoid. For a yarn of unit stretched length, this can be derived from Equation (13) as:

$$E_U = 1/2 \left[\frac{mT_R^2}{1 + mT_R} - \frac{mT_c^2}{1 + mT_c} \right]. \tag{19}$$

The corresponding expression for the case of a linear distribution of tension as given in Fig. 4 is approximately 1/3 of this amount, as shown in Equation (15).

$$E_D = 1/6 \left[\frac{mT_R^2}{1 + mT_R} - \frac{mT_c^2}{1 + mT_c} \right]. \tag{20}$$

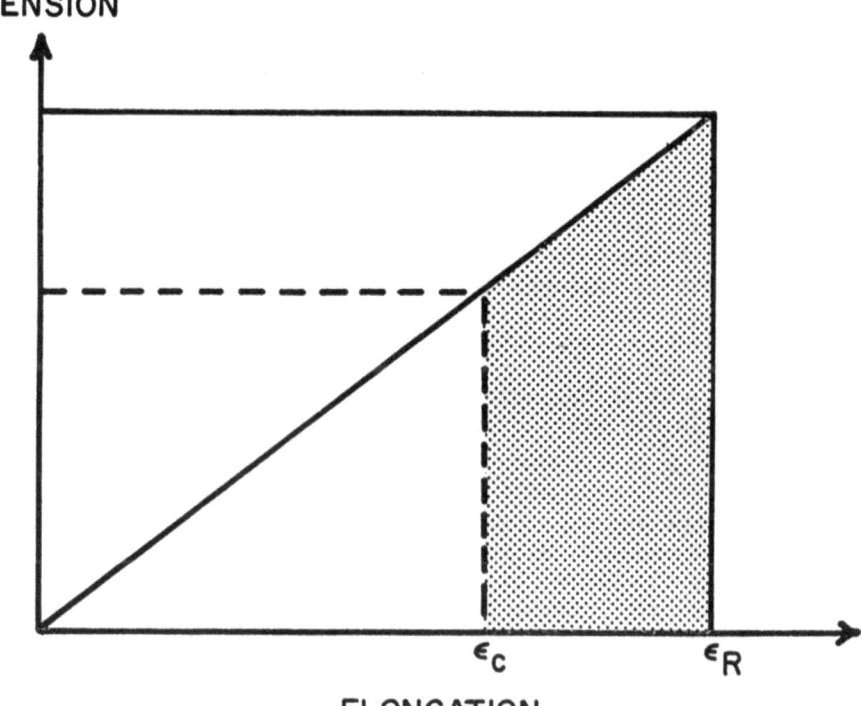

Fig. 6. Strain energy components in stressed yarn.

Thus, for a fabric in which all the yarns are identical, we may rewrite Equation (11) as:

$$E_R < K_C + 1/6 \left[\frac{mT_R^2}{1 + mT_R} - \frac{mT_C^2}{1 + mT_C} \right]. \quad (21)$$

Combining Equations (18) and (21) and simplifying, we obtain

$$\frac{mT_C^2}{1 + mT_C} < 6K_C + (1 - f) \frac{mT_R^2}{1 + mT_R}. \quad (22)$$

This equation gives the critical tension in terms of parameters which characterize the pertinent deformation properties of the fabric m, the tensile modulus; f, the energy resilience (recovery energy expressed as a fraction of strain energy to rupture for a yarn); T_R, the rupture tension; and K_C, the sum of the kinetic energy, heat energy and shear energy at stress T_C, the critical tension. Accurate values of all of these parameters are currently not available.

Equation (22) has not yet been useful for calculation of T_C. However the functional significance of the parameters can be understood, and this permits a useful insight into the nature of the phenomenon and the characteristics of the fiber or the structure on which it depends and work designed to measure or estimate some of the important parameters of the deformation is underway.

CONCLUSIONS

Two types of tearing behavior of woven fabrics are discussed corresponding to two distinct failure configurations. The first type simulates the tearing that occurs when an ever-increasing separation is imposed on the two sides of an extending tear, while the other is appropriate to the self-propagating failure that can occur when a tear is initiated in a fabric that is subject to a uniform distant stress field. Some theoretical considerations governing the two types of failure are described, and related to practical observations of the effects of variations in yarn and fabric elastic and constructural parameters.

REFERENCES

[1] C. M. Krook and K. R. Fox, Text. Res. J., 1945, 15, 389.
[2] N. A. Teixeira, M. M. Platt, and W. J. Hamburger, Text. Res. J., 1955, 25, 828.
[3] H. M. Taylor, J. Text. Inst. 1959, 50, T161.
[4] N. J. Abbott, T. E. Lannefeld, L. Barish and R. J. Brysson, J. Coated Fib. Mat., 1971, 1, 4.
[5] L. M. Olson, Technical Report DAAG53-76-0141, U. S. Army Mobility Equipment R&D Command, Ft. Belvoir, VA, 27060.
[6] N. J. Abbott and J. Skelton, J. Coated Fib. Mat., 1972, 1, 234.

CLASSICAL ELASTICA THEORY AND ITS GENERALIZATIONS

Milos Konopasek

School of Textile Engineering, Georgia Institute
of Technology, Atlanta, Georgia 30332

ABSTRACT. The mechanics of one-dimensional continuum belongs to the foundations of the theoretical and applied mechanics of textile structures and processes. The presentation of the statics of one-dimensional continuum is based on the concepts of planar and spatial elastica as developed respectively by Euler and Kirchhoff. This approach facilitates an analysis of practically important real situations which do not fit the assumptions of the classical elastica theory (these are summarized in the last part of the paper). It also fosters the numerical treatment, computation and applications described in the following papers of this volume.

1. INTRODUCTION

The title of this volume, 'Mechanics of flexible fibre assemblies', implies materials and structures exhibiting large bending deformations. The geometry of large deformations introduces nonlinearities into the formulations of the relationships between the original shape, loading conditions, geometric constraints, mechanical properties of the material, and the deformed shape even in cases of perfect elasticity and linear load-deformation characteristics. As if there was not enough trouble with this so called geometric nonlinearity, the real materials used in flexible structures (and, naturally, the structures themselves) also display nonlinear load-deformation relationships making particular problem physically nonlinear. In addition, these materials and structures usually do not exhibit perfectly elastic behaviour. All this makes the solution of any research or engineering design problems in this field a formidable task.

On the other hand, only fairly slender bodies can undergo large flexural deformations; high aspect ratio makes it possible to consider only one independent geometric variable (most conveniently the distance along the central line or longitudinal axis of the deformed body). Fig. 1 makes this point clear: the field of our interest is represented by boxes 1.2 and 1.3. Consequently, the general theory of large deflections and any particular problem may be formulated in terms of a set of ordinary differential equations, which, regardless of the degree of nonlinearity, may be solved by standard numerical procedures on a routine basis as opposed to problems in boxes 2.2, 2.3 and 3.3 which are represented in the general case by sets of unmanageable nonlinear partial differential equations.

Number of dependent geometric variables			
1	2	3	
1.1 Tensile or torsional deformations of fibres and yarns	1.2 Bending deformations of fibres and yarns in plane	1.3 Bending and torsional deformations of fibres and yarns in space	3
	2.2 Tensile and shear deformations of sheets (fabrics) in plane	2.3 Tensile, shear and bending deformations of sheets (fabrics) in space	2
		3.3 Complex deformations of fibres and fibre assemblies in space	3

Number of independent geom. variables

Fig. 1. Continuum models in mechanics of fibre assemblies

Keeping but one independent geometric variable does not necessarily mean that we must neglect the deformations in the other directions. We can, when solving the one-dimensional continuum problem, take into account the interdependencies of the curvature of the central line of a rod and the flattenning of its cross-section, the tensile modulus and the curvature, the torsion and the longitudinal contraction, etc. It does mean, however, that we have to obtain the information about these interdependencies separately, and feed it into the one-dimensional problem as an assumption, instead of generating it in the course of solving a three-dimensional problem.

During its nearly 300 year history, the mechanics of one-dimensional continuum attracted the attention of scores of prominent scientists, many of whom are better known for other contributions to mathematics, physics and mechanics, such as members of Bernoulli family, Euler, Poisson, Lagrange, Saint-Venant, Love, Born, Popov, Green, Truesdell - to mention just a few. It is not our aim to survey the exciting developments in this field; a comprehensive analysis of the work on large deflections up to 1788 was given by Truesdell [1] and later developments may be easily traced down from the most recent comprehensive account on the state of the art by Antman [2].

Antman's work represents the most general and rigorous treatment available of nonlinear problems in the mechanics of one-dimesional continua. However, the communication between the general theories of one-dimensional continua and the mechanics of fibres and fibre assemblies is yet to be established. At present the most useful aproach for us is to start with a classical Euler's planar and Kirchhoff's spatial elastica, and generalize from there.

2. EULER'S PLANAR ELASTICA

2.1 Bernoulli-Euler law

The starting point of our analysis is the Bernoulli-Euler law according to which the bending moment at any point of the central line (or longitudinal axis) of a slender body is proportional to the increment of the line curvature caused by that moment:

$$M = A \cdot k \qquad (1)$$

where M is the bending moment;
$k = 1/\rho = d\phi/ds$ is the central line curvature;
ρ is the radius of curvature;
$\tan\phi$ is the slope of the central line;
A = EI is the bending rigidity;
E is the Young modulus;
I is the moment of inertia of the cross-section of the body.

The Bernoulli-Euler law is equivalent to the assumption that the cross-section of the body remains unchanged, plane and normal to the central line and that only the body's resistance to extension and compression parallel to the central line is responsible for balancing the bending moment (i.e. the effects of the transverse tensile deformations and the shear deformations are neglected).

2.2 Small rotations

Let $y=f(x)$ define the shape of the central line of a deformed slender body in cartesian coordinates x,y. The curvature of the central line is:

$$k = \frac{d^2y/dx^2}{(1 + (dy/dx)^2)^{3/2}} \qquad (2)$$

If the slope of the tangent to the curve (or the rotation of the cross-section plane) dy/dx is very small, its square may be neglected and the substitution for k in (1) yields the linear differential equation:

$$M = A\, d^2y/dx^2 \qquad (3)$$

The solution of this equation in terms of a 2-, 3- or 4-order polynomial is easy to obtain; it is widely used in engineering calculations concerning relatively rigid beams, bars etc. The accuracy of the solution of the equation (3) deteriorates rapidly with increasing deflections as may be seen in Fig. 2 (from Bruggisser [3]).

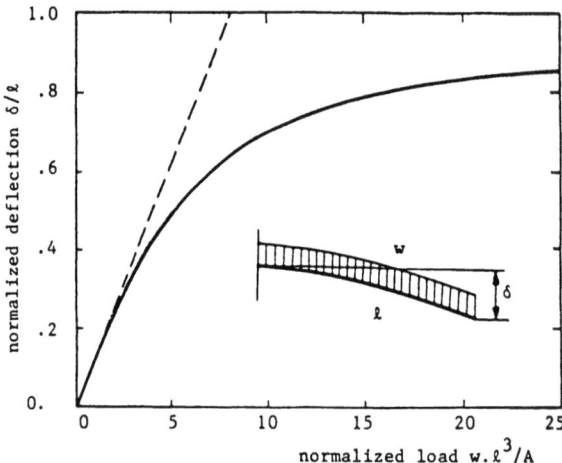

Fig. 2. Deflection of horizontal cantilever with distributed load w (——— heavy elastica, — — — small deflection theory)

2.3 Euler's solution

The first solution of equation (1) based on exact definition of the curvature and the principle of minimum energy of deformation is due to Euler [4]. He used the problem as a case study for the variational calculus he discovered. The energy of bending deformation in a slender body of length L is:

$$U = \frac{A}{2} \int_0^L k^2 ds \qquad (4)$$

The functional to be minimized was

$$J = \int_0^L k^2 ds = \int_0^X \frac{(d^2y/dx^2)^2}{[1+(dy/dx)^2]^{5/2}} dx = \int_0^X F(y'',y') dx \qquad (5)$$

for a given length

$$\int_0^L ds = \int_0^X \sqrt{1+(dy/dx)^2} dx = L \qquad (6)$$

Euler solved the problem by discovering a family of periodic planar curves called elasticas and introduced their classification recognizing 9 classes, the first and ninth class being a straight line and a circle, corresponding, respectively, to no deformation and pure bending (Fig. 3).

2.4 Analytical solution

Euler's techique of using the variational calculus and expressing the solution in terms of a power series is not easy to apply. Fig. 4 shows an alternative derivation of differential equations of elastica from the condition of moment equilibrium of a differential element ds:

$$\frac{d^2\phi}{ds^2} = -\frac{F}{A}\sin\phi \; ; \quad \frac{dx}{ds} = \cos\phi \; ; \quad \frac{dy}{ds} = \sin\phi \qquad (7)$$

The first integral of the first equation is

$$\phi' = 2\sqrt{F/A\,(C-\sin^2\phi/2)} \qquad (8)$$

where C is an integration constant.

The rest of the integration cannot be done in closed form. The slopes and coordinates of elasticas can be expressed in terms of elliptical integrals. The substitutions and resulting

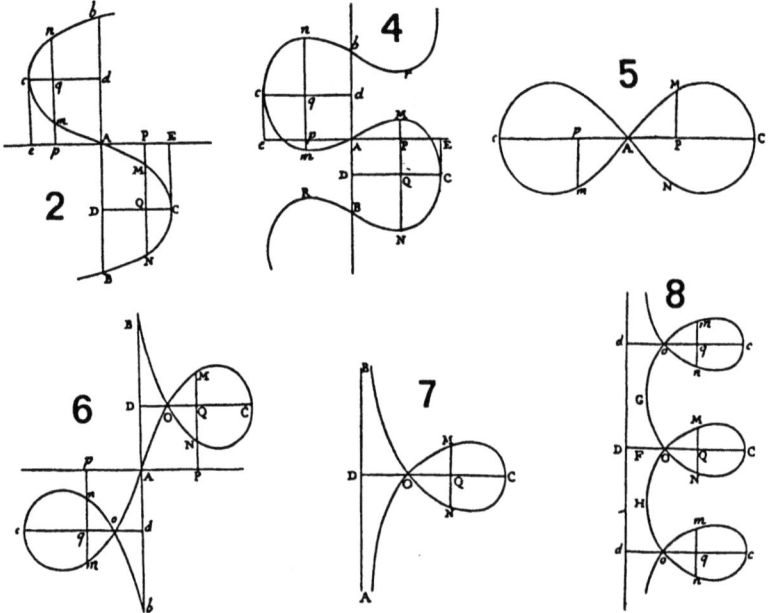

Fig. 3. Planar elasticas, from [4]
(numbers denote the class of elastica)

Moment equilibrium equation:

$-M + M + dM + Fds.\sin\phi = 0$

or

$$\frac{dM}{ds} = -F\sin\phi \qquad (*)$$

Bernoulli-Euler law:

$$M = A.k = A\frac{d\phi}{ds} \qquad \text{or}$$

$$\frac{dM}{ds} = A\frac{dk}{ds} = A\frac{d^2\phi}{ds^2} \qquad (**)$$

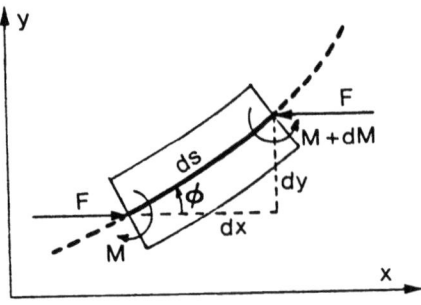

After substitution for dM/ds from (*) into (**): $\dfrac{d^2\phi}{ds^2} = -\dfrac{F}{A}\sin\phi$

Fig. 4. Derivation of the 1-st differential equation
of planar elastica

expressions are different for inflectional elasticas (i.e. elasticas of the 1-st through 6-th classes with an inflection point and alternating signum of curvature) and non-inflectional elasticas (i.e. elasticas of the 8-th and 9-th classes without inflection points and with constant signum, positive or negative, of the curvature). Only for the transition case of the elastica of the 7-th class with C=1 is an analytical solution in closed form available.

Any particular solution of a planar elastica may be defined by four parameters: 1. integration constant C expressing the position among the elasticas of the 1-st through 9-th class, 2. scale factor (amplitude), and 3.-4. two arc-length coordinates (the starting and terminating points of the elastica segment).

The use of analytical solutions for elastica in the form of elliptical integrals was common in the precomputer era and it still survives although it offers no advantage over the direct numerical solution of the nonlinear differential equations (7). Any generalization (e.g. distributed force, original curvature, nonlinear moment-curvature relationship) rules out analytical solution altogether and makes numerical integration of differential equations the only option available.

3. KIRCHHOFF'S SPATIAL ELASTICA

3.1 Definition of problem, coordinate systems

Let us consider a situation where a slender body is subjected to deformation by bending moments, the effects of which were discussed in a previous section, and also by torque, which will obviously cause the central line to become a spatial curve. The original formulation of this problem is due to Kirchhoff [5]. Love [6] was probably the most influential promoter of the concept of spatial elastica, although we found Popov's [7] interpretation to fit better the application and computation requirements. We will develop here from scratch a mathematical model expressing the relationship between the mechanical properties of the deformed body, external loads, geometric constraints, and the shape of its central line.

We introduce two orthogonal right-hand coordinate systems: fixed coordinate system x,y,z for definitionn of the shape of the curve representing the body, and coordinate system u,v,w with an origin moving along the elastica and with axes following the local directions of two main axes of inertia of the cross-section of the body (u,v), and tangent to the elastica (w).

3.2 Model composition, generalized Bernoulli-Euler law

The model will consist of the following components:

(a) The bending moment-curvature and torque-torsion equations constituting generalized Bernoulli-Euler law are

$$M_u = A \cdot p \quad ; \quad M_v = B \cdot q \quad ; \quad M_w = C \cdot r \qquad (9)$$

where M_u, M_v and A, B and p, q are respectively bending moment and rigidities and curvatures with respect to the two main axes of inertia of cross-section; M_w, C and r are respectively torque, torsional rigidity and torsion of the body.

(b) Equations of equilibrium of moments and forces acting on a differential element of the body (see Fig. 5)

$$\overline{M}(s+ds) - \overline{M}(s) + \overline{m} \cdot ds + (\overline{t} \times \overline{F}) ds = 0$$
$$\overline{F}(s+ds) - \overline{F}(s) + \overline{f} \cdot ds = 0 \qquad (10)$$

where $\overline{M}$ and $\overline{m}$ are the concentrated and distributed moments, $\overline{F}$ and $\overline{f}$ are the concentrated and distributed forces.

(c) Equations representing the curvature-to-orientation and orientation-to-coordinates transition.

An analysis and summary of differential-geometric aspects of (c) is given in the next sections.

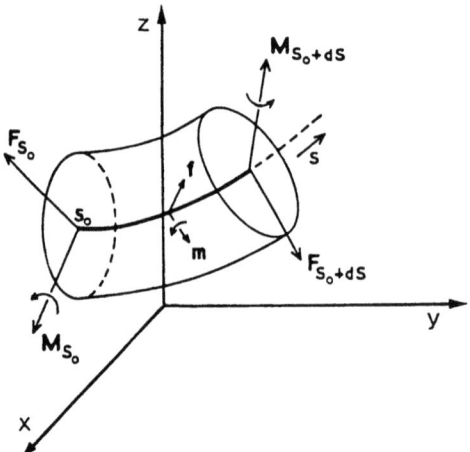

Fig. 5. Differential element of spatial elastica

3.3 Spatial curves, summary

The shape of a spatial curve (including any spatial elastica) may be defined by a radius-vector

$$\bar{r} = \bar{r}(s) \tag{11}$$

or by equivalent three scalar coordinate functions

$$x = x(s) \quad ; \quad y = y(s) \quad ; \quad z = z(s) \tag{12}$$

There are three important mutually orthogonal directions defined at any regular point of a spatial curve by a triad of unit vectors: tangent $\bar{t}$, principal normal $\bar{n}$ and binormal $\bar{b}$,

$$\bar{t} = \bar{r}' \quad ; \quad \bar{n} = t'/\kappa; \quad \bar{b} = \bar{t} \times \bar{n} \quad . \tag{13}$$

They form so called principal trihedron and the pairs of them define the following three distinct mutually orthogonal planes: normal, osculating and rectifying (see Fig. 6).

The vectorial equations of these lines and planes and their scalar equivalents are given in Table I.

The rates of rotation of the tangent and binormal, when moving along the curve, define, respectively, curvature κ and torsion τ of the curve:

$$\bar{t}' = \kappa \bar{n} \quad , \quad \bar{b}' = -\tau \bar{n} \tag{14}$$

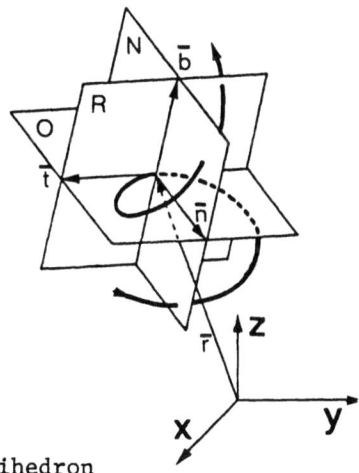

Fig. 6. Elements of principal trihedron

Table I
Attributes of spatial curves

	Vectorial equations	Scalar equations
Tangent	$\bar{R} = \bar{r} + \lambda\bar{r}'$	$\dfrac{X-x}{x'} = \dfrac{Y-y}{y'} = \dfrac{Z-z}{z'}$
Principal normal	$\bar{R} = \bar{r} + \lambda\bar{r}''$	$\dfrac{X-x}{x''} = \dfrac{Y-y}{y''} = \dfrac{Z-z}{z''}$
Binormal	$\bar{R} = \bar{r} + \lambda(\bar{r}' \times \bar{r}'')$	$\dfrac{X-x}{\begin{vmatrix}y' & z' \\ y'' & z''\end{vmatrix}} = \dfrac{Y-y}{\begin{vmatrix}z' & x' \\ z'' & x''\end{vmatrix}} = \dfrac{Z-z}{\begin{vmatrix}x' & y' \\ x'' & y''\end{vmatrix}}$
Normal plane	$(\bar{R} - \bar{r})\bar{r}' = 0$	$x'(X-x)+y'(Y-y)+z'(Z-z) = 0$
Osculating plane	$(\bar{R} - \bar{r})\bar{r}'\bar{r}'' = 0$	$\begin{vmatrix}X-x & Y-y & Z-z \\ x' & y' & z' \\ x'' & y'' & z''\end{vmatrix} = 0$
Rectifying plane	$(\bar{R} - \bar{r})\bar{r}'' = 0$	$x''(X-x)+y''(Y-y)+z''(Z-z) = 0$

R,X,Y,Z are radius-vector and coordinates on lines and planes;
λ is free parameter;
','' denote differentiations with respect to arc coordinate s.

These two equations and an equation for $\bar{n}'$ derived from $\bar{b} = \bar{t} \times \bar{n}$ constitute so called Serret-Frenet equations

$$\bar{t}' = \kappa\bar{n}, \quad \bar{n}' = -\kappa\bar{t}+\tau\bar{b}, \quad \bar{b}' = -\tau\bar{n} \tag{15}$$

which may be interpreted as differential equations of a curve. Because the $\bar{t},\bar{n},\bar{b}$ are othogonal unit vectors, the shape of the curve is defined uniquely by   and   as functions of arc length:

$$\kappa = \kappa(s) \quad \text{and} \quad \tau = \tau(s) \tag{16}$$

These are called natural equations of curve.

The rotation of the principal trihedron t,n,b moving with unit velocity along the curve may be expressed by so called Darboux's vector:

$$\bar{\Omega} = \kappa\bar{b} + \tau\bar{t} \tag{17}$$

This yields alternative expressions for the derivatives of the vectors $\bar{t},\bar{n},\bar{b}$:

$$\bar{t}' = \bar{\Omega} \times \bar{t} \; ; \qquad \bar{n}' = \bar{\Omega} \times \bar{n} \; ; \qquad \bar{b}' = \bar{\Omega} \times \bar{b} \; . \tag{18}$$

The equations (15) through (18) provide definitions of curvature and torsion in terms of radius-vector and its derivatives

$$\kappa = |\bar{t}'| = |r''| \; ; \qquad \tau = |\bar{b}'| = (r'.r''.r''')/(r''.r'') \tag{19}$$

and their equivalents in terms of coordinate functions:

$$\kappa = \sqrt{x''^2 + y''^2 + z''^2} \; ; \qquad \tau = \frac{1}{\kappa^2} \begin{vmatrix} x' & y' & z' \\ x'' & y'' & z'' \\ x''' & y''' & z''' \end{vmatrix} \tag{20}$$

3.4 Geometric vs. physical curvature and torsion

The relationships between the unit vectors $\bar{n}, \bar{b}, \bar{t}$ and the unit vectors $\bar{u}, \bar{v}, \bar{w}$ introduced earlier are

$$\bar{n}.\bar{u} = \bar{b}.\bar{v} = \cos\lambda \; ; \qquad \bar{t} \equiv \bar{w} \tag{21}$$

The vector of rotation of the u,v,w coordinate system, when moving along the elastica, may be expressed in a form similar to the Darboux's vector (17):

$$\bar{\omega} = \bar{p} + \bar{q} + \bar{r} = p\bar{u} + q\bar{v} + r\bar{w} \tag{22}$$

It also may be expressed as a sum of Darboux's vector and an additional rotation around the tangency vector

$$\bar{\omega} = \bar{\Omega} + \lambda'\bar{t} = \kappa\bar{b} + \tau\bar{t} + \lambda'\bar{t} \tag{23}$$

where λ, as follows from (21), is the angle between $\bar{n}$ and $\bar{u}$ or $\bar{b}$ and $\bar{v}$. Scalar intepretation of the equation (23) yields the relationship between geometric and physical curvatures

$$\kappa = \sqrt{p^2 + q^2} \tag{24}$$

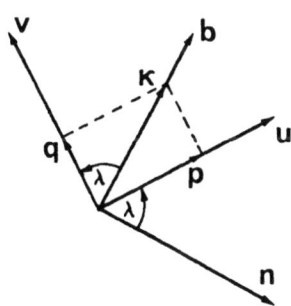

Fig. 7. Unit vectors of principal $(\bar{n}, \bar{b})$ and main $(\bar{u}, \bar{v})$ trihedron; geometric (κ) and physical (p,q) curvatures

and between geometric and physical torsion

$$\tau = r - \lambda' = r - \frac{p'q - q'p}{p^2 + q^2} \qquad (25)$$

The substitution for λ' is based on the equality $\tan\lambda = p/q$ (see Fig. 7).

We can define the curvature-orientation equation of elastica by analogy to the Serret-Frenet equations (15) or their interpretation (18):

$$\overline{w}' = \overline{\omega} \times \overline{w} \ ; \qquad \overline{u}' = \overline{\omega} \times \overline{u} \ ; \qquad \overline{v}' = \overline{\omega} \times \overline{v} \qquad (26)$$

and the orientation-coordinate equation from (13) and (21)

$$\overline{r}' = \overline{w} \qquad (27)$$

Considerations similar to those which have lead to the natural equations of a curve (16) yield physical equations of a curve

$$p = p(s) \ ; \qquad q = q(s) \ ; \qquad r = r(s) \ . \qquad (28)$$

These equations define uniquely not only the shape of a curve representing central line of a fibre, rod etc. (as natural equations do), but also the orientation of material points of the fibre cross-section.

3.5 Scalar equations of spatial elastica

Scalar equations of the elastica, as appeared in [8], are given in Table II. The equations are listed in the order maintained in the computer program subroutine BEC3D1 (see next paper). The differential equations for the energy of bending and torsional deformations and for total rotations around u,v,w axes are added.

3.6 Simplifications

The differential equations may be considerably simplified in case of initially straight central line, no distributed forces, and/or equal bending rigidities (A=B). If all the simplifications apply, the torsion equation turn into r'=0. This implies constant physical torsion r along spatial elastica regardless of the variability of the geometric torsion τ.

Table II
Scalar equations of spatial elastica

Direction cosine equations
(1) $u'_x = rv_x - qw_x$ (4) $v'_x = pw_x - ru_x$ (7) $w'_x = qu_x - pv_x$
(2) $u'_y = rv_y - qw_y$ (5) $v'_y = pw_y - ru_y$ (8) $w'_y = qu_y - pv_y$
(3) $u'_z = rv_z - qw_z$ (6) $v'_z = pw_z - ru_z$ (9) $w'_z = qu_z - pv_z$

Curvature equations
(10) $p' = [Br(q - q_0) - Cq(r - r_0) + F_u]/A + p'_0$
(11) $q' = [Cp(r - r_0) - Ar(p - p_0) - F_v]/B + q'_0$
(12) $r' = [Aq(p - p_0) - Bp(q - q_0)]/C + r'_0$

Internal force equations
(13) $F'_u = rF_v - qF_w - f_u$
(14) $F'_v = pF_w - rF_u - f_v$
(15) $F'_w = qF_u - pF_v - f_w$

Coordinate equations
(16) $x' = w_x$ (17) $y' = w_y$ (18) $z' = w_z$

Deformational energy equations
(19) $2U'_b = A(p - p_0)^2 + B(q - q_0)^2$
(20) $2U'_t = C(r - r_0)^2$

Angle deformation equations
(21) $\varphi'_u = p$ (24) $\varphi'_{0u} = p_0$
(22) $\varphi'_v = q$ (25) $\varphi'_{0v} = q_0$
(23) $\varphi'_w = r$ (26) $\varphi'_{0w} = r_0$

An elastica may be confined to the plane x,y by assuming q=r=0 and u_z=1 all along. The reduced system of differential equations of planar elastica in the order maintained in the subroutine BEC2D1 (see next paper) is as follows:

1. $2U' = A(p-p_0)$ 4. $w'_x = -pw_y$ 7. $F'_v = pF_w + f_y w_x$

2. $x' = w_x$ 5. $w'_y = pw_x$ 8. $F'_w = -pF_v - f_y w_y$

3. $y' = w_y$ 6. $p' = F_n/A + p'_0$ (29)

4. NONLINEAR MOMENT-CURVATURE RELATIONSHIP

4.1 Nonlinearly elastic rod

The nonlinear elastic behaviour of a rod may be accounted for by expressing the bending and torsional rigidities as functions of respective curvatures and torsions, i.e. A(p), B(q), C(r). The generalized Bernoulli-Euler law:

$$M_u = p \cdot A(p) \; ; \quad M_v = q \cdot B(q) \; ; \quad M_w = r \cdot C(r) \quad (30)$$

and its derivatives with respect to arc length s:

$$M'_u = [p \cdot (dA/dp) + A] \cdot p' \quad \text{and so on.} \quad (31)$$

The substitution for M' into the first scalar equation of moment equilibrium yields the following expression for p':

$$p' = [p \cdot (dA/dp) + rq(B-C) + F_v]/A \quad (32)$$

This and similar expressions for q' and r' constitute the differential equations of curvatures and torsion for originally straight nonlineraly elastic rod. The rest of the equations remains the same as in Table II.

For an originally curved rod, the equations have to be modified as follows:

$$p' = [(dA/dp)(p-p_o)+Br(q-q_o)-Cq(r-r_o)+F_v]/A + p'_o \qquad (33)$$

and so on.

4.2 Friction-elastic rod

An idealized moment-curvature relationship typical for a friction-elastic material is shown in Fig. 8. The most comprehensive treatment of friction-elastic planar elastica is due to Bruggisser [3]. The friction moment M_f (or the insensitivity zone) is accounted for by introduction of a virtual or actual original curvature. The former represents the friction resistance to the bending moment. The latter represents a residual curvature after the removal of the external loads at any stage of the deformation process.

The formulation of friction-elastic bending problems along these lines and the simulation of any non-trivial loading-deloading process is possible, but rather laborious.

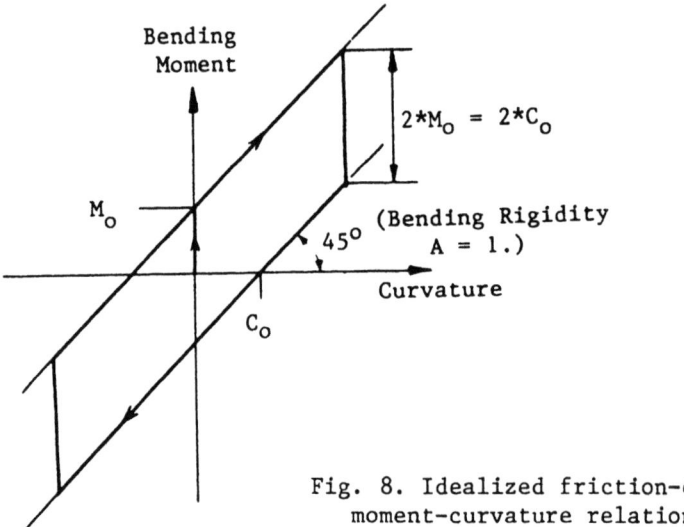

Fig. 8. Idealized friction-elastic moment-curvature relationship

4.3 Elasto-plastic rod

The graph of the simplest moment-curvature relationship of an elasto-plastic rod consist of a straight line passing through the origin with a slope proportional to the elastic bending rigidity, and a line parallel to the horizontal axis and implying no additional resistance above a certain limit of the bending moment.

The formulation of an elasto-plastic bending problem and the simulation of a load-deformation behaviour is not difficult. The elastoplastic moment-curvature relationship may be approximated by assuming the bending rigidity $A(p)=a-b.\exp(-c.p)$ and considering the rod to be nonlinearly elastic as discussed in section 4).1 (the a,b,c are positive constants. In case of spatial elastica the $B(q)$ and $C(r)$ have to be defined similar way.

4.4 Viscoelastic rod

The modification of the Hooke's law $\sigma=\varepsilon E$ as to cover the visco-elastic stress-strain behaviour yields, in a simplified case of a discrete 4-element model (Fig. 9), the following:

$$\sigma = \varepsilon_1 E_1 = \varepsilon_2 E + \eta_2 \frac{d\varepsilon_2}{dt} = \eta_3 \frac{d\varepsilon_3}{dt} \ ; \quad \varepsilon = \varepsilon_1 + \varepsilon_2 + \varepsilon_3 \qquad (34)$$

where σ is the stress;
  E_1, E_2 are the tensile moduli;
  $\varepsilon_1, \varepsilon_2, \varepsilon_3$ are the pure elastic, viscoelastic and plastic components of the total deformation;
  η_2, η_3 are the viscosity constants;
  t is the time (independent variable)

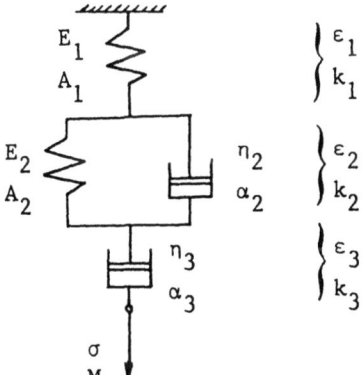

Fig. 9. Four-element model of viscoelastic behaviour

The models of viscoelastic bending and torsion of slender rods may be developed in a similar way through the generalization of the Bernouli-Euler law:

$$M = k_1 A_1 = k_2 A_2 + \alpha_2 \frac{\partial k_2}{\partial t} = \alpha_3 \frac{\partial k_3}{\partial t} \; ; \quad k = k_1 + k_2 + k_3$$

These equations together with the moment equilibrium equation, yield, after a series of differentiations, transformations and substitutions, a set of nonlinear partial differential equations with independent variables, arc length s and time t. Unfortunately, there is very little one can do with them.

We succeeded in bypassing the difficulties and simulating the bending and torsional behaviour of slender rods by using standard numerical procedures for integration of ordinary differential equations along the s-axis in combination with zero-order approximation along the t-axis. The procedure: (1) the ordinary differential equations of elastica are solved and the shape of the elastica at time t is obtained; (2) the incurred bending and torsional moments are let to relax during the time interval Δt; (3) the generated plastic bending and torsional deformation is interpreted as original curvature and torsion to enter another elastica solution, step (1), at time $t+\Delta t$, and so on.

This approach may be applied only in case of discrete viscoelastic model with finite set of relaxation times $\tau_i = \alpha_i/A_i$. Additional pairs of constants A_j, α_j may be easily incorporated into the computation scheme. It is not possible to work with a continuous distribution of relaxation times because this would require storage and continuous accounting for the whole history of deformations along the arc length coordinate.

5. PROBLEM FORMULATIONS

5.1 Normalization

It is convenient to normalize the problem by introducing a dimensionless independent variable $\tilde{s} = s/\ell$, normalized bending rigidities $\tilde{A} = 1$ and $\tilde{B} = B/A$, and normalized torsional rigidity $\tilde{C} = C/A$. The relationships between normalized and actual value are as follows:

coordinates: $\tilde{x} = x/\ell \; ; \quad \tilde{y} = y/\ell \; ; \quad \tilde{z} = z/\ell$,

curvatures and torsion: $\tilde{p} = p.\ell \; ; \quad \tilde{q} = q.\ell \; ; \quad \tilde{r} = r.\ell$,

moments: $\tilde{M}_u = M_u.\ell/A \; ; \quad \tilde{M}_v = M_v.\ell/A \; ; \quad \tilde{M}_w = M_w.\ell/A$, concentrated and distributed forces: $\tilde{F} = F.\ell^2/A$; $\tilde{f} = f.\ell^3/A$,

energy of bending and torsional deformation: $\tilde{U} = U.\ell/A$.

5.2 Single and compound elastica

The parameters of the differential equations of elastica, such as external forces, external moments, bending/torsional rigidities or original curvatures/torsion, may change along the curve continuously or stepwise. In the former case the constants in the differential equations representing the values of the parameters have to be replaced by functions of arc length. The stepwise changes have to be accounted for by redefinition of the parameters and/or the initial conditions.

If there are n points with stepwise change of any parameter including two end points) along the rod, the central line has to be looked at as a compound elastica consisting of n-1 single elastica segments with particular sets of initial conditions and parameters. Some of the initial conditions and parameters may be simply carried over from the previous elastica segment. The problems with compound elasticas usually imply increased number of a priori unknown initial conditions and parameters. On the other hand, there may be some boundary conditions defined at the ends of the elastica segments.

5.3 Boundary value problems

Formulation of a boundary value problem and finding a set of unknown initial conditions and parameters which gives the solution satisfying the boundary conditions is the essence of any particular application of the concept and theory of elastica.

To solve an elastica problem or to simulate a shape of the elastica usually means to integrate the relevant set of differential equations along the arc length s. There are always some of the parameters and values of dependent variables (or initial conditions) known or unknown at both ends. It is advisable to choose such a direction of integration as to reduce the number of unknown initial conditions and parameters (and equal number of boundary conditions) to a minimum.

5.4 Stability

The problem of stability of axially loaded rod or column was recognized by Euler who introduced the concept of critical force (in relation to length and bending rigidity) under which an originally straight rod buckles rather than compresses. There is a number of other stability problems, some of them discussed by Love in [6], concerned with the conditions of spontaneous collapse not only of a stright rod into a stable shape of planar elastica, but also a planar elastica into the shape of a spatial one.

If the potentially unstable shape may be described in closed form (like straight rod or circular arc), the chances are that the stability criteria may be formulated in closed form as well. Simple example: critical load required for buckling of a straight rod into a half-period of 2-nd class elastica, $F=\pi^2 A/\ell^2$.

If general case and especially in cases of out-of-plane instability of planar elastica, the explicit criteria are difficult or impossible to obtain. A universally applicable and safe procedure in any particular case is to examine the existence of multi-root solution of a boundary value problem within given geometric constraints and, given the choice, to select the one with minimum potential energy of the system (deformational and positional).

6. CONCLUDING REMARKS

The classical elastica theory reinforced by modern numerical and computational tools provides a reliable basis for analysis and simulation of any problem belonging to the wide class of problems of the mechanics (or, more precisely, statics) of thin rods, fibres, yarns etc., without any restrictions concerning the magnitude of deformation or its nonlinear and time-dependent nature. Nevertheless, there are several problems requiring further investigation. The most pressing, in our opinion, is the problem of the implications of finite cross-sectional area of the real slender bodies exhibiting large deflections, or the problem of actual bending rigidity of a real slender body.

The Bernoulli-Euler law approximates the behaviour of a real rod or fibre with a radius of cross-section vs. radius of curvature ratio R/ρ approaching zero. The reasons of convenience induce applications of the elastica approach to problems with R/ρ up to the limiting value 1/2. The errors introduced by assuming the validity of the Bernoulli-Euler law in the case of finite R/ρ should be reduced by linking the moment-curvature relationship to the behaviour of the fibre as a three-dimensional body and by incorporating into the analysis the five mechanical characteristics (two tensile and three shear moduli in a simple case) and their share in the resistance of the fibre to bending and torsional moments.

This should also provide a firm basis for exact treatment of the effects of extensibility and transverse shearability of the fibre or rod, which was neglected in our analysis.

The tangential and normal forces responsible for those deformations are available in the systems of differential equations for planar as well as spatial elastica. At least the influence of the extensibility of the elastica on its shape is easy to account for by introducing an additional variable, original arc length s_o, and a differential equation linking tensile modulus, tangential force, and local extension defined as a function of ds_o/ds. Accounting for the transverse shearability at this superficial level is less trivial.

The concept of extensible and shearable elastica may be useful in some very special cases of disproportionately high bending rigidity and low tensile modulus. For instance, bending of a helical spring (and the necessary stretching accompanying it) can result in the behaviour of the central axis of the imaginary rod, wrapped by the spring, close to that of an extensible elastica. However, in majority of situations where the bending rigidity is not far from EI, accounting for tensile and shear deformations can be omitted without introducing any appreciable error.

If the bending rigidity of a rod or fibre with circular cross-section is equal to EI, then, in the most severe case of the elastica of the 8-th class approaching the 7-th and the maximum curvature at the top of the loop being reciprocal to the fibre diameter, the extension of the tail of elastica is only 1.56%. If, on the other hand, the extending force suffices to cause higher extension, the shape of the fibre is governed almost entirely by the geometry of the interacting surfaces and there is no need (and no chance because of numerical instability) to use the elastica model.

There is little, if any, overlap between these two classes of problems which would justify the application of an extensible elastica model. Moreover, under severe deformation conditions, the simplifying assumptions concerning bending rigidity seem to loose ground well before the assumptions of inextensibilty do.

REFERENCES

[1] Antman, S.S., The Theory of Rods, in 'Handbuch der Physik' ed. by S. Flugge, Vol. VIa/2: 'Mechanics of Solids II', ed.by C. Truesdell, Springer-Verlag, 1972.
[2] Truesdell, C., The rational mechanics of flexible or elastic bodies 1638-1788, in 'L. Euleri Opera Omnia', Vol. II/11-2, Fussli, Zurich, 1960.
[3] Bruggisser, K., MS. Thesis, Georgia Institute of Technology, Atlanta, 1976.
[4] Euler, L., Methodus inveniendi lineas curvas maximi minimive proprietate gaudentes (Aditamentum I, De curvis elasticis), Lausanne, 1744; also in 'L. Euleri Opera Omnia', Vol. I/24, Fussli, Zurich, 1952.
[5] Kirchhoff, G., J.f.Mathematik (Crelle), 1859, 56, pp.285-313.
[6] Love, A.E.H., A treatise on the mathematical theory of elasticity, eds. 1-4, Cambridge University Press, 1892-1927; (reprinted by Dover Publications, New York, 1944).
[7] Popov, E.P., Nelineinyie zadatchi statiki tonkich sterzhney (Nonlinear problems of the statics of thin rods), Gostechizdat, Moscow, 1948.
[8] Konopasek, M. and Hearle, J.W.S., Fibre Sci. and Techn., 1972, 5, pp. 1-28.

COMPUTATIONAL ASPECTS OF LARGE DEFLECTION ANALYSIS
OF SLENDER BODIES

Milos Konopasek

School of Textile Engineering, Georgia Institute
of Technology, Atlanta, Georgia 30332

ABSTRACT. A set of FORTRAN subroutines called "Bending Curve Program Package" has proved to be an extremely useful instrument for solving a variety of problems in the area of large deflection analysis of slender bodies. The concept and the composition of the package is overviewed, the essential components are described, and extensions of the package are discussed.

1. INTRODUCTION

Analytical solutions of nonlinear differential equations for planar or spatial elasticas are unobtainable except in a few trivial or special cases (pure bending or bending/torsion, geometrical constraints forcing circular or helical shape, planar elastica of 7-th kind). Implicit analytical solutions in terms of elliptical integrals are available only for fairly simple problems and their use is usually inconvenient. Numerical integration of the differential equations performed by a computer is the only practical way to solve particular problems, i.e. to analyze relationships between mechanical properties of fibres or yarns, geometrical constraints, external and internal forces and moments, and the shape of the centre line (or longitudinal axis).

The need for an efficient computation tool in the large deflection analysis of slender bodies is amplified by the fact that the solutions of boundary value problems are required in practically all real situations. In such a highly nonlinear and analytically intractable environment only a carefully chosen, launched and controlled iteration procedure can lead to the satisfaction of a smaller or larger number of boundary conditions.

A set of FORTRAN subroutines called "Bending Curve Program Package" (BCPP) was developed at UMIST in the early seventies for the purpose of reducing programming chores, standardizing an approach to the solution of the whole class of elastica problems, and increasing problem-solving and computation efficiency (see [1] and sets of user notes [2,3] substituting for a BCPP manual).

2. BENDING CURVE PROGRAM PACKAGE

A typical block diagram for computation in the large deflection analysis of slender bodies is shown in Fig. 1.

The heart of the program is the subroutine BEC2D1 performing numerical integration of the differential equations of planar elastica. This subroutine may be substituted for by BEC3D1 in the case of spatial elastica or by some other subroutines reflecting various modifications of the differential equations of elastica. The input and output information represents, respectively, all initial values and parameters for the equations, and the values of the dependent variables at the end of the integration interval. The subroutines BEC2D1 or BEC3D1 sample the original curvatures and their derivatives by referring to function subprograms CIN2D or CIN3D. The referencing of this function may be omitted in the case of differential equations for originally straight elastica, or may quite involved in the case of friction-elastic or viscoelastic bending.

Control over the whole solution of a problem is exercised by the Main program which assigns the guessed values of the unknown initial conditions and parameters. Very often a series of problems is to be solved, in which case the sequence of values of a controlling parameter (e.g. deforming force or a coordinate value at the end of the integration interval) is also generated in the Main program and is communicated to the subroutine INTMED through a COMMON/GIVPAR/.

The last major parts of the scheme are a subroutine controlling the iteration process (in the simplest case it is called NERBEC) and the subroutine INTMED. The latter receives all the information about the unknown initial conditions and parameters from the subroutine NERBEC. It also possesses information about the fixed initial conditions and parameters and it has access to the controlling parameter through COMMON/GIVPAR/. All this information is used to generate a full set of input parameters required by the differential equation subroutine (e.g. BEC2D1) in order to produce a unique initial value solution to the problem.

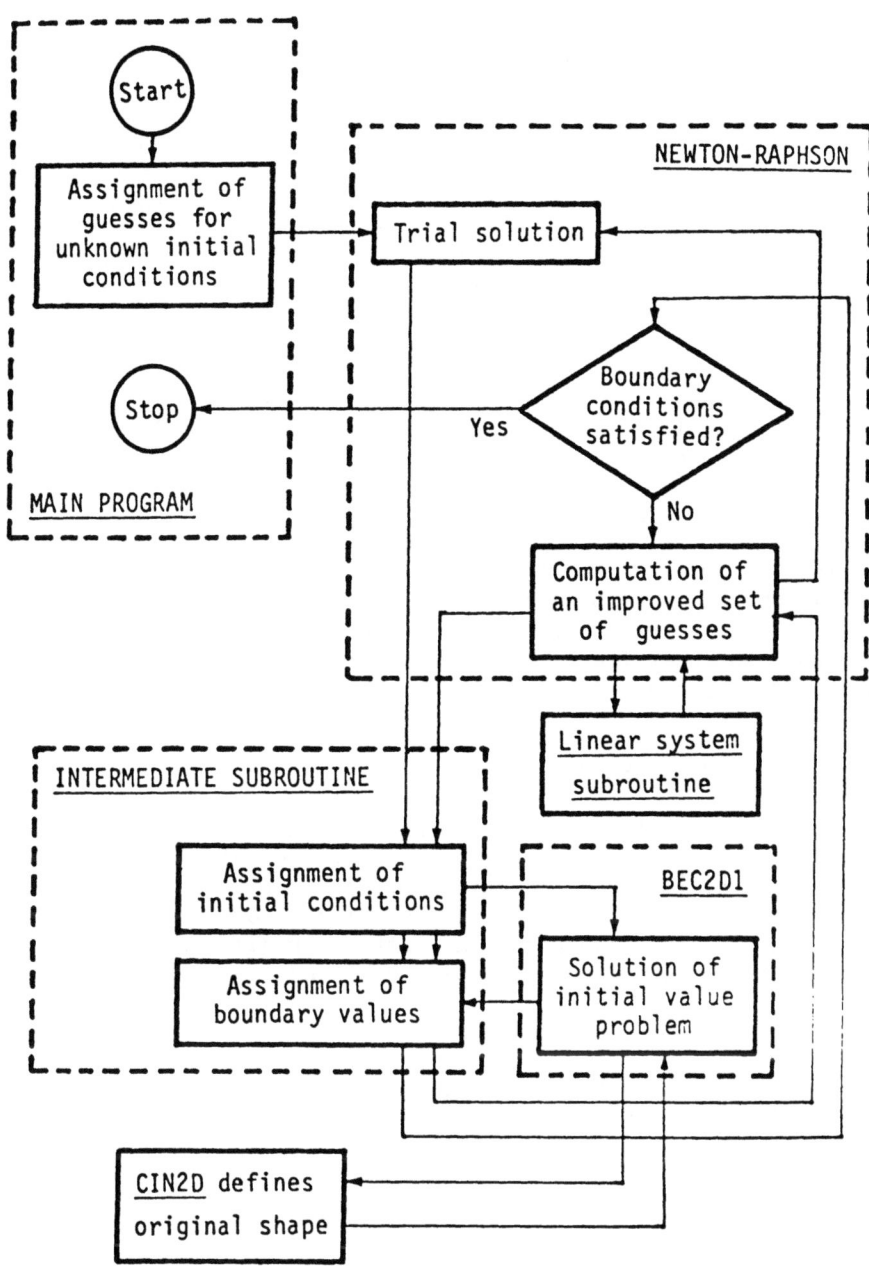

Fig. 1. Block diagram for computation in slender body mechanics
(special program segments are on the left side and
standard BCPP components are on the right)

The values of the dependent variables at the terminating point of the integration interval are fed back to the subroutine INTMED. Here the values constrained by boundary conditions (boundary values) are assigned and control is returned to the subroutine NERBEC. There the returned boundary values are compared with the target. If the discrepancy is below the chosen limit, the iteration process stops and control is returned to the Main program. If this is not the case, the iteration process continues, i.e. another set of improved guesses of the unknown initial values and parameters is calculated, the satisfaction of the boundary conditions is tested and so on (for more details see chapter 4).

The subroutine LINEL2 for solving systems of linear algebraic equations is used in the process of calculation of a new set of guesses.

The NERBEC, LINEL2 and BEC2D1 in Fig. 1 are standard BCPP subroutines. The Main program, the subroutines INTMED and the function CIN2D are problem-tailored. An example of the formulation of a problem and listing of the problem-tailored or user-supplied program segments are given in Fig. 2 for a case of a weightless pear-like shape made of originally straight fibre, and extended by a sequence of forces Fy.

3. INITIAL VALUE PROBLEM

Solving an initial value problem in the large deflection analysis of slender bodies is equivalent to numerical integration of nonlinear differential equations of elastica. Let us illustrate the concept of numerical integration of ordinary differential equations on an example of a first order differential equation involving a continuous function y(x) and its derivative:

$$dy/dx = f(x,y) \qquad (1)$$

Suppose we know the values x_o and $y_o = y(x_o)$ as well as x_o and we would like to find $y_1 = y(x_1)$ satisfying equation (1). According to Simpson's formula

$$y_1 = y_o + \int_{x_o}^{x_1} f(x,y)dx \approx \qquad (2)$$

$$\approx y_o + \frac{x_1 - x_o}{6}[f(x_o,y_o) + 4f(\frac{x_o+x_1}{2}, \frac{y_o+y_1}{2}) + f(x_1,y_1)]$$

Through averaging the slope along the curve y(x) in the interval $x_0 \leq x \leq x_1$, and approximating the changes in y_1, we obtain

$$y_1 = y_0 + (1/6) \sum_{i=1}^{4} \alpha_i k_1 \qquad (3)$$

where $k_i = h.f(x_0+\beta_i, y_0+\alpha_i k_{i-1})$
$k_0 = 0$

and

i	1	2	3	4
α_i	1	2	2	1
β_i	0	1/2	1/2	0

The values of $y_2, y_3, \ldots$ covering the whole integration interval, may be evaluated in a similar way step by step (it is a "single step method").

```
      DIMENSION PAR(3,20)
      COMMON /GIVPAR/GV1
      PAR(1,1)=2.
      PAR(1,2)=4.
      GV1=-.5
      DO 1 I=1,25
      GV1=GV1+.5
    1 CALL NEREBEX(PAR,1.E-6,2,2,I)
      STOP
      END

      SUBROUTINE INTMED(PAR,BOV,KEY)
      DIMENSION PAR(20),BOV(20),A(8),B(12)
      COMMON /GIVPAR/GV1
      DATA B/.5,1.,5*0.,-1.3*0.,.0001/
      B(3)=PAR(1) ....... unknown initial condition k(0)
      B(4)=PAR(2) .......  - " -   parameter $F_x$
      B(5)=GV1 .............. controlling parameter $F_y$
      CALL BEC2D1(B,A,40,KEY)
      BOV(1)=A(2) .......... boundary condition x($\ell$)=0
      BOV(2)=A(5) ..........  - " -   - " -   w ($\ell$)=0
      RETURN
      END
```

half length $\ell = B(1) = .5$
bending rigidity $A = B(2) = 1.$
curvature $k(0) = B(3) = ?$
concentr. forces $F_x = B(4) = ?$
                   $F_y = B(5) = 0, .5, \ldots, 12$
distrib. force $f = B(6) = 0.$
direct. cosine $w_x(0) = B(7) = 0.$
sign $w_y(0) = B(8) = -1.$
init. arc length $s(0) = B(9) = 0.$
" x-coordinate $x(0) = B(10) = 0.$
" y-coordinate $y(0) = B(11) = 0.$
error limit $\varepsilon = B(12) = .001$

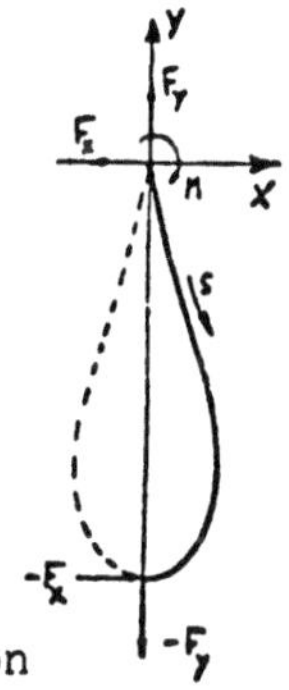

Fig. 2. Computer program set-up for simulation
of load/deformation behaviour of pear-like loop
(Main routine and Intermediate routine INTMED on the left,
formulation of the problem on the right)

Approximations of the integral (2) other than by using Simpson's rule yield a variety of Runge-Kutta methods of various orders, different from the classical 4-th order version (3). The accuracy of integration may be improved by using a higher order method or a smaller integration step h - both at the expense of increased computing time. The difference between the results from higher and lower order gives an estimate of the error and may be used for an automatic step control.

For this reason we use in BCPP the Merson's version of 5-th order Runge-Kutta method, which is defined as follows:

$$y_1 = y_0 + (k_1 + 4k_4 + k_5)/6 \qquad (4)$$

where

$$k_1 = h \cdot f(x_0, y_0)$$

$$k_2 = h \cdot f(x_0 + h/3, y_0 + k_1/3)$$

$$k_3 = h \cdot f(x_0 + h/3, y_0 + (k_1 + k_2)/6)$$

$$k_4 = h \cdot f(x_0 + h/2, y_0 + (k_1 + k_3)/8)$$

$$k_5 = h \cdot f(x_0 + h, y_0 + (k_1 - 3k_3 + 4k_4)/2)$$

An estimate of the error at the end of the step

$$\varepsilon = (2k_1 - 9k_3 + 8k_4 - k_5)/30 \qquad (5)$$

The subroutines BEC2D1 and BEC3D1 use a fixed number of integration steps, in multiples of 20, because the output generator is set up for displayng and printing results at 21 equidistant points. There is a message printed in case the estimated error is larger than the given limit; then the number of integration steps has to be increased for another 20, or multiple of 20, steps.

The Runge-Kutta methods are well known for their reliability and are widely used. New improved versions are still being developed. One of them due to Fehlberg and referred to in [4], p. 377, is becoming increasingly popular and should be tried in BCPP.

There is probably even more effort by numerical analysts put into the development of so called multistep predictor-corrector methods. These methods use computer resources more efficiently, but they are more difficult to start when only information concerning the initial point is available and they pose numerical stability problems. They do not seem to fit the needs of BCPP at this point.

The initial value assignment and integration part of the subroutine BEC2D1 is set up as follows:

```
      NPI=NI/20
      DS=B(1)/FLOAT(NI)
      S=B(9)                              .
      A(1)=0.                             .
      A(2)=B(10)                          .
      A(3)=B(11)                          .  initial
      A(4)=B(7)                           .  value
      A(5)=B(8)*SQRT(1.-A(4)*A(4))        .  assignment
      A(6)=B(3)                           .
      A(7)=B(5)*A(4)-B(4)*A(5)            .
      A(8)=-B(4)*A(4)-B(5)*A(5)           .
      DO 10 I=1,21
      IF (I.EQ.1) GO TO 6
      DO 5 IP=1,NPI
      DO 5 J=1,5
      S=S+DS*CS(J)            ............... arc-length update
      C(1,J)=(A(6)-CIN2D(S,1,I))**2 .... double energy
      C(2,J)=A(4)             ............... x-coordinate
      C(3,J)=A(5)             ............... y-coordinate
      C(4,J)=-A(6)*A(5)       ............... direction cosine $w_x$
      C(5,J)=A(6)*A(4)        ...............   - " -   - " - $w_y$
      C(6,J)=A(7)/B(2)+CIN2D(S,2,I) .... curvature
      C(7,J)=A(6)*A(8)+A(4)*B(6)  ...... normal force
      C(8,J)=-A(6)*A(7)-A(5)*B(6) ..... tangential force
      DO 5 K=1,8
      C(K,J)=C(K,J)*DS
      DO 5 L=1,J
    5 A(K)=A(K)+C(K,L)*CV(J,L)
    6 IF (KEY.NE.1) GO TO 10
        .............................. output preparation
   10 CONTINUE
```

The integration steps totalling NI are controlled by two DO loops "DO 10 I=1,21" which enables the collection of intermediate results for standard output when the success flag KEY=1, and "DO 5 IP=1,NPI" where NPI=NI/20. The "DO 5 J=1,5" loop gives all the $k_{i,j}$ necessary for evaluation of the function values at each integration interval (they are stored in array C). The assignment statements for C(1,J) through C(8,J) reflect the defined set of the differential equations of planar elastica. These statements are subject to modification when the structure of differential equations changes.

This is the integration part of the subroutine BEC3D1 for spatial elastica:

```
      DO 5 J=1,5
      S=S+DS*CS(J) .................... arc-length update
      DO 4 K=1,3                       rotations of main trihedron
      C(K+23,J)=CIN3D(S,K,I) ........... from original curvatures
      AA(K)=B(K+1)*(A(K+9)-C(K+23,J)) .... moments M_u,v,w
      C(K,J)=A(12)*A(K+3)-A(11)*A(K+6) ..... direction cosines u_x,y,z
      C(K+3,J)=A(10)*A(K+6)-A(12)*A(K) .....   - " -     - " -   v_x,y,z
      c(k+6,j)=a(11)*a(k)-a(10)*a(k+3) .....   - " -     - " -   w_x,y,z
      C(K+15,J)=A(K+6) ................. coordinates x,y,z
    4 C(K+20,J)=A(K+9) ................. rotations of main trihedron
      C(10,J)=(AA(2)*A(12)-AA(3)*A(11)+A(14))/B(2)+CIN3D(S,4,I)... p
      C(11,J)=(AA(3)*A(10)-AA(1)*A(12)-A(13))/B(3)+CIN3D(S,5,I)... q
      C(12,J)=(AA(1)*A(11)-AA(2)*A(10))/B(4)+CIN3D(S,6,I) ........ r
      C(13,J)=A(12)*A(14)-A(11)*A(15)-B(11)*A(3) .......... force F_u
      C(14,J)=A(10)*A(15)-A(12)*A(13)-B(11)*A(6) .......... -"-  F_v
      C(15,J)=A(11)*A(13)-A(10)*A(14)-B(11)*A(9) .......... -"-  F_w
      C(19,J)=AA(1)*AA(1)/B(2)+AA(2)*AA(2)/B(3)..dble bending energy
      C(20,J)=AA(3)*AA(3)/B(4) ............... double torsion energy
      DO 5 K=1,26
      C(K,J)=C(K,J)*DS
      DO 5 L=1,J
    5 A(K)=A(K)+C(K,L)*CV(J,L)
```

The programs are set up differently from (4) for reasons of efficiency in sequential computation. The values of each function A and error ε accumulate during the five steps within the integration interval as follows:

$$A_j = A_{j-1} + \sum_{\ell=1}^{j} C_\ell CV_{\ell,j} \; ; \quad \varepsilon_j = \varepsilon_{j-1} + C_j CE_j \qquad (6)$$

for $j=1,2,\ldots,5$

where $C_\ell = DS \cdot f(S_\ell, A_{\ell-1})$
$S_\ell = S_{\ell-1} + DS \cdot CS$
S,DS are the independent variable and its increment,
A_o, S_o are the values of function and independent variable at the beginning of an integration interval,

$$CV = \begin{bmatrix} 1/3 & 0 & 0 & 0 & 0 \\ -1/6 & 1/6 & 0 & 0 & 0 \\ -1/24 & -1/6 & 0 & 0 & 0 \\ 3/8 & 0 & 15/8 & 0 & 0 \\ -1/3 & 0 & 3/2 & -4/3 & 1/6 \end{bmatrix}$$

$$CS = [\; 0 \quad 1/3 \quad 0 \quad 1/6 \quad 1/2 \;]$$

$$CE = [\; 1/15 \quad 0 \quad -3/10 \quad 4/15 \quad -1/30 \;]$$

4. BOUNDARY VALUE PROBLEM

Finding those values of the unknown initial conditions and parameters, which would bring about the satisfaction of the boundary conditions, is equivalent to finding the roots of a system of nonlinear equations:

$$f_i(p_1, p_2, \ldots, p_n) - b_i = 0, \quad i=1,2,\ldots,n \tag{7}$$

where $p_1, p_2, \ldots, p_n$ are the unknown parameters;
$f_1, f_2, \ldots, f_n$ represent the dependency of boundary conditions on parameters $p_1, p_2, \ldots, p_n$. They are defined implicitly by elástica equations;
$b_1, b_2, \ldots, b_n$ are constants or boundary conditions;
n is the number of unknown parameters.

Two iterative procedures for solving nonlinear systems are available in the BCPP: the "false position method" and the Newton-Raphson procedure. The former is applicable only to problems with one unknown parameter and is rarely used. The latter has proved to fit the conditions and requirements of solving elastica boundary value problems extremely well. It converges rapidly providing the initial guesses are not far off. The sensitivity of the Newton-Raphson procedure to initial guesses does not help. However, we need good guesses anyway in order to prevent the convergence to wrong roots and consequently we prefer not to take advantage of some less sensitive method.

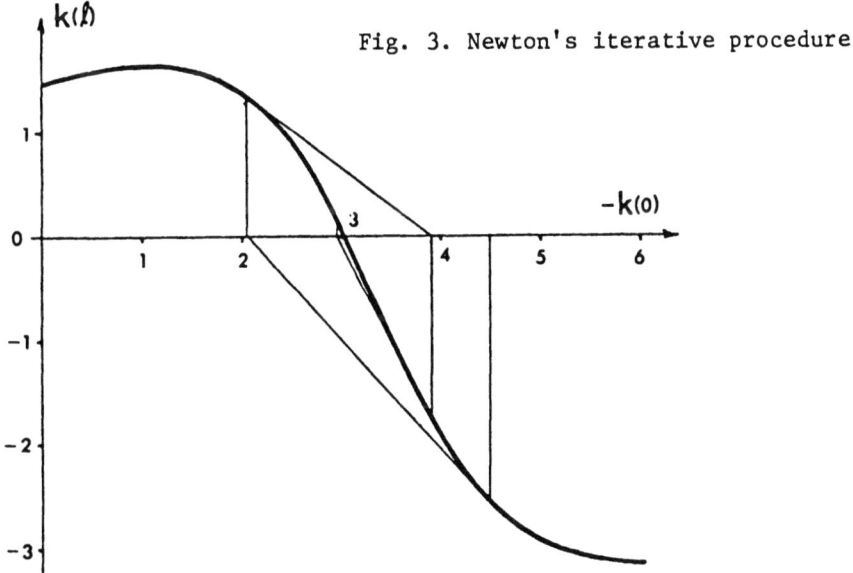

Fig. 3. Newton's iterative procedure

4.1 Newton-Raphson procedure

The Newton's iterative procedure for one unknown is illustrated in Fig. 3 which shows a graph of the curvature at the end of a horizontal cantilever as a function of the initial curvature at the pinched end. Our guess for the initial value of the curvature was -4.5. The next guess, -2.08, was found at the cross-section of the tangent to the graph at the point of the initial guess. Repetition of the same trick gave the value of the initial curvature as -3.85, the following was -3.004, the next -3.0626 (not shown) and the final was -3.0618 with the corresponding boundary value -0.000000508, i.e. below an error limit set as $|\epsilon| \leq 0.000001$.

In a generalized Newton-Raphson procedure applicable to any number of unknowns the intersection of the tangent to the curve is replaced by the intersection of the traces of tangential hyperplanes to the hypersurfaces of boundary values. The iterative procedure is based on an assumption that

$$f_i + \sum_{j=1}^{n} \frac{\partial f_i}{\partial p_j} (p_j^{[k+1]} - p_j^{[k]}) = 0 \tag{8}$$

where $k=0,1,2,\ldots$ is a counter of iteration steps. It was possible to omit the reference to the boundary conditions $b_1, b_2, \ldots$ due to the convention in the BCPP to formulate the boundary conditions so that $b_1 = b_2 = \ldots = 0$.

The partial derivatives in (8) are replaced by easy-to-compute partial differences and an improved set of parameters at each iteration step is calculated by solving a system of linear equations, in matrix form,

$$\left[\frac{\Delta f_i}{\Delta p_j}\right] \times \left[p_i^{[k+1]}\right] = \left[\sum_{j=1}^{n} \frac{\Delta f_i}{\Delta p_j} p_j^{[k]} - f_i\right] \tag{9}$$

where
$$\Delta f_i / \Delta p_j =$$
$$= [f_i(p_1^{[k]}, p_2^{[k]}, \ldots, p_n^{[k]}) - f_i(p_1^{[k]}, p_2^{[k]}, \ldots, p_j^{[k]} - \lambda, \ldots, p_n^{[k]})]/\lambda$$

A simplified listing of the essential part of the subroutine NERBEC follows

```
            1 CALL INTMED(PAR,BOV)
              KEY=1
              DO 2 I=1,N
              IF (ABS(BOV(I)).GT.PRECIS) KEY=0
            2 CONTINUE
```

```
            IF (KEY.EQ.1) RETURN
            DO 3 I=1,N
            PAR(I)=PAR(I)-PRECIS
            CALL INTMED(PAR,BOD)
            PAR(I)=PAR(I)+PRECIS
            DO 3 J=1,N
          3 PD(J,I)=(BOD(J)-BOV(J))/PRECIS
            DO 4 I=1,N
            DO 4 J=1,N
          4 BOV(I)=BOV(I)+PD(I,J)*PAR(J)
            CALL LINEL2(PD,BOV,PAR,N)
            GO TO 1
```

First, the subroutine INTMED is called in order to provide a response (array BOV) to the current set of parameters (array PAR). The following DO loop serves to test the satisfaction of the boundary conditions. The components of the BOV array are defined in the subroutine INTMED in such a way as to become zero when the boundary conditions are satisfied. Consequently the check amounts to transferring control to a segment of the subroutine which would generate another, hopefully improved, set of parameters PAR whenever an absolute value of any one of the elements of array BOV happens to be greater then the error limit PRECIS (see DO 3 loop). If all the BOV elements pass the test, control is returned to the program the subroutine was called from.

If the iteration continues in the following double DO 4 loop the parameters are modified one at a time by substracting a small number (usually the same PRECIS), and the Jacobian matrix of partial differences is built column by column. Next, a right hand side column matrix is built inside the DO 5 loop and a linear system subroutine LINEL2 is called to generate an improved set of parameters PAR.

A boundary value problem with n unknown parameters solved in m iteration steps requires $m(n+1)+1$ calls to subroutine INTMED and executions of initial value problems.

The real subroutine NERBEC is protected against catastrophic divergency (the values of PAR or BOV becoming enormous) or failure to converge within a certain number of iteration steps. It also offers a choice of four levels of output of the trace of iteration process ranging from none to printing the PAR and BOV arrays as well as the Jacobian matrix and the right side colum matrix. The value of the parameter increment for generating the Jacobian may be individually adjusted if the Jacobian threatens to ill-condition the linear system because of uneven sensitivity of the problem to different parameters.

4.2 Improving the convergence by extrapolation

As mentioned previously, the most common pattern of solving problems in large deflection analysis of slender bodies is an expansion of the pilot solutions in various directions by stepping up or down certain parameters of a problem. The simplest way to generate guesses of unknown initial values and parameters for each next step in the sequence of solutions is to use the terminal values of the previous step. This technique is shown in the leftmost section of Table I (see the description below).

The graph of the correct initial values of curvature as a function of force in Fig. 4 shows that the function, although nonlinear, is monotonic and smooth. This is typical for most of the elastica problems. It implies the possibility of extrapolating the curves in order to obtain better guesses and faster convergence. The BCPP subroutine NERBEX is put between the main routine and the subroutine NERBEC. The subroutine NERBEX passes over to the subroutine NERBEC the initial guess for the first problem in the sequence (k=1); the following guesses or predictions $\tilde{P}$ are generated from the previous final values P as follows:

$$\tilde{P}_2 = P_1 \; ; \quad \tilde{P}_3 = P_2 + \Delta_{1,3} = 2P_2 - P_1$$

$$\tilde{P}_k = P_{k-1} + \Delta_{1,k} + \Delta_{2,k} = 3(P_{k-1} - P_{k-2}) + P_{k-3} \quad \text{for } k \geq 4$$

(10)

The critical situation in this extrapolation scheme usually occurs during the solution of the second problem in the series when there is no better guess available than the first final value. The chosen increment of the controlling parameter must be small enough to ensure the convergence of this second solution although it may be wasteful and unnecessary to keep it through the whole series of solutions.

There is a subroutine NERBIN which, when put between the Main routine and the NERBEX, subdivides or interpolates the first increments of the controlling parameter in order to ease the convergence at the beginning of the solution series. The first, second,...,n-th level of interpolation may be used with the effect of subdividing the first increment into two, four,...,2^n subincrements. A particular finer grid is maintained only for 5 solutions; then the 1-st, 3-rd and 5-th solutions are taken as forerunners for extrapolation when shooting for two more solutions (k=7 and k=9) with a coarser grid, and so on until the desired coarse grid is reached and the solution of the whole series can go on from there.

Table I
Extrapolation procedures (ring example)

F O R C E F_y	NERBEC			NERBEX+NERBEC			NERBIN+NERBEX+NERBEC		
	Prediction	True value	Iter. steps	Prediction	True value	Iter. steps	Prediction	True value	Iter. steps
0	2.000	2.000	0	2.000	2.000	0	2.000	2.000	0
2	2.000	2.607	3	2.000	2.697	3	2.000	2.607	3
4	2.607	3.159	3	3.214	3.159	2	3.214	3.159	2
6	3.159	3.664	3	3.657	3.664	2	3.657	3.664	2
8	3.664	4.127	3	4.120	4.127	2	4.120	4.127	2
10	4.127	4.557	3	4.550	4.557	2			
12	4.557	4.957	3	4.952	4.957	2	4.904	4.957	2
14	4.957	5.332	4	5.328	5.332	2			
16	5.332	5.682	4	5.682	5.686	2	5.647	5.686	2
18	5.686	6.021	4	6.018	6.021	2			
20	6.021	6.340	4	6.338	6.340	2			
22	6.340	6.645	5	6.643	6.645	2			
24	6.645	?		6.935	6.937	2	6.675	6.937	4
26	iteration failed			7.216	7.218	1			
28	to converge to			7.487	7.488	1			
30	correct solution			7.749	7.750	1			
32				8.002	8.003	1	7.881	8.003	3
Total			39			29			20

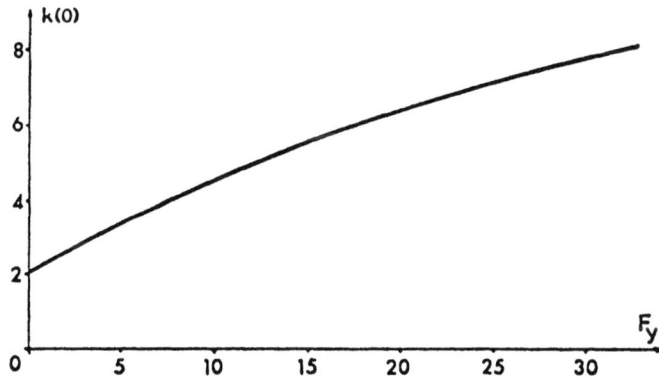

Fig. 4 Dependence of unknown initial condition k(0) on varying parameter F_y in Table I

The advantages of this procedure may be seen from Table I. We wanted a solution of an elastic ring for both diameter and bending rigidity equal to 1, extended by concentrated forces $F_y = 0, 8, 16, 24, 32$. The largest first step converging to the right solution was for $F_y = 2$. The solutions with NERBEC only, when final parameter value from the previous step is used as a guess for the next, failed to converge for $F_y > 22$. The extrapolation of guesses with NERBEX helped to reach the end of the investigated interval at the price of generating 12 unsought solutions in addition to the 5 required ones. Twenty three iteration steps were needed in order to reach the $F_y = 22$ as compared with 39 when using NERBEC only. The use of NERBIN+NERBEX+NERBEC at the 2-nd interpolation level produced only 5 unwanted solutions and covered the whole interval with 20 iteration steps in total as compared with 29 for NERBEX+NERBEC.

5. REPRESENTATION OF ORIGINAL CURVATURE

The load-free shape of a crimped or heat-set fibre, yarn or general slender body has to be expressed in terms of curvatures and torsion as functions of arc length, i.e. $k_o(s)$ or $p_o(s)$, $q_o(s), r_o(s)$ in the planar or spatial case respectively. Within the BCPP framework these functions are defined by the FORTRAN function subprograms CIN2D or CIN3D. The values of k_o or p_o, q_o, r_o and their derivatives are sampled by referencing these subprograms from the subroutines BEC2D1 or BEC3D1 (see Fig. 1). For an originally straight elastica the CIN2D or CIN3D are set to 0.

The original curvature is often defined not as an analytical function of s, but is obtained directly or indirectly from another solution of elastica. In that case we have values of curvature at 20 points of elastica (or multiple of 20 for a compound elastica) to interpolate through during the integration. The cubic spline interpolation technique has proved to fit our purpose. For any value of s

$$k = c_{1,i} + c_{2,i}s + c_{3,i}s^2 + c_{4,i}s^3$$

$$k' = c_{2,i} + 2c_{3,i}s + 3c_{4,i}s^2, \quad s_{i-1} \leq s \leq s_i.$$

(11)

Two of the 80 cubic coefficients $c_{1,4} = c_{20,4} = 0$ and the remaining 78 are found from a set of 78 linear equations. Forty of them ensure that each cubic passes through both terminal points of the interval

$$k_{i-1} = c_{1,i} + c_{2,i}s_{i-1} + c_{3,i}s_{i-1}^2 + c_{4,i}s_{i-1}^3$$
$$k_i = c_{1,i} + c_{2,i}s_i + c_{3,i}s_i^2 + c_{4,i}s_i^3$$

(12)

At each point s_i, k_i the first and the second derivatives of both left and right cubic have the same value, i.e.

$$c_{2,i} + 2c_{3,i}s_i + 3c_{4,i}s_i = c_{2,i-1} + 2c_{3,i-1}s_i + 3c_{4,i-1}s_i^2$$
$$2c_{3,i} + 6c_{4,i}s_i = 2c_{3,i-1} + 6c_{4,i-1}s_i$$
(13)

The choice of spline cubic for interpolation offers the sufficient smoothness and precision required in order to prevent oscillation when solving boundary value problems. The solution of 78 cubic coefficients for a single elastica with curvature values given at 21 equidistant points is trivial: it is performed by subroutine SPLBEC and it takes advantage of the inverse of a normalized matrix of linear system coefficients defined in a BLOCK DATA segment in COMMON/SPLBAR/.

In the case of a compound elastica the cubic spline coefficients have to be evaluated by solving the linear system from scratch using subroutines SPLIN2 (which generates linear systems for up to 101 arbitrary points s_i, k_i) and LINEL3 (sparse matrix version of subroutine LINEL2 for solving linear systems using Gaussian elimination with partial pivoting).

6. INTERACTIVE FACILITY

The Bending Curve Program Package was originally designed for batch processing environments. Although advantage was taken later of interactive editing when reformulating the problems and initiating jobs from a remote terminal, the need to re-edit and recompile special parts of programs makes the solution of complicated problems a frustrating experience.

The interactive and conversational versions QBC2 and QBC3 (for planar and spatial elastica respectively) considerably speeded up searches for pilot solutions, and the operation of BCPP in general. In particular, the QBC3 was instrumental in finding the solution of twist-textured fibres (see section 3.3 of the next paper).

The QBC2 or QBC3 dialogue alternates between three-number user entries and computer responses. An unsophisticated list of commands (see Table II) provides for individual assignment of the components of the array of initial conditions and parameters B, specification of output information (printing of selected components of the result array A, writing complete numerical or graphical output information into a file for future disposal, specifying the level of iteration trace), formulating the boundary value problem (by specifying free parameters and boundary conditions), launching a single or a series of solutions with NERBEC, NERBEX or NERBIN, enquiring about the status of conversation, and

Table II
List of QBC commands

Command	Action
0,0,0	stop the program
1,0,0	list input array B
1,i,0	print B(i)
2,0,0	list output assignments
3,1,0	switch numerical output on
3,2,0	switch numerical and graphical output on
3,3,0	switch output off, copy accumulated output into file defined in response to FILE NAME ?
4,0,0	list indeces of unknown parameters and amplification factors
5,0,0	list indeces and boundary value array BOV
6,0,0	print NI,PRECIS
8,0,0	save current problem under defined filename
9,0,0	restore problem saved under defined filename
10,i,p	assign B(i)=p (default: all B(1,2,...,12)=0)
21,i,0	assign A(i) as output
22,i,0	remove A(i) as output
30,0,0	solve initial value problem and print elements of array A assigned to output
41,i,k	assign B(i) as unknown parameter and and 10^k as amplification factor
42,i,0	remove B(i) as unknown parameter
51,i,p	assign A(i)=p as boundary condition
52,i,0	remove A(i) as boundary condition
61,n,0	assign number of integration intervals NI=n (multiple of 20)
62,k,p	assign error limit PRECIS=p.10^k
7j,0,0	launch single Newton-Raphson iteration
8j,i,Δ	launch series of Newton-Raphson iterations with controlling parameter B(i) and increment Δ; additional question: NUMBER OF STEPS?
9j,i,Δ	launch series of iteration through NERBIN with controlling parameter B(i) and increment Δ; additional questions: NUMBER OF STEPS ? INTERPOLATION LEVEL ?

j=0,1,2 or 3 indicates the level of iteration trace

saving and restoring the status of conversation (important when exploring the conditions of solving difficult problems).

As an example of the QBC2 use we reproduce the user entries and computer responses during the solution of a series of the ring deformation problems, the results of which are reported in the right part of Table I. The user's part is underlined and the explanations are given on the right:

```
/RUNQBC2
? 10,1,.7853983     assign length ℓ=B(1)=π/4
? 10,2,1              "    bending rigidity A=B(2)=1
? 10,3,2              "    curvature k(0)=B(3)=2
? 10,7,1              "    direction cosine w_x(0)=B(7)=1
? 10,12,.001          "    integr. error limit  =B(12)=.001
? 21,2,0           request output coordinate x(ℓ)=A(2)
? 21,3,0              "      "      - " -    y(ℓ)=A(3)
? 21,6,0              "      "     curvature k(ℓ)=A(6)
? 61,40,0      define number of integration intervals NI=40
? 41,3,0       select curvature k(0)=B(3) as unknown parameter
? 51,4,0       define dir. cosine w_x(ℓ)=A(4)=0 as boundary value
? 62,-6,1      define boundary value error limit PRECIS=.000001
? 91,5,-8              launch series of solutions via NERBIN
   NUMBER of steps ?  5        with control. parameter F_y=B(5)
   INTERPOLATION level ?  2    and incrementΔ =-8

    NS =  0                    no of iter. steps and final va-
2.00000000000E+00  -7.914E-07  lues of k(0)=B(3) and w_x(ℓ)=A(4)
    5     0.                   controlling parameter F =B(5)
    2     5.000000E-01         coordinate x(ℓ)=A(2)
    3     5.000000E-01         coordinate y(ℓ)=A(3)
    6     2.000000E+00         curvature k(ℓ)=A(6)

    NS =  3
2.607108257E+00  -3.564E-12
    5    -2.000000E+00
    2     4.673404E-01
    3     5.332208E-01
    6     1.672427E+00
.........................
    NS =  4
6.936845667E+00  -4.012E-07
    5    -2.400000E+01
    2     2.746140E-01
    3     6.666785E-01
    6     3.461097E-01
                                  ⎫
    NS =  3                       ⎬ the last two problems of 9
8.003041771E+00  -3.018E-07       ⎭ in the right part of Table I
    5    -3.200000E+01
    2     2.432064E-01
    3     6.821106E-01
    6     2.204385E-01
? 0,0,0
   4.800 CP SECONDS EXECUTION TIME
```

7. CONCLUSION

The concept of the Bending Curve Program Package and the design of its components represent a good compromise between computing efficiency and user convenience from the viewpoint of current needs of the mechanics of flexible fibre assemblies and large deflection analysis of slender bodies. The interactive amendment described in chapter 4 speeds up the process of formulating new problems, developing pilot solutions and expanding pilot solutions over the multidimensional space of variable parameters.

There are two main avenues for further enhancement of existing computer programs and techniques:

1. The accumulated expertise in formulating the problems and developing pilot solutions should be generalized and a substantial part of the operation should be performed automatically. For instance, given the structure topology, an intelligent program would automatically determine a structural repeat, select relevant known and unknown initial conditions and parameters, define boundary conditions, and, by a carefully controlled trial and error method, find a pilot solution.

2. The numerical solution of the differential equations of elastica has a tendency to become unstable after reaching certain critical loading conditions. Consequently, the iteration procedure fails to converge to a solution satisfying the boundary conditions. The interval of solvability may be extended through an automatic switching to one of the finite element methods the performance record of which is opposite to that of elastica: they are difficult to use within the interval of true large bending/torsion or in a post-buckling situation, and their reliability increases with increasing loads or increasing role of geometric constraints. A similar positive effect may be expected from the combination of the elastica approach, being essentially a force method, with the generalized energy method developed and described in this volume by de Jong.

REFERENCES

[1] Konopasek, M., "Program package for large deflection analysis of thin rods and their asemblies", in 'Proceedings of International Conference CAD 74', London, 1974.
[2] "Bending Curve Package on 1906A - 7600" (Users' manual), UMIST, Manchester, 1976.
[3] "Bending Curve Program Package", Users' notes 1 - 5, Georgia Tech, Atlanta, 1976.
[4] Atkinson, K.E., "An Introduction to Numerical Analysis", John Wiley, 1978.

TEXTILE APPLICATIONS OF SLENDER BODY MECHANICS

Milos Konopasek

School of Textile Engineering, Georgia Institute
of Technology, Atlanta, Georgia 30332

ABSTRACT. The application of elastica theory in the field of the mechanics of flexible fibre assemblies and the use of the Bending Curve Program Package (BCPP) is illustrated by a number of examples. The need and design of a more general treatment covering the statics, kinematics and dynamics of fibres or slender bodies with finite bending and torsional rigidity is outlined.

1. INTRODUCTION

Nonlinear and time-dependent stress-strain characteristics of textile fibres in combination with large tensile, bending and shear deformations of fibres and fibre assemblies are notoriously difficult to handle in both engineering practice and theoretical analyses.

The reduction of the number of independent geometrical variables to one (coinciding with longitudinal axis of the fibre or yarn, or with the orthogonal projection of a cylindrical sheet) is usually the most significant and the least harmful simplification. It turns the problems of textile mechanics into those of the mechanics of one-dimensional continua.

Two classes of nontrivial problems may be interpreted in terms of ordinary differential equations with solutions easily obtainable by numerical, or, in some cases, analytical means:
(a) problems of static equilibrium of elastic fibres with finite bending and torsional rigidity, and
(b) problems of statics, kinematics and dynamics of the elastic and perfectly flexible fibres.

The first class covers all the textile applications of Euler's and Kirchhoff's elastica as pioneered by Peirce and followed by an increasing number of textile scientists in the last two decades. The second class stretches from capstan friction and the catenary to the phenomena of ballooning and whipping - all taking into account the inertial mass of the slender body, but neglecting its finite bending and torsional rigidity. The single independent variable in the first case is the arc length of the longitudinal axis, and in the second case the time.

The problems of the first class can be taken good care of by the generalized theory of elastica and by the computation framework described earlier in this volume. We will display a number of formulations and solutions of the problems belonging to this class in order to demonstrate the versatility and uniformity of our approach. By doing so we will concentrate on the methodology of application of the available theoretical and computational apparatus rather than the details of particular problems.

Current developments in fibre manufacturing and processing as well as in intermediate and end use of fibres and fibre assemblies bring about a pressing need for expansion of the above two classes and their fusion under a generalized approach covering differential changes along both space and time axes. The resulting nonlinear partial differential equations with respect to both arc length and time do not have analytical solutions and do not submit themselves easily to numerical treatment.

No framework, comparable with that for solving the problems of (a), is available for analysis and simulation of the dynamics of fibres or slender bodies with finite bending and torsional rigidity, and even the class of problems (b) alone seems to have been neglected in the past. Consequently, we cannot but limit ourselves to just a few notes on the subject in the last chapter.

2. PLANAR ELASTICA

The first two of the following examples deal with large deflections of a yarn in a plane. The first is a case of yarn embedded in a textile structure (plain woven fabric), and the second is an analysis of a free yarn segment in a textile process (embroidery loop). The last three examples represent cylindrical bending of sheet material interpreted in terms of planar elastica.

2.1 Plain woven fabric

The repeat of plain woven fabric consists of one period of both warp and weft yarn. The analysis interval may be reduced to quarter periods because of symmetry. If the fabric is sufficiently loose, the quarter-periods may be represented by horizontal cantilevers.

Three unknown parameters of the system are two maximum curvatures at the crown of the waves (or corresponding moments M_1 and M_2, see Fig. 1) and an interlacing force F_y. The first two boundary conditions are zero values of the curvatures at the inflection points of the waves. The third boundary condition requires both inflection points to belong to a single plane parallel to the plane of the fabric due to symmetry. The problem formulation:

given lengths of quarter-periods $\quad \ell_1 = L_1/2$
$\qquad\qquad\qquad\qquad\qquad\qquad\qquad \ell_2 = L_2/2$
initial curvatures $\quad k_1(0) = ?$
$\qquad\qquad\qquad\quad k_2(0) = ?$
interlacing force $\quad F_y = ?$
initial direction cosines $\quad w_{x1}(0) = w_{x2}(0) = 1$

Boundary conditions: $\quad k(\ell_1) = 0$
$\qquad\qquad\qquad\qquad\quad k(\ell_2) = 0$
$\qquad\qquad\qquad\qquad\quad y(\ell_1) + y(\ell_2) = (d_1 + d_2)/2$

where $\quad L_1, L_2$ are yarn lengths in a fabric cell,
$\qquad\quad d_1, d_2$ are yarn diameters.

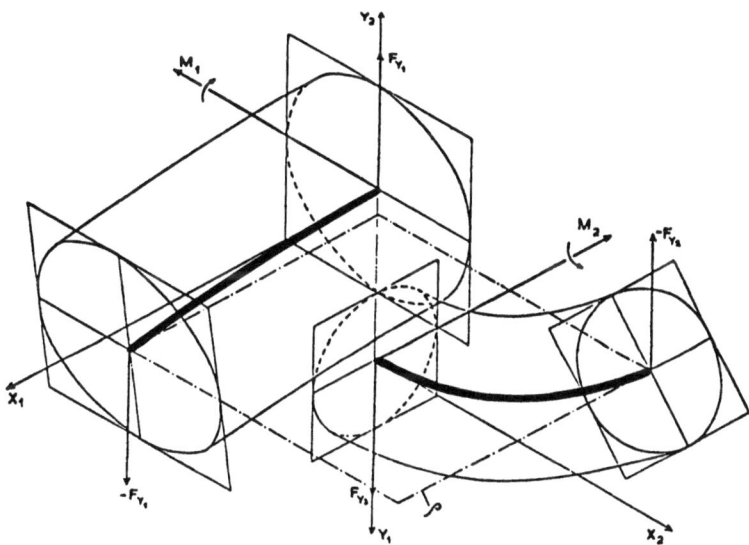

Fig. 1. Elastica model of plain woven fabric (ρ is the plane of inflection points, see the last boundary condition)

In a more tight structure the finite lengths of yarns are wrapped around the yarn surfaces, the free lengths of yarns subject to the solution of the elastica equations are reduced and the problem is to be reformulated as follows:

$\ell_1 = (L_1-(d_1+d_2) \cdot \text{arcos } w_{x1}(0))/2$ $F_{y1} = F_{y2} = ?$

$\ell_1 = (L_2-(d_1+d_2) \cdot \text{arcos } w_{x2}(0))/2$ $w_{x1}(0) = ?$

$k_1(0) = k_2(0) = 2/(d_1+d_2)$ $w_{x2}(0) = ?$

The iterative solution of practical problems often oscilates around the boundary line between the above defined loose and tight structures. It would be inconvenient to keep manually reformulating the problems during the solution. Both cases may be unified, as shown in [1], by using an exponential transformation and automatic selection of unknown $w_x(0)$ or $k(0)$ whichever is applicable:

$k_i(0) = \min(P_i, 2/(d_1+d_2))$

$w_{xi}(0) = 1/\exp((\max(P_i(d_1+d_2)/2, 2)+2)^2)$ $i=1,2$

$\ell_i = (L_i-(d_1+d_2) \cdot \text{arcos } w_{xi})/2$

The formulation may be easily modified for simulating fabric under uniaxial or biaxial tension, and under bending moment acting along a weft or warp direction as discussed elsewhere in this volume by Leaf and Grosberg. In the first case all one needs to do is to assign finite values of F_{x1}, or F_{x2}, or both.

2.2 Embroidery loop

In the embroidering or sewing process a needle with a thread alternately penetrates the fabric and withdraws. During the needle withdrawal one branch of the thread withdraws as well when pulled back by the tensioning mechanism. The other branch the end of which is clamped between the needle and the fabric, buckles. Bruggisser [2] analysed the buckling forces and the shape of the buckled loop as depending on the length of the thread and the position of the needle.

The initial conditions and parameters are:

length	ℓ - given
curvature	$k(0) = ?$
slope	$w_x(0) = 0$
force component	$F_x = ?$
" "	$F_y = ?$

The boundary conditions are:

$$k(\ell) = 0 \; ; \quad x(\ell) = 0 \; ; \quad y(\ell) - \text{given}.$$

Typical results of a computational experiment are shown in Fig. 2.

2.3 Pierce's ring, pear and heart

Pierce's classical study on the measurement of fabric hand [3] involved an analysis of the load/deformation characteristics of ring-, pear- and heart-shaped samples (Fig. 3). Pierce suggested amazingly accurate approximate analytical relationships as a substitute for nonexisting analytical and unobtainable (50 years ago!) numerical solutions. The formulation and solution of these problems using BCPP is very simple.

It is sufficient to integrate along one quarter of the ring only because of symmetry. The initial conditions and parameters are:

length	ℓ - given
curvature	$k(0) = ?$
slope	$w(0) = 1$
horizontal force	$F_x = 0$
vertical force	F_y - given

The boundary condition is: $w_x(\ell) = 0$.

Single line symmetry in cases of pear or heart specimens requires integrating over half of the length. The initial conditions and parameters for a pear are:

length	ℓ - given
curvature	$k(0) = ?$
slope	$w(0) = 0$
"	sign $w_y(0) = -1$
horizontal force	$F_x = ?$
vertical force	F_y - given

The boundary conditions are: $x(\ell) = 0 \; ; \quad w(\ell) = 0$.

For the heart shape the initial conditions and parameters as well as boundary conditions are the same except for the sign $w_y(0) = +1$.

Areal density or own weight of the fabric may be easily accounted for in all three cases by assigning a non-zero value of distributed force f_y.

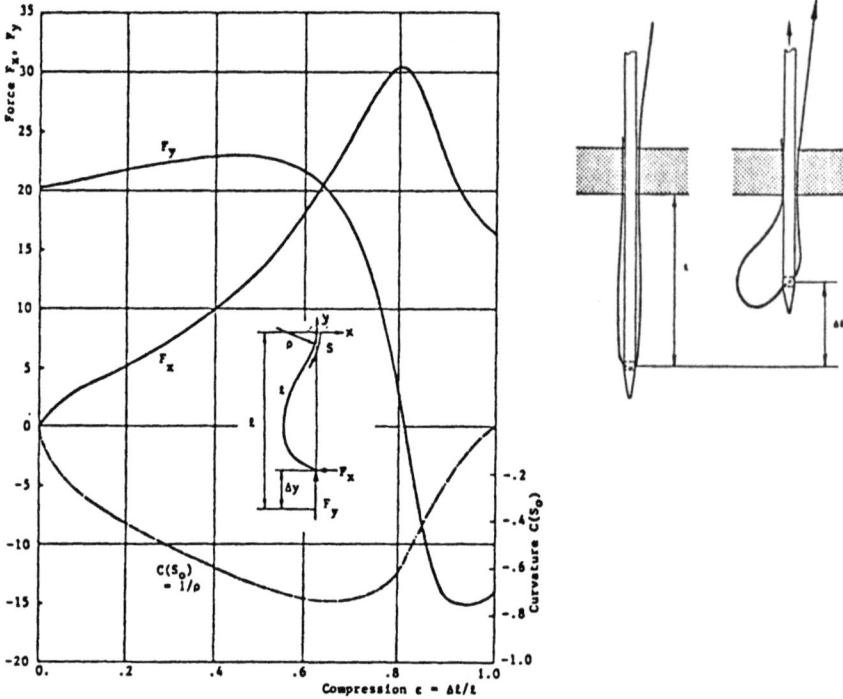

Fig. 2. Model of embroidery loop and typical computation results

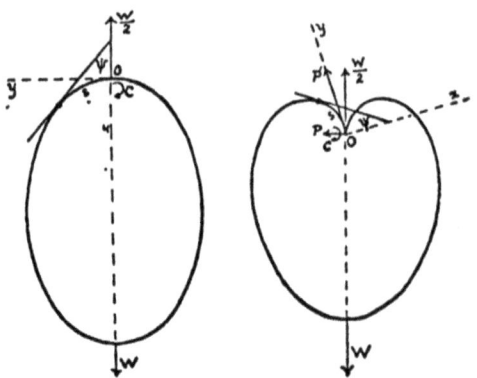

Fig. 3. Pierce's ring and heart from [3]

2.4 Folding of heavy fabric sheets

The folding of heavy fabric sheets analyzed by Lloyd et al. [4] represents a multistage compound curve situation, which is not unusual in the large deflection analysis of textile structures.

As shown on Fig. 4, the fabric is fed vertically in such a way that it buckles and eventually forms folds on one side and then on the other. There are altogether 6 stages which differ in the definition of their initial values and boundary conditions.

At the first stage, which represents a simple problem of buckling a vertical column under its own weight, the unknown parameters are the curvature and forces at the feeding point and the boundary conditions are 0,0 coordinates and zero curvature at the terminating point. The stage 2 starts at the moment when the slope w( )=-1 and the sheet starts rolling on the surface. A new boundary condition is w( )=0 instead of x( )=0.

Stage 3 takes place and stage 4 begins when the lowest point of the sheet touches the horizontal surface splitting one bending curve into two. At stage 4 there are 5 unknown initial conditions and 5 boundary conditions; there is only a single contact point in the middle, which rolls to the left when more

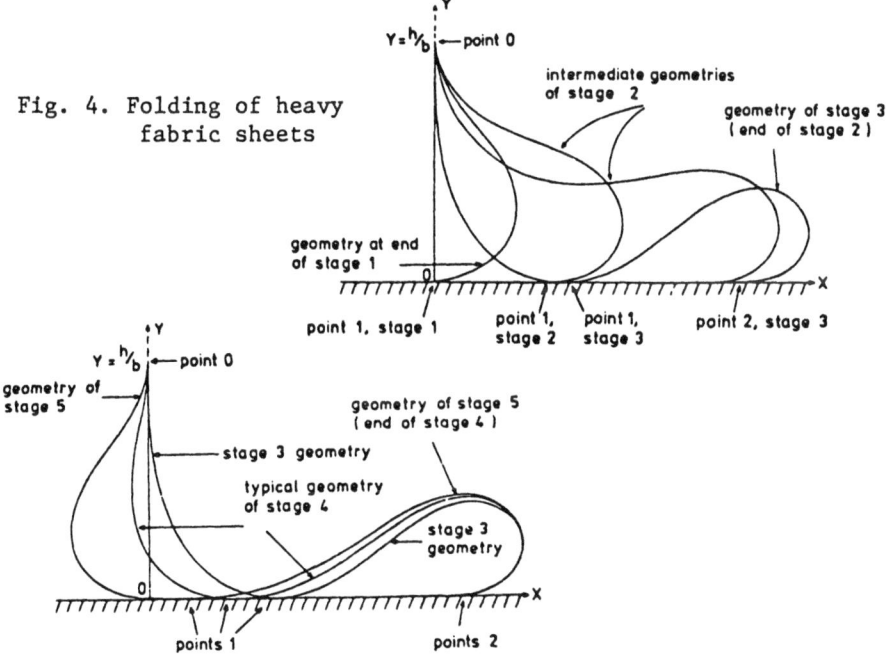

Fig. 4. Folding of heavy fabric sheets

fabric is fed in and which, having a nonzero curvature, affects the shape of both the fed-in left branch and the closed right branch. The curvature at this point continues to decrease. When it reaches zero (stage 5), the fed-in and closed branches become separate and the shape of the closed branch stabilizes. Then a brief sixth stage brings the rolling point of the fed-in branch to the coordinate origin so that it becomes equivalent to the previous stage 3 and the process repeats at the left side of the coordinate system.

The whole process may easily be simulated by assigning and reassigning the initial values and boundary conditions in the subroutine INTMED (see previous paper).

2.5 Simulation of fluid filled flexible walled tank

The ratios of length to perimeter of free standing flexible walled fluid tanks are usually large enough so that they can be considered as cylindrical surfaces and their contour cross-section analyzed using the bending curve approach. In an analysis by Olson [5] the hydrostatic pressure acting on the tank walls was expressed in terms of a distributed load. This required a small modification in the subroutine BEC2D1. The boundary conditions were similar to those discussed in a previous paragraph. An example of a solution of the fluid filled flexible tank is shown in Fig. 5.

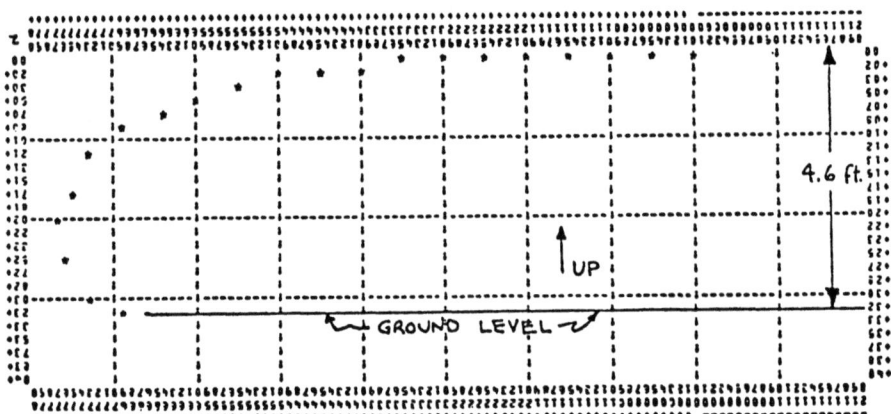

Fig. 5. One half of the cross-section contour of a fluid-filled tank (see the dots representing 21 equidistant points along the curve, BCPP output option)

3. SPATIAL ELASTICA

The first of the following sections deals with helices seen as trivial solutions of the equations of spatial elastica. The second and third sections cover examples from the field of fibre buckling and crimp. The last two sections deal with elastica interpretation of yarn embedded in textile structures: non-plain woven fabrics and plain knitted fabrics.

3.1 Helices in fibre assemblies

Helices play an important role in the mechanics of fibre assemblies. It is useful to realize that a helix is a trivial case of spatial elastica and that a helix equation represents an analytical solution of the elastica equation. Indeed, the differential equations for physical curvatures and torsion (see the general paper on elastica in this volume) in the absence of forces and original curvature, and when the bending rigidities A=B, simplify as follows:

$$p' = (A-C) q.r \; : \quad q' = (C-A).p.r \; : \quad r' = 0 \; ,$$

and their solution:

$$p = p(0).\cos(\alpha s) \; ; \quad q = p(0).\sin(\phi s) \; ; \quad r = r(0)$$

where $\phi = (C-A).r(0)$.

Substitution into the equations linking geometric and physical curvatures and torsion yields

$$K = (\sin^2\phi)/R = p(0) \quad \ldots\ldots\ldots\ldots \text{geometric curvature}$$

$$\tau = (\sin\phi.\cos\phi)/R = r(0).(C-A+1) \quad \ldots \text{geometric torsion}$$

and consequently

$$\tan \phi = p(0)/(r(0).(C-A+1)) \; ; \quad R = 1/(p(0).(\cotan \phi +1))$$

where ϕ is the helical angle,
       R is the radius of the generating cylinder.

A generalized analytical solution of spatial elastica in helical form was applied to an analysis of two-ply twisted yarn reported in [6]. This analysis yielded, for instance, an equation for yarn torsional moment

$$M_y = [8A.\sin^3\phi + 2C.\cos 2\phi (2\sin\phi + r_o D/\cos\phi)]/D + DF_y \tan(\phi/2)$$

and for distributed normal forces acting between singles

$$f = 2\sin\phi \cdot \tan\phi [2\sin\phi (Cr_o D - \sin\phi (A-C))/D^2 - F_y/2]/D$$

where F_y is the yarn tension,
D is the yarn diameter,
r_o is the original single's torsion.

Another example of analytical interpretation of a helix derived immediately from the differential equations of elastica is given in [7] dealing with helical models of collapsed twist-textured fibres. If p_o and r_o are heat set curvature and torsion respectively, and $q_o = p_o' = q_o' = r_o' = 0$, the only meaningful differential equation

$$q' = [C \cdot p(r-r_o) - A \cdot r(p-p_o) - F_u]/B + q_o'$$

turns into

$$F_u = F_z \sin\phi = C \cdot p(r-r_o) - A \cdot r(p-p_o)$$

where ϕ is the helix angle of the collapsed helix,
F_z is the extending force.

3.2 Bicomponent and twist-textured fibres

Helical models of bicomponent or twist-textured fibres approximate only very roughly the real situation in a fibre the original load-free shape of which is, respectively, a circle or a helix. If the rectifying moments are applied to the ends of such fibres, causing the circle or the helix to open, the instability of the rectified shape leads to spontaneous collapse into quasi-helical branches separated by reversal points.

There are two (left and right) quasi-helical segments in a full repeat of a collapsed bicomponent fibre. Each of these two segments is flanked by two reversal points. Owing to plane symmetry around the middle reversal point and the point symmetry around the point of inflection in the middle between two reversal points, only one quarter of the repeat has to be interpreted and solved as spatial elastica. Let it be the quarter between a reversal point (s=0) and an inflection point (s= ℓ).

In the process of formulating the problem we choose the stretching force, and consequently the tangency vector $\overline{w}$ at the reversal point to be oriented along the y-axis, and the vector $\overline{u}$ to be oriented along the x-axis. We assume the original load-free curvature p_o =const and $q_o = r_o = 0$. Thus the initial conditions and parameters

$p(0) = ?$ $F_x = 0$ $u_x(0) = 1$
$q(0) = 0$ F_y - given $w_x(0) = 0$
$r(0) = 0$ $F_z = 0$ $w_y(0) = 1$

The boundary conditions

$$q(\ell) = 0 ; \quad v_y(\ell) = 0$$

follow from symmetry considerations requiring an opposite sign of the curvature q in two halves of a branch. For the same reason the vector $\overline{V}$ must be perpendicular to the y-axis.

The problem of a twist-textured fibre differs from the above by a non-zero value of the original torsion r reflecting the helical load-free shape. The plane symmetry around the reversal point does not apply any more and we have to integrate over one half of the repeat from one inflection point to another. The initial conditions are:

$$p(0) = ? \qquad F_x = 0 \qquad u_x(0) = 0$$
$$q(0) = 0 \qquad F_y - \text{given} \qquad w_x(0) = 0$$
$$r(0) = ? \qquad F_z = 0 \qquad w_y(0) = ?$$
$$\text{sign } w_z(0) = 1$$

The boundary conditions are:

$$q(\ell) = 0 ; \quad v_y(\ell) = 0 ; \quad v_z(\ell) = 0 .$$

The last two boundary conditions imply orientation of the vector $\overline{V}$ along the x-axis, i.e. the same as at the beginning point. The initial and boundary conditions are equivalent to holding the collapsed fibre on pins inserted at inflection points. The pins are oriented along the x-axis allowing free rotation of the fibre cross-section around the vector $\overline{V}$ implying zero moment M_V and curvature q (Fig. 6).

The solution of the twist-textured fibre was found to be extremely sensitive to the guesses of the values of unknown initial conditions. The dramatic search for a pilot solution is described in detail in [8]. It took the full power of the interactive extension of BCPP to move gradually from a known bicomponent fibre configuration to the first twist-textured one. Once this first or pilot solution was reached, it was not difficult to expand it over a multidimensional space of varying ℓ, F_z, C/A, p_0 and r_0, although the convergency of the iterative procedure of solving the boundary value problems did not extend over very slack or very strained cases.

3.3 Plane-restricted buckling

As opposed to the planar elastica intepretation of buckled and folded sheets described in section 2.4, the buckle of a fibre or yarn fed against a solid plane turns from a planar to a spatial one (see Fig. 7). The diameter of a circular buckle depends on the distance H of the buckling plane from the feeding

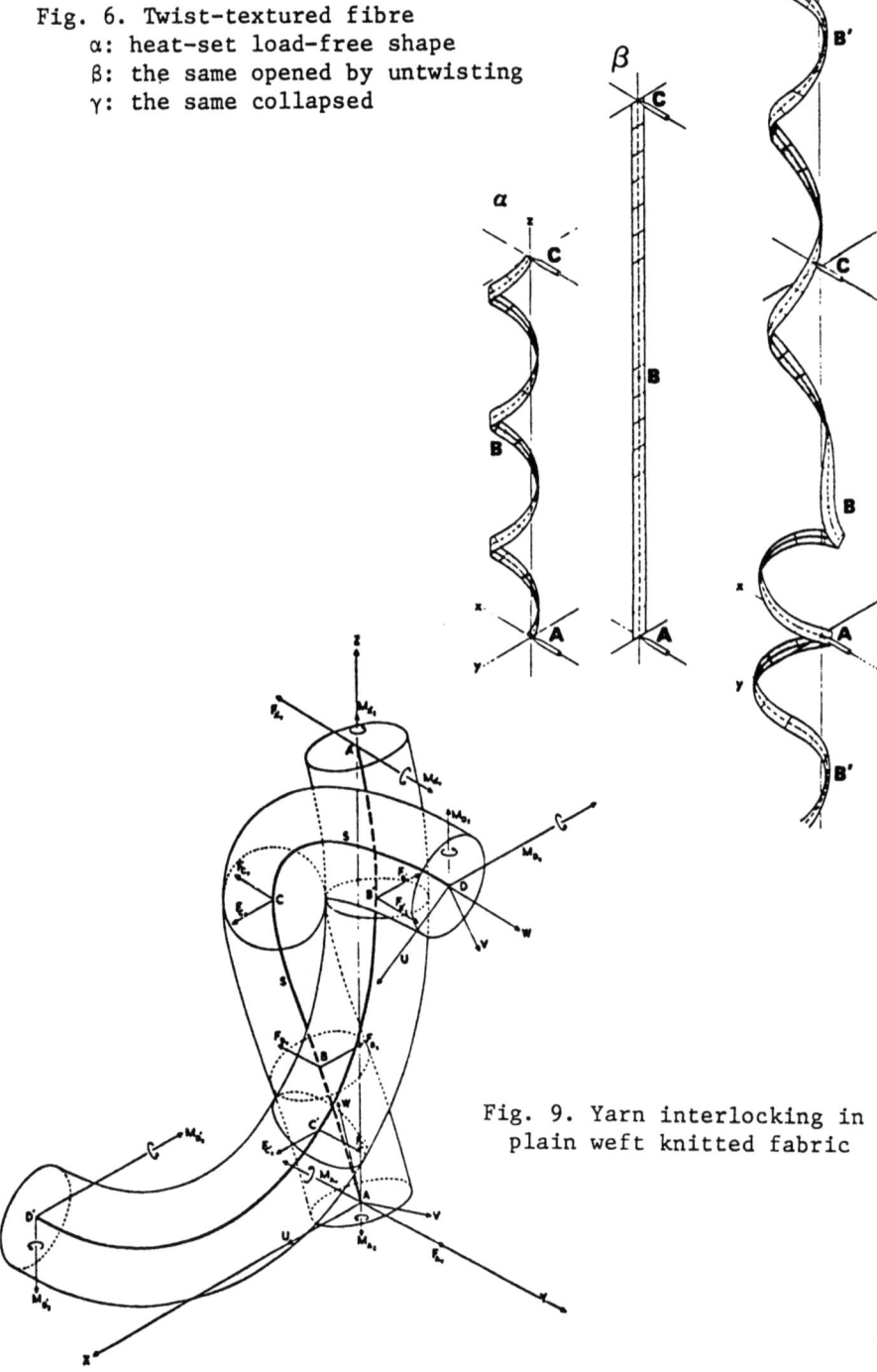

Fig. 6. Twist-textured fibre
α: heat-set load-free shape
β: the same opened by untwisting
γ: the same collapsed

Fig. 9. Yarn interlocking in plain weft knitted fabric

point, bending and torsional rigidity, and, in case the gravitational forces count, on the fibre or yarn linear density.

The differential equations of elastica may be integrated either from the fibre-plane contact point to the feeding point or vice versa. In the former case the initial conditions and parameters are as follows:

length:	$\ell = ?$				
curvatures:	$p(0)=?$	;	$q(0)=0$	;	$r(0)=-p(0)$
direction cosines:	$u_y(0)=0$	;	$w_x(0)=1$	;	$w_y(0)=0$
forces:	$F_x= ?$	;	$F_y= ?$	;	$F_z= ?$

(The equality of the torsion and curvature at the supporting plane $r(0)=-p(0)$ follows from the fact that there is one full twist generated for each complete coil.)

There are five boundary conditions for five unknown initial conditions and parameters:
1.-2. The tangent to the axis at the feeding point is parallel to the z-axis, i.e. $w_x(\ell)=w_y(\ell)=0$;
3. the feeding point belongs to the y,z plane, i.e. $x(\ell)=0$;
4. the distance between the s=0 point and the projection of the feeding point on the x,y plane must be $y(\ell)=1/p(0)$;
5. $z(\ell)=H$ for the given distance between the feeding point and supporting plane.

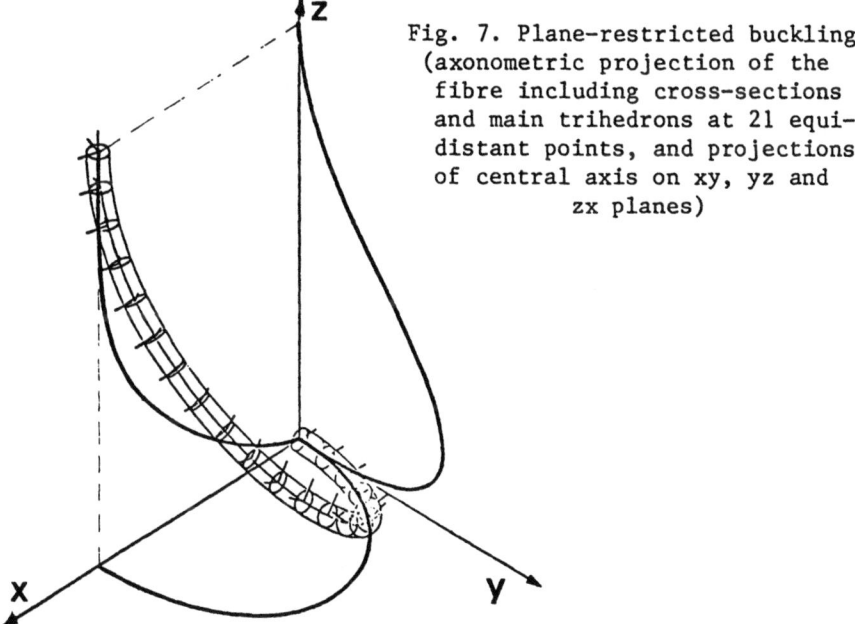

Fig. 7. Plane-restricted buckling (axonometric projection of the fibre including cross-sections and main trihedrons at 21 equidistant points, and projections of central axis on xy, yz and zx planes)

The solution of this problem may help in the analysis of the stuffer-box texturing process, and it also may be used for the simulation of the deposition of yarns or fibres in the fabrication of nonwovens.

3.4 Non-plain woven fabrics

Any departure from plain woven structure (or basket weave which is only a minor modification of the latter) means loss of symmetry and consequent deviation of the yarn axes out of planes perpendicular to the fabric plane. This situation was exemplified in an analysis of the square-set 2/2 twill in [9]. The repeating unit of the fabric was represented by two segments of spatial elastica, one half of an alternating segment and one half of a nonalternating segment. The integration was carried out from the inflection point (middle of the alternating segment, $s=0$) to the interlacing point ($s=\ell_1$) and it continued (after reassignment of forces) from the interlacing point to the middle of the nonalternating segment. The initial conditions and parameters were:

length	$\ell_1 = ?$	$\ell_2 = L-\ell_1$, L is given
u-comp. of curvature	$p(0)=0$	$p(\ell_1)$ carried over
v-comp. of curvature	$q(0)=0$	$q(\ell_1)$ carried over
physical torsion	$r(0)=0$	$r(\ell_1)=0$
direction cosines	$u_x(0)= 1-w(0)$	$u_x(\ell_1)$ carried over
" "	$w_x(0)= ?$	$w_x(\ell_1)$ " "
" "	$w_y(0)= ?$	$w_y(\ell_1)$ " "
force components	$F_x = ?$	$F_x = ?$
" "	$F_y = ?$	$F_y = ?$
" "	$F_z = ?$	$F_z = ?$

The eight unknown initial conditions and parameters may be reduced to six owing to the symmetry of the crossing warp and weft yarns, and accounting for known shear forces or biaxial tension.

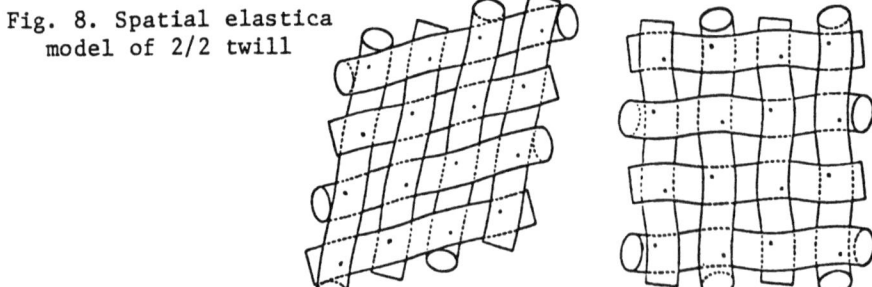

Fig. 8. Spatial elastica model of 2/2 twill

There are the following six boundary conditions (we skip the elaborate symbolic expressions given in [9]): 1. the projection of the yarn axis on the fabric plane must have a point of inflection at the point $s=\ell_2$; 2. the tangent to the yarn axis at that point must be parallel to the fabric plane; 3. the vector of the interlacing force must be perpendicular to the tangent of the yarn axis at the point $s=\ell_1$; 4. by symmetry the contact point and the point $s=\ell_2$ must belong to the same plane parallel to the fabric plane; 5. the contact point must belong to the line of symmetry of the warp and weft yarns; 6. the position of the contact point with respect to the position of the point $s=\ell_2$ must satisfy the given skew angle.

These are certainly more complicated boundary conditions than those we dealt with in previous examples (this is characteristic for segments of spatial elasticas embedded in textile structures). The analysis of 2/2 twill structure confirmed an observed asymmetry of shear modulus and revealed the inherent instability of this fabric (see Fig. 8).

3.5 Plain weft knitted fabric

A true mechanical model of the plain weft knitted fabric developed simultaneously by Hepworth and Leaf [10] and Konopasek [1] may be considered as one of major feats of the applications of the elastica in the mechanics of textile structures.

Thanks to symmetry it is possible again to consider only one quarter of a knitted stitch - from the middle of a stitch wall to the middle of the loop top (see Fig. 9). There are two interlacing points breaking this interval into three segments. The initial conditions and parameters were defined as follows:

length	$\ell_1 = ?$	$\ell_2 = ?$	$\ell_3 = L-\ell_1-\ell_2$, L given
curvature	$p(0)=0$	$p(\ell_1)$ C.O.	$p(\ell_2)$ C.O.
- " -	$q(0)=0$	$q(\ell_1)$ C.O.	$q(\ell_2)$ C.O.
torsion	$r(0)=0$	$r(\ell_1)=0$	$r(\ell_2)=0$
slope	$u_x(0)=1$	$u_x(\ell_1)$ C.O.	$u_x(\ell_2)$ C.O.
"	$w_x(0)=0$	$w_x(\ell_1)$ C.O.	$w_x(\ell_2)$ C.O.
"	$w_y(0)=?$	$w_y(\ell_1)$ C.O.	$w_y(\ell_2)$ C.O.
force	$F_{x1}=0$	$F_{x2}=F_{y2}w_y(\ell_1)/w_z(\ell_1)$	$F_{x3}=0$
"	$F_{y1}=0$	$F_{y2}=F_{y1}/2$	$F_{y3}=0$
"	$F_{z1}=0$	$F_{z2}=0$	$F_{z3}=0$

(C.O. means "carried over from previous segment")

The five boundary conditions were derived from the conditions of symmetry of two interlacing quarter-loops (1,2), geometric constraints in the interlocking points (3), and continuity of loop tops (4,5):

1. $w(\ell_2)/w_x(\ell_2) + w_y(\ell_1)/w_x(\ell_1) = 0$

2. $[x(\ell_1)-x(\ell_2)]/[y(\ell_1)-y(\ell_2)] - w(\ell_1)/w(\ell_2) = 0$

3. $[x(\ell_1)-x(\ell_2)] + [y(\ell_1)-y(\ell_2)]^2 - 1 = 0$

4. $w(\ell_3) = 0$

5. $w(\ell_3) = 0$

All the properties of this model showed good general agreement with real structures and it also finally dispelled the myth about the twist believed to be imposed by the topology of knits (an outgrowth of the confusion of geometric for physical torsion). The definition interval of the model, however, covered only rather loose structures with yarn diameter over stitch length ratio (d/ℓ) smaller than 0.0313 as compared with realistic values around or greater than 0.06. The reasons were (a) no provision was made for contact between the yarns other than in interlacing points, (b) the assumption about circular and non-deformable yarn cross-section, and (c) disregard for the tendency of the yarn to develop plastic bending deformation (the yarn was assumed to be straight in the load-free situation). Konopasek showed in [1] that removal of assumption (c) alone

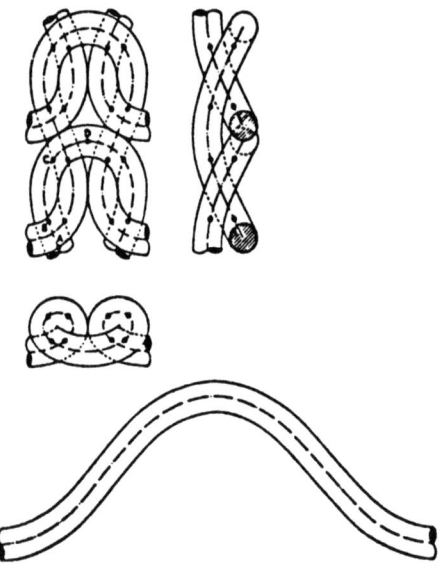

Fig. 10. Spatial elastica model of partially relaxed plain weft knitted fabric

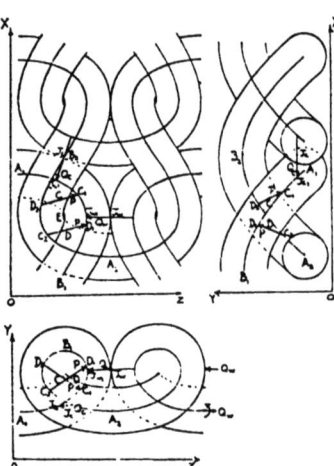

Fig. 11. Spatial elastica model of plain weft knitted fabric with wale and course jamming from [11]

widens the definition interval for d/ℓ up to ca 0.055 (see Fig. 10). Hepworth and Leaf [11] reached d/ℓ=0.07 by accounting for jamming in the course and wale direction (see Fig. 11). De Jong and Postle using a generalized energy method got a tighter structure with distributed reaction forces in interlacing points and a deformable yarn cross-section [12]. These authors as well as Hepworth in this volume expand their models further in order to simulate tensile and other deformations of knitted fabrics.

4. STATICS, KINEMATICS AND DYNAMICS OF SLENDER BODIES

Studies of the behaviour and mechanical response of fibres and fibre assemblies during processing and use require a variety of approaches depending on conditions and aims. We have statics, kinematics and dynamics dealing respectively with the equilibrium of fibres at rest, fibre motion, and motion under the influence of inertial masses. According to another important criterion we may assume perfectly flexible fibres or we may account for fibre finite bending and torsional rigidity. The elastica problems discussed in chapters 2. and 3. belong to the latter class.

The simplest cases of static equilibrium of perfectly flexible and extensible fibres may be described by the following vectorial differential equation:

$$\frac{\varepsilon + 1}{\lambda_o} \cdot \frac{dT}{ds} \bar{t} + \frac{T}{\rho} \bar{n} + \bar{F} = 0$$

where λ_o is the original linear density of fibre,
 T is the fibre tension ($T = \varepsilon E$),
 ε is the strain,
 s is the arc length (independent variable),
 ρ is the radius of fibre curvature,
 $\bar{n}, \bar{t}$ are normal and tangency vectors,
 $\bar{F}$ is the body force (force/mass).

From this equation may be derived, for instance, Euler's formula for capstan friction, linking the input tension T_0, output tension T, friction coefficient μ, and capstan angle ϕ, $T = T_0 \cdot \exp(\mu\phi)$.

A comprehensive theory covering the most complex problems of the dynamics of fibres with finite bending and torsional rigidity may be built on and around the following differential equations of the translational and rotational motion of a fibre element, in vector form:

$$\frac{1 + \varepsilon(F_w, t, \ldots)}{\lambda_o} \cdot \frac{\partial \bar{F}}{\partial s} = \frac{\partial^2 \bar{r}}{\partial t^2} - \bar{f}$$

$$\frac{1 + \varepsilon(F_w, t, \ldots)}{\lambda_o} \cdot \frac{\partial \overline{M}}{\partial s} + \overline{w} \times \overline{F} = [I] \frac{\partial \overline{\omega}}{\partial t} - \overline{m}$$

where $\overline{F}, \overline{M}$ are vectors of concentrated forces and moments,
   $\overline{f}, \overline{m}$ are vectors of distributed forces and moments (from interacting solids, fluids or fields; they may be functions of location, time etc.),
   $\overline{r}$ radius vector of the element, stands for the element coordinates; its derivatives with respect to time represent translational velocity and acceleration,
   $\overline{w}$ is the tangency vector,
   is the original linear density,
   $\varepsilon(F_w, t, \ldots)$ is the strain as a function of the tangential component of concentrated force, time, loading history, temperature etc.,
   $\overline{\omega}$ is the rotational velocity of the element,
   $[I]$ is the generalized rotational moment of inertia,
   s,t are arc length and time (independent variables).

It would be desirable to develop such a unified theory and supporting computer program packages which would make it easier to simulate the mechanical response and behaviour of fibres and yarns under the widest variety of conditions and circumstances.

REFERENCES

[1] Konopasek, M., PhD. Thesis, UMIST, Manchester 1970.
[2] Bruggisser, K., MS. THESIS, Georgia Institute of Technology, Atlanta, 1976.
[3] Peirce, F.T., J.Text.Inst, 1930, 21, T377.
[4] Lloyd, D.W., Shanahan, W.J. and Konopasek, M., Int. J. of Mech. Sci., 1978, 20, p. 521.
[5] Olson, L.H., 'Simulation of skin loading in a fluid filled, flexible walled tank' (report), Georgia Institute of Technology, Atlanta, 1977.
[6] Konopasek, M., and Hearle, J.W.S., J.Text.Inst., 1974, 65, p. 217.
[7] Konopasek, M., Text.Res.J., 1976, 46, p. 278.
[8] Konopasek M. and Bruggisser, K., J.Appl.Pol.Sci., Applied Polymer Symposium 33, 1978, p. 203.
[9] Konopasek, M., and Shanahan, W.J., J.Text.Inst., 1975, 66, p. 351.
[10] Hepworth, B. and Leaf, G.A.V., in 'Studies in Modern Fabrics (edited by P.W. Harrison), the Textile Institute, Manchester, 1970, p. 181.
[11] Hepworth, B. and Leaf, G.A.V., J.Text.Inst., 1976, 67, p.241.
[12] De Jong, S. and Postle, R., J.Text.Inst., 1977, 68, p. 307.

THE ANALYSIS OF COMPLEX FABRIC DEFORMATIONS*

D.W. Lloyd

Department of Textile Industries, University of Leeds,
Leeds, England[+]

ABSTRACT. The complex deformations of fabrics, whilst of great
technological interest, present formidable difficulties of
analysis. These difficulties are discussed and the types of
deformations most amenable to analysis are reviewed. A theoretical
framework for describing the mechanical properties of fabrics is
formulated within the context of approximating fabrics as two-
dimensional continua. The finite element method is outlined with
its ability to handle the various nonlinearities exhibited by
deformations of fabrics. Applications of finite element methods
within textile mechanics are suggested, and illustrative examples
are described.

1. INTRODUCTION

Although a large part of the textile industry is concerned
with imparting desirable physical properties to fabrics, theoretical understanding of the effects of these properties on fabric
response is strictly limited. This apparent neglect of an area
of important technological interest stems from the analytical
difficulties inherent in any but the simplest models of fabric
deformations, when compared to models of fabric structure, or
studies of yarn mechanics.

The difficulties are easily understood. A fabric of any
appreciable area, as a structure, consists of perhaps many thousands

---

*The author acknowledges the support of the Courtaulds Educational
 Trust and the Science Research Council.
[+]Present address.

of yarns with a consequent vast number of mutual contact points, each with several degrees of freedom. Since modelling a fabric as such a collection of yarns leads to an impossibly large number of equations, even for modern computers, it is tempting to model fabrics as continuous sheets. Deferring the discussion of the assumption of continuity till later, the problem becomes that of modelling the deformations in two and three dimensions of a two dimensional sheet. Any deformation of a fabric sheet will almost inevitably be complicated, and hence difficult to model, involving in general large displacements and large strains of a sheet with complex mechanical properties. Leaving aside for the moment the question of the mechanical properties of fabrics, it is helpful to consider the different types of deformations that can occur.

1. Planar deformations. The simplest class of problems are those in which an initially flat sheet is deformed in its own plane, i.e., no transverse displacements occur. This class of problem can be modelled, at least in principle, as suitable definitions of strain are available, the appropriate mechanical properties can be formulated and measured, and the governing equations are well known. The strains developed in the plane of the sheet are known as membrane strains, a term that will be retained for the in-plane strains in more complex situations involving transverse displacements.

2. 'Tension membranes'. Where transverse displacements of the fabric occur, but the fabric has negligible bending stiffness, and the boundary conditions are such that the fabric is everywhere in tension, then the problem is that of a 'tension membrane', and can be solved by suitably extended methods from planar problems.

3. 'Plate' and 'shell' problems. A fabric with significant bending stiffness which is subject to transverse displacements is, in engineering terms, a 'plate' if initially flat, or a 'shell' if naturally curved. The plate and shell theories developed for use in engineering are limited to "large" transverse displacements of the order of the plate or shell thickness, and so have only limited application to fabrics where "very large" transverse displacements are common. There is as yet no large displacement, large strain, shell theory suitable for general textile applications at least to the author's knowledge. Consequently, very few of these shell-type problems are amenable to analysis, though useful simplifications and approximations are available in particular examples, e.g., cylindrical bending with no membrane strains (1).

4. Buckling. It is necessary to distinguish between two cases. Buckling can occur after only small membrane strains: the 'initial stability' problem in engineering, familiar in textile mechanics as Euler buckling of an elastica. More common in fabric deformations is the case where buckling occurs only

after large membrane strains. Examples in this category would be buckling in fabric shear tests, or the drapemeter. Whilst it is possible to investigate buckling of both types theoretically, further discussion must await an examination of the methods available.

5. Post-buckling behaviour. Clearly once buckling has occurred, the problem is transformed into a shell-type problem with all its attendant difficulties.

Thus, although some areas of technological interest still present formidable problems of analysis, others appear more amenable to study. Before considering methods of analysis, it is convenient to formulate a framework for expressing the mechanical properties of fabrics.

2. GENERAL FABRIC STRESS-STRAIN RELATIONSHIPS

2.1 Fabric structure and the assumption of continuity

The impracticability of modelling a fabric as an assemblage of its constituent yarns has already been pointed out. A convenient assumption for the purposes of analysis is that the fabric constitutes a two-dimensional continuum, whose actual structure is irrelevant except insofar as it affects the form and values of the sheet's mechanical properties (2). In homogeneous elastic materials the load-deformation behaviour is a special case of three-dimensional elastic behaviour. However, in textile materials the structure is not homogeneous, and there is no simple recourse to three dimensional theory. It is therefore important to define the behaviour of sheet materials as a purely two-dimensional system, without any simplifications arising from homogeneity assumptions. The fabric is hence idealised by a representative plane whose thickness does not enter the formulation. The mechanical properties of such fabric sheets have been examined in the literature for the case of membrane strains in planar deformations (3,4), but has not been widely discussed for cases involving bending, or combined bending and membrane strains.

Although an engineering material exhibits structure and inhomogeneity (crystallite grains, inclusions, molecular structure, etc.) textile materials have a macrostructure several orders of magnitude larger than engineering materials. This leads to a restriction in the use of the assumption of continuity. The classical measure of strain for a continuum material is obtained by considering the relative displacement of two neighbouring material points. The original distance separating the two points can be made arbitrarily small, leading to strain measures related to the first derivatives of displacement with respect to coordinates. The macrostructure of textile materials however, places a lower limit to the size of area element that can be considered. Below this limit, which is of the order of a few "unit cells" of the structure, the material, and hence the strain, can no longer be considered as being continuously distributed.

Thus if the continuity assumption and the classical definition of strain is to be retained, attention must be confined to those situations where the nonuniformity of the deformation of the material treated as a continuum is not significant within areas of the same order of size as the structural unit cell. Thus the size of subdivisions that need to be made for numerical methods of solution to the required accuracy should be large compared to the unit cell; if accuracy requires smaller subdivisions then the continuum approximation must be abandoned and at least part of the system replaced by an explicit model of fabric structure.

2.2 Features of fabric deformations

It is useful to examine the different approaches to representing the deformations and mechanical properties of fabrics. The simplest approach is that of small strains and small displacements, typified by traditional engineering stress analysis methods. Next in complexity is small strains with large displacements, as for example, in elastica theory. This is an adequate representation for some fabric deformations, but in general a full large strain, large displacement approach is needed.

The simplest representation of mechanical properties is linear elasticity. In general the assumption of isotropy common in engineering is inadequate, and an anisotropic formulation must be used. Linear elasticity provides a useful basis for understanding the more complex representations which are more realistic descriptions of real textile behaviour. The simplest of these is nonlinear elasticity, followed in both complexity and realism by viscoelastic models, and inelastic behaviour, involving both time and history dependence (5).

The situation is additionally complicated where dynamic (inertial) effects are important, as the mechanical properties are strongly dependent on strain-rate, and additional information (damping coefficients, etc.) is required. It is convenient to examine the small strain, linear elastic case to illustrate the important features of the stress-strain relations of sheet materials, and as a basic framework which can be extended to more complex models.

The basic assumptions of engineering plate and shell theory require that the sheet thickness be small compared to its other dimensions and its minimum radius of curvature. In addition, even for "large" deflections, if small strains are to be retained and membrane strains and bending are to remain uncoupled, the displacements may be only of the same order as the thickness and small compared to the plate dimensions. (True large deflections would produce finite membrane strains as coupling occurs between membrane strains and bending through the changes of geometry.) Elements normal to the plate or shell are assumed to remain undeformed and normal during deformations, and normal components of stress to produce negligible strains.

2.3 Strain measures

The strain in the fabric may be described in terms of the representative plane (or surface, for shells). If the undeformed reference plane is the xy plane, points on whose surface have displacements u,v,w, in the x,y,z directions, then the strain measures for small deflection theory are

$$\epsilon_1 = \frac{\partial u}{\partial x}$$

$$\epsilon_2 = \frac{\partial v}{\partial y} \tag{1}$$

$$\epsilon_{12} = \frac{1}{2}\left(\frac{\partial u}{\partial y} + \frac{\partial v}{\partial x}\right)$$

where ϵ_1 and ϵ_2 are the tensile strains in the x and y directions, and ϵ_{12} is the shear strain. It is worth comparing these with strain measures for large strains.

$$\epsilon_1 = \frac{\partial u}{\partial x} + \frac{1}{2}\left\{\left(\frac{\partial u}{\partial x}\right)^2 + \left(\frac{\partial v}{\partial x}\right)^2 + \left(\frac{\partial w}{\partial x}\right)^2\right\}$$

$$\epsilon_2 = \frac{\partial v}{\partial y} + \frac{1}{2}\left\{\left(\frac{\partial u}{\partial y}\right)^2 + \left(\frac{\partial v}{\partial y}\right)^2 + \left(\frac{\partial w}{\partial y}\right)^2\right\} \tag{2}$$

$$\epsilon_{12} = \frac{1}{2}\left(\frac{\partial u}{\partial y} + \frac{\partial v}{\partial x}\right) + \frac{1}{2}\left\{\frac{\partial u}{\partial x}\frac{\partial u}{\partial y} + \frac{\partial v}{\partial x}\frac{\partial v}{\partial y} + \frac{\partial w}{\partial x}\frac{\partial w}{\partial y}\right\}$$

in an initial, or Lagrangian, reference frame. In a current or Euleran frame the plus sign outside the curly brackets becomes a minus.

Returning to small deflections, the curvatures can be expressed as (2)

$$K_1 = -\frac{\partial^2 w}{\partial x^2}$$

$$K_2 = -\frac{\partial^2 w}{\partial y^2} \qquad (3)$$

$$K_{12} = -2\frac{\partial^2 w}{\partial x \partial y}$$

where K_1 and K_2 are the curvatures in the x and y directions, and K_{12} is the twist. In the case of shells, analogous, but more complex expressions exist (6).

2.4 Measures of stress

In the engineering treatment, the actual stress distribution through the thickness is replaced by resultant forces and couples

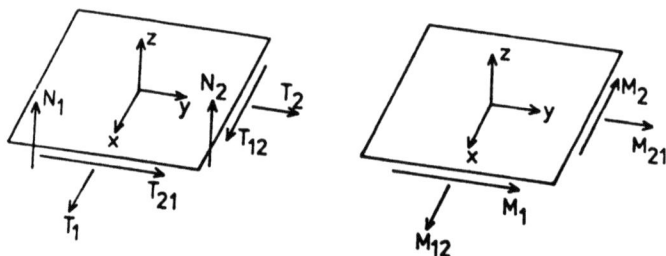

Fig. 1. Stress-resultants and stress-couples

("stress-resultants" and "stress-couples") regarded as acting on the reference surface, and obtained by integration over the material thickness, figure 1. For flat plates the shear components T_{12} and T_{21} become identical, as do M_{12} and M_{21}, the twist-couples. Thus in a strictly two-dimensional treatment, the "stresses" will be the forces per-unit-length familiar in textile mechanics. Clearly a conventional stress is obtained by dividing by the material thickness. Also, since by definition they produce negligible strains, the transverse shear components N_1 and N_2 do not enter the stress-strain relationships.

2.5 Stress-strain relationships

As a result of the basic assumptions, six stress and six strain measures remain. In the most general case (7) these are related linearly by

$$\begin{Bmatrix} T_1 \\ T_2 \\ T_{12} \\ M_1 \\ M_2 \\ M_{12} \end{Bmatrix} = \begin{bmatrix} A_{11} & A_{12} & A_{13} & B_{14} & B_{15} & B_{16} \\ & A_{22} & A_{23} & B_{24} & B_{25} & B_{26} \\ & & A_{33} & B_{34} & B_{35} & B_{36} \\ & & & D_{44} & D_{45} & D_{46} \\ & \text{symmetrical} & & & D_{55} & D_{56} \\ & & & & & D_{66} \end{bmatrix} \begin{Bmatrix} \epsilon_1 \\ \epsilon_2 \\ \epsilon_{12} \\ K_1 \\ K_2 \\ K_{12} \end{Bmatrix} \qquad (4)$$

Here the submatrices A_{ij} and D_{ij} represent the membrane and bending (and twisting) rigidities respectively. The B_{ij} are coupling rigidities that connect the membrane and bending modes of deformation.

In general for a skew fabric having a 'face' and 'back', all the terms in equation 4 would exist, though symmetry arguments for particular fabric structures would lead to simplifications. In contrast to a layered material in engineering, however, the submatrix B_{ij} cannot be assumed to be symmetrical, though fundamental energy considerations dictate that the complete matrix should be so. Thus in the general case 21 independent rigidities are required to specify the elastic behaviour of a sheet: 6 for membrane deformations, 6 for bending and twisting, and 9 for coupling between the two modes.

2.6 Special cases

The special case of anisotropy most likely to be of interest in fabric applications is that of an orthotropic sheet, i.e., one which possesses a line of symmetry, such as the warp or weft direction in a non-skew woven fabric. The symmetry leads to the elimination of certain stiffness terms so that equation 4 reduces to

$$\begin{Bmatrix} T_1 \\ T_2 \\ T_{12} \\ M_1 \\ M_2 \\ M_{12} \end{Bmatrix} = \begin{bmatrix} A_{11} & A_{12} & 0 & B_{14} & B_{15} & 0 \\ & A_{22} & 0 & B_{24} & B_{25} & 0 \\ & & A_{33} & 0 & 0 & B_{36} \\ & & & D_{44} & D_{45} & 0 \\ & \text{symmetrical} & & & D_{55} & 0 \\ & & & & & D_{66} \end{bmatrix} \begin{Bmatrix} \epsilon_1 \\ \epsilon_2 \\ \epsilon_{12} \\ K_1 \\ K_2 \\ K_{12} \end{Bmatrix} \quad (5)$$

where the x and y axes are assumed to coincide with the principal directions of orthotropy, i.e., the warp and weft directions in a woven fabric. This has 13 independent rigidities, reducing to 12 if the coupling matrix is symmetric; to 8 if the fabric is symmetrical about its central plane so that the B_{ij} disappear; to 6 for a square fabric such as a plain weave with the same yarns in each direction; to 4 for an isotropic sheet with bending behaviour unrelated to planar behaviour; and to 2 plus the thickness for an isotropic solid sheet. If the fabric is skew, but the fabric is symmetrical about its central plane, equation 4 is reduced by the B_{ij} vanishing, leaving 12 independent parameters.

The interpretation of the parameters can best be illustrated using the special case of an orthotropic fabric, initially flat, with no elastic coupling between membrane strains and bending/twisting modes. The relationship between membrane stresses and strains can be written as either

$$\begin{Bmatrix} \epsilon_1 \\ \epsilon_2 \\ \epsilon_{12} \end{Bmatrix} = \begin{bmatrix} \frac{1}{E_1} & \frac{-\nu_2}{E_2} & 0 \\ \frac{-\nu_1}{E_1} & \frac{1}{E_2} & 0 \\ 0 & 0 & \frac{1}{G} \end{bmatrix} \begin{Bmatrix} T_1 \\ T_2 \\ T_{12} \end{Bmatrix} \quad \text{(compliance)} \quad (6)$$

or

$$\begin{Bmatrix} T_1 \\ T_2 \\ T_{12} \end{Bmatrix} = \begin{bmatrix} \dfrac{E_1}{1-\nu_1\nu_2} & \dfrac{\nu_2 E_1}{1-\nu_1\nu_2} & 0 \\ \dfrac{\nu_1 E_2}{1-\nu_1\nu_2} & \dfrac{E_2}{1-\nu_1\nu_2} & 0 \\ 0 & 0 & G \end{bmatrix} \begin{Bmatrix} \epsilon_1 \\ \epsilon_2 \\ \epsilon_{12} \end{Bmatrix} \qquad (7)$$

(stiffness)

where E_1, E_2 are tensile rigidities in the principal directions, ν_1 and ν_2 are the corresponding Poisson ratios for extension, and G is the shear rigidity. Symmetry requires that

$$\nu_1 E_2 = \nu_2 E_1 . \qquad (8)$$

The bending behaviour of a fabric can be expressed by either

$$\begin{Bmatrix} M_1 \\ M_2 \\ M_{12} \end{Bmatrix} = \begin{bmatrix} F_1 & \sigma_2 F_1 & 0 \\ \sigma_1 F_2 & F_2 & 0 \\ 0 & 0 & \tau \end{bmatrix} \begin{Bmatrix} K_1 \\ K_2 \\ K_{12} \end{Bmatrix} \qquad (9)$$

(stiffness)

or

$$\begin{Bmatrix} K_1 \\ K_2 \\ K_{12} \end{Bmatrix} = \begin{bmatrix} \dfrac{1}{F_1(1-\sigma_1\sigma_2)} & \dfrac{-\sigma_2}{F_2(1-\sigma_1\sigma_2)} & 0 \\ \dfrac{-\sigma_1}{F_1(1-\sigma_1\sigma_2)} & \dfrac{1}{F_2(1-\sigma_1\sigma_2)} & 0 \\ 0 & 0 & \dfrac{1}{\tau} \end{bmatrix} \begin{Bmatrix} M_1 \\ M_2 \\ M_{12} \end{Bmatrix} \qquad (10)$$

(compliance)

where F_1 and F_2 are the principal flexural rigidities, τ is the twisting rigidity, and σ_1 and σ_2 are analogous to Poisson's ratios and give rise to anticlastic curvature, i.e., bending in a perpendicular direction to the imposed direction of bending. As before symmetry requires that

$$\sigma_1 F_2 = \sigma_2 F_1 \quad . \tag{11}$$

The ratios σ_1 and σ_2 cannot be identified with ν_1 and ν_2 as for homogeneous materials, as the mechanism of bending for textile structures is completely different.

2.7 Tests of validity

The validity of this formulation in textile applications can be partially tested by comparing the theoretical transformation behaviour of the various rigidities and compliances (8) with the known behaviour of fabrics. For membrane strains the theoretical expression for tensile rigidity in a direction θ to the x axis is

$$\frac{1}{E_\theta} = \frac{1}{E_1} \cos^4\theta + \left(\frac{1}{G} - \frac{2\nu_1}{E_1}\right)\cos^2\theta\sin^2\theta + \frac{1}{E_2}\sin^4\theta \tag{12}$$

Kilby has noted this formula (4), finding that it corresponded reasonably with experimental measurements of initial tensile rigidity of woven fabrics.

The flexural rigidity in the θ-direction is given by

$$F_\theta = F_1\cos^4\theta + (4\tau + 2\sigma_2 F_1)\cos^2\theta\sin^2\theta + F_2\sin^4\theta \tag{13}$$

which, although different from a formula given by Peirce (9), is of the same form as one derived by Cooper (10) from a structural model of a woven fabric. In Cooper's expression the coefficient of $\cos^2\theta\sin^2\theta$ was related to the torsional rigidities of the yarns. Equation 13 can be fitted reasonably well to experimentally determined polar plots of fabric bending rigidity by an appropriate choice of the coefficient of $\cos^2\theta\sin^2\theta$. If the middle coefficient $(4\tau + 2\sigma_2 F_1)$ is less than the sum of the other two $(F_1 + F_2)$ then a polar plot of F_θ would exhibit minima between the principal directions, as is observed for some fabrics (11). The more usual type of fabric behaviour (9) with a smooth transition without pronounced maxima or minima between principal directions will occur if $(4\tau + 2\sigma_2 F_1) \simeq (F_1 + F_2)$, giving a polar plot with an approximately elliptical shape.

Since the polar plots of equations 12 and 13 must show four maxima in the principal directions, approximately elliptical form, or circular shape, triaxial fabrics with three principal directions might appear to contradict equations 12 and 13. The only way of reconciling the equations to three principal directions is for the fabric to yield a circular polar plot, i.e., to be isotropic. Measurements by Skelton (12) on such fabrics have indeed revealed isotropic behaviour.

2.8 Difficulties in the experimental measurement of rigidities

The practical difficulties of experimental techniques for measuring the rigidities in equation 4 are beyond the scope of these lectures: more fundamental theoretical difficulties exist. Also the flexible nature of fabrics, the difficulty of determining in-plane strains directly, and the large displacements and finite areas needed to obtain measurable effects, appear to rule out the measurement techniques normally employed for anisotropic materials (7). The usual approach is to impose a mode of deformation as far as possible in isolation, whilst monitoring the loads and deformations produced. Thus it is possible to measure the diagonal terms A_{ii}, the membrane tensile and shear rigidities. The off-diagonal terms, incorporating the Poisson ratios, await the development of a suitable biaxial test method (2,13).

Flexural rigidity is determined from a knowledge of the applied moment required to produce large bending deflections in one direction. The anticlastic curvature predicted by equations 9 and 10 would induce finite membrane strains. Typically the A_{ij} are of greater magnitude than the D_{ij}, so membrane stresses provide the balancing moment needed to prevent bending in the other direction. Only when the membrane stiffnesses are small can anticlastic curvature occur for large deflection bending (14). Twisting produces an additional difficulty: it is not possible to impose a twist K_{12} in isolation over a finite area, as twisting changes the length of lines on the fabric surface, leading to membrane stresses. Neither can the twist rigidity or Poisson ratios be obtained from the transformation behaviour - only the combined term $(4\tau + 2\sigma_2 F_1)$ can be deduced.

The off-diagonal bending rigidities, D_{ij}, relate to double curvatures. It is usually said that double curvature causes shear. This is not strictly true: double curvature over a finite area requires changes in area, which in woven fabrics is achieved by the easy shear and consequent area reduction. Other mechanisms may exist, for example, sheet rubber can accommodate area changes through changes in thickness. Where no such mechanism exists, double curvature can occur only over infinitesimal areas, hence the sharp peaks formed in crumpled paper.

Double curvatures imposed over finite areas bring another difficulty. Double curvature cannot be imposed over an arbitrarily large area as a uniform deformation, for, as the area becomes larger distances on the fabric surface get smaller, until eventually a large circumference would collapse to a point. This has an important consequence. Although for simplicity it has been assumed up to this point that fabrics are usually initially flat, many fabrics, including most knitted fabrics, have an unstrained state which is initially doubly curved (15,16). Such fabrics cannot adopt their unstrained state without going into double curvatures over large areas, and hence producing impossibly large membrane strains. The minimum energy state of an element of such a fabric depends on how much double curvature it can adopt, and is thus a function of the size of the piece in which it is contained. This presents a considerable problem to any analysis which should start from the unstrained state.

The area reduction caused by length changes during shear raises another problem. This might be expected to be obtained from the A_{13} term. However in a square fabric symmetry requires that this be zero. This area reduction which is independent of the direction of shear must be a large second-order term suggesting that equation 2 must be used at comparatively small strains.

This leads to the coupling rigidities B_{ij}, which couple membrane and bending modes as a property of the material structure, and independently of the geometrical coupling already mentioned.

In any fabric with a definite face and back, some at least of these terms will be non-zero, although in practice they may be small and unimportant. The possibility of the importance of the B_{ij} must throw doubt on the validity of methods of measuring the A_{ij} based on cylindrical specimens.

2.9 Extensions to basic framework

The restriction of the treatment to small-strain, linear elastic deformations is unrealistic in relation to textile materials. However, the framework outlined above opens up more realistic possibilities. Time independent nonlinearities can be introduced by making the terms in equation 4 functions of their associated strains. For example

$$T_1 = A_{11}(\epsilon_1) + A_{12}(\epsilon_2) + A_{13}(\epsilon_{12}) + B_{14}(K_1)$$
$$+ B_{15}(K_2) + B_{16}(K_{12}) \quad . \quad (13)$$

If interaction between the modes is stronger, more complex strain dependence can be used, involving cross-terms such as $\epsilon_1\epsilon_2$, or worse. In these circumstances, the formulation provides a procedural framework that is readily compatible with computational methods.

A similar approach can be used for time dependence; the terms in the matrix become functions not only of the strains, but of time as well. Again this lends itself to computational methods, with the functional dependence calculated from structural models of fabrics and the mechanical properties of yarns, or stored as tables of experimental results. The details of the way in which values for the terms of the matrix are obtained from these relationships will depend on the particular method of analysis which is chosen. This will be discussed later in the context of finite element methods.

3. FINITE ELEMENT METHODS IN TEXTILE MECHANICS

3.1 Introduction

A fabric represented as a large continuum will have an infinite number of degrees of freedom, so any analysis of the system and its deformations requires some method of reducing the number of degrees of freedom to manageable proportions. This necessarily involves some measure of approximation. The finite element method, developed for difficult engineering structures, follows the intuitive approach of formulating the governing equations for small representative pieces of the system, each with a limited number of degrees of freedom, and then using the equations for all the small pieces together to represent the complete system. This is most conveniently expressed as a series of matrix equations which are readily programmed for solution on modern digital computers.

The matrix equations themselves can be formulated in a variety of ways (17), with more or less complicated notations. The notation used by Zienkiewicz (18) will be followed here, as it is easy to understand, and is widely used. Space considerations allow only a brief outline of the method and its capabilities here; greater detail will be found in the appropriate texts (see for example, 19,20).

3.2 Formulation of the basic method

The most common approach to the finite element method, the 'displacement method', will be illustrated here for small strains and linear elasticity, and then generalised. Since the main concern is with fabrics, attention will be confined initially to two-dimensional continua. The two-dimensional continuum is first divided into pieces - the finite elements - which are considered to be joined only at a limited number of points called nodes, figure 2. Each node has prescribed degrees of freedom, expressed

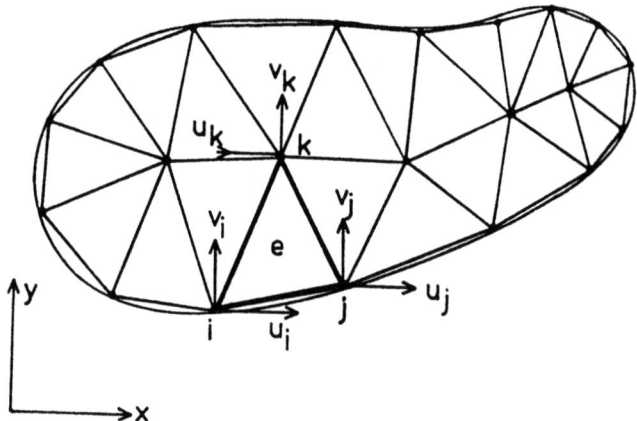

Fig. 2. Finite element model of 2 dimensional continua

in terms of its displacements. Thus, in the figure, element e (lying in the xy plane) has nodes i,j,k, each with displacements u,v, in the x and y directions (i.e., membrane deformations).

The displacements at any point within the element, $u(x,y)$ and $v(x,y)$ are approximated in terms of the displacements at the nodes.

$$\{u\} = \begin{Bmatrix} u(x,y) \\ v(x,y) \end{Bmatrix} \simeq \begin{bmatrix} [N_i], [N_j], [N_k] \end{bmatrix} \begin{Bmatrix} \{a_i\} \\ \{a_j\} \\ \{a_k\} \end{Bmatrix} = [N]\{a\}$$

(14)

where $\{a_i\}^T = \{u_i v_i\}$ are the displacements of node i; and the $[N_i]$ are functions of position which give the appropriate nodal displacements when the nodal coordinates are substituted in equation 14. The functions $[N]$ are called "shape functions" and play a fundamental role in finite element analysis, enabling the properties of the elements to be calculated by approximating the unknown distribution of displacements with a known function.

From equation 14 the displacements at any point in the element can be obtained and hence the strains at any point can be found by suitable differentiation. Using equation 1 gives

$$\{\epsilon\} = [L]\{u\} \qquad (15)$$

where $\{\epsilon\}$ are the strains, and $[L]$ is an operator given by

$$[L] = \begin{bmatrix} \frac{\partial}{\partial x}, & 0 \\ 0, & \frac{\partial}{\partial y} \\ \frac{1}{2}\frac{\partial}{\partial y}, & \frac{1}{2}\frac{\partial}{\partial x} \end{bmatrix} \qquad (16)$$

Approximating $\{\epsilon\}$ in the same way as $\{u\}$ gives

$$\{\epsilon\} = [B]\{a\} \qquad (17)$$

where

$$[B] = [L][N] \qquad (18)$$

Clearly, once the strains are known, it is possible to calculate the stresses. If the material is linear elastic the stresses are given by

$$\{\sigma\} = [D](\{\epsilon\} - \{\epsilon_o\}) + \{\sigma_o\} \qquad (19)$$

where $[D]$ is a matrix containing the appropriate material properties, equivalent to equation 4, $\{\epsilon\}$ are the elastic strains, and $\{\epsilon_o\}$ are initial strains caused by shrinkage, etc. The stresses $\{\sigma\}^T = \{\sigma_x, \sigma_y, \tau_{xy}\}$ are either conventional stresses, if the material thickness is included in the formulation, or the stress-resultants discussed in section 2. The $\{\sigma_o\}$ are initial stresses, and can be used to include information on the previous deformation history of the material.

Equilibrium requires that these stresses be balanced by the loads on the element boundary. As the element is only joined through the nodes, these loads must be expressed as interelement forces at the nodes $\{q\}$. If the distributed loads (in the form of body forces) are $\{b\}$, and virtual displacements exist at the nodes, $\delta\{a\}$, then the external work done is equal to the internal work done by the stresses and distributed forces; thus for an element of volume V

$$\delta\{a\}^T\{q\} = \delta\{a\}^T(\int_V [B]^T\{\sigma\}dV - \int_V [N]^T\{b\}dV) \qquad (20)$$

and hence

$$\{q\} = \int_V [B]^T\{\sigma\}dV - \int_V [N]^T\{b\}dV . \qquad (21)$$

Substituting from equation 19 gives

$$\{q\} = [K]\{a\} + \{f\} \qquad (22)$$

where

$$[K] = \int_V [B]^T[D][B]dV \qquad (23)$$

is the element stiffness matrix, and

$$\{f\} = -\int_V [N]^T\{b\}dV - \int_V [B]^T[D]\{\epsilon_0\}dV + \int_V [B]^T\{\sigma_0\}dV \qquad (24)$$

are the nodal forces due to body forces, initial strains, and initial stresses. Clearly, if stress-resultants are used instead of stresses the integration will be over the element area. As the nodes may also be subject to concentrated loads, an additional vector of nodal loads $\{r\}$ is required for nodal equilibrium.

The equilibrium of the whole continuum can now be expressed in terms of the overall equilibrium at the nodes. Assuming that the elements all have the same coordinate system (or have had suitable transformations applied to bring them to a single global coordinate system), the contributions for each element from equation 22 can be summed, together with the vector of all the concentrated nodal loads. The interelement forces balance out overall, leaving a set of equations for the whole system

$$\{r\} = [K]\{a\} + \{f\} \qquad (25)$$

where the vectors now include all the nodes. The matrix $[K]$ is the system or global stiffness matrix, and is obtained by summing at common nodes the stiffness contributions from all the elements. It is convenient to use the global equivalents to the element expressions, thus

$$[K] = \int_{V_T} [B]^T[D][B]dV_T , \qquad (26)$$

and

$$\{f\} = -\int_{V_T} [N]^T \{b\} dV_T - \int_{V_T} [B]^T [D] \{\epsilon_0\} dV_T$$

$$+ \int_{V_T} [B]^T \{\sigma_0\} dV_T - \int_{A_T} [N]^T \{t\} dA_T \qquad (27)$$

where all the matrices now refer to the complete system, the surface forces $\{t\}$ have been included for completeness, and V_T and A_T are the total volume and area respectively.

The process of obtaining equation 25 from the contributions of the individual elements is known as assembly, and is another fundamental feature of finite element methods. The element matrix $[N]$ can be found in terms of the nodal coordinates of the element, so from equations 16 and 18, matrix $[B]$ can be found, and the stiffness matrix $[K]$ calculated using the mechanical properties of the element, $[D]$. The element stiffness is then added into the global stiffness matrix. The processes of finding the element stiffness, and assembly of the element stiffnesses into the global stiffness matrix are thus separate, and can be programmed separately. This permits different finite elements, with different shape functions, different shapes (triangles, quadrilaterals, etc.), and different behaviour (bending, membrane modes, etc.) to be incorporated into a standard program, and, if desired, mixed in a single problem. The element size, type, position, etc., are input as data, while the assembly and solution procedure are standard. This allows great flexibility in the choice of elements (new or special elements can be incorporated in a standard manner), and in the geometry and type of problem.

The solution of equation 25 requires that a sufficient number of the nodal parameters are given in advance. Clearly, if all the $\{f\}$ are known, then a solution is possible if either all the nodal loads $\{r\}$ are known, or all the nodal displacements $\{a\}$ are known. The usual situation is that some displacements $\{a\}$ are given, and that the concentrated forces are specified for all remaining components, the majority of applied forces commonly being zero.

After the solution, by Gaussian elimination or some equivalent numerical method, all the forces $\{r\}$ and displacements $\{a\}$ are known. These can be substituted back into the elements to obtain the stresses and strains from

$$\{\sigma\} = [D][B]\{a\} - [D]\{\epsilon_0\} - \{\sigma_0\} \qquad (28)$$

where the matrices refer to the individual element, and from equation 17 respectively.

The solution thus obtained approximates the distribution of displacements in the actual continuum by a piecewise approximation made up of the individual element shape functions. If the shape functions have been properly chosen the approximation will converge to the real distribution as the number of elements increases. If the shape functions satisfy two basic requirements, compatibility and completeness, then convergence is assured. Compatibility requires that the element sides should remain in contact - more formally that only finite strains (and hence finite strain energy) should occur at the element boundaries. Completeness requires that the shape functions can produce all the rigid body and constant strain states for the element. Thus the simplest shape functions are linear in displacements, and give constant strains. Higher order shape functions, implying elements with extra nodal points, are used, as these often require fewer elements to obtain a given accuracy.

So far the only displacements and forces mentioned have been those associated with membrane deformations. The concept can be broadened to include other generalised forces and displacements; moments and curvatures, torques and twists, etc. The maximum number of degrees of freedom of a node is thus six - three translations and three rotations.

3.3 Extensions to the basic theory

The basic theory requires modification to deal with the non-linearities common in fabric deformations: nonlinear material properties, large strains and large displacements. The effect of these nonlinearities is to make the stiffness matrix $[K]$ a function of the displacements $\{a\}$, through the terms in the matrix of material properties $[D]$ being functions of the strains, and hence of the displacements; through the strains being nonlinear functions of displacement, from equation 2; or through changes in the equilibrium equations as the system geometry changes with displacement.

Equation 2 was obtained from the definition of the Lagrangian strain tensor

$$e_{ij} = \frac{1}{2}\left(\frac{\partial u_i}{\partial x_j} + \frac{\partial u_j}{\partial x_i} + \frac{\partial u_t}{\partial x_i}\frac{\partial u_t}{\partial x_j}\right) \tag{29}$$

which may be rewritten

$$e_{ij} = \epsilon_{ij} + \eta_{ij} \qquad (30)$$

where $\epsilon_{ij} = \frac{1}{2}(u_{i,j} + u_{j,i})$ \qquad (31)

and $\eta_{ij} = \frac{1}{2} u_{t,i} u_{t,j}$, \qquad (32)

and where a Cartesian basis has been assumed and the notation is that of Cartesian tensors. Equations 31 and 32 may be written in the form of equation 17

$$\{\epsilon\} = [B]\{a\} \qquad (33)$$

$$\{\eta\} = \frac{1}{2}\{a\}^T[B_L]\{a\} . \qquad (34)$$

The stress-strain relations may be written as

$$S_{ij} = D_{ijkl} e_{kl} \qquad (35)$$

where S_{ij} are the Kirchhoff stresses

$$S_{ij} = |J| \sigma_{rs} \frac{\partial x_i}{\partial X_r} \frac{\partial x_j}{\partial X_s} \qquad (36)$$

and $|J|$ is the Jacobian determinant $\partial x_i/\partial x_j$, and X_i are the current state coordinates. D_{ijkl} is a general functional dependent on the current state and the past deformation history.

Applying the variational principle as before yields

$$[K_T] = ([K] + [K_\sigma] + [K_L]) \qquad (37)$$

where $[K_T]$ is the total or tangential stiffness matrix, and $[K]$ is the small displacements stiffness matrix of equation 26.

$$[K_\sigma]d\{a\} = \int_V d[B_L]^T\{\sigma\}dV , \qquad (38)$$

where $\{\sigma\}$ is the stress in the deformed state, $d\{a\}$ is a small change in the displacements, and $[K_\sigma]$ is known as the geometric stiffness matrix or initial stress matrix.

$$[K_L] = \int_V ([B]^T[D][B_L] + [B_L]^T[D][B_L] + [B_L]^T[D][B])dV \qquad (39)$$

is the initial displacement matrix, or large displacement matrix, and has precisely the effect of a small strain approach with the current element coordinates used in the calculation of stiffness.

Neglecting any initial stresses or strains, the nodal forces due to body and surface forces become

$$\{F\} = -\int_V \{F_B\}^T dV - \int_A \{F_S\}^T dA \qquad (40)$$

where $\{F_B\}$ and $\{F_S\}$ are the body and surface forces in the current state transformed to the Lagrangian frame, and are functions of the displacements, and V and A are the original volume and surface area. It is convenient to rewrite equation 40 as

$$\{F\} = -\int_V [N]^T \{b\} dV - \int_A [N]^T \{t\} dA$$
$$- [Q]\{a\} \qquad (41)$$
$$= \{f\} - [Q]\{a\}$$

where $[Q]$ is a matrix which expresses the changes of the loads with displacement, and is known as the initial load stiffness matrix or load correction matrix (21). Clearly $[Q]$ is only non-zero for non-conservative loads, such as pressure loading. The equivalent expression to equation 25 is now

$$\{r\} = ([K_T] - [Q])\{a\} + \{f\}, \qquad (42)$$

where $[D]$ in $[K_T]$ contains the general form of D_{ijkl}.

Numerical analysis has two complementary approaches to solving sets of nonlinear equations such as this, iterative techniques and incremental methods. In principle, an iterative solution should be possible for a given load case; however large departures from the initial state are likely to lead to divergence of the iterative procedure. Thus it is normal in nonlinear finite element analyses to use an incremental method with the loads or displacements applied in a series of small steps. This has the advantage of modelling the loading process.

More formally, if only conservative forces are considered so that $[Q]$ is zero, and $\{r\}$ and $\{f\}$ are summed into a single vector of nodal forces $\{f\}$, equation 42 is equivalent to

$$[K_T]\{a\} + \{f\} = 0 . \qquad (43)$$

If the loads are applied in increments $\{\Delta f\}$, giving displacement increments $\{\Delta a\}$, the n+1 th increment is represented as

$$\{\Delta a\} = \{a_{n+1}\} - \{a_n\} = -[K_T]_n^{-1} \{\Delta f_n\} \qquad (44)$$

in the simplest (Euler) method. $[K_T]_n$ is calculated using the displacements $\{a_n\}$ at the end of the previous load step, and is assumed to be linear within the increment. Usually the loads and displacements are known for the start of the problem, often with $\{a\}$ and $\{f\}$ zero. If sufficiently small steps of $\{f\}$ are used the system will be well-behaved and convergent, though subject to possible drift. This is normally corrected by an additional iterative procedure within a load increment to ensure equilibrium, i.e., satisfaction of equation 43.

Particular nonlinearities are readily handled within this framework. Nonlinear material properties are expressed by the terms in the $[D]$ matrix being functions of the strains. This dependence may be input as known functions, experimentally derived data, or numerical data from other theoretical modelling, usually in the form of stress-strain laws. The matrix $[D]$ now becomes

$$[D_T] = \frac{d\{\sigma\}}{d\{\epsilon\}} \qquad (45)$$

the tangential elasticity matrix.

Alternatively, if $[D]$ is kept constant, the resulting linear elastic solution will require corrections to the stresses calculated from equation 19. If the initial stresses are zero at zero displacement, then the nonlinearities can be contained in $\{\sigma_0\}$ and used to apply the necessary correction. This is known as the initial stress method. Similarly, the initial strain method uses the vector $\{\epsilon_0\}$ to carry the necessary information on material nonlinearities. If the material is viscoelastic, the incremental approach can be used to model stress relaxation or creep problems, with the increments now representing steps in time, and with the changes in stress or strain calculated from the appropriate viscoelastic law.

A nonlinear phenomenon of particular interest in fabric deformations is that of buckling. In some cases such as that of Euler buckling of an elastica, buckling occurs after only infinitesimal deformations, and hence $[K_L]$ is zero in equation 37. After a small first increment producing displacements $d\{a\}$, the stresses $\{\sigma\}$, and hence $[K_\sigma]$, can be calculated. If further increase in the load by a factor λ will result in neutral stability, a classical eigenvalue problem arises

$$([K] + \lambda[K_\sigma])d\{a\} = 0 \qquad (46)$$

where for non-zero solutions the determinant

$$\left| [K] + \lambda[K_\sigma] \right| = 0 \quad . \qquad (47)$$

Solution of equation 47 yields n values for λ where $[K]$ and $[K_\sigma]$ are nxn matrices, and the resulting eigenvectors of displacements give the corresponding buckling modes.

Many fabric deformations, such as occur with the drapemeter (14), result in buckling only after relatively large deformations, so that $[K_L]$ can no longer be taken as zero. In this case a full incremental nonlinear analysis can be made, with the eigenvalue analysis carried out after each increment. Alternatively instability can be detected by checking the second variation of potential energy, $d^2\Pi$, where

$$d^2\Pi = d\{a\}^T [K_T] d\{a\} \quad . \qquad (48)$$

If this is positive, the system is stable, and $[K_T]$ is positive definite. If the system is unstable, and subject to buckling, $d^2\Pi$ is negative and $[K_T]$ is negative definite. Thus the onset of buckling can be detected by testing the sign of the determinant of $[K_T]$ as the incremental solution progresses.

Up to this point only static equilibrium has been considered. To deal with dynamic problems equation 25 must be extended to include inertial and damping effects (22,23). The extended equation takes the form

$$[M]\{\ddot{a}\} + [C]\{\dot{a}\} + [K_T]\{a\} + \{f\} = \{r\} \qquad (49)$$

where $[M]$ is a mass matrix which is discussed below, $\{\ddot{a}\}$ is the second derivative of displacements with respect to time, i.e., the nodal accelerations, $[C]$ is a damping matrix, also discussed below, $\{\dot{a}\}$ is the first derivative of displacements with respect to time, i.e., the nodal velocities, and the other quantities have the same meanings as before.

Two forms of mass matrix are used. The consistent mass matrix assigns the distributed mass of an element to its nodes in an analogous manner to the displacements, i.e.,

$$[M] = \int_V [N]^T \rho [N] \, dV \qquad (50)$$

where ρ is the density of the element material. The alternative, the lumped mass matrix, assigns the mass to the nodes in a way which always leads to a diagonal matrix. As this leads to computational savings without introducing serious inaccuracies, lumped mass matrices are widely used. The element mass matrices are assembled in the same way as the element stiffness matrices.

It might be expected that the damping matrix $[C]$ would be calculated in a similar way to the consistent mass matrix. However, the quantities equivalent to ρ are not usually known, and $[C]$ is normally calculated from

$$[C] = \alpha[M] + \beta[K_T] \qquad (51)$$

where α and β would be determined experimentally (24) and are known as the Rayleigh damping coefficients.

Equation 49 is, as before, solved by an incremental approach, with the increments representing time steps. $[M]$ and $[C]$ normally remain constant during a solution, and in addition to the force and displacement boundary conditions, initial values are required for the nodal velocities and accelerations. Once again equilibrium iteration is used within increments to prevent drifting of the solution.

3.4 Applications in textile mechanics

In the twenty five years or so that the finite element method has been developing, many different types of element have appeared in the literature. The two-dimensional elements already mentioned have counterparts in one-dimension and three dimensions. This leads to the idea of an integrated approach to problems in textile mechanics, using the framework of finite element methods, and the different types of element to model problems in yarn mechanics, fabric structures, and fabric deformations.

Similarly, although only continuum deformations have been dealt with here, the same discretization procedures can be applied to other steady-state and time-dependent field problems. Thus other problems, not normally considered as part of textile mechanics might logically come within the scope of an integrated approach based on finite elements, for example thermal insulation properties of items such as duvets using elements developed for engineering heat transfer problems, and examples in extrusion and dyeing using elements developed for fluid flow and porous seepage.

Alternatively, elements can be developed for specifically textile applications. The best example of these is the cruciform element developed by Torbe (25) for calculating the shape of sails and other inflated fabric structures. This element approximates a fabric not as a continuous membrane but as a coarser representative grid than that formed by the yarns.

Although an integrated approach through finite elements has its attractions, finite element methods tend to be expensive in computer time, the cost rising dramatically with the number of elements, the different nonlinearities included, and the length of the incremental and iterative processes. Consequently the finite element method represents a brute force technique which is complementary to the established analytical and computational methods in textile mechanics, but which, with its ability to handle different types of nonlinearity simultaneously, is probably the only currently available way of making progress in the more intractable areas of textile mechanics that also tend to be the most interesting.

4. ILLUSTRATIVE EXAMPLES OF FINITE ELEMENT SOLUTIONS

The following examples of finite element solutions in textile applications have been produced using the NONSAP finite element package (22,23). The intention has been to select interesting real applications which illustrate some of the points made earlier, and to demonstrate simply how they may be modelled using finite elements.

The first example concerns a common type of planar deformation, tensile testing of a woven fabric. In the test a rectangular fabric specimen has its ends clamped, and is extended longitudinally, the clamps preventing any changes in width at the ends. Such a specimen has two axes of symmetry, so only one quadrant needs to be modelled. The deformed shape obtained from a finite element model for 10% extension is shown in figure 3, and demonstrates the limited amount of waisting expected of a woven fabric. The elements are two-dimensional eight node isoparametric elements, i.e., the same shape functions are used to generate the curved shape of the element as are used to approximate the displacements. These elements were chosen as fewer elements are required to obtain a solution of given accuracy than would be needed if lower order shape functions were used. Intuitively, elements with curved edges are better at representing structures with curved boundaries than elements with straight edges. A full tangent stiffness formulation was incorporated into an incremental solution with equilibrium iteration at each step, appropriate to the large strains and displacements encountered. The anisotropy of woven fabrics was modelled by orthotropic material properties, based on experimental measurements, though the values were kept constant for

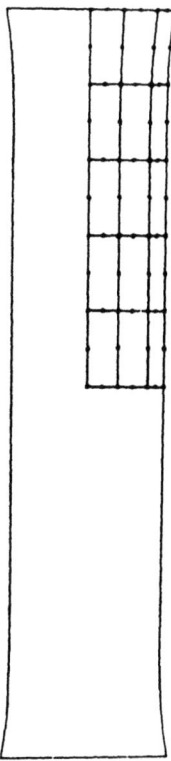

Fig. 3. Model of tensile test

simplicity. The boundary conditions are input as zero displacements at nodes along the clamp, zero transverse displacements along the longitudinal centreline of the specimen, and specified, incrementally increasing longitudinal displacements along the transverse centreline. All remaining components are specified as zero applied forces. Smaller elements were used at the edges where steeper stress gradients would be expected. The solution was obtained in 10 increments, required a maximum of 17 equilibrium iterations within an increment, and took 9.4 seconds execution time on a CDC 7600.

The second example involves a different kind of element, a one-dimensional element capable of carrying tension loads and with zero bending stiffness. Ropes are tested for safety by drop-testing, i.e., by using a falling weight to impose shock-loads on the rope. The specimen of rope, 2.5 metres long, is fixed at one end to a beam and at the other end to a heavy weight, figure 4. The weight is dropped from above the beam, and runs on vertical guide wires. The guide wires are displaced a short

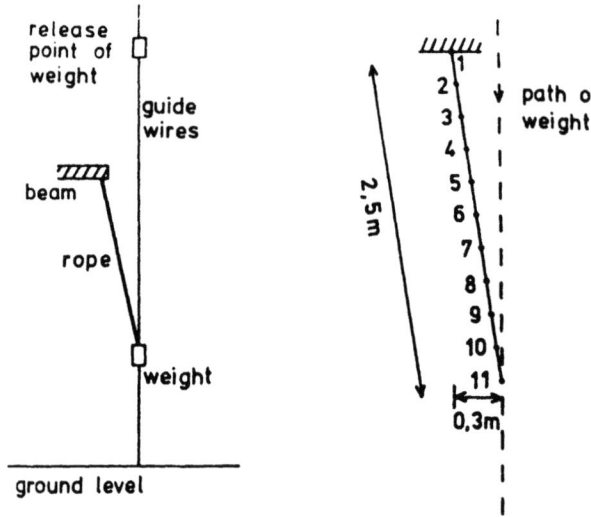

Fig. 4. Drop test   Fig. 5. Finite element model

distance horizontally from the beam to ensure a free fall, and prevent dangerous horizontal movements of the weight. The finite element model is shown in figure 5 where the rope is represented by 10 elements of equal length. The solution commences from the point at which the falling weight starts to load the rope. This state is modelled as follows. Node 1 is fixed, i.e., has zero displacements, and represents the anchorage point. The rope is assumed to be at rest initially, so all nodes except node 11 have zero initial velocity and acceleration, and nodes 2 to 10 have boundary conditions specified as zero applied nodal forces. A consistent mass formulation was used to generate inertial loadings, with a linear density for the rope of 180g/m, and a cross-sectional area of 1cm^2. The falling weight is represented by giving node 11 an additional concentrated mass of 80Kg and velocity and acceleration appropriate to the length of fall of the weight. The guide wires are represented by constraining node 11 to zero horizontal displacement.

It is particularly convenient to give one-dimensional elements nonlinear material properties, as only a single stress-strain curve is needed, the idealised curve used in this example being shown in figure 6. The solution required time increments of one microsecond, with equilibrium iterations every second step, thus to model the first 11 milliseconds of the loading and produce a 10.9cm vertical displacement of the weight, required 11,000 increments. The program had to be restarted twice using information stored in files, and took 139 seconds execution time on the CDC 7600, illustrating the high cost of some solutions. The ability to stop and restart a solution permits another feature to be modelled. The Rayleigh damping coefficients α and β were both assumed to be constant at 0.2 in this example. However the character of a rope changes during a test, and this could be partly modelled by changing the Rayleigh damping coefficients at restarts.

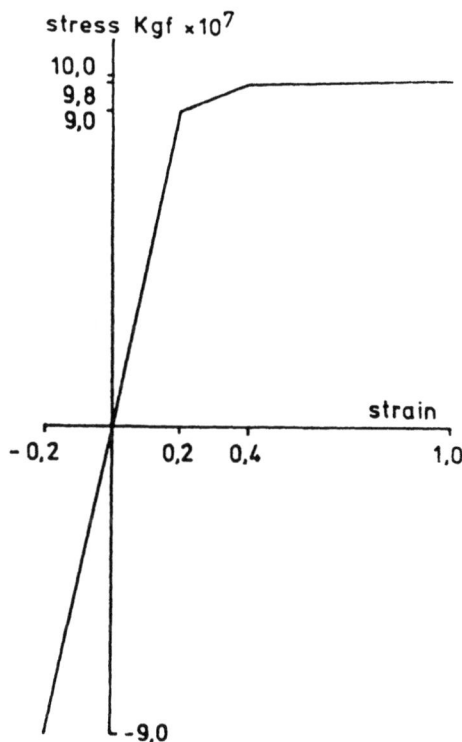

Fig. 6. Idealised stress-strain curve

In the drop-test the inertia of the rope does not substantially change the deformed geometry. However, in the case of ballistic penetration of a membrane the fabric inertia dramatically affects the deformation. Normally such a penetration example would be modelled using planar elements; however some simplifications are available in special cases. If the fabric is isotropic, a three-dimensional treatment is possible with the same mechanical properties being used through the thickness. If in addition the problem has rotational symmetry about some axis, then the deformation can be modelled as a fabric cross-section, using fewer elements and fewer degrees of freedom per node, with consequent computational savings. The example models ballistic penetration of a circular fabric sheet by a projectile moving along the central axis. The undeformed model is shown in figure 7 where the thickness of the fabric is modelled using six node isoparametric elements which have a "thickness" of 1 radian. The clamped edge of the fabric is represented by nodes 19 and 38 being fixed, and the constraint of the remaining fabric at the axis by nodes 1 and 20 being constrained to zero radial displacements. The remaining nodes have zero applied forces. The impact of the projectile is represented by giving nodes 1, 2, 20 and 21 concentrated masses

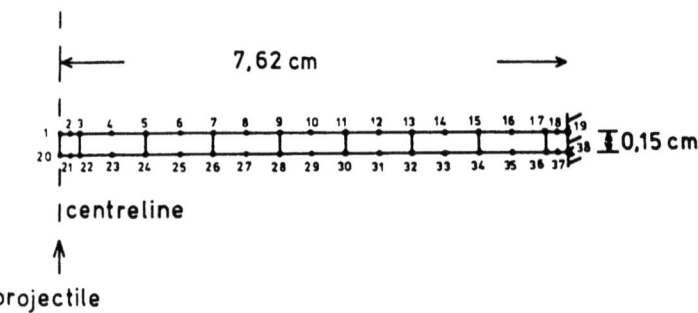

Fig. 7. Finite element model for ballistic penetration

339

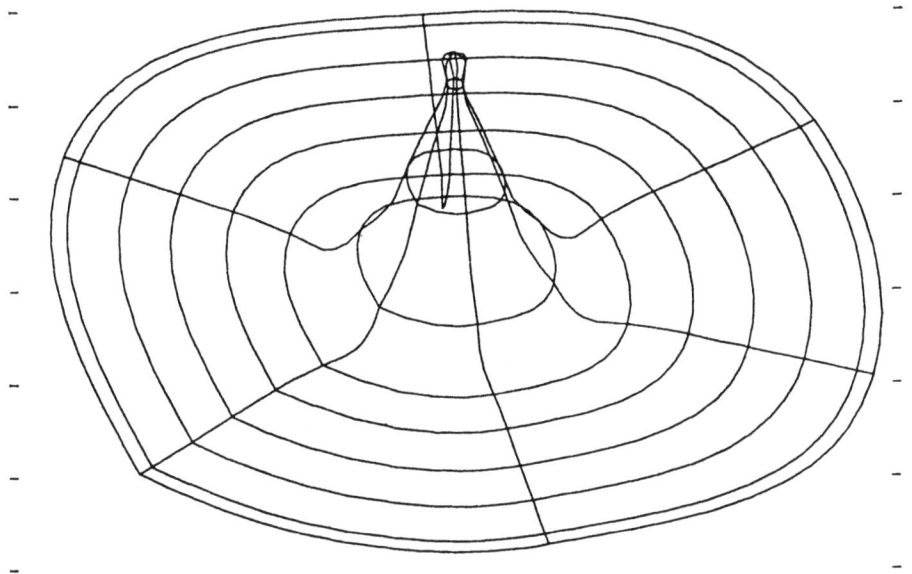

Fig. 8. Deformed geometry - model result

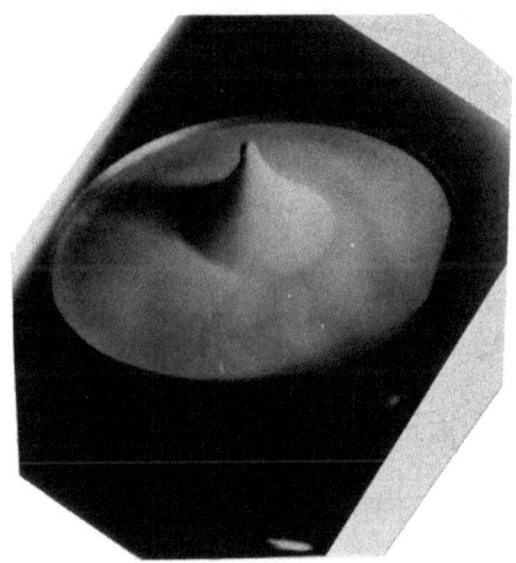

Fig. 9. Deformed geometry - experimental

equal in total to half the projectile mass, and initial transverse
nodal velocities equal to that of the projectile. The inertial
properties of the fabric are calculated from a consistent mass
formulation using the assumed fabric density. As with the previous
example, 1 microsecond increments were required with equilibrium
iterations every second step. The first 1 millisecond (i.e.,
1000 increments) required 8.9 seconds execution time. The deformed
shape is most easily interpreted when plotted in an isometric
projection for the complete fabric disc, i.e., repeating the
deformed cross-section at one radian intervals, and drawing the
remaining element boundaries. This is shown in figure 8, where
only the top surface of the membrane is shown for clarity, and
the slight distortions in the closed curves are caused by the
rather primitive graphics that were used. It is interesting to
compare figure 8 with a photograph of an experiment on a knitted
fabric, figure 9, which was kindly supplied by Mr. Colin Cork at
UMIST.

The effect of fabric inertia in controlling the deformed
shape can be seen by comparison with figure 10, which plots the

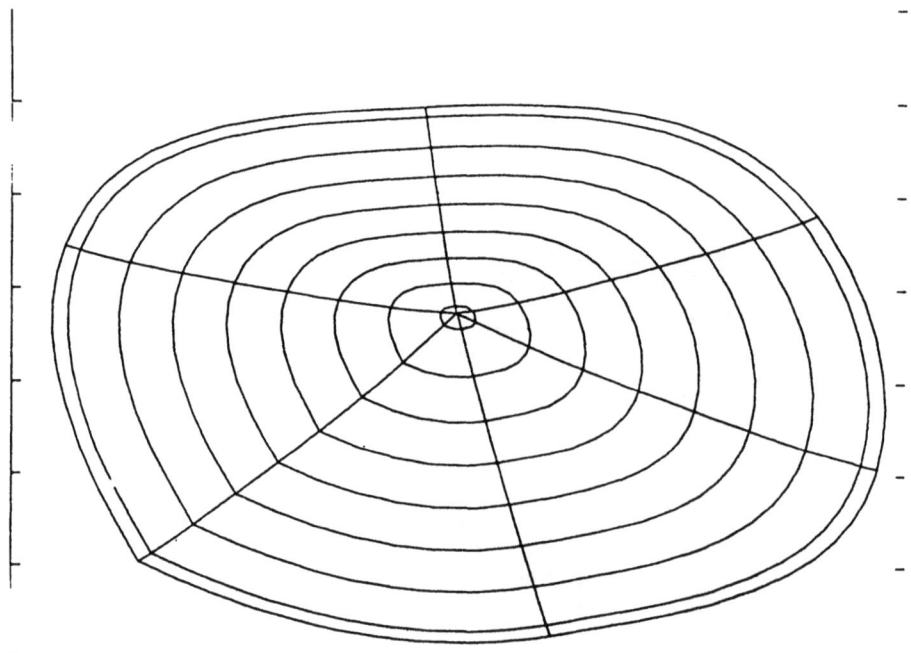

Fig. 10. Deformed geometry of quasi-static penetration

deformed shape of the same model as before, but where the dynamic element is removed, i.e., a static analysis, with the inertial loads at nodes 1, 2, 20 and 21 replaced by incrementally increased forces.

REFERENCES

1. Lloyd, D.W., Shanahan, W.J., and Konopasek, M., Int.J.Mech. Sci., 20, 521-527, (1978).
2. Shanahan, W.J., Lloyd, D.W., Hearle, J.W.S., Text.Res.J., 48, 495-505, (1978).
3. Buckley, W.R., McCullough, J.R., Sowerby, C., and Tennyson, R.C., in "Stresses and Strains in Textile Structures", Shirley Institute Publication S10, Shirley Institute, Manchester, 1974, pp.Buckley 1-18.
4. Kilby, W.F., J.Text.Inst., 54, T9-27, (1963).
5. Hearle, J.W.S., Konopasek, M., and Newton, A., Text.Res.J., 42, 613-626, (1972).
6. Novozhilov, V.V., "Thin Shell Theory", P. Noordhoff, Groningen, 1964.
7. Ambartsumyan, S.S., "Theory of Anisotropic Shells", NASA TTF-118, 1964.
8. Hearmon, R.F.S., "An Introduction to Applied Anisotropic Elasticity", Oxford University Press, Oxford, 1961.
9. Peirce, F.T., J.Text.Inst., 21, T377-416, (1930).
10. Cooper, D.N.E., J.Text.Inst., 51, T317-335, (1960).
11. Jilla, D.A., "On the Mechanics of Deformation of Cotton Fabrics", U.S.D.A. Report FP175-68, No. 5, 1968.
12. Skelton, J., Text.Res.J., 41, 637-647, (1971).
13. Lloyd, D.W., Hearle, J.W.S., J.Text.Inst., 68, 299-302, (1977).
14. Cuisick, G.E., "A Study of Fabric Drape", Ph.D. Thesis, University of Manchester, 1962.
15. Hepworth, B., and Leaf, G.A.V., in "Studies in Modern Fabrics", Papers of the Diamond Jubilee Conference of the Textile Institute, Textile Institute, Manchester, 1970, pp. 181-196.
16. Konopasek, M., "Improved Procedures for Calculating the Mechanical Properties of Textile Structures", Ph.D. Thesis, University of Manchester, 1970.
17. Robinson, J., "Integrated Theory of Finite Element Methods", John Wiley and Sons, 1973.
18. Zienkiewicz, O.C., "The Finite Element Method", third edition, McGraw-Hill, 1977.
19. Martin, H.C., Carey, G.F., "Introduction to Finite Element Analysis", McGraw-Hill, 1973.
20. Przemieniecki, J.S., "Theory of Matrix Structural Analysis", McGraw-Hill, 1968.

21. Oden, J.T., "Finite Elements of Non-linear Continua", McGraw-Hill, 1971.
22. Bathe, K.J., Wilson, E.L., Iding, R.H., "NONSAP - A Structural Analysis Program for Static and Dynamic Response of Nonlinear Systems", Report No. UCSESM 74-3, University of California.
23. Bathe, K.J., Ozdemir, H., Wilson, E.L., "Static and Dynamic Geometric and Material Nonlinear Analysis", Report No. UCSESM 74-4, University of California.
24. Clough, R.W., and Penzien, J., "Dynamics of Structures", McGraw-Hill, 1975.
25. Torbe, I., Proc.Inst.Maths. Conference on Finite Element Methods, Brunel University, 1975, pp. 359-367.

THE DYNAMICS OF FLEXIBLE FILAMENT ASSEMBLIES

C. M. Leech

Department of Mechanical Engineering, University
of Manchester Institute of Science and Technology,
Manchester, England.

ABSTRACT. The dynamics of flexible filament assemblies are
examined by considering first the motion of a filament, second
the interaction of interfering filaments and finally by
synthesizing the assembly from filament arrays. The filament
behaviour is considered from the point of view of wave
propagation and when assembled these behavioural characteristics
can be exploited for the prediction of the system dynamics.
Finally, various continuum models for dense assemblies are
considered. Finite deformations and Green strains are used in
the model formulation and Hamilton's (variational) principle is
employed to generate the necessary equations of motion.

1. INTRODUCTION

For the modelling and prediction of the dynamics of high
speed large deformation processes of systems of multiply connected
filaments which together when densely packed form a cloth or
fabric and when sparsely assembled form a net or grillage, it is
necessary firstly to specify the dynamic principles assumed and,
secondly, to define a coordinate representation and hence an
appropriate strain measure [1-5].

The dynamic principle chosen for this development is Hamilton's
principle, [6-10], a variational energy principle based upon a
closed system or a fixed aggregate of matter or mass particles.
This then suggests a Lagrangian formulation in which the dynamics
of labelled points are followed through space rather than the
Eulerian formulation in which motion through a specific space is
examined. Following the Lagrangian formulation leads quite

naturally to the Green strain tensor as a means of measuring deformation. All motions are referred to a coordinate system defined in a reference state, chosen here to be the rest state, FIG. 1.1

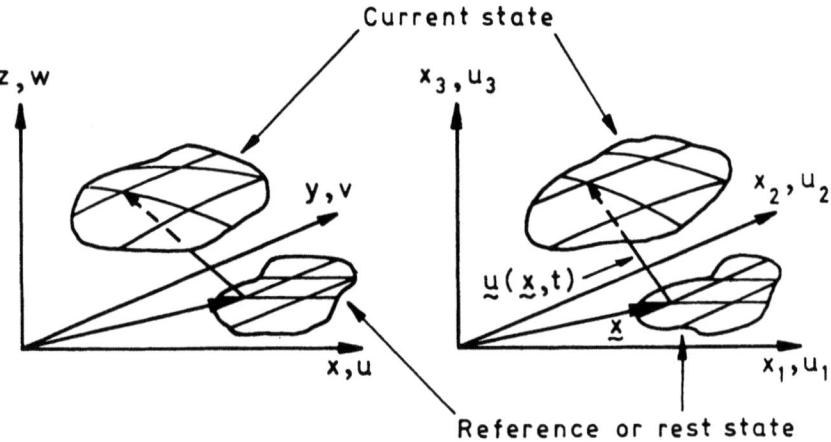

Fig. 1.1   Coordinate system

1.1 Hamilton's principle [6-8]

In its most general form, Hamilton's principle states that "of all possible motions the actual motion of a dynamic system is such that"

$$\delta \int_{t_1}^{t_2} (T-U) \, dt + \int_{t_1}^{t_2} \delta W dt = 0 \qquad (1.1)$$

where the variations of the kinematic variables vanish at the arbitrary times t_1 and t_2. T is the kinetic co-energy of the system, U is the potential or strain energy and δW is the virtual work associated with non-conservative elements typified by slippage and internal friction.

The principle is applied to the system 'as a whole' and scans through all admissible or compatible states for the actual motion which satisfies equation (1.1).

1.2 Strain Measure [1-3]

The exploitation of the energy principles must lead to the consideration of hyperelastic or hypoelastic systems which are constitutively based upon work or energy functions. The strain measure used is the Green strain tensor E which conforms to the Lagrangian formulation and the application of Hamilton's principle since it "follows the motion".

It is defined as follows:

$$E_{ij} = \frac{1}{2}\left[u_{i,j} + u_{j,i} + u_{k,i}\, u_{k,j}\right] \tag{1.2}$$

where $u_i\,(x_1, x_2, x_3, t)$ is the material deformation.

The variation of strain energy then results in a consistent stress tensor, namely the Kirchhoff or 2nd Piola-Kirchhoff stress tensor T, [9-10].

That is

$$T_{ij} = \frac{\partial U^*}{\partial E_{ij}} \tag{1.3}$$

for hyperelastic (conservative) systems, or

$$\delta W^* = -T_{ij}\,\delta E_{ij} \tag{1.4}$$

for hypoelastic (non-conservative) systems, where U^* and W^* are the energy and work densities [3,5].

1.3 System Analysis

The system will be assumed to be composed of a large number of one dimensional elements (filaments) arranged in such a fashion so as to form a surface. The denseness of the surface as in cloth systems must naturally suggest that a continuum mechanics approach be considered. However, the routing of any information through the surface must be such as to follow the filament lines since transmitting matter does not exist elsewhere.

The analysis of the system dynamics will be pursued firstly by considering the dynamics of filaments, secondly by accounting for the interference of crossing filaments, and finally by considering an integrated effect of many interfering filaments to make up a cloth continuum.

Before proceeding the following definitions will be made FIG. 1.2;

　i) a node is the crossing or interference of two or more filaments,

　ii) a cell is the empty space bounded by interfering filaments.

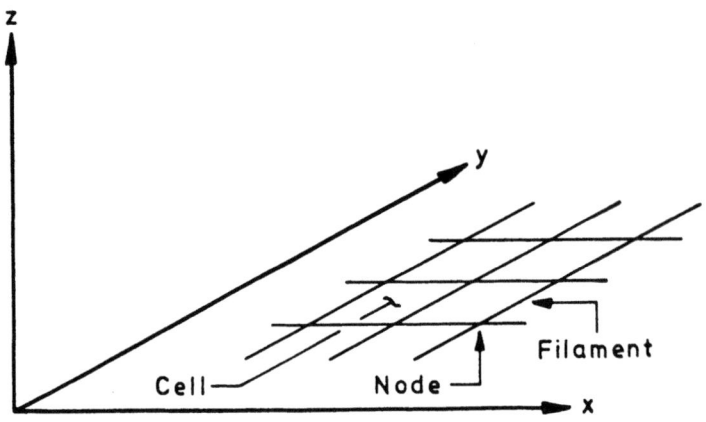

Fig. 1.2   Assembly definitions

1.4 Modelling Assumptions

In order to keep the following developments within reasonable bounds, the following will be assumed:

i) the filaments are hyperelastic,
ii) the filaments do not resist bending, shear, or torsion,
iii) nodes are fixed on each interfering filament precluding the effects of slip or separation,
iv) interference only occurs at nodes and not, for example, by the fibrous excursions to parallel filaments,
v) the system assumed throughout this development is orthogonally woven thus generating rectangular cells.

No restrictions are placed as to the largeness of the deformations or to the speed of the deformation process; however, there will be no folding of the fabric so as to form multiply connected surfaces as can occur, for example, in tangled nets. The surface defined by nodes is assumed to be single valued and continuous but it need not be developable.

2. FILAMENT DYNAMICS

Many investigators have derived the equations of motion for flexible filaments [11-17]. The equations will now be derived using variational principles and finite strain measures; the equilibrium conditions applicable at nodes or points of interference between filaments will also be generated.

In the previous section, the Green strain tensor has been introduced; however, although this has been defined in the reference or rest coordinates, it is possible to define strain tensors in, for example, unstrained coordinates. In many materials there is very little difference between these two coordinates systems since the maximum strain may be of the order of 0.5%. However, filaments, especially polymer chain filaments can withstand strains typically of about 20% and in many cases, much higher strains can be supported. It is thus essential that the distinction between rest and unstrained coordinates be made where they exist.

The Green strain tensor $\underset{\sim}{E}$ has been defined, and for a filament with the reference state being a strained rest state, only one component of the tensor is considered, that is

$$E_{11} = u_{,x} + \frac{1}{2}(u^2_{,x} + v^2_{,x} + w^2_{,x}). \qquad (2.1)$$

Now the classical strain ε being defined as (new length - old length)/(old length) is then as follows:

$$\varepsilon = \sqrt{1+2E_{11}} - 1. \qquad (2.2)$$

However, the classical strain could have been defined in the unstrained coordinate system. If ε is this classical strain, ε_p is the prestrain in the unstrained coordinate system and ε_p is the prestrain (positive) measured in the reference coordinate system, then the following relationships can be deduced:

$$\varepsilon = \frac{\varepsilon - \varepsilon_p}{1+\varepsilon_p} \quad \text{and} \quad \varepsilon = \frac{\varepsilon + \varepsilon_p}{1-\varepsilon_p} \qquad (2.3)$$

2.1 Equations of Motion

Using the reference (rest) state and the Green strain tensor, the equations of motion for the filament will now be derived. The kinetic co-energy is given as follows:

$$T = \int_0^L \frac{\rho A}{2}(u^2_{,t} + v^2_{,t} + w^2_{,t}) \, dx$$

and the strain energy for a hyperelastic filament is as follows:

$$U = \int_0^L AU^*(E_{11}) \, dx$$

where A is the cross section area of the filament, L is its length and ρ is the filament material density. The functional $U^*(E_{11})$ is the strain energy/unit volume of the filament material given a Green strain E_{11}, FIG. 2.1.

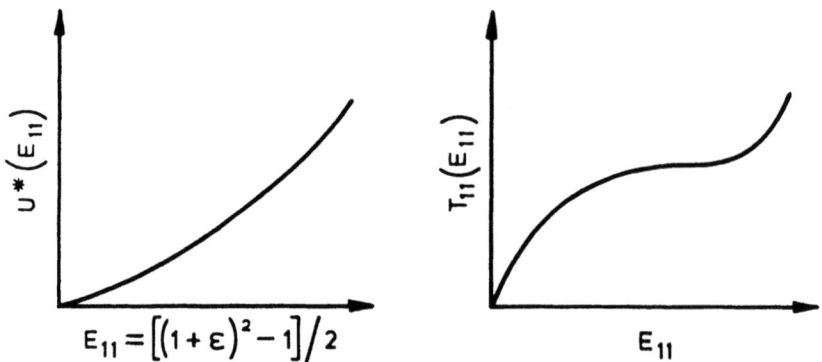

Fig. 2.1 Strain energy density and Kirchhoff stress functions of Green strain

Applying Hamilton's principle with the above energies yields the following equations of motion

$$\rho u_{,tt} - \{T_{11}(1+u_{,x})\}_{,x} = 0, \qquad (2.4)$$

$$\rho v_{,tt} - \{T_{11} v_{,x}\}_{,x} = 0, \qquad (2.5)$$

$$\rho w_{,tt} - \{T_{11} w_{,x}\}_{,x} = 0, \qquad (2.6)$$

where T_{11} the Kirchhoff stress is given as follows

$$T_{11} = \frac{dU^*}{dE_{11}}.$$

Note that the ordinary differential operator has replaced the partial differential operator since the strain energy density is a function only of one component of the strain tensor, namely E_{11}.

These equations are conveniently written in first order form by defining the following terms,

$$p = u_{,t}, \qquad q = v_{,t}, \qquad r = w_{,t}$$

and $\theta = u_{,x}, \qquad \phi = v_{,x}, \qquad \psi = w_{,x}.$

Then the following equations follow:

$$p_{,t} - (a\,\theta_{,x} + b\,\phi_{,x} + c\,\psi_{,x}) = 0,$$

$$q_{,t} - (b\,\theta_{,x} + d\,\phi_{,x} + e\,\psi_{,x}) = 0,$$

$$r_{,t} - (c\,\theta_{,x} + e\,\phi_{,x} + f\,\psi_{,x}) = 0, \qquad (2.7)$$

$$\theta_{,t} - p_{,x} = 0,$$

$$\phi_{,t} - q_{,x} = 0,$$

$$\psi_{,t} - r_{,x} = 0,$$

where $a = \{T_{11} + \dfrac{dT_{11}}{dE_{11}}(1+\theta)^2\}/\rho,$

$b = \dfrac{dT_{11}}{dE_{11}}(1+\theta)\,\phi/\rho,$

$c = \dfrac{dT_{11}}{dE_{11}}(1+\theta)\,\psi/\rho,$

$d = \{T_{11} + \dfrac{dT_{11}}{dE_{11}}\phi^2\}/\rho,$

$e = \dfrac{dT_{11}}{dE_{11}}\phi\,\psi/\rho,$

and $f = \{T_{11} + \dfrac{dT_{11}}{dE_{11}}\psi^2\}/\rho.$

The equations (2.7) represent a system of six simultaneous quasilinear hyperbolic partial differential equations. Characteristic theory [18-22] will be used to develop these equations into their characteristic speeds and the associated transport equations.

Returning to the equations (2.4), (2.5) and (2.6) and considering highly stressed systems where

$$\dfrac{dT_{11}}{dE_{11}} >> T_{11} >> T_{11}\,u_{,x}^2,\ T_{11}\,v_{,x}^2\ \text{or}\ T_{11}\,w_{,x}^2.$$

These equations can then be reduced to the following,

$$\rho\,u_{,tt} - \dfrac{dT_{11}}{dE_{11}} u_{,xx} = 0, \qquad (2.8)$$

$$\rho\,v_{,tt} - T_{11}\,v_{,xx} = 0, \qquad (2.9)$$

and $\rho w_{,tt} - T_{11} w_{,xx} = 0.$ (2.10)

These are again three hyperbolic systems but in this case they are both linear and decoupled. At this point it is sensible to consider these equations, representing the dynamics of tightly drawn filaments and examining their solution in terms of the propagation characteristics.

2.2 Propagation in Filaments.

The equations (2.8), (2.9) and (2.10) are hyperbolic and a consequence of this is that they model the propagation of signals through the filament, [23,24]. To illustrate this, consider solely equation (2.8) and let

$$v^2 = \frac{1}{\rho} \frac{dT_{11}}{dE_{11}}.$$

It is noted here that v has the dimensions of velocity. Now writing equation in first order form with $p = u_{,t}$ and $\theta = u_{,x}$ and substituting v in the equations leads to the following:

$$p_{,t} - v^2 \theta_{,x} = 0 \quad \text{and}$$
$$\theta_{,t} - p_{,x} = 0. \quad (2.11)$$

Consider now a solution of the form $p(\eta)$ and $\theta(\eta)$ where $\eta = \eta(x, t)$; for this solution to be admitted it must follow that

$$\eta_{,t} = \pm v \eta_{,x} \quad \text{or} \quad \eta = t \pm \int^x \frac{dx}{v}$$

There are thus two possible functions $\eta(x, t)$ which will allow the solutions $p(\eta)$ and $\theta(\eta)$. To generate these solutions let $\xi = t + \int^x \frac{dx}{v}$, and $\zeta = t - \int^x \frac{dx}{v}$ and let $p(\xi,\zeta)$ and $\theta(\xi,\zeta)$ be solutions to the equations (2.10).
They must then satisfy the following "transport" equations

$$p_\xi - v \theta_\xi = 0 \quad \text{along } \zeta \text{ constant}$$
$$\text{and} \quad p_\zeta + v \theta_\zeta = 0 \quad \text{along } \xi \text{ constant} \quad (2.12)$$

The ξ and ζ lines are characteristics of the differential equations (2.11), FIG. 2.2.

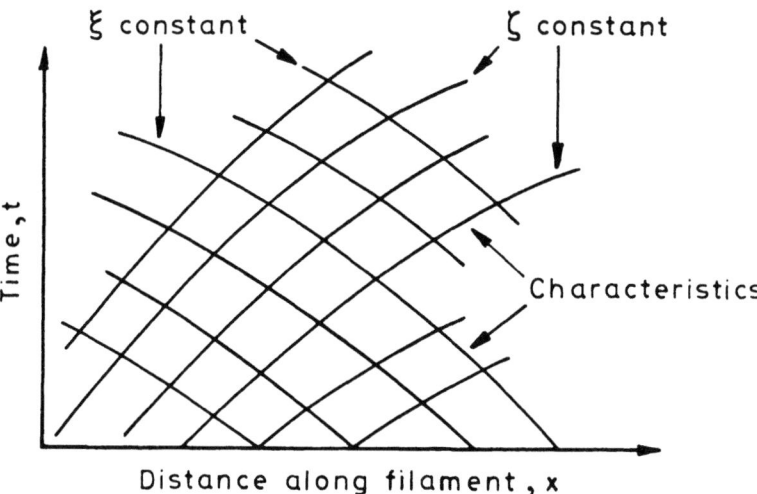

Fig. 2.2 Characteristic mesh

The transport equations are thus the mode of propagation of signals through the filament, and the characteristic equations dictate the speed of propagation. Returning now to the equations (2.8), (2.9) and (2.10), it is noticed that they are all of the form of (2.11). Equation (2.8) is the model for extension along the filament and these extensions thus propagate with the extensional propagation speed c_e $(= \sqrt{\frac{1}{\rho} \frac{dT_{11}}{dE_{11}}})$.

Equations (2.9) and (2.10) model the propagation of transverse signals (deflections perpendicular to the filament axis), and these signals propagate with the transverse propagation speed c_t $(= \sqrt{T_{11}/\rho})$.

2.3 Propagation Order

Noting that T_{11} the Kirchhoff stress, and $\frac{dT_{11}}{dE_{11}}$ the secant modulus, can be observed on a stress-strain curve, then for a typical "S" curve obtained from fishing nylon, FIG. 2.3, it can be seen that in the majority of the curve the extensional propagation speed is larger than the transverse propagation speed; however, in the vicinity of the softening point the two speeds become equal and beyond that point the transverse speed is larger than the extensional speed. However, for most applications, the extensional speed is very much larger than the transverse speed since the tensile stress must be very much lower than the secant modulus for sub yield stresses.

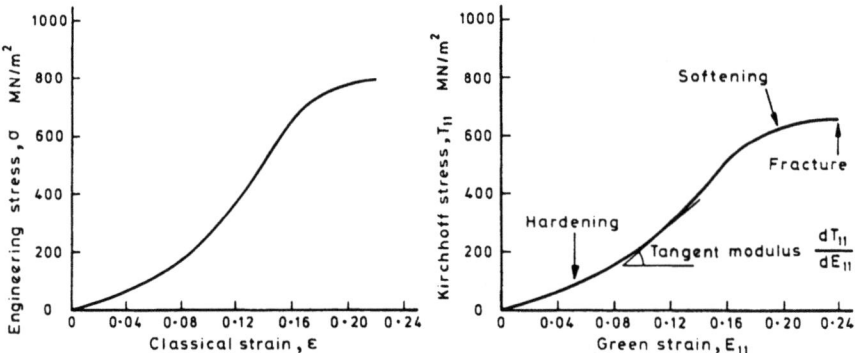

Fig. 2.3   Stress-strain curve for nylon

	Material Density (Kg/m^3)	Secant Modulus (GN/m^2)	Extensional Propagation Speed (m/s)
Steel	7800	210	5200
Glass	2500	70	5300
Rubber	930	0.02	147
Nylon [at Green Strain E_{11} = .02]	1135	1.5	1150
Nylon [at Green Strain E_{11} = .12]	1135	4.8	2050

Table 1. Propagation Velocities for some common materials

2.4 Dispersion and Shocking

For non-linear stress-strain behaviour, the propagation speed c_e is governed by the local state of strain of the filament. Again, referring to the "S" curve, in the vicinity of the hardening point, the higher the strain the faster the propagation; similarly in the vicinity of the softening point higher strains lead to slower propagation speeds.

Now consider firstly a ramp signal, spanning the hardening point, FIG. 2.4; small strains propagate slower than the high strains which must ultimately catch up with the slower signals. However, they cannot overtake and must result in a steep fronted signal or shock.

Alternatively, in the vicinity of a softening point a sharp fronted signal will degenerate into a rounded signal since high strains propagate slower than the low strains. This is dispersion where the small front distance is dispersed to a larger front. Dispersion can also arise from internal reflections from internal inclusions, flaws and defects and is thus a real effect in many natural and composite materials, whereas in many manufactured materials (nylon and polymeric filaments) where great care is taken to remove these defects and attain near perfect-homogeneous materials, dispersion only arises from material softening.

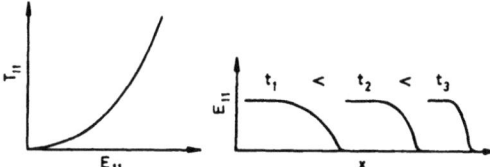

(a) Hardening (shocking) stress-strain behaviour

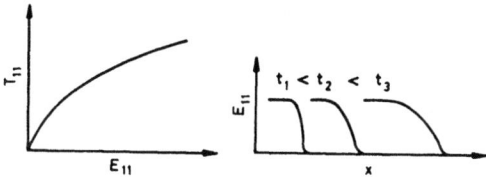

(b) Softening (dispersive) stress-strain behaviour

Fig. 2.4   Signal fronts for hardening and softening materials

2.3 Reflection

In the earlier section, it has been shown that the signals or strains and velocities propagate along characteristic lines (in the time-distance or $t \sim x$ plane) according to the equations (2.12). However, when a boundary or free edge is encountered, the equations must be modified to account for the appropriate conditions at the boundary.

For a free edge, the strain θ is zero; then noticing that the two solutions $\theta_1(\xi)$ and $\theta_2(\zeta)$ are solutions to the differential equations (2.11), then the general solution can be written as:

$$\theta(x, t) = \theta_2(\zeta) + \theta_1(\xi).$$

Then setting the origin at the boundary so that at this point $\xi = \zeta$ for all times, FIG. 2.5, it must follow that the reflected signal $\theta_1 = -\theta_2$, the incident signal. Thus a compressive strain reflects as a tensile strain.

For a fixed edge, the velocity p is zero since the boundary is immobile. Thus $p_2(\zeta) = -p_1(\xi)$. Also along the incident characteristic (ξ is constant), $\theta_\zeta = -p_\zeta/\nu$ and along the reflected characteristic (ζ is constant), $\theta_\xi^{\zeta} = p_\xi/\nu$. It then follows that the "rate" θ_ξ equals θ_ζ and the strain signal thus reflects unchanged.

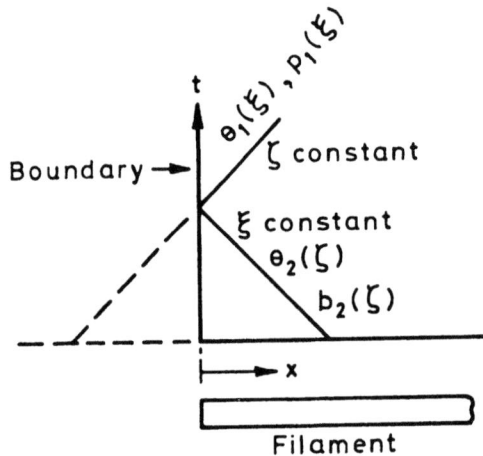

Fig. 2.5 Reflection from a boundary

2.6 Non-Linear Propagation in Filaments

For the non-linear model, equations (2.7), there are three sets of characteristics since the model accounts for extensional and two transverse motions [11,18]. The propagation velocities are as follows:

$$c_e = \sqrt{\left[T_{11} + (1+2E_{11})\frac{dT_{11}}{dE_{11}}\right]/\rho} \qquad (2.13)$$

for the extensional signals and

$$c_t = \sqrt{T_{11}/\rho} \qquad (2.14)$$

for both (v and w) transverse signals.

It is noticed that the extensional propagation speed is always greater than the transverse propagation speeds unless $\dfrac{dT_{11}}{dE_{11}} < 0$

which corresponds to tensile instability and is precluded from this development. The transport equations are given by using the following shorthand,

$$Z \equiv \{p, q, r, \theta, \phi, \psi\}$$

$$U^n \equiv \{1 + \theta, \phi, \psi, -\lambda_n(1 + \theta), -\lambda_n \phi, -\lambda_n \psi\} \quad n = 1, 2$$

$$\equiv \{\phi, -(1 + \theta), 0, -\lambda_n \phi, \lambda_n(1 + \theta), 0\} \quad n = 3, 5$$

$$\equiv \{\psi, 0, -(1 + \theta), -\lambda_n \psi, 0, \lambda_n(1 + \theta)\} \quad n = 4, 6$$

where the characteristic directions are given:

$$\frac{dx}{dt} = \lambda_n$$

and $\lambda \equiv \{+ c_e, - c_e, + c_t, + c_t, - c_t, - c_t\}$.

Finally the transport equations are as follows,

$$U_n \cdot Z = 0 \quad \text{along the } \lambda_n \text{ direction.} \tag{2.15}$$

This set of characteristic equations gives the characteristic mesh, with the characteristic lines

$I^{\pm}$ corresponding to $\frac{dx}{dt} = \pm c_e$ and

$II^{\pm}$ corresponding to $\frac{dx}{dt} = \pm c_t$.

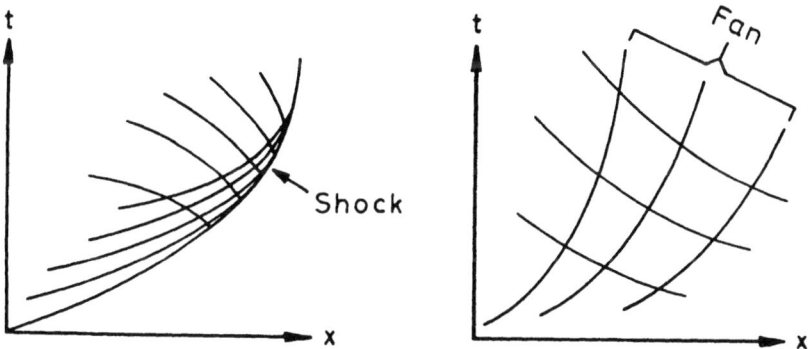

Fig. 2.6   Characteristics for non-linear filaments

356

The characteristic speeds are not necessarily constant, reflecting the non-linearity of the system; the characteristic lines are not then straight and could coalesce into fronts (shocking) or form fans (dispersion), FIG. 2.6.

2.7 Propagation Through Nodes (Knots)

When a signal, extensional or transverse, propagates along a filament and encounters a node it is firstly in part reflected and in part transmitted through the node, and secondly some of the signal is transmitted along the interfering filament. In a previous section, reflection from strain free nodes and fixed nodes was discussed. It is thus argued that the reflection from a node produced by interfering filaments must lie between ±1. These limits will be tightened in the following development. It is pointed out that there is a plane containing both the input filament and the interfering filament and this plane will contain the u and v deformations. Any deformations out of this plane will be called "out of plane" deformations and any deformations in this plane will be "in plane" deformations, FIG. 2.7.

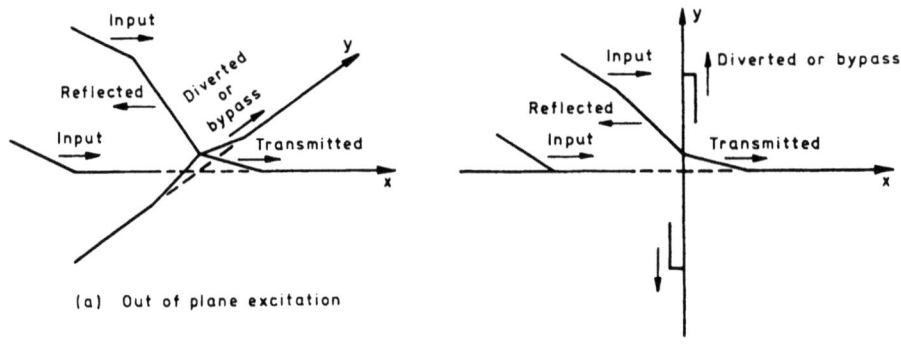

Fig. 2.7 Nodal motion arising from an incident pulse

From Hamilton's principle, the natural boundary conditions for the node are obtained as follows:

$$\sum_{i=1}^{n} A_i \frac{\partial U_i^*}{\partial E_{11}} \left\{ \frac{\partial E_{11}}{\partial u_{i,x}} \delta u_i + \frac{\partial E_{11}}{\partial v_{i,x}} \delta v_i + \frac{\partial E_{11}}{\partial w_{i,x}} \delta w_i \right\} \Bigg|_{x_i=0} = 0$$
(2.16)

where there are n intersecting filaments, <u>each originating from</u> the node. Since the node is assumed to remain fixed on each filament then for compatibility:

$$\delta u_i = \delta u \cos \chi_i + \delta v \sin \chi_i,$$

$$\delta v_i = -\delta u \sin \chi_i + \delta v \cos \chi_i,$$

and $\delta w_i = \delta w \quad$ for $i = 1, 2 \ldots n,$

where δu, δv, and δw are the virtual displacements of the node and χ_i is the angle made by the ith filament with the reference coordinate system, FIG. 2.8.

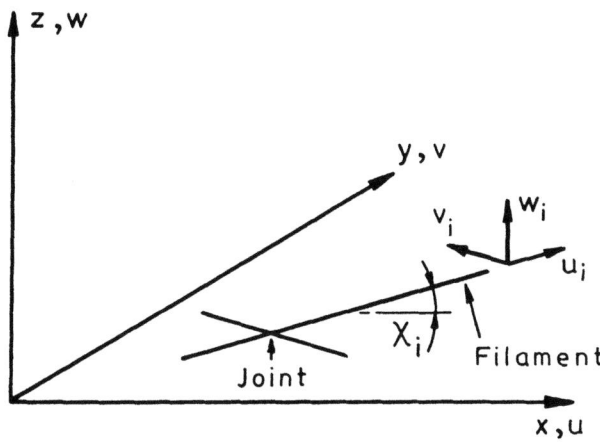

Fig. 2.8   Nodal coordinate system

It then follows after suitable simplification that

$$\delta u \sum_{i=1}^{n} A_i \frac{dU_i^*}{dE_{11}} \left\{ (1 + \theta_i) \cos \chi_i - \phi_i \sin \chi_i \right\} = 0,$$

$$\delta v \sum_{i=1}^{n} A_i \frac{dU_i^*}{dE_{11}} \left\{ (1 + \theta_i) \sin \chi_i + \phi_i \cos \chi_i \right\} = 0, \quad (2.17)$$

and $\delta w \sum_i A_i \dfrac{dU_i^*}{dE_{11}} \psi_i = 0.$

Since the nodes are not fixed in space, it follows that the multiplicands of δu, δv and δw are zero. These are thus the nodal equations for dynamic equilibrium.

For small deformations in tightly strung orthogonally constructed nets, the reflected, transmitted and diverted signals are as follows for firstly an out of plane transverse signal:

$$\theta_r = \frac{c_2}{2(c_1+c_2)} \left(\frac{p_i}{c_1} + \theta_i\right),$$

$$\theta_t = -\frac{c_1}{2(c_1+c_2)} \left(\frac{p_i}{c_1} + \theta_i\right),$$

$$\theta_d = -\frac{c_1^2}{2c_2(c_1+c_2)} \left(\frac{p_i}{c_1} + \theta_i\right), \qquad (2.18)$$

where p_i and θ_i are the input or incident velocities and strains. For inplane disturbances, similar impedance expressions are determined [25]. Nodal impedances can then be identified, these arising from the attenuation of signals through the nodes in interfering strings. For an input entering the node along an x directed filament, and a y directed filament causing the interference, the impedance δ is $\dfrac{T_y A_y c_x}{T_x A_x c_y}$ where T is the Kirchhoff stress, A is the cross-section area and c is the propagation velocity of transverse signals. Optimum energy partitioning occurs when $\delta = 1$, in which case equal amounts of energy are reflected, transmitted through the node, and passed to each interfering filament.

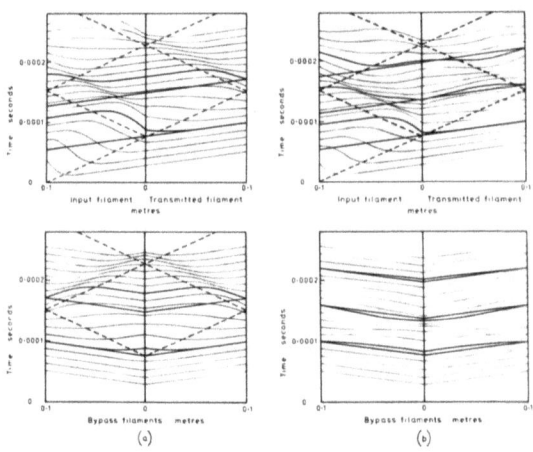

Fig. 2.9   Characteristic solutions for the effect of (a) an out of plane and (b) an extensional incident pulse

Finally, the non-linear model showing the effect of a node on the input signal can be solved using the characteristic theory and equations (2.15). The computer plotted output, FIG. 2.9, shows the propagation of slope through the constituent filaments.

2.8 Closure

In this chapter, the equations of motion for a filament, using finite strain theory, are developed. They are hyperbolic with two distinct propagation velocities, namely the propagation speed of extensional (fast) and of transverse (slow) signals. The latter characteristic speed is repeated, in that there are two associated transport equations. Reflection and transmission is then discussed with the intention of incorporating the derived attenuation and impedance factors into the modelling of coarse and dense nets and cloths.

3. DYNAMICS OF FILAMENT ASSEMBLIES

Various investigations [25-33] have been conducted into the vibrations and dynamics of nets. These investigators all commenced from the premise that the systems were sufficiently sparse so as to preclude any assumptions that the systems could be considered as continua. These ideas will be pursued here.

The dynamics of various filaments is considered by an extension of the ideas presented in the previous chapter. Cloths if sufficiently dense could cursorily be considered from continuum concepts; however, in this chapter they will be considered from the filament assembly ideas and it will be seen that various conclusions can be deduced, these conclusions being exclusive to the assembly rather than the continuum models.

3.1 Dispersion and Routing

For the majority of this chapter the filament assembly will be assumed to be constructed of homogeneous highly tensioned filaments, with the nodes fixed on interfering filaments. As previously discussed, for small signals transverse "out of plane" deformations upon encountering a node will in part be transmitted through the node, in part reflected along the incident path and part will be diverted to the interfering filaments. Out of plane transverse signals cause out of plane transverse signals whereas "in plane" transverse signals or extensional signals will cause, depending upon the inclination of the interfering filaments, both in plane, transverse and extensional signals, FIG. 3.1.

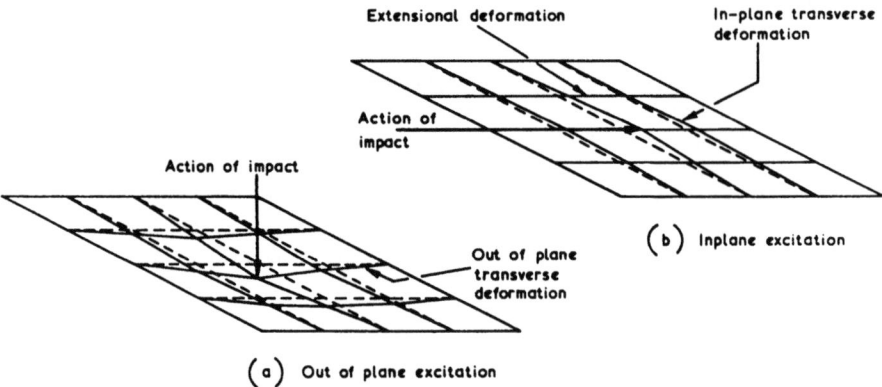

Fig. 3.1   Assembly deformations

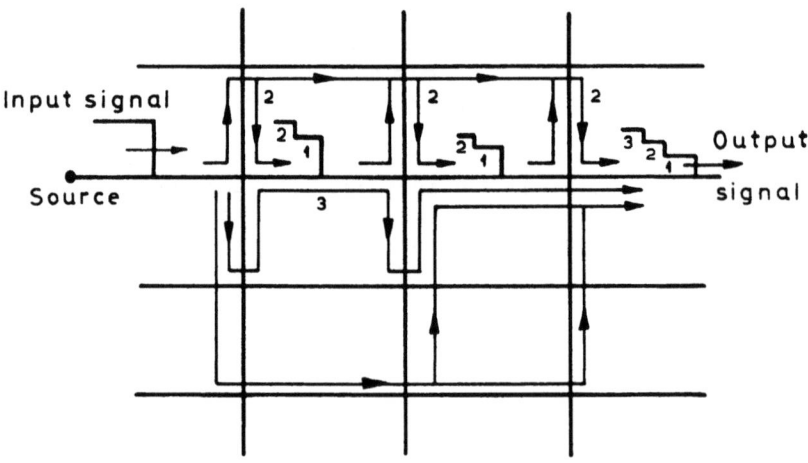

Fig. 3.2   Signal dispersion by routing

Considering solely the out of plane transverse signals, although the following arguments will be applicable to the other excitations, the signal upon encountering a node is only in part transmitted through the node. Here then is the first distinction between nodal assemblies and continuum since in the latter the total energy packet is transferred through all points. The reflected signal returns along the incident path and could be reflected back to the excitation area, FIG. 3.2. However, in returning to this zone it likely encounters other nodes and thus will be again reflected back to the incident node. Again part will be transmitted through the incident node, but obviously at a later time to the first transmitted signal. This then is the mechanism for dispersing a signal; the signal which initially occupies a small space along a filament now occupies a larger space. Thus even if the dispersion of energy by reflection and diversion is ignored, the energy density and thus the associated strains must be lower. This then is another distinction between the behaviour of filament assemblies and continua; in the former the front is dispersed over a large time whereas in the latter the front occupies the same time base as the signal source.

3.2 Zones of Influence

Signals propagating through filament assemblies must travel along the filaments and branch at the nodes. At a time t after the initiation of an excitation at some point in the assembly, this point being taken as the origin, the effect of this disturbance will be felt at points remote from this excitation point. These points can be joined, ignoring for the moment the sparsity of the assembly, to form a front. The zone of influence will thus be defined as that zone lying between the signal source and signal front and all points outside this zone will be at rest or "quiet".

Consider the action of a disturbance at some point in an assembly of orthogonally interfering filaments, this disturbance exciting the assembly in "out of plane" transverse motion. By evaluating the time taken for a signal to arrive at a point (x, y) on the front, it is seen that the zone of influence is limited by the following expression:

$$t = \pm \frac{x}{c_{tx}} \pm \frac{y}{c_{ty}} \qquad (3.1)$$

where c_{tx} and c_{ty} are the propagation speed of transverse signals in the x and y directions. The zone of influence is thus rhomboidal, the quiet region lying outside the rhomboid, FIG. 3.3a.

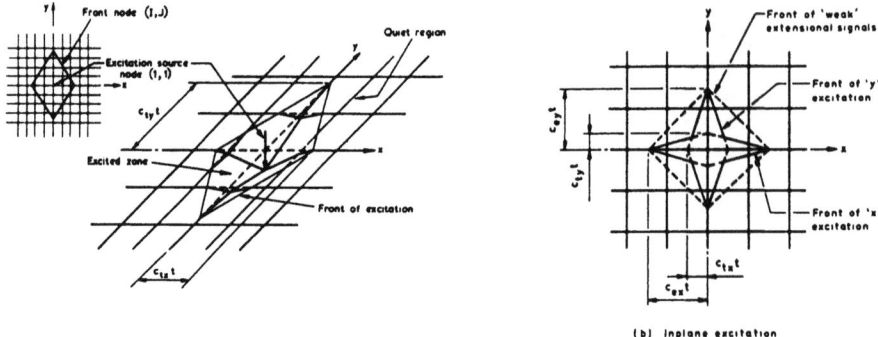

Fig. 3.3   Zones of influence

The action of "in plane" signals gives rise to similar zones of influence. The zone of influence of a general in plane excitation is given by the outermost extent of the two rhomboids

$$t = \pm \frac{x}{c_{ex}} \pm \frac{y}{c_{ty}} \quad \text{and}$$

$$t = \pm \frac{x}{c_{tx}} \pm \frac{y}{c_{ey}} \tag{3.2}$$

where c_{ex} and c_{ey} are the propagation speeds of extension signals in the x and y directions, FIG. 3.3b. This results in the star front; however, the transverse signals give rise to small extensional signals and these form a weak front (rhomboid)

$$t = \pm \frac{x}{c_{ex}} \pm \frac{y}{c_{ey}}$$

containing the previous two rhomboids. These signals, however, are very small and are usually ignored.

3.3 Front Signals [29]

In an earlier section it was shown that the signal passing through a node is attenuated, the attenuation for each node being characterised by a nodal impedance. For "out of plane" transverse excitation, the nodal impedances are

$$\lambda_x = \frac{c_y T_x}{c_x T_y + c_y T_x} \quad \text{for emanating signals in the } x \text{ direction}$$

and $\lambda_y = \dfrac{}{c_x T_y + c_y T_x}$ for emanating signals in the y direction.

Thus for an input signals, $w_o(t - x/c_x - y/c_y)$, the attenuated signals in the x and y directions are:

$w = w_o \lambda_x$ leaving along an x filament, and

$w = w_o \lambda_y$ leaving along the y filaments.

Thus a signal originating from node (1,1) and leaving the frontal node (i,j) will have a signal strength

$$w = w_o \lambda_x^{i-1} \lambda_y^{j-1}$$

since it will have passed through $i-1$ nodes in the x direction and $j-1$ nodes in the y direction. However, the route to the fronts is not unique for $i > 1$ and $j > 1$. The number of possible primary routes to the front $R(i,j)$ can be evaluated by noting that

$$R(i,j) = R(i-1,j) + R(i,j-1),$$

and from elementary combinational theory it can be seen that

$$R(i,j) = \frac{\Gamma(i+j-1)}{\Gamma(i)\,\Gamma(j)} = \frac{1}{(i+j-1)\,B(i,j)},$$

where $\Gamma(i)$ is the Gamma function and $B(i,j)$ is the Beta function.

The signal thus leaving the front node (i,j) is then

$$w = w_o \lambda_x^{i-1} \lambda_y^{j-1} R(i,j).$$

For dense assemblies it is logical to convert from the node number (i,j) to the node coordinates (x,y). This is achieved by noting that $x \simeq (i-1)\Delta_x$ and $y \simeq (j-1)\Delta_y$ where Δ_x and Δ_y are the cell sizes. Also the attenuations λ_x and λ_y are converted to logarithmic attenuations α_x and α_y, where

$$\alpha_x = -\frac{1}{\Delta_x} \log_e \lambda_x \quad \text{and} \quad \alpha_y = -\frac{1}{\Delta_y} \log_e \lambda_y.$$

Then for an "assembly continuum" where the cells are limiting small but still finite the front signal w is given

$$w = \Lambda w_o$$

where the attenuation $\Lambda(x,y)$ is as follows:

$$\Lambda(x,y) = \exp\left(-\alpha_x x - \alpha_y y\right) \left\{ \left(\frac{x}{\Delta_x} + \frac{y}{\Delta_y} + 1\right) B\left(\frac{x}{\Delta_x} + 1, \frac{y}{\Delta_y} + 1\right) \right\}. \quad (3.3)$$

The assembly (cloth) slopes at the front are thus

$$w,_x = -\frac{1}{c_x} \dot{w}_o \Lambda \quad \text{and} \quad w,_y = -\frac{1}{c_y} \dot{w}_o \Lambda.$$

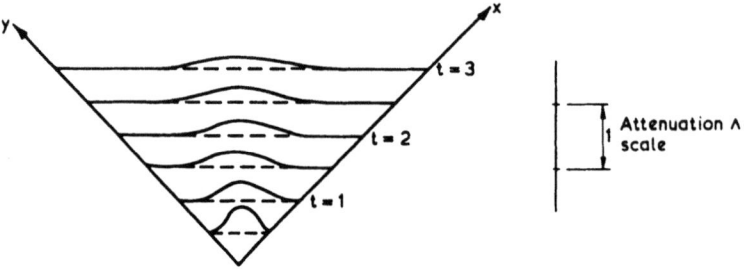

$\Delta x = \Delta y = 0.1; \quad Cx = Cy = 1; \quad \lambda x = \lambda y = 0.5$

Fig. 3.4 Signal attenuation in an orthogonal assembly

In these latter expressions, the differentials $w,_x$ and $w,_y$ are obtained by holding Λ constant and finding the signal slope $\dot{w}_o$ since the attenuation is <u>constant</u> between nodes, even in the limit as Δ_x and $\Delta_y \to 0$. The attenuation for an orthogonal filament assembly is shown in FIG. 3.4 and it can be seen that maximum signals occur in the quadrant and not on the orthogonal axis.

3.4 Final Element Nodal Model [30-33]

In the introduction, Hamilton's principle has been introduced, eq. (1.1). This considers the total energy contents (kinetic and strain) of the system and states that an extreme is produced for the actual motion of the system. This then implies that one can search through all possible motions of the system and for each, one assembles the Lagrangian, performs the integration and tests the integral value against that produced for other possible motions.

This is indeed laborious; an alternative is to scan through these motions and for each find a necessary condition for the attainment of the extreme. Again this is impossible unless the complete range of possible motions is expressed in some functional or collapsed means. However, a viable alternative is to select one of these possible motions and find the numerical conditions for this admissible configuration to give a local extreme. This extreme can then be checked against other local extremes and whilst never obtaining the exact motion, it does give reasonable engineering solutions.

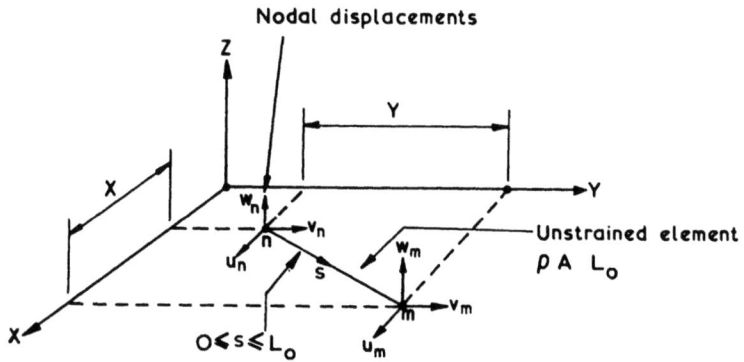

Fig. 3.5   Element and nodal coordinate systems

A possible motion of a filament assembly, especially when the cells are small as occurring in cloths, is for each filament to be linear between the nodes. The nodal motions are then still unknown and the variational principle will yield the necessary conditions for the extreme, within the constraints of the selected compatible or possible motion, namely no flexure or bending of the filaments between nodes. For the filament connecting node m to node n, the displacement of the filament material is given, FIG. 3.5,

$$\underline{u} = \underline{u}_n(1 - s/L_o) + \underline{u}_m \, s/L_o$$

where the displacement vector $\underline{u} \equiv (u,v,w)$, s is a distance coordinate along the rest or reference position of the filament and L_o is the reference length between nodes m and n. The kinetic energy of the filament connecting node m to node n is then

$$T_{mn} = \frac{\rho A L_o}{6} \left\{ \underline{u}_{m,t} \circ \underline{u}_{m,t} + \underline{u}_{m,t} \circ \underline{u}_{n,t} + \underline{u}_{n,t} \circ \underline{u}_{n,t} \right\}.$$

The classical strain ε for the filament connecting the nodes m and n is given by

$$\varepsilon = (L - L_o)/L_o$$

where L is the current distance between nodes m and n. That is

$$L = \sqrt{(X + u_m - u_n)^2 + (Y + v_m - v_n)^2 + (w_m - w_n)^2}$$

where X and Y are the x and y distances between these two nodes, such that

$$L_o = \sqrt{X^2 + Y^2}.$$

The Green strain is simply, as follows,

$$E_{11} = \left\{ X(u_m - u_n) + Y(v_m - v_n) + (u_m - u_n)^2/2 \right.$$
$$\left. + (v_m - v_n)^2/2 + (w_m - w_n)^2/2 \right\} \bigg/ L_o^2 .$$

The strain energy of each filament is then evaluated using the Green strain E_{11} or more conventionally the classical strain ε and the Lagrangian for the filament connecting node m to node n is as follows,

$$\mathcal{L}_{mn} = T_{mn} - U_{mn}(\varepsilon),$$

where $U_{mn}(\varepsilon)$ is the strain energy/unit volume. The total system Lagrangian is simply the sum of the constituent filament Lagrangians and the necessary equations of motion are simply obtained.

$$\sum_m \frac{(\rho A L_o)_m}{6} \left(2 u_{n,tt} + u_{m,tt} \right) + \frac{dU_{mn}}{d\varepsilon} \frac{A_m}{L_{mn}} (u_n - u_m - X_{mn}) = 0$$

$$\sum_m \frac{(\rho A L_o)_m}{6} \left(2 v_{n,tt} + v_{m,tt} \right) + \frac{dU_{mn}}{d\varepsilon} \frac{A_m}{L_{mn}} (v_n - v_m - Y_{mn}) = 0$$

$$\sum_m \frac{(\rho A L_o)_m}{6} \left(2 w_{n,tt} + w_{m,tt} \right) + \frac{dU_{mn}}{d\varepsilon} \frac{A_m}{L_{mn}} (w_n - w_m) = 0 \quad (3.4)$$

where the $\sum_m$ denotes summation over all m filaments interfering with node n, and L_{mn} is the current length of the filament connecting node m to node n.

Completing the model, it is observed that the mass matrix $[M]$ associated with this system is applicable for each of the component directions. It can thus be evaluated and inverted before the marching solution is attempted with a finite difference scheme. That is

$$\{u\}^{t+\Delta t} = 2\{u\}^t - \{u\}^{t-\Delta t} + \Delta t^2 [M]^{-1} \{P_u\},$$

$$\{v\}^{t+\Delta t} = 2\{v\}^t - \{v\}^{t-\Delta t} + \Delta t^2 [M]^{-1} \{P_v\},$$

$$\{w\}^{t+\Delta t} = 2\{w\}^t - \{w\}^{t-\Delta t} + \Delta t^2 [M]^{-1} \{P_w\}, \qquad (3.5)$$

where $\{u\}$, $\{v\}$ and $\{w\}$ are now the vectors of the nodal displacements and $\{P_u\}$, $\{P_v\}$ and $\{P_w\}$ are the x, y and z components of the generalised forces.

This procedure is quite acceptable for small assemblies and has been successfully applied for various weaves in a sixteen node assembly, FIG. 3.6. The results are summarised in Table 3.1. However for larger systems the problems associated with the inversion of the matrix necessitate formulations not so convenient as (3.5). For example, an assembly of 10 x 10 nodes requires a mass matrix of (100 x 100) which even for this coarse system will stretch many medium size computers. This storage problem is overcome for example by not assembling the mass matrix but by evaluating the mass components as they are required and using a Gauss-Seidel or Successive Over Relaxation method to solve the displacements, [31].

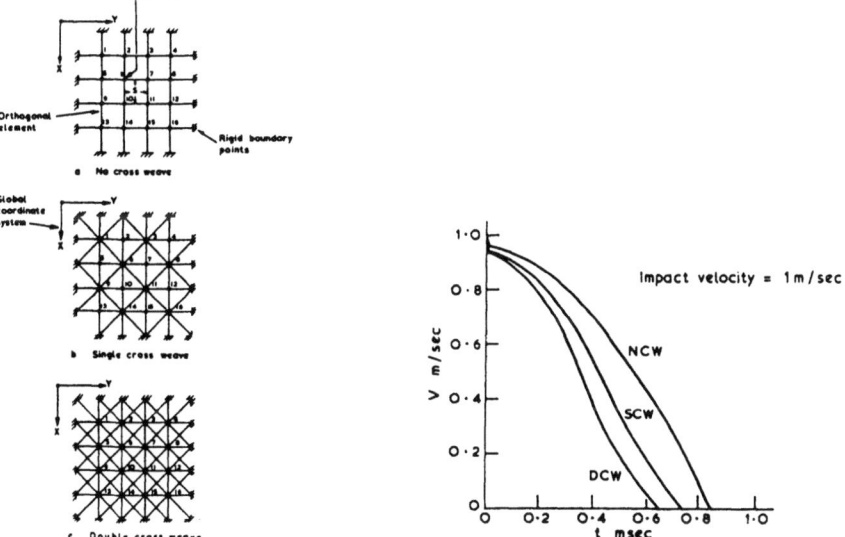

Fig. 3.6 4 x 4 filament assemblies and arrest characteristics

Velocity m/sec	Cross weave parameter	Maximum over stress MN/m²	Arrest time m/sec	Node 6 final displ. mm
1	1	4.8×10	8.3	3.4
1	2	2.9×10	7.1	3.4
1	3	3.7×10	6.3	3.4
10	1	6.4×10^2	3.1	19.5
10	2	4.7×10^2	2.7	14.9
10	3	5.6×10^2	2.5	13.0
50	1	3.2×10^3	1.4	44.1
50	2	2.6×10^3	1.2	33.9
50	3	2.8×10^3	1.18	29.4
100	1	6.6×10^3	1.0	62.5
100	2	5.2×10^3	0.87	48.2
100	3	5.8×10^3	0.83	41.8
1000	1	9.8×10^4	0.34	210.8
1000	2	6.0×10^4	0.30	162.4
1000	3	5.9×10^4	0.28	141.7
10000	1	1.1×10^6	0.18	1025.8
10000	2	5.4×10^5	0.14	744.8
10000	3	6.5×10^5	0.13	683.3

Notes:
(a) The maximum overstress quoted is for element (2)-(6) over the period from the initial impact to the projectile arrest.
(b) The cross weave parameter is as follows:
    No cross weave   1
    Single cross weave   2
    Double cross weave   3
(c) Material constants:
    Youngs Modulus   = 200 GN/m²
    Material Density   = 10^4 Kg/m³
    Cross Section Area = 10^{-6} m²
(d) Assembly Properties:
    Cell rectangular dimension = 10^{-2} m
    Assembly dimension   = 5×10^{-2} m
    Filament prestress   = 30 MN/m²

Table 3.1. 4 × 4 rectangular assembly impact characteristics

3.5 Characteristic Nodal Model

In section 2.6, the characteristic solution for the dynamics of filaments was discussed. In this section this technique is extended to find solutions of dense assemblies. Firstly, since the assembly is dense it is sensible to consider only the nodal velocities and displacements. The characteristic velocity variables (p,q,r) are defined at each node and the gradient variables (θ,ϕ,ψ) are defined for each interfering filament at each node. Thus for a node with m interfering filaments that are $3(m+1)$ unknown quantities. Entering each node there are $3m$ characteristic equations, FIG. 3.7, and to complete the formulation there are 3 force equations at the node arising from nodal compatibility. Thus the maximum matrix to be inverted is rank $3(m+1)$ and is governed by the node with the maximum number of interfering filaments, whereas in the finite element model the matrix to be inverted has a rank equal to the number of nodes.

The three force equations, applicable at the nodes, are obtained from the natural boundary conditions arising from Hamilton's principle. These equations have been previously derived, eq. (2.16) and eq. (2.17).

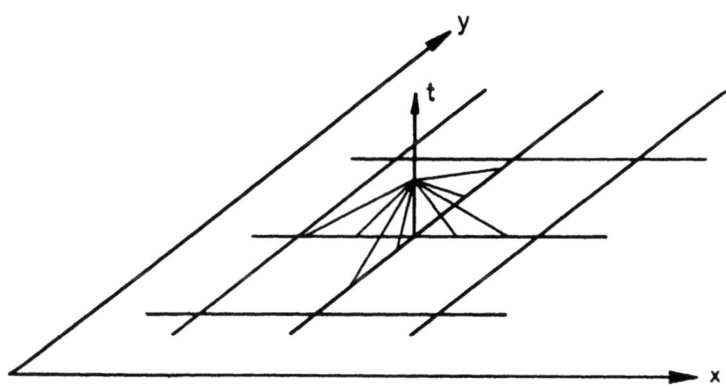

Fig. 3.7 Characteristic scheme for dense assemblies

These three equations together with the 3m input character-equations can then be solved for the $3(m+1)$ unknowns. However this last set are non-linear equations since $\frac{dU_i}{dE_{11}}$ is in general a function of θ_i, ϕ_i and ψ_i. However these equations hold for all times and can be used to generate a vertical characteristic; this is done by differentiating the equations (2.17) analytically with respect to time and integrating them numerically.

These equations become

$$\sum_{i=1}^{m} (a_{ij} \theta_{i,t} + b_{ij} \phi_{i,t} + c_{ij} \psi_{i,t}) = 0, \quad j = 1, 2, 3,$$

where a_{ij}, b_{ij} and c_{ij} are obtained from the differentials of the coefficients in equation (2.17). In a general three dimensional assembly it is of course necessary to define the filament axis with reference to the global coordinate system by introducing the filament direction cosines.

The advantages of such a formulation are two-fold. Firstly the maximum matrix to be inverted is local to the node under consideration and is governed by the maximum number of interfering filaments. Secondly since the solution is based upon characteristic theory and the characteristics propagate along the filament lines, the signals will propagate with finite propagation velocities along the filaments and the zones of influence (sec. 3.2) will result

These disadvantages are again two-fold. Firstly the solution is tied to the nodes since it is not possible because of signal branching and reflections to skip nodes and secondly the time interval in the iterative process must be small enough so that all characteristics impinging on the node must originate only on filaments interfering with that node.

3.6 Closure

In this chapter various models have been considered for filament assemblies. The models have evolved from filament models and been modified so that solutions of assemblies rather than filaments have been sought.

In the first place, filament behaviour has been used to deduce behaviour of assemblies; it was shown that such assemblies are more dispersive than continuous structures, a behavioural characteristic useful in shock isolation and protection. The zones of influence

are not elliptical as are those for anisotropic homogeneous
continuum; for orthogonal networks they are rhomboidal. The
evaluation of front signals show that on the front the maximum
occurs within the quadrant and not on the direct filament paths.
This is because of the multiplicity of routes available to the
front.

The finite element method, with the finite element grid
conforming to the assembly nodes, is then described in outline as
are some of the numerical (storage) problems. Finally a characteristic nodal model is described. Again this is tied to the
assembly nodes but it will generate shock solutions. It also
does represent a bridge between nodal models and continuum models
where characteristic theories have been used but abandoned in
favour of the easily formulative finite element and finite
difference models.

4. DYNAMICS OF DENSE FILAMENT ASSEMBLIES

In the previous sections the modelling of filaments and
filament assemblies have been considered; however for assemblies
which are so dense as to be considered a continuum, other modelling
techniques must be used and developed if only to remove the book-
keeping of each node and each filament. Three such models are
described here, the first being a membrane analogy, the second
follows from a rigid application of Hamilton's principle and the
third is an approximation and depends upon known fronts.

4.1 The Membrane Analogy

This analogy, as suggested by the title, consists of replacing
a surface defined by a dense arrangement of filaments by a uniform
or homogeneous membrane [34, 35]. The material properties of this
membrane must be connected in some way with the constituent filament material properties and also must account for the interaction
of interfering filaments.

In order to illustrate this analogy and to make sensible
comparisons, consider a regular orthogonally woven cloth or net
and held by high edge tensions. The system is considered only for
small deformations so that only "out of plane" deformations are
considered. Again, for the purpose of comparison, it will be
considered regular, which in this context will be taken to imply
that all parallel filaments have the same material properties,
support the same tension and along each filament the nodes are
regularly spaced, FIG. 4.1.

Now, by assuming that the system is dense, the finite element model with linear elements 31-33 generates the following differential equations for each node,

$$\frac{\rho_x A_x \Delta_x}{6} \{w_{i+1,j} + w_{i-1,j}\}_{,tt} + \frac{\rho_y A_y \Delta_y}{6} \{w_{i,j-1} + w_{i,j+1}\}_{,tt}$$

$$+ \frac{2}{3} \{\rho_x A_x \Delta_x + \rho_y A_y \Delta_y\} w_{i,j,tt} - \frac{\sigma_x A_x}{\Delta_x} \{w_{i+1,j} + w_{i-1,j}\}$$

$$- \frac{\sigma_y A_y}{\Delta_y} \{w_{i,j+1} + w_{i,j-1}\} + 2\left\{\frac{\sigma_x A_x}{\Delta_x} + \frac{\sigma_y A_y}{\Delta_y}\right\} w_{i,j} = F_{ij} , \quad (4.1$$

where ρ is the material density, A is the filament cross section area and w_{ij} is the deflection of the node (i,j). The subscripts x and y allocate the parameters to the x and y running filaments and Δ_x and Δ_y are the pitch or distance between nodes along the x and y running filaments respectively. $F_{ij}(t)$ is the force as a function of time applied to the node (i,j).

Now, considering the motion of the surface containing all the nodes and ignoring the action of the impressed force F, it can be readily deduced that the eigen functions of this system (modes of vibration) are simply

$$w_{ij}(t) = \bar{w}(t) \sin \frac{\alpha(i-1)\pi}{m_x} \sin \frac{\beta(j-1)\pi}{m_y}$$

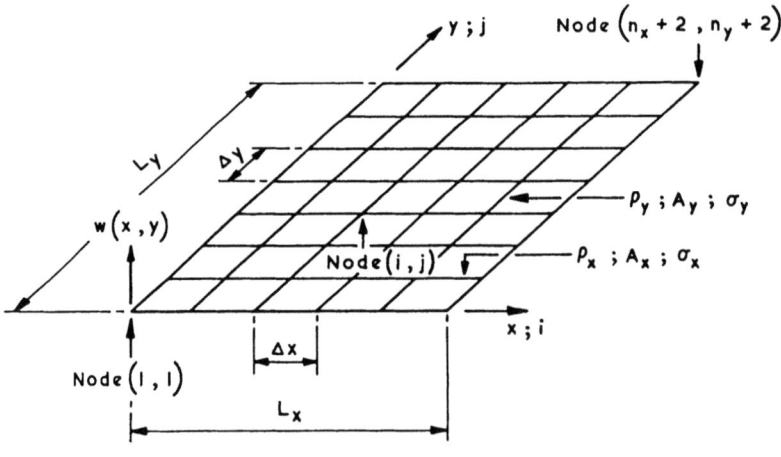

Fig. 4.1   Regular orthogonal assembly

where $\bar{w}(t)$ is an harmonic function of time, α and β are integers identifying the mode of vibration and m_x and m_y are the number of cells in the x and y directions; it also follows that including the boundary nodes there are $m_x + 1$ nodes in the x direction.

The natural frequency Ω_1 associated with this mode is given by

$$\Omega_1^2 = \frac{6\left\{\dfrac{\sigma_x A_x}{\Delta_x}\left(1 - \cos\dfrac{\alpha\pi}{m_x}\right) + \dfrac{\sigma_y A_y}{\Delta_y}\left(1 - \cos\dfrac{\beta\pi}{m_y}\right)\right\}}{\left\{\rho_x A_x \Delta_x\left(2 + \cos\dfrac{\alpha\pi}{m_x}\right) + \rho_y A_y \Delta_y\left(2 + \cos\dfrac{\beta\pi}{m_y}\right)\right\}} \qquad (4.2)$$

The first mode is given when α and β are both unity; higher modes will follow when other integer values of α and β are inserted. For very dense systems, when m_x and m_y are large, the natural frequencies become

$$\Omega_2^2 = \frac{\pi^2\left\{\dfrac{\sigma_x A_x}{\Delta_x}\dfrac{\alpha^2}{m_x^2} + \dfrac{\sigma_y A_y}{\Delta_y}\dfrac{\beta^2}{m_y^2}\right\}}{\left\{\rho_x A_x \Delta_x\left(1 - \dfrac{\alpha^2\pi^2}{6m_x^2}\right) + \rho_y A_y \Delta_y\left(1 - \dfrac{\beta^2\pi^2}{6m_y^2}\right)\right\}} \qquad (4.3)$$

Now, for the equivalent membrane, the differential equation is

$$\rho h\, w_{,tt} - \tau_x w_{,xx} - \tau_y w_{,yy} = f(x,y,t), \qquad (4.4)$$

where ρ is the membrane material density, h is the thickness of the membrane, τ is the tension/unit width and $f(x,y,t)$, the analogy of $F_{ij}(t)$ is the distribution of force loading across the surface of the membrane. As before the impressed force $f(x,y,t)$ is ignored and natural frequencies and modal shapes are sought. For the rectangular domain, whose dimensions are given by $L_x (= m_x A_x)$ and $L_y (= m_y A_y)$ the modal shapes

$$w(x,y,t) = \bar{w}(t) \sin\frac{\alpha\pi x}{L_x} \sin\frac{\beta\pi y}{L_y}$$

are introduced, and the natural frequency follows

$$\Omega_3^2 = \frac{\pi^2}{\rho h}\left(\frac{\tau_x \alpha^2}{L_x^2} + \frac{\tau_y \beta^2}{L_y^2}\right) \qquad (4.5)$$

Now, noting that (i) $\rho h = \dfrac{\rho_x A_x}{\Delta_y} + \dfrac{\rho_y A_y}{\Delta_x}$

(ii) $\tau_x = \dfrac{\sigma_x A_x}{\Delta_y}$ and $\tau_y = \dfrac{\sigma_y A_y}{\Delta_x}$

it then follows that

$$\Omega_2^2 > \Omega_1^2 > \Omega_3^2 \text{ for } m_x \text{ and } m_y \to \infty$$

Thus, in summarising, the modal shapes for the membrane analogy are identical to the assembly surface and as the denseness of the surface is increased the analogy frequencies tend to the assembly frequencies from below (lower bound).

The general solution to such a system can then be found by following the conventional eigenfunction expansion, that is

(i) finding the eigen functions,

(ii) expanding the surface force in terms of these eigen functions,

(iii) solving the resulting decoupled second order ordinary differential equations.

4.2 The Continuum Model

In the previous section the membrane analogy is introduced by suggestion rather than argument; the introduction is then justified by the examination of the system behavioural characteristics (eigen functions and eigen frequencies) and by comparison of these characteristics with those deduced by finite element

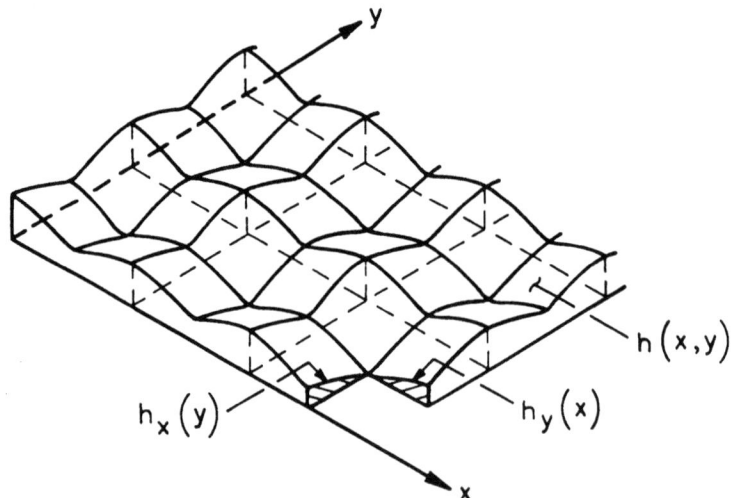

Fig. 4.2   Thickness functions $h_x(y)$, $h_y(x)$ and $h(x,y)$

arguments. However a rigorous approach would be to assume that the filaments are continuously distributed in both directions. Then the material density is a continuously variable function and can be introduced in a continuum model into Hamilton's principle. Before applying the principle to a regular orthogonal system of interfering filaments, it is first necessary to consider the area distribution of material throughout the assembly. To do this, the thickness functions $h_x(y)$ and $h_y(x)$ are introduced, FIG. 4.2.

The kinetic energy of the system in then simply given as

$$T = \frac{1}{2} \int_0^{L_x} \int_0^{L_y} (\rho_x h_x + \rho_y h_y) w^2_{,t} \, dy\, dx \,,$$

and the strain energy, for transverse and small deformations, is

$$U = \frac{1}{2} \int_0^{L_x} \int_0^{L_y} (\sigma_x h_x w^2_{,x} + \sigma_y h_y w^2_{,y}) \, dy\, dx \,.$$

The application of Hamilton's principle then leads to the following differential equation,

$$(\rho_x h_x + \rho_y h_y) w_{,tt} - \sigma_x h_x w_{,xx} - \sigma_y h_y w_{,yy} = 0 \qquad (4.6)$$

since both $h_{x,y}$ and $h_{y,x}$ are zero.

This is a hyperbolic linear partial differential equation with variable coefficients which for filaments with zero width but finite area are generalised distribution functions. For real filaments with finite width and used in a regular assembly, these coefficients are periodic in x or y and a Fourier fit to the thickness distribution is an obvious step. These series are as follows,

$$h_x(y) = a_0 + \sum_{k=1}^{\infty} a_k \cos \frac{2k\pi y}{\Delta_y} \,, \text{ and}$$

$$h_y(x) = b_0 + \sum_{k=1}^{\infty} b_k \cos \frac{2k\pi x}{\Delta_x} \,, \qquad (4.7a)$$

where a_k and b_k are Fourier coefficients. For filaments with a rectangular cross section area, area A and half width d, these coefficients are

$$a_0 = A_x/\Delta_y \,, \quad b_0 = A_y/\Delta_x$$
$$a_k = (A_x/\Delta_y)(\Delta_y/k\pi dx) \sin(2k\pi dx/\Delta_y) \qquad (4.7b)$$
$$\text{and } b_k = (A_y/\Delta_x)(\Delta_x/k\pi dy) \sin(2k\pi dy/\Delta_x) \,.$$

An appropriate Rayleigh function to the differential equation (4.6), giving an upper bound to the fundamental natural frequency, is as follows,

$$\Omega^2 = \min \frac{\int_0^{L_x}\int_0^{L_y}[\sigma_x h_x w{,}_x^2 + \sigma_y h_y w{,}_y^2]\,dy\,dx}{\int_0^{L_x}\int_0^{L_y}[\rho_x h_x + \rho_y h_y]w^2\,dy\,dx} \qquad (4.8)$$

where $w(x,y)$ is any admissible function and is thus zero at $x=0$, $x=L_x$, $y=0$ and $y=L_y$. A possible admissible function is $\sin\frac{\pi x}{L_x}\sin\frac{\pi y}{L_y}$, and upon substituting this function for $w(x,y)$ into the quotient leads to the upper bound of the fundamental natural frequency Ω,

$$\Omega^2 \leq \frac{\pi^2\left(\dfrac{\sigma_x A_x}{\Delta_x m_x^2} + \dfrac{\sigma_y A_y}{\Delta_y m_y^2}\right)}{\rho_x A_x \Delta_x + \rho_y A_y \Delta_y}$$

which is the same as that implied by the lowest frequency of the membrane analogy (4.5).

4.3 Rays and Wave Fronts [21, 36-38]

Tha hyperbolic partial differential equation (4.6) admits a propagation solution and with this there must be fronts separating quiet and excited zones. A hyperbolic equation with constant coefficients (homogeneous system) will generate elliptic fronts which, for isotropic systems, become circular. The signal paths connecting the front to the point of excitation are straight lines. However, if the coefficients are spatially dependent as for an inhomogeneous system, the rays become bent; they will usually bend around a point of slowness since minimum travel time will follow from routes passing through fast regions. Equation (4.6) has such coefficients, namely $h_x(y)$ and $h_y(x)$.

To find these rays the Eiconal equation must be found, and is as follows,

$$(\phi,_t)^2 - \frac{\sigma_x h_x}{\rho_x h_x + \rho_y h_y}(\phi,_x)^2 - \frac{\sigma_y h_y}{\rho_x h_x + \rho_y h_y}(\phi,_y)^2 = 0.$$

The propagation velocities are now, following

$$c_x^2 = \frac{\sigma_x h_x}{\rho_x h_x + \rho_y h_y} \quad \text{and} \quad c_y^2 = \frac{\sigma_y h_y}{\rho_x h_x + \rho_y h_y}$$

and these velocities are obviously spatially dependent.

The ray paths are then obtained by the extremisation of the ray length integral

$$\int_{\tau_2}^{\tau_1} \left\{ \left(\frac{X,_\tau}{c_x}\right)^2 + \left(\frac{Y,_\tau}{c_y}\right)^2 \right\} d\tau$$

where τ_1 and τ_2 are time functions evaluated at t_1 and t_2.

This then results in the following necessary differential equations,

$$X,_\tau = c_x^2 p \; ; \quad p,_\tau = -\frac{1}{2}\left[(c_x^2),_x p^2 + (c_y^2),_x q^2\right], \quad \text{and}$$

$$Y,_\tau = c_y^2 q \; ; \quad q,_\tau = -\frac{1}{2}\left[(c_x^2),_y p^2 + (c_y^2),_y q^2\right], \quad (4.9)$$

where $X(\tau)$ and $Y(\tau)$ are the coordinates of the ray at "time" τ.

In this above development it has been assumed that the thickness functions $h_x(y)$ and $h_y(x)$ are well behaved functions, whereas earlier it has been implied that they could be generalised distributions (Dirac delta functions). In most applications, that is for real filaments, truncated Fourier series are employed thus side-stepping the mathematical arguments that could evolve from such usage of generalised functions.

The solutions to equation (4.9) require a set of initial or "launch" conditions, which are namely at $t = 0$, $X = Y = 0$, $p = c_y/\sqrt{c_y^2 + c_x^2 \tan^2 \alpha}$, and $q = c_x/\sqrt{c_y^2 + c_x^2 \tan^2 \alpha}$ where α is the launch angle of the ray.

The rays are the routes taken by signals propagating to the front and for assemblies composed of interfering filaments they must follow the filaments or cell edges since in the cell interiors there is no material to support the propagation. The rays must, in the continuum model, go around these points of slowness.

Evaluation of the position on a set of rays originating from the same launch point, but with different launch angles, at a fixed "time" τ, then gives a set of points on the front for "time" τ.

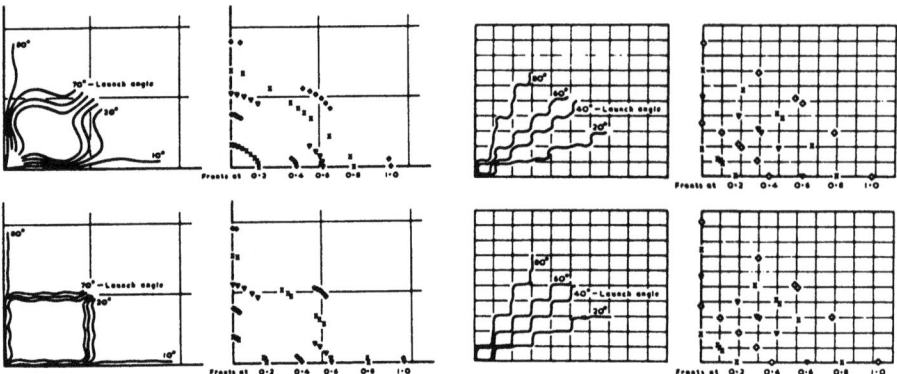

Fig. 4.3   Rays and fronts for 4 and 100 cell assemblies
(one periodic and three periodic functions)

Connection of these points then generates the fronts and, as has been previously shown, the fronts should be rhomboidal for regular assembly composed of orthogonally interfering filaments, FIG. 4.3.

4.4 Travelling Waves [35, 39-42]

For a homogeneous continuum, the wave form of a travelling wave is preserved through the domain and propagates through the continuum with a constant velocity which is either frequency independent, or is frequency dependent, as in the case of a dispersive media. In this case, however, because of the periodicity of the medium for a regular disposition of interfering filaments, the wave form which will be preserved will be governed by the periodic nature of the differential equation (4.6).

Looking for harmonic solutions to this equation (4.6) where $w(x,y,t) = e^{j\Omega t} w(x,y)$ and $j = \sqrt{-1}$, it then follows that

$$\sigma_x h_x w,_{xx} + \sigma_y h_y w,_{yy} + \Omega^2 \left(\sigma_x h_x + \sigma_y h_y\right) w = 0$$

and with a uniform filament material ($\rho_x = \rho_y = \rho$)

$$\left(\frac{\sigma_x w,_{xx}}{w} + \rho \Omega^2\right)\frac{1}{h_y} + \left(\frac{\sigma_y w,_{yy}}{w} + \rho \Omega^2\right)\frac{1}{h_x} = 0 .$$

This is amenable to separation of variables: that is, with $w(x,y) = X(x)\,Y(y)$ then

$$\left(\frac{\sigma_x X,_{xx}}{X} + \rho\Omega^2\right)\frac{1}{h_y} = -\left(\frac{\sigma_y Y,_{yy}}{Y} + \rho\Omega^2\right)\frac{1}{h_x}$$

where the expression on the left of the equality is a function of x only and that on the right is a function solely of y. Both sides of the equality must then be constant, λ, and hence the following equations can be written

and
$$\sigma_x X,_{xx} + \left(\rho\Omega^2 - \lambda h_y\right)X = 0$$
$$\sigma_y Y,_{yy} + \left(\rho\Omega^2 + \lambda h_x\right)Y = 0 \ .$$
(4.10)

These two equations are Hill's equations, generally written as follows [43],

$$X,_{xx} + \left\{\theta_o + 2\sum_{\nu=1}^{\infty}\theta_\nu \cos 2\nu x\right\}X = 0$$

where the boundaries between stable and unstable solutions are given approximately by $\theta_o = \nu^2 \pm \theta_\nu$, [43], for small values of θ_ν.

These boundaries give the conditions for the perpetuation of oscillations; they thus separate the damped and divergent oscillations and the conditions obtained at these boundaries must apply for the existence of unattenuated travelling waves. Applying this criterion to the separated equations (4.10) for condition on frequency "Ω" and wavelength "λ" leads to the following equations for rectangular filaments,

$$\rho\Omega^2 - \frac{\lambda A_y}{\Delta_x}\left\{1 \pm \frac{\sin\mu}{\mu}\right\} = \frac{\sigma_x \pi^2 i^2}{\Delta_x^2} \ ,$$

$$\rho\Omega^2 + \frac{\lambda A_x}{\Delta_y}\left\{1 \pm \frac{\sin\nu}{\nu}\right\} = \frac{\sigma_y \pi^2 j^2}{\Delta_y^2} \ ,$$
(4.11)

where $\mu = \dfrac{2i\pi d_y}{\Delta_x}$ and $\nu = \dfrac{2j\pi d_x}{\Delta_y}$,

For wide filaments (webs) where dx and dy are significant the terms $\sin\mu/\mu$ and $\sin\nu/\nu$ can be ignored. For example when webs are touching parallel webs so that $dy = 0.5\,\Delta_x$, and with $i = 3$, then $\sin\mu/\mu \sim 0.1$. Similarly, consideration of the higher frequency Fourier components will also warrant ignoring $\sin\mu/\mu$.

For these wide filaments the travelling waves have a "frequency" given by

$$\Omega^2 = \frac{\pi^2}{\rho} \frac{\left[\frac{\sigma_x A_x}{\Delta_x} i^2 + \frac{\sigma_y A_y}{\Delta_y} j^2 \right]}{A_x \Delta_x + A_y \Delta_y},$$

and a corresponding "wavelength"

$$\lambda = \pi^2 \frac{\left[\sigma_y \frac{\Delta_x}{\Delta_y} j^2 - \sigma_x \frac{\Delta_y}{\Delta_x} i^2 \right]}{A_x \Delta_x + A_y \Delta_y}.$$

For thin filaments when μ and ν are small compared to unity, there are four possible combinations of frequency and wavelength associated with these travelling waves. The two equations (4.11) plotted λ against $\hat{\Omega}^2$ in FIG. 4.4 form two pairs of straight lines, the four intersecting points giving the conditions for the propagation of travelling waves. It is pointed out that this suggests a small wave packet with the four discrete waves propagating through the assembly.

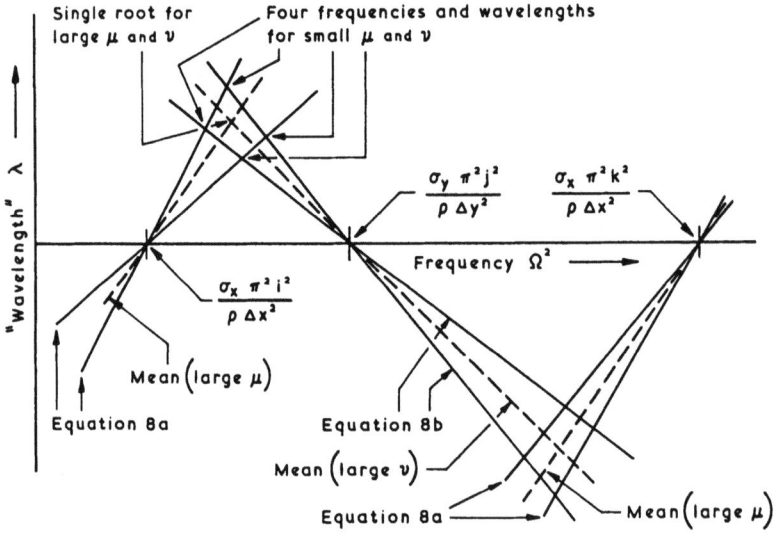

Fig. 4.4 Frequency spectrum and wavelengths for travelling waves

4.5 The Variational Model

Various researchers [44-46] have considered the arrest by textile structure of high speed low mass projectiles. In this section the use of variational principles is explored in order to generate an approximate solution for the arrest dynamics of such dense filament assemblies. It is applied to regular orthogonal systems, although there are applications in other types of system and indeed for various continuum systems. It is demonstrated here for the cloth arrest of a normally impacting projectile.

It has been shown consistently that for the above assemblies the fronts are rhomboidal; this piece of information (both theoretical and experimental), FIG. 4.5, will now be exploited by applying Hamilton's principle solely to the excited zone. A new coordinate set is now introduced so that a single parameter will define the front and stations behind the front. This similarity parameter η is defined by the following

$$\eta = \frac{x}{c_x t} + \frac{y}{c_y t} .$$

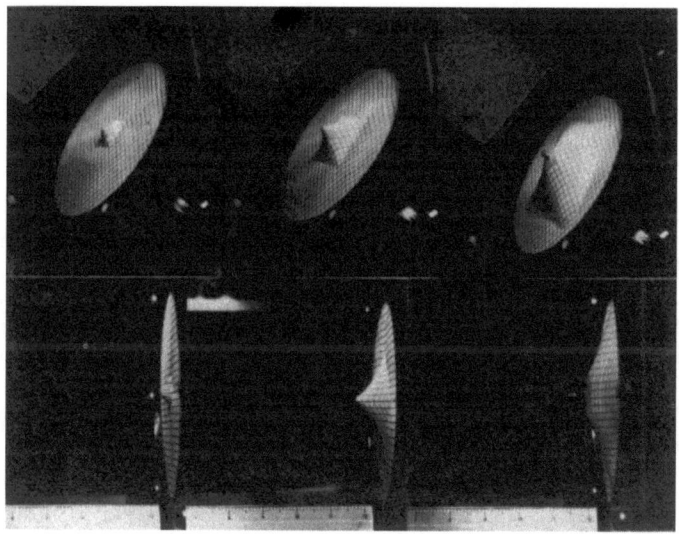

Fig. 4.5   Deformation sequence for impact in a regular orthogonal filament assembly (nylon)

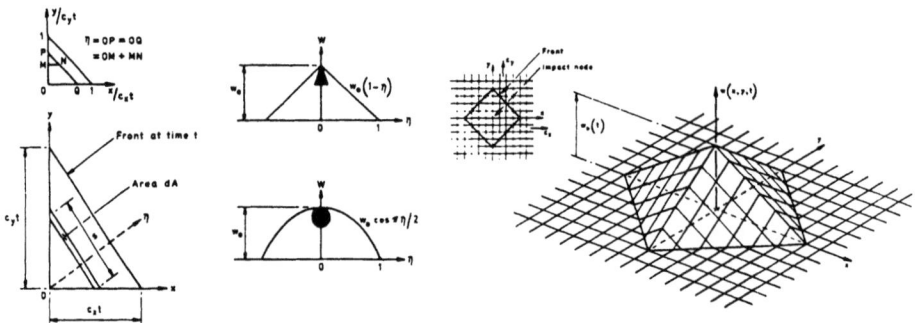

Fig. 4.6 Similarity function and admissible deformation

This parameter is zero at the centre of the excited zone and is unity at the front. The increment of area dA, FIG. 4.6, is then given by

$$dA = \eta \, t^2 \, c_x \, c_y \, d\eta$$

and a similarity solution sought so that

$$w(x,y,t) = W_o(t) \, \overline{W}(\eta)$$

The function $\overline{W}(\eta)$ is any <u>kinematically</u> admissible function, that is it must be zero at the front and be continuous so that the required differentials exist. Possible functions are the linear function $\overline{W} = 1 - \eta$, and the cosine function $\overline{W} = \cos \pi\eta/2$. The latter could be significant for blunt impaction where the centre of the excited zone must remain flat.

Hamilton's principle for a conservative system can now be introduced, that is

$$\delta \int_{t_1}^{t_2} (T - U) \, dt = 0 \, .$$

The total kinetic energy of the material behind the front, assuming that all filament particles move solely in the z and not x and y directions is then

$$T = 2 \int_A (\rho_x \, h_x + \rho_y \, h_y) \, w^2_{,t} \, dA \, .$$

Then substituting for $w = W_o \bar{W}$, performing the area integral and defining $\rho h \; (= \rho_x h_x + \rho_y h_y^o)$ as the average assembly mass/unit area leads in the following

$$T = 2 \rho h \, c_x c_y \, t^2 \{\dot{W}_o^2 I_1 + 2 \dot{W}_o W_o I_1/t + W_o^2 I_2/t^2\} \quad (4.12)$$

where

$$I_1 = \int_0^1 \bar{W}^2 \eta \, d\eta \quad \text{and} \quad I_2 = \int_0^1 \bar{W}^{1\,2} \eta^3 \, d\eta \, .$$

The strain energy due to the rotation of the prestress σ_x and σ_y is as follows, for highly tensioned systems

$$U = 2 \int_A \{\sigma_x h_x w_{,x}^2 + \sigma_y h_y w_{,y}^2\} \, dA$$

$$= 2 c_x c_y \left(\frac{\sigma_x h_x}{c_x^2} + \frac{\sigma_y h_y}{c_y^2} \right) W_o^2 I_3 \, , \quad (4.13)$$

where

$$I_3 = \int_0^1 \bar{W}^{1\,2} \eta \, d\eta \, .$$

Then by accounting for an impacting mass M_p on this assembly and by applying Hamilton's principle leads to the following differential equation

$$\left[M_p + 4 \rho h \, c_x c_y \, t^2 I_1 \right] \ddot{W}_o + 8 \rho h \, c_x c_y \, t \, I_1 \dot{W}_o$$

$$+ 4 c_x c_y \left\{ \left(\frac{\sigma_x}{c_x^2} + \frac{\sigma_y}{c_y^2} \right) I_3 + \rho h (I_1 - I_2) \right\} W_o = 0 \quad (4.14)$$

This is a linear ordinary differential equation with varying coefficients and can be solved by numerical integration. However there is an analytic solution for $M_p = 0$ or for $t \to \infty$, namely

$$W_o = t^{-\frac{1}{2}} \; A \cos(\nu \log_e t) + B \sin(\nu \log_e t)$$

where A and B are constants of integration and the "frequency" is given by

$$\nu = \frac{1}{2} \sqrt{3 + 4(I_3 - I_2)/I_1} \, .$$

This solution predicts damped oscillations and it can be seen that a the time t increases the amount of material involved in the dynamics increases and that the mass of the impactor could become insignificant. This is of course the damping effect afforded by large hanging drapes against small impact loads.

The differential equation (4.14) is sensibly non-dimensionalised by letting $\tau = \alpha t$, where $\alpha^2 = 4 \rho h\, c_x c_y\, I_1/M_p$, and it now becomes

$$(1 + \tau^2)\, W_{o,\tau\tau} + 2\tau\, W_{o,\tau} + \delta\, W_o = 0 \qquad (4.15)$$

where $\delta = (I_3 + I_1 - I_2)/I_1$ and the propagation velocities for transverse waves are

$$c_x^2 = \sigma_x/\rho_x \quad \text{and} \quad c_y^2 = \sigma_y/\rho_y.$$

The values of the integrals I_1, I_2 and I_3 and the parameter δ are shown in Table 4.1 for both the linear function $\bar{W} = 1 - \eta$ and the cosine function $\bar{W} = \cos \pi\eta/2$. Initial conditions are required, that is at $t = 0$, $W_o = 0$ and $W_{o,t} = V_{imp}$ or at $\tau = 0$, $W_{o,\tau} = V_{imp}/\alpha$. Since the equation (4.15) is linear and since the system is assumed initially at rest then any deflections are proportional to the non-dimensional impact velocity. Solutions of (4.15) for various values of δ are shown, FIG. 4.7, and it is concluded that the solution is quite insensitive to the kinematically assumed function $\bar{W}(\eta)$.

TABLE 4.1

Integrands	Linear function $\bar{W} = 1 - \eta$	Cosine function $\bar{W} = \cos \pi\eta/2$
$I_1 = \int_0^1 \bar{W}^2 \eta\, d\eta$	0.08333	0.1487
$I_2 = \int_0^1 \bar{W}^{1^2} \eta^3\, d\eta$	0.25	0.4571
$I_3 = \int_0^1 \bar{W}^{1^2} \eta\, d\eta$	0.5	0.8668
$\delta = \dfrac{I_3 + I_1 - I_2}{I_1}$	4	3.755

Integrand Values for Linear and Cosine Similarity Functions

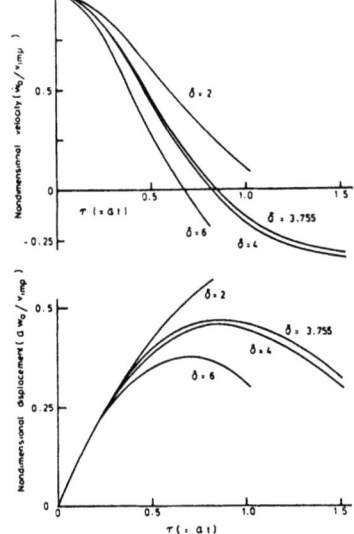

Fig. 4.7 Effect of indentation shape on system dynamics

However, upon comparison with experimental results (46) the solution, whilst giving basic trends, is quite inaccurate. A dramatic improvement in accuracy can be obtained by accounting for the stretch of each constituent filament. In this case the strain energy (4.13) becomes

$$U = c_x c_y t^2 \left\{ E_y h_y \left[(1 + \epsilon_y) \sqrt{1 + \frac{W_o^2}{c_y^2 t^2}} - 1 \right]^2 \right.$$
$$\left. + E_x h_x \left[(1 + \epsilon_x) \sqrt{1 + \frac{W_o^2}{c_x^2 t^2}} - 1 \right]^2 \right\}$$

for the linear similarity function $\overline{W} = 1 - \eta$, and where ϵ is the prestrain.

The non-dimensional equation of motion becomes

$$(1 + \tau^2) W_{o,\tau\tau} + 2\tau W_{o,\tau} + 6 W_o \left\{ \frac{\beta_x(1+\epsilon_x)}{\epsilon_x} \left[1 + \epsilon_x - \frac{1}{\sqrt{1 + \left(\frac{\alpha W_o}{c_x \tau}\right)^2}} \right] \right.$$
$$\left. + \frac{\beta_y(1+\epsilon_y)}{\epsilon_y} \left[1 + \epsilon_y - \frac{1}{\sqrt{1 + \left(\frac{\alpha W_o}{c_y \tau}\right)^2}} \right] - \frac{1}{3} \right\} = 0$$

where as before $\alpha^2 = \frac{\rho h \, c_x \, c_y}{3 M_p}$ and β_x and β_y are the mass fractional contributions to the assembly of the x and y running filaments and where the preload $\sigma_x = E_x \epsilon_x$ and the transverse wave speed is given by $c_x^2 = \dfrac{\sigma_x}{\rho_x}$.

The linear model, non-linear model and experimental results [46] shown in FIG. 4.8 demonstrate the significant increase in accuracy of prediction by incorporating the filament stretching. This does suggest that for such a system, the prestress is relatively unimportant except in so far as it controls the propagation velocity of transverse waves. It does also suggest that it is the filament constitutive properties and material densities that are important.

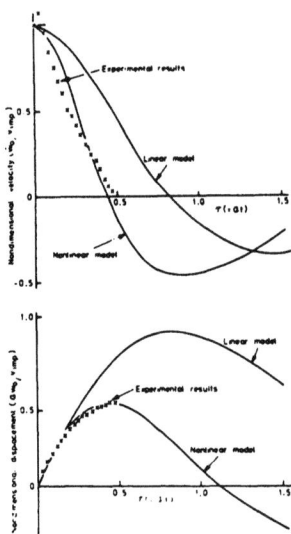

Fig. 4.8 Comparison of theoretical solutions with experimental results

4.6 Closure

In this chapter various ideas on the modelling of dense assemblies of filaments are presented whilst the emphasis has been on regular orthogonal systems. As a suggestion yet unexplored, the use of probability functions and the variational models § 4.2 and § 4.5 could be applied to random aggregate continuum.

The three models described are all sensible in that firstly the analog model is a good "a priori" model. However it does not predict the proper wave fronts and is only really applicable to long time transient problems. The second model is not really useful in the engineering application since it is tangibly solved only for regular linear systems. It does have application in its predictive ability and for wave front analysis. The third model does depend initially upon some experimental information, namely the wave front and propagation speeds. However, from that point it does appear to be a very powerful and quickly applied engineering model.

REFERENCES

1. Y. C. Fung, Foundations of Solid Mechanics, Prentice Hall Inc., 1965.

2. I. N. Sneddon and D. S. Berry, Classical Theory of Elasticity, in Enclyopaedia of Physics, Vol. VI: Elasticity and Plasticity, ed. by S. Flügge, Springer Verlag, 1958.

3. C. Truesdell and R. A. Toupin, Classical Field Theories, in Encyclopaedia of Physics, Vol. III/1: Principles of Classical Mechanics and Field Theory, ed. by S. Flügge, Springer Verlag, 1960.

4. A. E. Green and J. E. Adkins, Large Elastic Deformations, Clarendon Press, Oxford, 1970.

5. C. Truesdell, The Non-linear Field Theories in Mechanics, in Topics in Non-linear Physics, ed. by N. J. Zabusky, NATO ASI, Munich, 1966, from the International School of Non-linear Mathematics and Physics, Springer Verlag, 1968.

6. E. J. Saletan and A. H. Cromer, Theoretical Mechanics, John Wiley, 1971.

7. L. A. Pars, A Treatise on Analytical Dynamics, Heinemann, 1965.

8. R. M. Rosenberg, Analytical Dynamics of Discrete Systems, Mathematical Concepts and Methods in Science and Engineering, Vol. 4, Plenum Press, 1977.

9. S. Nemat-Nasser, On Variational Methods in Finite and Incremental Elastic Deformation Problems with Discontinuous Fields, Q. Appl. Math., Vol. 30, July 1972.

10. S. Nemat-Nasser, General Variational Principles in Non-linear and Linear Elasticity with Applications, Mechanics Today, Vol. 1, 1972.

11. N. Cristescu, Dynamic Plasticity, Series in Applied Mathematics and Mechanics, North Holland, 1967.

12. Y. Kim and B. Tabarrok, On the Non-linear Vibrations of Travelling Strings, J. Franklin Inst., Vol. 293, No. 6, 1972.

13. W. F. Ames, S. Y. Lee and J. N. Zaiser, Non-linear Vibrations of a Travelling Threadline, *Int. J. Non-linear Mech.*, Vol. 3, No. 4, 1968.

14. G. F. Carrier, On the Non-linear Vibration Problem of the Elastic String, *Q. Appl. Math.*, Vol. 3, No. 1, 1945.

15. J. P. Roos, C. Schweigmann and R. Timman, Mathematical Formulation of the Laws of Conservation of Mass and Energy and the Equations of Motion for a Moving Thread, *J. Eng. Maths.*, Vol. 7, No. 2, 1973.

16. L. J. F. Broer, On the Dynamics of Strings, *J. Eng. Maths.*, Vol. 4. No. 3, 1970.

17. L. Y. Shih, Motion of Elliptic Ballooning for Travelling String, *Int. J. Non-linear Mech.*, Vol. 10, No. 3, 1975.

18. A. Jeffrey and T. Taniuti, *Non-linear Wave Propagation*, Mathematics in Science and Engineering, Vol. 19, Academic Press, 1964.

19. P. C. Chou and R. W. Mortimer, Solution of One Dimensional Wave Problems by the Method of Characteristics, *J. Appl. Mech.*, Vol. 34, Series E, 1967.

20. Y. Mengi and H. D. McNiven, Analysis of the Transient Excitation of an Elastic Rod by the Method of Characteristics, *Int. J. Solids Structures*, Vol. 6, 1970.

21. R. Courant and D. Hilbert, *Methods of Mathematical Physics*, Vol. II, 1966, Interscience Publishers Inc.

22. R. D. Richtmyer and K. W. Morton, *Finite Difference Methods for Initial Value Problems*, Second Edition, Interscience Publishers Inc., 1967.

23. H. Kolsky, *Stress Waves in Solids*, Dover Publications, New York, 1963.

24. K. F. Graff, *Wave Motion in Elastic Solids*, Oxford Engineering Science Series, Clarendon Press, 1975.

25. C. Leech and J. Mansell, Some Aspects of Wave Propagation in Orthogonal Nets, *Int. J. Mech. Sci.*, Vol. 19, No. 2, 1977.

26. H. Afshari and A. I. Solar, Vibration of Cable Gridworks with Small Initial Deformation, *J. Appl. Mech.*, Vol. 41, No. 1, Series E, 1974.

27. A. I. Solar and H. Afshari, On the Analysis of Cable Network Vibrations using Galerkin's Method, J. Appl. Mech., Vol. 41, No.1, Series E, 1974.

28. N. Subramonian and D. V. Reddy, Frequency Analysis of Cable Networks using Finite Difference Calculus, J. Appl. Mech., Vol. 41, Series E, 1970.

29. C. Leech and J. Mansell, Propagation Fronts Arising from Impulsive Excitation of Orthogonally Woven Cloth and Nets, J. Appl. Mech., Vol. 45, Series E, 1978.

30. B. Noble, Variational Finite Element Methods for Initial Value Problems, Academic Press, 1973.

31. O. C. Zienkiewicz, The Finite Element Method, Third Edition, McGraw-Hill, 1977.

32. C. S. Desai and J. F. Abel, Introduction to the Finite Element Method, Van Nostrand, 1972.

33. C. Leech and J. Shanks, Dynamics of Reinforced Coarse Nets using a Finite Element Analysis, Int. J. Mech. Sci., Vol. 21, 1979.

34. C. Sun and T. Y. Yang, A Continuum Approach towards the Dynamics of Gridworks, J. Appl. Mech., Vol. 40, No. 1, 1973.

35. C. Leech, Modelling of Net and Cloth Dynamics, to be published in J. Franklin Inst., 1979.

36. B. A. Auld, Acoustic Fields and Waves in Solids, Wiley-Interscience, 1973.

37. F. C. Karal and J. B. Keller, Elastic Wave Propagation in Homogeneous and Inhomogeneous Media, J. Acoust. Soc. Am., Vol. 31, No. 6, 1959.

38. R. K. Luneberg, Mathematical Theory of Optics, University of California Press, 1966.

39. E. H. Lee, A Survey of Variational Methods for Elastic Wave Propagation in Composites with Periodic Structures, in Dynamics of Composite Materials, ed. by E. H. Lee, ASME, 172.

40. E. H. Lee, Wave Propagation in Composites with Periodic Structures, Proc. 5th Can. Congress App. Mech.

41. L. Brillouin, Wave Propagation in Periodic Structures, Dover, 1965.

42. A. Bedford, D. S. Drumheller and H. J. Sutherland, On Modeling the Dynamics of Composite Materials, *Mechanics Today*, Vol. 3, 1976.

43. C. Hayashi, *Non-linear Oscillations in Physical Systems*, McGraw-Hill, 1964.

44. D. Roylance, A. Wilde and G. Tocci, Ballistic Impact into Textile Structures, *Textile Res. J.*, Vol. 43, No. 1, 1973.

45. J. R. Vinson and J. A. Zukas, On the Ballistic Impact of a Textile Body, *J. Appl. Mech.*, Vol. 42, No. 2, 1975.

46. C. R. Maheux, G. M. Stewart, D. R. Petterson and F. A. Odell, *Dynamics of Body Armor Materials under High Speed Impact, Part 1*, U.S. Army Chemical Warfare Laboratories, CWLR 2141, 1957.

VISCOELASTIC, FRICTIONAL AND STRUCTURAL EFFECTS IN
FABRIC WRINKLING

B.M. Chapman

Physical Technology Unit,
CSIRO, Institute of Earth Resources

ABSTRACT.

Wrinkling of fabrics is explained in terms of the viscoelastic behaviour of individual fibres, the frictional interactions between them and their arrangement within the assembly. Topics covered include: 1. Introduction 2. Single fibre viscoelasticity 2.1. Fibre bending stress, strain and moduli. 2.2 Nominal stress strain and moduli 2.3 Generalised linear viscoelasticity 3. Fabric bending 3.1 Two-dimensional assembly of GLVE fibres 3.2 Bending hysteresis loops 3.3 Wrinkling 3.4 Blend fabrics.

1.   INTRODUCTION

The phenomenon of wrinkling which is observed in fabrics is manifest principally as an arrangement of incompletely recovered bending deformations. Typically a wrinkle is produced during, say, sitting when a fold of cloth is pressed against the body to produce a bend of higher curvature for a period of time during which the temperature and regain may rise. When the wearer resumes a standing position the fold falls away and commences recovery under ambient conditions again. Thus the fabric is subjected not only to a sequence of mechanical constraints but simultaneously to a cycle of temperature and humidity variations. During this fabric bending-and-recovery cycle, individual fibres are subjected to both bending and torsional deformations and will move longitudinally relative to one another, giving rise to a frictional loss of energy as they slide over one another. Any mechanical analysis of the problem therefore must take into account the time-dependent viscoelastic behaviour of fibres under conditions of changing temperature and relative humidity and the effects of interfibre frictional interactions.

2. SINGLE FIBRE VISCOELASTICITY

2.1 Fibre bending stress, strain and moduli

According to the theory of simple bending of elastic beams [1] the bending moment M required to hold a member of modulus E and moment of inertia of cross-section I bent to a radius of curvature Γ is given by:

$$M = E I/\Gamma \tag{1}$$

These quantities will now be reduced with respect to specimen dimensions to define bending stress, strain and modulus [2]. Let the semidimension of the specimen in the plane of bending be b. For a cross-section symmetrical about the neutral axis, the maximum tensile and compressive strains occur at the inner and outer portions of the cross-section respectively and are equal to b/Γ. If we define bending strain ϵ_B as (b/Γ) equation (1) for linear elastic materials can be rearranged to give:

$$[M/(I/b)] = E (b/\Gamma) \tag{2}$$

The quantity on the L.H.S. has the dimensions of stress and we define it as the bending stress F_B. This relation shows bending stress to be proportional to bending strain, with the constant of proportionality equal to the modulus of elasticity. For viscoelastic materials we define bending stress and bending strain in the same way but the relation between them will be more complex than (2).

2.2 Nominal stress, strain and moduli

It is convenient to measure the transverse dimensions of a fibre at one standard temperature and humidity only. All stress, strain and modulus calculations are then defined as nominal values [2] in terms of these dimensions even though measurements are made under other conditions. Actual values of F_B, ϵ_B and E_B are related to their nominal values (indicated by an asterisk) through the fractional lateral swelling δ relative to the original conditions of measurement:

$$\begin{aligned} F_B &= F_B^* (1+\delta)^{-3} \\ \epsilon_B &= \epsilon_B^* (1+\delta) \\ E_B &= E_B^* (1+\delta)^{-4} \end{aligned} \tag{3}$$

2.3 Generalised linear viscoelasticity (GLVE)

Linear viscoelastic theory has been used successfully in the study of the mechanical behaviour of certain materials at low strains (usually<1%). The theory can be based on the assumption that the time-dependent stress responses to strain inputs or vice versa are linearly additive (Boltzmann superposition principle). This can be expressed in the following Stieltjes integral forms:

$$f(t) = \int_0^t G(t-\tau)\,d\varepsilon(\tau) \quad \text{(superposition of stress)} \quad (4)$$

$$\varepsilon(t) = \int_0^t J(t-\tau)\,df(\tau) \quad \text{(superposition of strain)} \quad (5)$$

where f and ε are the stress and strain at any time t and G(t) and J(t) are the stress relaxation modulus and creep compliance which are unique for the material under the specified conditions. These functions are related through the equations

$$\int_0^t G(t-\tau)\,\frac{dJ(\tau)}{d\tau}\,d\tau = 1 \quad (6)$$

$$\int_0^t J(t-\tau)\,\frac{dG(\tau)}{d\tau}\,d\tau = 1. \quad (7)$$

This theory can be applied to certain materials at equilibrium with a constant environment. In our general situation, however, fibres are not in equilibrium with their environment which may even be changing.

The simplest example is the phenomenon of ageing [4]. If a fibre is subjected to a thermal or hygral shock and then allowed to come to moisture and thermal equilibrium with a constant environment, it will exhibit mechanical properties which change with time (e.g. gradually become stiffer over a period of years). A wool fibre, previously wet out and allowed to recondition in air at constant conditions will age in this way. Stress relaxation curves, obtained by imposing a step strain at increasing times after reconditioning, show a shift to greater relaxation times. (see Fig.1). Thus the stress relaxation function is not invariant with time of testing. It is in fact a function not only of time t elapsed after imposition of step strain but also of the absolute time t_A at which the step strain was imposed; i.e. $G=G(t_A, t)$.

In the more complex situation where temperature and relative humidity are changing with absolute time the stress relaxation modulus will change as a result [5]. For a given "environmental sequence" e of temperature and R.H. as a function of absolute time, the stress relaxation modulus $G_e(t_A,t)$ and creep compliance $J_e(t_A,t)$ will be functions of absolute time t_A as well as elapsed

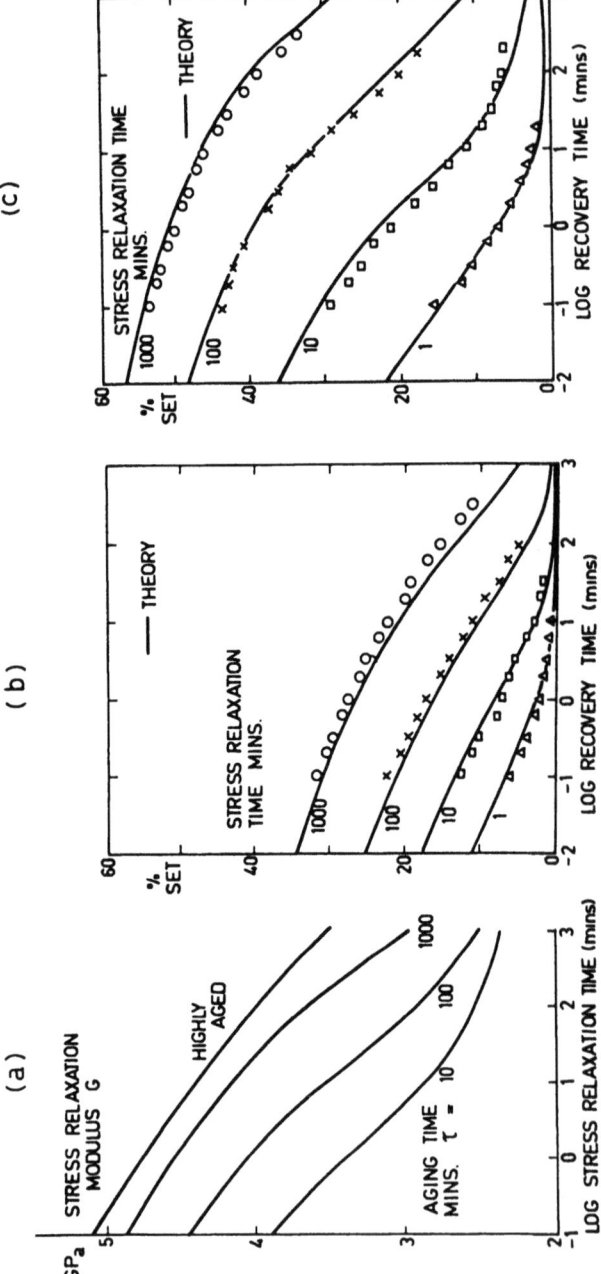

Figure 1: (a) Stress-relaxation modulus curves for a wool fibre aged for various times. Recovery for (b) highly aged fibre (c) ten minute aged fibre following release after stress relaxation times indicated.

time t. For convenience, f, ε, G_e and J_e are defined as nominal values calculated from the dimensions of the specimen measured under some standard conditions. Thus complex changes in specimen dimensions will not have to be continually compensated for.

During ageing, or during an environmental sequence, changes may occur in the strain parameter of a stress-free specimen. For example, increasing R.H. will increase the length of a wool fibre in the absence of tensile stress. This is equivalent to an increase in the tensile strain parameter. Alternatively if the specimen were held at constant length (zero strain), a compressive stress would be built up during the course of the experiment. Assuming these "background" effects are additive with the viscoelastic effects equations (4) and (5) generalise to

$$f(t) = \phi_e(t) + \int_0^t G_e(\tau, t-\tau) \, d\varepsilon(\tau) \qquad (8)$$

$$\varepsilon(t) = \eta_e(t) + \int_0^t J_e(\tau, t-\tau) \, df(\tau) \qquad (9)$$

where ϕ_e and η_e are the background stress and strain functions respectively. They represent the stress or strain changes that would be observed in an environmental sequence with the specimen maintained at zero strain or zero stress respectively.

For bending experiments on single straight fibres, structurally symmetrical in cross-section, the functions ϕ_e and η_e are identically zero [5]. In Fig.1 equations (4) and (8) have been used to calculate recovery of the same wool fibre both highly aged and ageing. Per cent set is the percentage of unrecovered strain remaining in the fibre at any time.

3. FABRIC BENDING

3.1 Two-dimensional assembly of GLVE fibres

Assume we have a two dimensional assembly of long thin straight fibres of n distinct types arranged in various directions within the plane of the assembly [6]. Within a circle of unit diameter let there be N_{oj} fibres of type j with direction distribution function $P_j(\phi)$ where ϕ is the clockwise angle between a given fibre direction and an arbitrary but fixed direction $\phi = 0$. Then the number of fibres of type j in unit circle having direction between ϕ and $\phi + \Delta\phi$ is given by:

$$N_{\phi j} \Delta\phi = N_{oj} P_j(\phi) \Delta\phi \qquad (10)$$

For fibres GLVE in bending (k=1) and torsion (k=2) the bending and torsional relaxation moduli for fibre type j are given by

$G_{jk}(t_A,t)$. Therefore bending or torsional stress f_{jk} is related to bending or torsional strain ε_k by:

$$f_{jk}(t) = \int_0^t G_{jk}(\tau,t-\tau)\, d\varepsilon_k(\tau) \qquad (11)$$

The bending or torsional moment for a fibre of appropriate moment of inertia I_{jk} is given by:

$$m_{jk}(t) = \int_0^t G_{jk}(\tau,t-\tau)\, I_{jk}\, d\eta_k(\tau) \qquad (12)$$

where η_k is the bending curvature or torsional twist.

In addition to the viscoelastic moment m_{jk} for any fibre there is a frictional moment $m_{jk}^*(\eta_k, t_A)$ associated with each fibre in bending and in torsion which will in general be a function of strain and also of absolute time.

If the assembly is bent to curvature K such that the generators of the cylindrical portion make a counter-clockwise angle θ with the direction $\phi=0$ then any fibre with direction ϕ will be subjected to curvatures and twists of

$$\eta_k = K_{\phi k} = D_{\phi k}\, K \qquad (13)$$

where $D_{\phi k} \equiv \sin(\theta+\phi)\sin(\theta+\phi+(k-1)\pi/2)$

The total moment $m_{\phi jk}$ for any fibre of type j in direction ϕ for strain mode k contains both frictional and viscoelastic components and is given by

$$m_{\phi jk}(K,t) = m_{jk}(D_{\phi k}K,t) + \dot{K}/|\dot{K}|\, m_{jk}^*(D_{\phi k}K,t) \qquad (14)$$

The factor $\dot{K}/|\dot{K}|$ indicates that the frictional moment operates in a direction such to oppose the movement. The total bending moment per unit length of the assembly bent along direction θ,

$$m_\theta(K,t) = \sum_{j,k} \int_0^\pi N_{\phi j} D_{\phi k} m_{\phi jk}(K,t)\, d\phi \qquad (15)$$

From equations (12)-(15)

$$m_\theta(K,t) = \int_0^t B_\theta(\tau,t-\tau)\, dK(\tau) + \dot{K}/|\dot{K}|\, \Omega_\theta(K,t) \qquad (16)$$

where $B_\theta(t_A,t) = \sum_{j,k} \int_0^\pi N_{\phi j} D_{\phi k}^2\, I_{jk} G_{jk}(t_A,t)\, d\phi \qquad (17)$

and $\Omega_\theta(K,t_A) = \sum_{j,k} \int_0^\pi N_{\phi j} D_{\phi k}\, m_{jk}^*(D_{\phi k}K,t_A) \qquad (18)$

In general therefore, the assembly will behave as an anisotropic generalised linear viscoelastic sheet with "bending relaxation rigidity" $B_\theta(t_A,t)$ and internal frictional moment $\Omega_\theta(t_A,t)$. These functions will depend of course on the environmental sequence of interest.

For woven fabrics the fibre/direction distribution function is discrete with say $\phi=0$ for warp and $\phi=\pi/2$ for weft. If the number of yarns per unit length in warp and weft are n_1 and n_2, with each yarn containing ν_{1j} and ν_{2j} fibres of type j the bending relaxation rigidity is therefore given by:

$$B_\theta(t_A,t) = \sum_{j=1}^{m}\sum_{k=1}^{2} [n_1 \nu_{1j} \sin^2\theta \sin^2(\theta+(k-1)\pi/2)$$
$$+ n_2 \nu_{2j} \cos^2\theta \cos^2(\theta+(k-1)\pi/2)] I_{jk} G_{jk}(t_A,t) \quad (19)$$

3.2 Bending hysteresis loops

If a non-ageing fabric is bent under constant temperature and R.H. conditions and if the frictional moment is constant with curvature and equal to M_o equation (16) becomes:

$$M = \int_0^t B(t-\tau)dK(\tau) + \dot{K}/|\dot{K}| M_o \quad (20)$$

The fabric will behave according to the model in Fig.2(a) which is a linear viscoelastic element in parallel with a constant frictional element.

Fabric bending hysteresis curves [7] where the fabric is cycled between curvatures K^* and $-K^*$ gives rise to a hysteresis loop which makes an intercept W on the bending moment axis as in Fig.2(b). This intercept has been interpreted [7] as being entirely due to the frictional moment and equal to $2 M_o$. However if the fibres are viscoelastic, portion of this intercept will be due to viscoelastic effects [8]. The viscoelastic bending moment per unit length M of a fabric subjected to a constant rate of change of curvature ρ is given by

$$M = \rho \int_0^t B(z)dz$$

Using the Boltzmann superposition principle to add the effects caused by the component strain rates for each portion of the hysteresis curve of the viscoelastic component we obtain for the moments at points X, Y and Z of Fig.2(b)

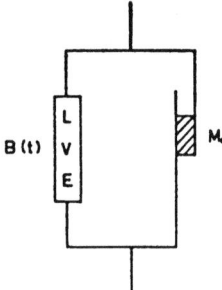

Fig.2(a) Model with parallel, linear viscoelastic element plus frictional element for fabric bending.

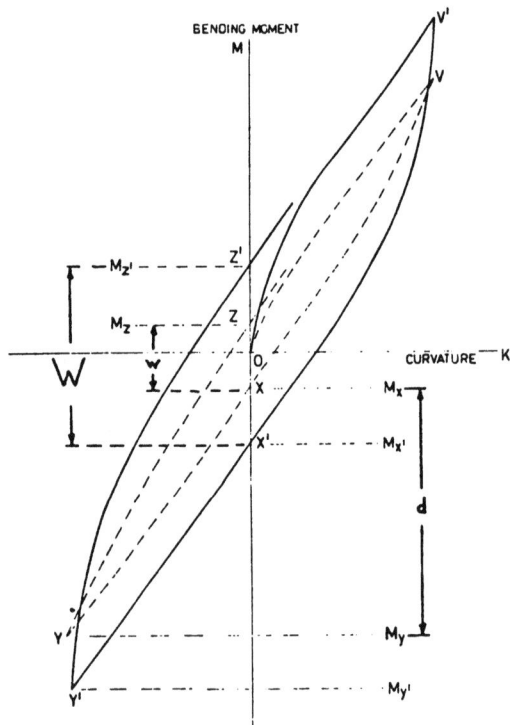

Fig.2(b) Fabric hysteresis curve showing relationship with underlying viscoelastic hysteresis. The difference in span between the two curves $W = M_z' - M_x'$ and $w = M_z - M_x$ is due to friction.

$$M_X = \rho(\int_0^{2t^*} B(z)dz - 2\int_0^{t^*} B(z)dz),$$

$$M_Y = \rho(\int_0^{3t^*} B(z)dz - 2\int_0^{2t^*} B(z)dz),$$

$$M_Z = \rho(\int_0^{4t^*} B(z)dz - 2\int_0^{3t^*} B(z)dz + 2\int_0^{t^*} B(z)dz),$$

where $t^* = K^*/\rho$. (22)

For many fibres the relaxation function $B(t)$ is linear with log time so that:

$$B(t) = a - b \log_{10} t$$

If we define the fractional relaxation rate per decade R_d so that

$$R_d = \frac{b}{a - b \log_{10} t^*},$$

then the width to depth ratio w/d (see Fig.2(b)) for the viscoelastic hysteresis can be shown to be from eqn. (22).

$$\frac{w}{d} = \frac{1.056}{(1/R_d) - 0.059} = 1.06 R_d \qquad (23)$$

to a good approximation since $1/R_d > 10$ for most fibres.

3.3 Wrinkling

The generalised model containing a GLVE element in parallel with a constant frictional element as in Fig.3. is the more useful one to employ for wrinkling studies [9,10]. The bending relaxation rigidity will be defined for the particular environmental sequence e of interest, thus bending moment per unit length of fabric is related to curvature by

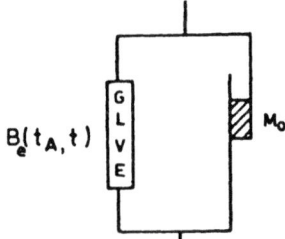

Fig.3 Rheological model used to study fabric wrinkling.

$$M = \int_0^t B_e(\tau, t-\tau) dK(\tau) + \dot{K}/|\dot{K}| M_o. \qquad (24)$$

A mechanical and environmental sequence to simulate conditions during wrinkling is as follows. A fabric at ambient conditions is bent to a fixed curvature K_o at time $t=t_1$ for a fixed time during which the air temperature and R.H. are increased. The fabric is then released at $t=t_2$ and allowed to recover as conditions are returned to ambient.

During recovery the applied moment M is zero and eqn. (24) becomes

$$0 = B_e(t_1, t-t_1)K_o + \int_{t_2}^t B_e(\tau, t-\tau) dK(\tau) - M_o \qquad (25)$$

The quantity $(1-K/K_o)$ is the fractional recovered strain (or recovery) R. Equation (25) can be easily solved numerically [11]. However Denby [12] has found a simple, convenient but close approximation where:

$$R(t) = \frac{B_e(t_1, t-t_1) - M_o/K_o}{B_e(t_2, t-t_2)} \qquad (26)$$

For fixed t_1 and t_2, the exact solution of eqn. (25) requires a knowledge of the single variable function $B_e(t_1, t)$, as well as the two-variable function $B_e(\tau, t-\tau)$ over the range $t_2 < \tau < t$. However the approximation (26) only requires a knowledge of two single-variable functions, viz $B_e(t_1, t)$ and $B_e(t_2, t)$. The function B_e is determined by subjecting unstrained fabric specimens to the environmental sequence over a period of time during which at time $t=t_A$, a step strain is imposed and maintained until the end of the sequence. By varying the value of t_A for a number of new fabric specimens the function B_e can be built up from the series of bending moment relaxation curves so obtained.

The use of equation (26) is illustrated by the following example. Consider the standard multiple pleat wrinkle test [13] where a fabric, initially at 65% R.H. 20°C (at time t_o) is preconditioned at 85% R.H. 30°C for 30 min before being bent to and held at a fixed curvature (at time t_1) for 15 min. After this time the fabric is allowed to recover (at time t_2) for 15 min under the original 65% R.H. 20°C conditions. The percentage of the original bending angle which has recovered after 15 min (at time t_3) is termed the "wrinkle recovery". This sequence of events is illustrated in Fig.4(a).

If we were to separately carry out a stress-relaxation test for fabrics in bending by firstly subjecting an unbent sample to the same sequence of temperature and R.H. changes as for the wrinkle test and then rapidly bending it (at time t_1) to a fixed

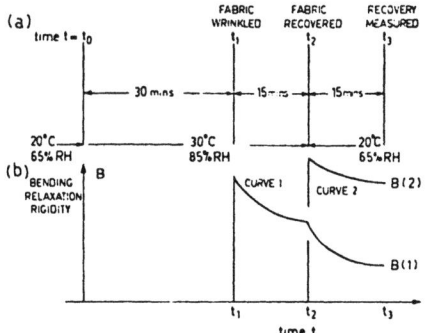

Fig.4(a) Temperature and humidity sequence for the standard multiple-pleat wrinkle test.
(b) Typical bending stress-relaxation curves obtained by imposing step strains at t_1 (curve 1) or t_2 (curve 2) on a fabric subjected to the concurrent temperature-humidity sequence in (a).

curvature, we would obtain a plot of decaying bending moment against time such as curve 1 in Fig.4(b). Similarly, if another stress relaxation test is then performed on a different sample of the same material such that, while undeformed, it is subjected to the humidity-temperature sequence up to time t_2 when it is bent simultaneously with the R.H. and temperature conditions being changed back to 65% R.H. 20°C as in Fig.4(a), then the plot of decaying bending moment against time will be as in curve 2 in Fig.4(b).

In the absence of friction equation (26) gives:

$$R = \frac{B(1)}{B(2)}, \qquad (27)$$

where $B(1)$ and $B(2)$ are the bending moments at time t_3 on curves 1 and 2 respectively.

3.4 Blend fabrics

Consider a series of blend fabrics of the same weight containing a range of different proportions of component fibres Y and Z [13]. If for any given sequence of temperature and R.H. the bending relaxation rigidity of the 100% Y fabric is B_Y and for the 100% Z fabric is B_Z, then the relaxation rigidity for a blend fabric containing a proportion b of fibre Y is:

$$B_b = bB_y + (1-b)B_z. \qquad (28)$$

For a wrinkle test sequence such as that illustrated in Fig.4(a) let the bending relaxation rigidities for fabric Y required to calculate recovery at a given point in time be $B_Y(1)$ and $B_Y(2)$ similarly to that illustrated in Fig.4. Thus recovery for fabric Y in absence of friction at this point is given by:

$$R_Y = \frac{B_Y(1)}{B_Y(2)}. \qquad (29)$$

Similarly, for fabric Z:

$$R_Z = \frac{B_Z(1)}{B_Z(2)}. \qquad (30)$$

Similarly, the corresponding recovery for a blend with proportion b of fibre type Y is:

$$R_b = \frac{B_b(1)}{B_b(2)} \qquad (31)$$

where $B_b(1)$ and $B_b(2)$ correspond to the two points, on the stress relaxation curves for the blend, required to calculate recovery.

From Equations (28) and (31) we obtain:

$$R_b = \frac{bB_Y(1) + (1-b)B_Z(1)}{bB_Y(2) + (1-b)B_Z(2)},$$

whence from Equations (29) and (30) we obtain:

$$R_b = R_Z + \frac{1}{1 + \frac{1}{f}(\frac{1}{b} - 1)}(R_Y - R_Z), \qquad (32)$$

where $f = B_Y(2)/B_Z(2)$.

For a series of blends not necessarily of constant weight where the densities of fibres Y and Z are ρ_Y and ρ_Z, their respective radii are r_Y and r_Z, and their appropriate material moduli at $t = t_3$ for curve 2 are $G_Y(2)$ and $G_Z(2)$, it can be shown that equation (32) still holds with

$$f = \frac{\rho_Z}{\rho_Y}(\frac{r_Y}{r_Z})^2 \frac{G_Y(2)}{G_Z(2)} \qquad (33)$$

For any given series of blends and a given wrinkle test R_Y, R_Z, and f will be constant. Thus, except in the special case where $B_Y(2) = B_Z(2)$, there will be a nonlinear relation between wrinkle recovery and blend proportion b. The ratio f can be considered as the ratio of the bending stiffnesses (defined under certain specified experimental conditions) of the two pure fabrics.

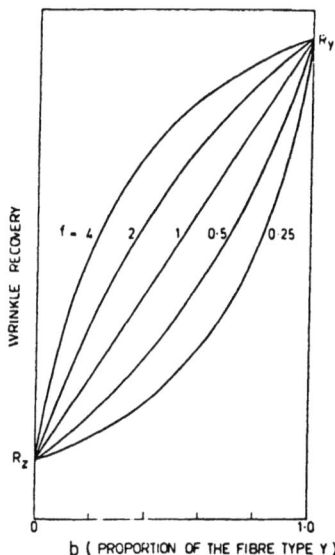

Fig.5 Theoretical wrinkle recoveries of a blend fabric as a function of the proportion of one component, for several values of the stiffness ratio f.

Fig.5 gives some idea of how the recovery of a fabric can vary with blend proportion for different values of f equal to 0.25, 0.5, 1, 2, and 4.

When the stiffness ratio is unity, it can be seen that a linear relation is obtained. However, by suitable choice of this ratio it is possible to obtain a disproportionately large increase in recovery for a low percentage of the high-recovery fibres. Conversely, for other stiffness ratios even a large proportion of the high-recovery fibres will effect only a moderate improvement in recovery.

In general terms Equation (32) indicates that the wrinkle recovery of the blend will be biased away from linearity towards the recovery determined by the stiffer fibre.

For a given experimental curve of wrinkle recovery against blend proportion, it is possible to calculate the ratio f in the following way. If $b_{\frac{1}{2}}$ is the blend proportion giving a wrinkle recovery midway between R_Y and R_Z, then substitution in Equation 32 yields:

$$f = \frac{1}{b_{\frac{1}{2}}} - 1.$$

If it is necessary to keep the proportion of high-recovery fibres to a minimum (either because of cost consideration or because they may possess certain undesirable properties), we can beneficially affect the stiffness ratio by constructing the blend from higher-modulus high-recovery fibres and/or increasing their diameter. Since the foregoing treatment is strictly correct only for linear viscoelastic fibres within the linear region (usually < 1% bending strain), the presence of higher-modulus, larger-diameter, high-recovery fibres may actually detract from the recovery of high-curvature wrinkles where the possibility of plastic flow exists. However, it should be possible to determine an optimum set of fibre characteristics in any particular case.

REFERENCES

1. A.E.H.Love, A Treatise on the Mathematical Theory of Elasticity, Dover Publication, New York 1944.
2. B.M.Chapman, J.Text.Inst.64, 312, 1973.
3. J.D.Ferry, Viscoelastic Properties of Polymers, Wiley, New York, 1961.
4. B.M.Chapman, Rheol.Acta 14, 466, 1975.
5. B.M.Chapman, J.App.Polymer Sci. 18, 3523, 1974.
6. B.M.Chapman, Text.Res.J. 44, 306, 1974.
7. P.Grosberg, Text.Res.J. 36, 205, 1966.
8. B.M.Chapman, Text.Res.J. 45, 137, 1975.
9. B.M.Chapman, Text.Res.J. 44, 531, 1974.
10. B.M.Chapman, Text.Res.J. 46, 113, 1976.
11. B.M.Chapman, J.App.Polymer Sci. 17, 1693, 1973.
12. E.F.Denby, Rheol.Acta. 14, 591, 1973.
13. G.M.Abbott, J.Delmenico, J.D.Leeder and D.S.Taylor, Appl.Polymer Symposium 18, 963, 1971.
14. B.M.Chapman, Text.Res.J. 46, 711, 1976.

Acknowledgement

The author would like to thank the Textile Research Institute for permission to reproduce Figs. 4 and 5 and corresponding sections of the text.

EXAMINATION OF EFFECT OF BASIC MECHANICAL PROPERTIES OF FABRICS
ON FABRIC HAND*

S. Kawabata

Department of Polymer Chemistry,
Kyoto University, Kyoto, 606, Japan

ABSTRACT. Fabric hand are related closely with the mechanical
properties of the fabric. In order to control or improve the
fabric quality, we have to know how the hand is related with these
mechanical properties. A large scale research project about this
subject has been carried out recently. The progress in this investigation will be introduced in this lecture as well as the result
of the examination of the effect of the basic mechanical properties
on fabric hand.

1. INTRODUCTION

It is important whether a fabric posseses the suitable properties
for clothing or not. The fabric for clothing should be in conformity with function of clothing and with human sense. The mechanical property of the fabric are closely related with this conformity. For example, the fabric must be moderately flexible for
bending and deformable for stretching and shearing in plane to
cover the three dimensional surface of human body with a beautiful
silhouette. The perfect elastic behaviour like a vulcanized rubber
is not suitable for wearing because of lack of damping. By contraries, the lack of moderate elasticity causes an uncomfortable
clothing. Surface of the fabric must be smooth and soft for human
skin especially for winter suit or heavy fabric. This is also an
important property for the fabric being high quality.

---

* This work has been supported in part by the grant of aid for
  scientic research, 1978, from Japanese government and carried
  out as a part of the activity of the Hand Evaluation and Standardization Committee of The Textile Machinery Society of Japan.

All of these properties of fabric are mainly related with the mecanical properties of fabric directly or, in some cases, indirectly, and have judged by the experts in factories and by marchants, by means of so called hand judgement. This judgement is done by hand. Although the judgement by the experts is very sensitive in many cases, this judgement is essentially sensory. The judged result is usually shown by sensory expressions. This makes the engineering design of fabric very difficult and have interrupted to establish the scientific design system of textile products in textile industry.

In 1972, the Hand Evaluation and Standardization Committee (HESC) was organized by author for the purpose of standardization of the hand evaluation. Although the hand judgement is essentially based on the individual criterion of experts, there are some of common expressions of hand which are usually used in the experts. For example, KOSHI (stiffness) is one of them. This common expression is however not standardized. Each expert, for example, has not always a perfectly common concept of the KOSHI feeling. Moreover, there is no common scale of the intensity of the feeling, too.

The main works which were carried out by the committee are: [1-2]
1) Selection of the important hand expressions which are considered to be primary hand expressions and are common among the experts in factories.
2) Definition of each of these selected hand expressions.
3) Numerical expression of the intensity or degree of these hand feelings.
4) Settlement of the standard samples for each of the primary hand expressions.

Based on the standardization of the primary hand and the development of the numerical expressions of them, the research on connecting between these hands and the mechanical properties of fabric was started by author and his coworkers. As seen from the non-linearity in the mechanical properties of fabrics and from the type of investigation, following two were important works to be done.
1) Determination of the characteristic values which characterize the non-linear mechanical properties concerned with the hand.
2) Development of the quick measuring system of the fabric mechanical properties.

In this lecture, the result of these investigations will be introduced especially with respect to men's winter suit fabric.

2. THREE PRIMARY HANDS AND THEIR NUMERICAL EXPRESSIONS

2.1 The Three Primary Hand

From many expressions of fabric hand, following three were sellected as the primary hand for winter suit fabric by the ten experts who were elected as the committee from companies.
  1) KOSHI (stiffness):

A feeling related with bending stiffness. Springy property promotes this feeling. The fabric having compact weaving density and woven by springy and elastic yarn makes this feeling strong.
  2) NUMERI (smoothness):
A mixed feeling come from smooth, limber and soft feeling. The fabric woven from cashimere fiber gives this feeling strongly.
  3) FUKURAMI (fullness and softness):
A feeling come from bulky, rich and well formed feeling.

For summer suit, the KOSHI and SHARI were selected as the primary hand. The SHARI (crispness) is the feeling come from crisp and rough surface of fabric. This word means a crisp, dry and sharp sound arisen when the fabric is rubbed with itself. The SHARI feeling brings us a cool feeling and is brought by hard and highly twisted yarn. In this lecture, however, only the winter suit fabric will be discussed and analysed to avoid the scatter of the subject. The word "primary" comes from their important and basic character. Many of the other hand expressions can be decomposed into these three primary hand and expressed by a combination of them.
These primary hand expressions are useful to examining the performance of the fabric such that,
  1) KOSHI's usefulness:
The fabric having the moderate KOSHI plays an important role to keep moderate space between the clothing and the body. This enables the movement of the body to be easy. And KOSHI is related closely with beautiful silhouette of clothing. Too much strong KOSHI, however, bring us an uncomfortable feeling.
  2) NUMERI's usefulness:
Smooth surface with moderate softness gives us a comfortable and rich feeling.
  3) FUKURAMI's usefulness:
Soft feeling come from FUKURAMI also bring a comfortable and warm feeling. The fabrics having high FUKURAMI can be deformed easily in small deformation region. This also promotes the comfortability for wearing.

2.2 Numerical Expression of Intensity of Hand Feeling

About a thousand samples which were commercially produced for men's winter suit were collected and individual of the ten experts evaluated these samples concerning the three primary hands by following manner. For example, regarding KOSHI, a expert classifies the samples to three groups, 1) the group having relatively strong, 2) medium and 3) low KOSHI, then he repeats the same procedure in each of those groups. Then the nine groups of samples are obtained in order of the feeling intensity of KOSHI. The samples having extremely high feeling intensity are seperated from the highest group and extremely low ones from the lowest group. Now he obtains eleven groups of samples classified with the feeling intensity. And the

Table 1. Classification of the intensity of hand feeling by eleven degrees. The intensity is expressed numerically by "Hand Value".

Group	XH	A(strong)			B(medium)			C(low)			XL
		A-1	A-2	A-3	B-1	B-2	B-3	C-1	C-2	C-3	
H.V.	10	9	8	7	6	5	4	3	2	1	0

samples which belong to the groups are labeled by the number from 10 to 0 where 10 indicates the samples belong to the highest and 0 the lowest group as shown in Table 1. We call the number "Hand Value". The same procedure is repeated for NUMERI and FUKURAMI, and other experts also carried out the same procedure. All samples had evaluated their hand values for each of the three primary hands by the experts individually. However, the values are different in general between judges. therefore the mean values were taken as the representative value. Some of the samples of which hand values were not so scattered were picked up as the standard samples for the range from 10 to 0.

By this procedure, the scale of the hand judgement has been coordinated. After this, the judgement has been done by means of comparing a sample with these standards to determine its hand values. Now we can express the hand of a fabric such that, HV(KOSHI)=5, HV(NUMERI)=4 and HV(FUKURAMI)=5.5. The value 5.5 is the case when the hand value is judged being between 5 and 6 of the standards.

3. CHARACTERIZATION OF THE MECHANICAL PROPERTIES

The mechanical properties of fabircs are not so simple as of the continuous body having linear property. Firstly, the fabrics are largely deformable and show a typical non-linear behaviour with considerably large amount of hysteresis. We have to determine which conditions and deformations are appropriate for correlating the hand of experts. The preliminary and fundamental investigation concerning this problem had been carried out by author and his group and following 16 of the characteristic values of six blocks were selected as shown in Table 2.

These values have been selected from the consideration as following.
1) Deformation region must be nearly in the deformation region of the fabric under the usual wearing condition.
2) The mode of deformation must be as simple as possible to connect those easily with the fabric structure, and yarn property.
3) The smaller number of the characteristic values is desirable but those values must well-cover the fabric property related with the hand.

The definition of the characteristic values are shown in Fig. 1. These sixteen characteristic values can be easily and quickly measured by KES measuring system which has developed by author.

Table 2. Characteristic values of fabric mechanical property

Blocked Properties	Characteristic Values	Remarks	
Tensile	LT WT RT	Linearity Tensile energy Regilience	- g.cm/cm^2 %
Bending	B 2HB	Rigidity Histeresis	g.cm^2/cm g.cm/cm
Shearing	G 2HG 2HG5	Shear stiffness Histeresis at $\phi=0.5°$ Histeresis at $\phi=5°$	g/cm.degree g/cm g/cm
Compression	LC WC RC	Linearity Energy Regilience	- g.cm/cm^2 %
Surface	MIU MMD SMD	Coefficient of friction Mean deviation of MIU Geometrical roughness	- - micron
Weight & Thickness	W T	Weight per unit area Thickness	mg/cm^2 mm

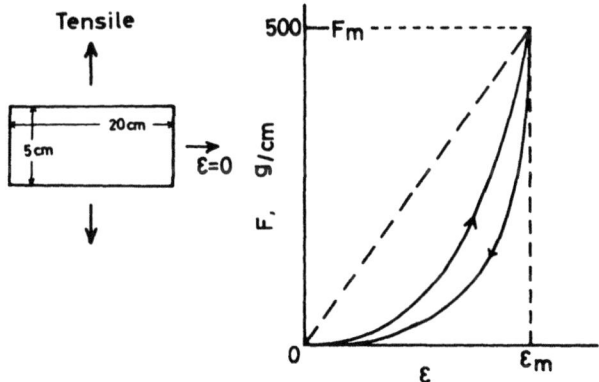

Fig. 1-1. Tensile property, LT, WT, RT.

$\underline{LT}; = 2WT/\epsilon_m F_m$, $\quad \underline{WT}; = \int_0^{\epsilon_m} F \, d\epsilon$, $\quad \underline{RT}; = \int_0^{\epsilon_m} F(recovering) d\epsilon/WT$,

strain rate; 400×10^{-3} sec, F_{max}; 500 g/cm.

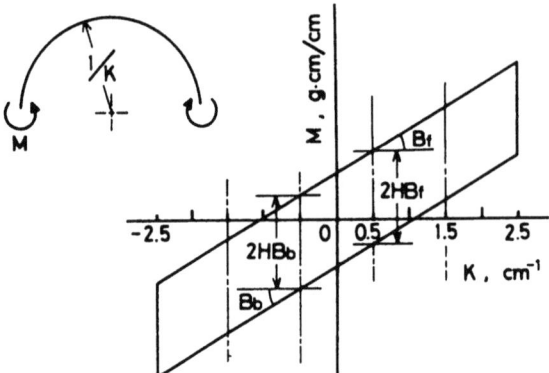

Fig.1-2. Bending property, B, $2HB$.
bending rate; $0.5\ cm^{-1}/sec$, K_{max}; $\pm 2.5\ cm^{-1}$, bending specimen; 1 cm long.

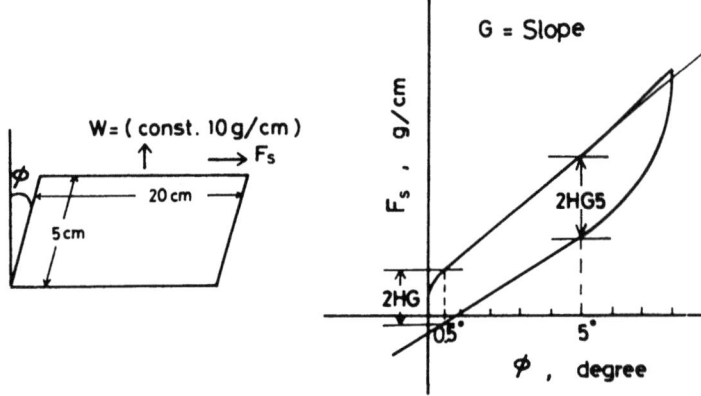

Fig.1-3. Shearing property, G, $2HG$, $2HG5$.
rate; $0.417\ mm/sec$, ϕ_{max}; $8°$.

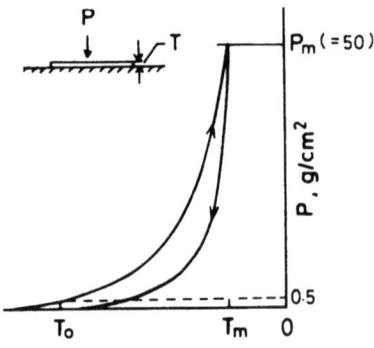

Fig.1-4. Compressional property, LC, WC, RC.

$\underline{LC}\colon = 2WC/(T_0 - T_m)P_m, \quad \underline{WC}\colon = \int_{T_m}^{T_0} P\, dT, \quad \underline{RC}\colon = \int_{T_m}^{T_0} P(recovering)\, dT/WC,$

rate; 20 micron/sec, compressed area; 2 cm^2, P_{max}; 50 g/cm^2.

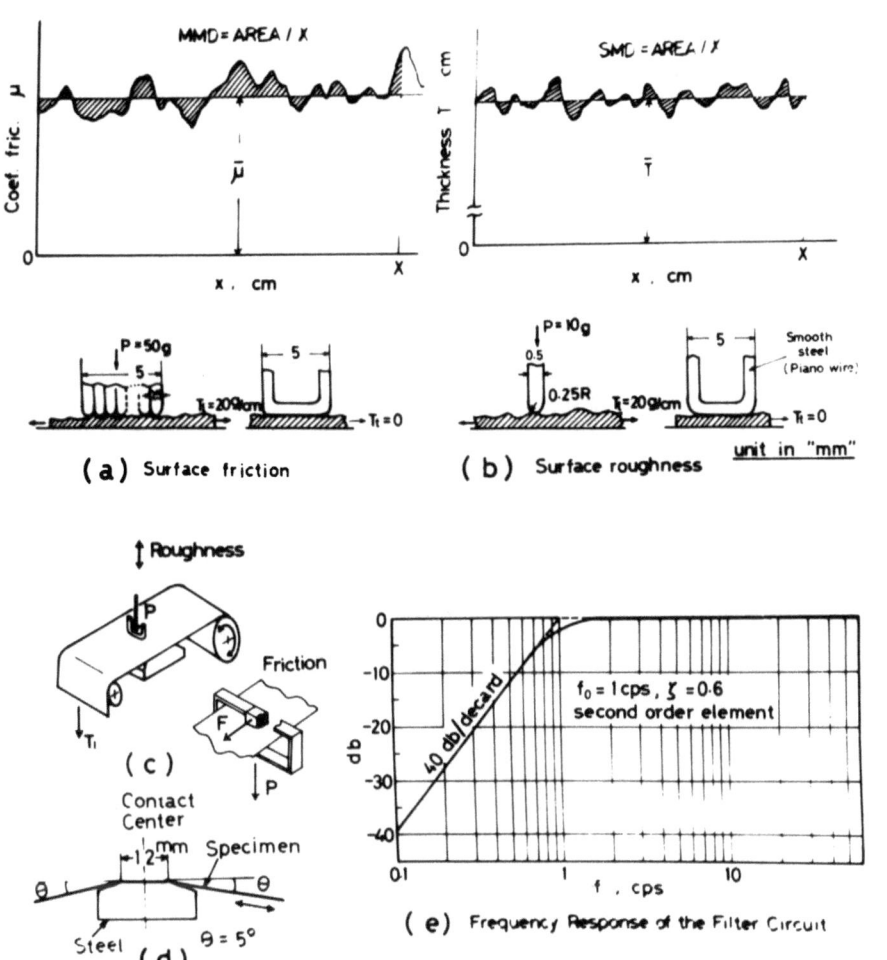

Fig.1-5. Surface property, MIU, MMD, SMD.

$\underline{MIU}\colon = \bar{\mu}, \quad \underline{MMD}\colon = \frac{1}{X}\int_0^X |\mu - \bar{\mu}|\, dX, \quad \underline{SMD}\colon = \frac{1}{X}\int_0^X |T - \bar{T}|\, dX,$

rate; 0.1 cm/sec, pressure for roughness is given by a spring of which constant is 25±1 g/mm.

Thickness and weight are defined as $\underline{T}$; Thickness under 0.5 g/cm^2 pressure, $\underline{W}$; weight/unit area, 20°C, 65%RH.

4. THE EFFECT OF BASIC MECHANICAL PROPERTIES ON THE FABRIC HAND

One of the purpose of the regression analysis is to obtain a good equation by which the hand values can be calculated from the mecanical properties. Another purpose is to see how the hand is related with the mechanical properties. The ordinary multi-variable regression, however, is not always appropriate for this latter purpose in the case when the variables are mutually correlated. Therefore, a stepwise block regression method has been developed. As a preliminary study, we have examined how many samples are necessary at least to obtain a good regression equation, because that use of too much samples makes the judgement by experts so difficult and leads to error in the judgement. It must be noted that the smaller number of samples is, the better regression is obtained at the process obtaining the regression equation. For example, in the case when the numbers of variables and of samples are equal, we can obtain a complete regression equation. However, if we use the equation obtained from such a small number of samples against new samples to calculate their hand values, a large error will be occured. The large number of samples at the regression analysis leads better result in this case.

After this examination, we have found that more than 150 samples are necessary, however, the accuracy will be saturated at about 200 samples. Based on this result, 214 samples from the one thousand samples were selected to use them for the regression analysis and these samples were again evaluated their hand values by the experts based on the standards which had been already settled. In addition, 66 samples were prepared for the examination of the accuracy of the regression equation apart from the 214 samples used for obtaining the regression equation.

As described before, the characteristic values of the fabric are grouped to six blocks, and the values in a block are treated to be inseperable throughout the regression analysis. The stepwise block regression analysis is carried out by following way.
We assumed the regression equation as follows.

$$Y = C_0 + C_i \sum_{i=0}^{16} Z_i \quad (1)$$

Y = Hand values
Z_i = Normalized valiables ($i=1\sim 16$) and are reduced by

$$Z_i = \frac{X_i - \bar{X}_i}{\sigma_i} \quad (2)$$

X_i = The i th characteristic values
$\bar{X}_i$ = Mean of the i th X
σ_i = Standard deviation of the i th X
C_0, C_i = Constant coefficients

For some of characteristic values, following normalization is used instead of eq.(2)

$$z_i = \frac{\log X_i - \overline{\log X_i}}{\sigma_{\log X_i}} \qquad (3)$$

The stepwise block regression is carried out as following.
1) Firstly, the regression equations between Y and each of the block are obtained seperately using ordinary multivariables regression method. Therefore, we have six regression equations. The regressed hand values [Y] and the experimental hand values (that is the value obtained by experts' judge) [Y_E] are examined by the correlation between them for each of the equation. Here, [ ] means the matrix representation of the values for 214 samples. Then, the block having the highest correlation is picked up as the first block, and named as the first regression equation. That is,

$$Y_1 = C_{01} + C_k z_k + C_{k+1} z_{k+1} + C_{k+2} z_{k+2} \qquad (4)$$

1st block

where Y_1 denotes the regressed value by the first regression equation.
2) The same procedure is repeated between the residuals [Y_E]-[Y_1] and the each of the remaining five blocks. Thus, the following regression equation is obtained.

$$(Y - Y_1) = C_{02} + C_\ell z_\ell + C_{\ell+1} z_{\ell+1} + C_{\ell+2} z_{\ell+2} \qquad (5)$$

2nd block

The first regression equation and the eq.(5) are composed to lead the 2nd regression equation

$$Y = Y_1 + (Y - Y_1) = (C_{01}+C_{02}) + (C_k z_k+..) + (C_\ell z_\ell +..) \qquad (6)$$

1st block   2nd block

This procedure is continued to built up regression equation which includes full characteristic values.
3) Finally, the stepwise regressions are again applied to the characteristic values within the first block to rearrange them in this block. Then the rearrangement in the second block is done based on the rearranged equation of the first block. This procedure is continued for following blocks without reordering of the blocks. Fig.2 shows the increasing of the correlation coefficient R between [Y_E] and the regressed hand values. In this figure, the R is calculated with increasing terms of characteristic values. The coefficients of the regression equations obtained by this stepwise block regression analysis are shown in Table 3, 4 and 5. The X_i and the σ_i which are necessary for the normalization are shown in Table 6. The strong point of this method is firstly that the contribution of each of the characteristic values to the hand value is shown by the coefficient without constrained relation with the other characteristic values. Second, the saturation of R with increasing

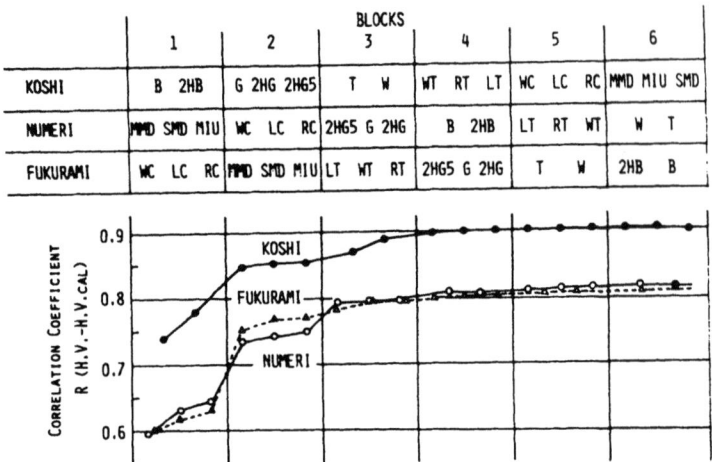

	BLOCKS					
	1	2	3	4	5	6
KOSHI	B 2HB	G 2HG 2HG5	T W	WT RT LT	WC LC RC	MMD MIU SMD
NUMERI	MMD SMD MIU	WC LC RC	2HG5 G 2HG	B 2HB	LT RT WT	W T
FUKURAMI	WC LC RC	MMD SMD MIU	LT WT RT	2HG5 G 2HG	T W	2HB B

Fig. 2. Correlation coefficient between experimental and the regressed hand values are shown with increasing number of the terms of the regression equation.

Table 3. Coefficient for KOSHI (Stiffness)

Block	i	X_i	C_i	R
		C_0	5.7093	
1	1	logB	0.8459	0.740
	2	log2HB	-0.2104	0.780
2	3	logG	0.4268	0.849
	4	log2HG	-0.0793	0.854
	5	log2HG5	0.0625	0.854
3	6	logT	-0.1714	0.868
	7	logW	0.2232	0.889
4	8	logWT	-0.1345	0.896
	9	RT	0.0676	0.898
	10	LT	-0.0317	0.899
5	11	logWC	-0.0646	0.900
	12	LC	0.0073	0.900
	13	RC	-0.0041	0.901
6	14	logMMD	0.0307	0.901
	15	MIU	-0.0254	0.901
	16	logSMD	0.0009	0.901

terms of X_i suggests that the behind terms of after the saturation can be ignored. This is profitable for the application of this equation into the fabric design.

Table 4. Coefficients for NUMERI (Smoothness)

Block	i	X_i	C_i	R
		c_0	4.7537	
1	1	logMMD	-0.9270	0.595
	2	logSMD	-0.3031	0.633
	3	MIU	-0.1539	0.645
2	4	logWC	0.5278	0.734
	5	LC	-0.1703	0.742
	6	RC	0.0972	0.749
3	7	log2HG5	-0.3702	0.794
	8	logG	-0.0263	0.794
	9	log2HG	0.0667	0.792
4	10	logB	-0.1658	0.807
	11	log2HB	0.1083	0.803
5	12	LT	-0.0686	0.808
	13	RT	-0.1619	0.812
	14	logWT	0.0735	0.813
6	15	logW	-0.0122	0.813
	16	logT	-0.1358	0.812

Table 5. Coefficients for FUKURAMI (Softness & Fullness)

Block	i	X_i	C_i	R
		c_0	4.9798	
1	1	logWC	0.8845	0.600
	2	LC	-0.2042	0.616
	3	RC	0.1879	0.630
2	4	logMMD	-0.5964	0.754
	5	logSMD	-0.1702	0.768
	6	MIU	-0.0569	0.770
3	7	LT	-0.1558	0.782
	8	logWT	0.2241	0.793
	9	RT	-0.0897	0.795
4	10	log2HG5	-0.0657	0.799
	11	logG	0.0960	0.800
	12	log2HG	-0.0538	0.802
5	13	logT	0.0837	0.807
	14	logW	-0.1810	0.805
6	15	log2HB	0.0848	0.805
	16	logB	-0.0337	0.806

Table 6. $\bar{X}_i$ and σ_i

X_i	$\bar{X}_i$	σ_i
LT	0.6082	0.0611
logWT	0.9621	0.1270
RT	62.1894	4.4380
logB	-1.0084	0.1267
log2HB	-1.3476	0.1801
logG	-0.0143	0.1287
log2HG	0.0807	0.1642
log2HG5	0.4094	0.1441
LC	0.3703	0.0745
logWC	-0.7080	0.1427
RC	56.2709	8.7927
MIU	0.2085	0.0215
logMMD	-1.8105	0.1233
logSMD	0.6037	0.2063
logT	-0.1272	0.0797
logW	1.4208	0.0591

According to the definition made by the experts, KOSHI is related with the bending stiffness with springy behaviour. The analysis shown in Table 3 agrees well with this definition. It is interesting that "thin and heavy" property promotes the KOSHI feeling.
The NUMERI and the FUKURAMI are very similar feeling and the NUMERI is firstly related with smoothness of surface and secondly, with softness of compressional deformation (the larger value of WC corresponds to the deformable behaviour under compression). On the other hand, FUKURAMI is firstly correlated with softness under compression. The deformability against the extension also contributes to the FUKURAMI.
Fig. 3 shows the correlation for between the calculated hand values and the experimental values (mean of the experts' judge) for KOSHI of 66 samples which prepared for examination of the accuracy of the regression equation. R is the correlation coefficient between them. SDE is the standard deviation of the difference between them.

5. CONCLUSION - TOTAL HAND VELUE -

The primary hand expressions have been analysed here, however, one of the other important hand evaluations is the judgement of "good" or "poor" fabric. This is a final hand judgement and also a kind of quality judgement. We have also studied this kind of hand.

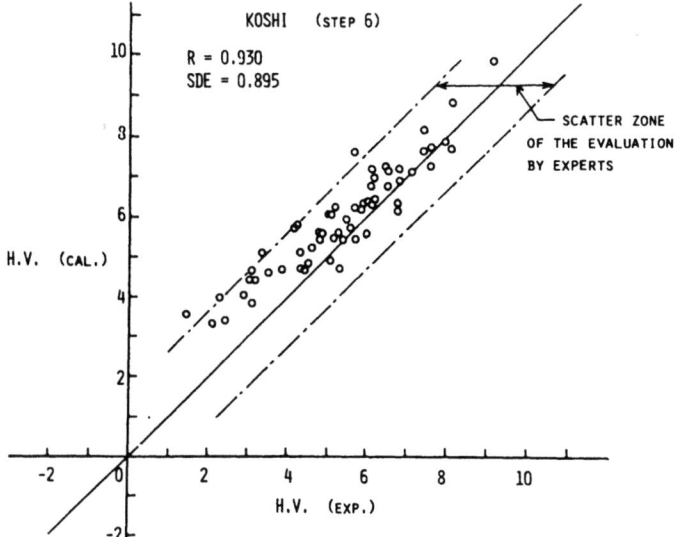

Fig. 3. Correlation between the hand values of KOSHI evaluated by experts (experimental hand value) and these calculated from mechanical properties using the regression equation samples used here are the new samples prepared for examination of the accuracy of the regression equation.

The "good" hand is classified by the number between 5 and 0 in order of the "good" hand where 5 means excellent. This number was named "Total Hand Value". And the T.H.V. are considered to be a function of the primary hand values. We have constructed a equation which connects the primary hand values with the Total Hand Value. This equation contains the square terms of the valiables because that an optimum combination of the hand values is expected. The mechanical properties are translated firstly to the three primary hand values, then the values are translated into the Total Hand Value. This two steps method is the most accurate method for the judgement of the Total Hand Value from the mechanical properties in this stage. We have learned this sequence of calculation from the sequence which is probably taken in the experts' brain at the judgement.

REFERENCES

1. S. Kawabata, Editor HESC Standard of Hand Evaluation, The Hand Evaluation and Standardization Committee, The Textile Machinery Society of Japan, Osaka, 1975.
2. S. Kawabata, The Standardization and Analysis of Hand Evaluation, The Hand Evaluation and Standardization Committee, The Textile Machinery Society of Japan, Osaka, 1975.

THE ELASTIC MODULUS OF PAPER —*
THE CONTROLLING MECHANISMS

D.H. Page, R.S. Seth and J.H. De Grâce

Pulp and Paper Research Institute of Canada,
Pointe Claire, Quebec, Canada

ABSTRACT. The elastic modulus of paper varies considerably with pulping and papermaking conditions. Experiments which demonstrate the factors that control elastic modulus are presented. The main factors are the elastic modulus of fibres, the degree of bonding, and the presence of curl, kinks, crimps and microcompressions in the fibres. Mechanisms by which these factors operate are discussed. Based on these, the observed large variation in the elastic modulus of paper can now be explained.

1. INTRODUCTION

When paper is subjected to tensile stress, it behaves, for small strains, as an elastic material. The elastic modulus, generally determined from the initial slope of the load-elongation curve (Figure 1), is an important mechanical property of paper. It is a critical end-use property for certain products, and in addition provides an understanding of other strength properties. Over the past twenty-five years several theories relating the modulus of paper to the properties of fibres and paper structure have been put forward. None of these theories are, however, sufficiently well developed to provide understanding of particular practical problems.

The work carried out here demonstrates what the papermaking factors are that control elastic modulus and the mechanisms by which they operate. In this report, we consider these mechanisms

---

* To be published in Tappi (1979)

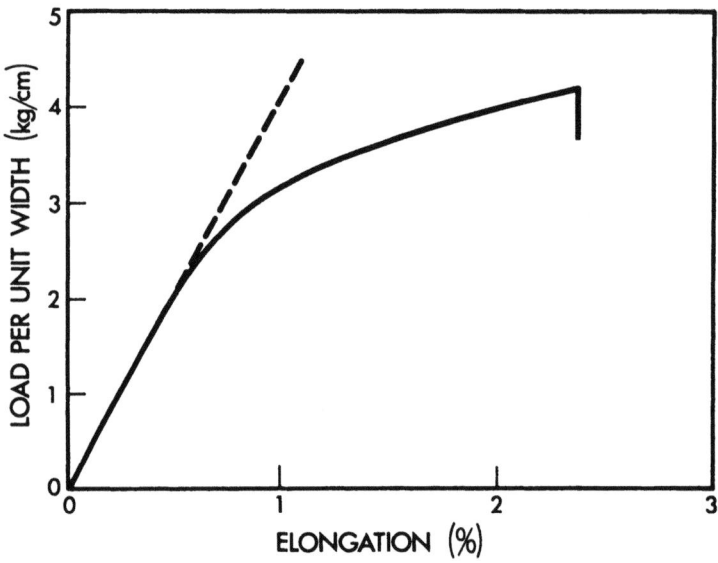

Fig. 1. A typical load-elongation curve for paper.

in qualitative terms. In subsequent reports, a mathematical theory is developed based on these mechanisms, culminating in an explicit theoretical equation for the modulus of paper.

2. OBSERVED VARIATION IN MODULUS

The elastic modulus of a material is usually expressed as a stress per unit strain. For paper it is particularly convenient, because of the fact that the stress is borne only by the solid fraction of the sheet, to divide the load per unit sample width by the basis weight of the sheet. The elastic modulus is then expressed as a length analogous to breaking length.

The variation in elastic modulus that is found in paper samples is illustrated in Table I. Modulus varies from around 300-600 km for mechanical pulps and lightly beaten dried chemical pulps, to 1,000-1,200 km for never-dried chemical pulps or well beaten dried chemical pulps.

It is the source of this variation that is discussed here.

3. NETWORK THEORY OF COX

The starting point for all theories of the elastic modulus of

TABLE I

ELASTIC MODULUS OF STANDARD HANDSHEETS OF VARIOUS PULPS

PULP	TREATMENT	C.S. FREENESS (ml)	ELASTIC MODULUS (km)
Commercial stone groundwood of spruce	-	84	449
Commercial refiner mechanical pulp of spruce	-	119	417
Commercial thermomechanical pulp of spruce	-	104	431
Commercial, dried, bleached softwood sulphite pulp	unbeaten	675	320
Commercial, dried, bleached hardwood kraft pulp	20 min in Valley beater	416	528
Commercial, dried, bleached softwood kraft pulp	unbeaten	668	600
Commercial, dried, bleached softwood kraft pulp	40 min in Valley beater	315	953
Commercial, never-dried, unbleached softwood sulphite pulp	unbeaten	687	1020
Unbleached, never-dried softwood laboratory kraft pulp	unbeaten	708	1039
Unbleached, never-dried softwood laboratory kraft pulp	6000 rev. in PFI mill	507	1149

paper is the work of Cox[1]. He considered paper as a two-dimensional network of randomly oriented long straight linearly elastic fibres. He assumed, initially, that the distribution of stress within the paper sheet was homogeneous, so that the strain in every fibre was equal to the local strain of the sheet in the fibre direction. From an analysis of strains and resolved stresses in fibres, he concluded that the elastic modulus of paper is simply equal to one-third of the elastic modulus of its component fibres.

Using this theory we can estimate the modulus of paper from published data on fibre moduli. Page et al.[2] have shown that the elastic modulus of pulp fibres of softwoods, in the most common range of fibril angles 0°-20°, is about 4.5-5.5 GPa. This value is surprisingly independent of pulping process and yield in the yield range generally used. In units of length, this gives a modulus of 3,000-3,600 km. Using Cox's factor of one-third, the modulus of paper is expected to be 1,000-1,200 km. It seems from Table I that the highest moduli of paper indeed correspond to this theoretical value, but often the modulus falls short, for reasons that merit attention.

4. VARIATION IN MODULUS WITH WET-PRESSING AND BEATING

An insight into the mechanism of modulus comes from an investigation of the effect of different wet-pressing pressures and beating degrees on sheet properties. It is instructive to plot elastic modulus against light scattering coefficient, since this is generally accepted as a measure of the degree of bonding. Typical plots for a number of commercial, dried bleached kraft pulps are shown in Figures 2-4. The pulps were beaten in the Valley beater and at each beating level, handsheets were made at different wet-pressing pressures.

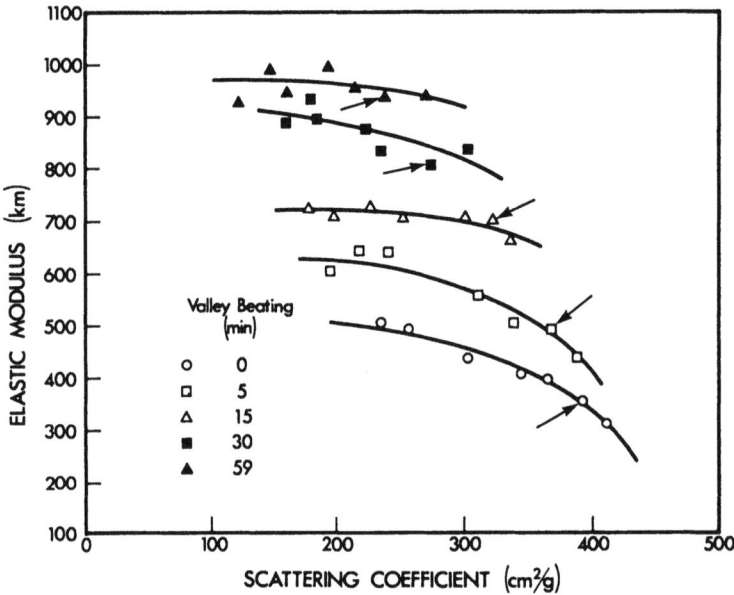

Fig. 2. Plot of elastic modulus against scattering coefficient for sheets made from a bleached kraft pulp of Eastern Canadian white pine. The lines are for constant beating but varying wet-pressing pressures. The arrows in this and in subsequent plots indicate data for the standard pressure of 345 kPa.

Without exception, the data in Figures 2-4 show two features. First, an increase in wet-pressing increases sheet modulus. Second, the increase in modulus levels off to a plateau that is different

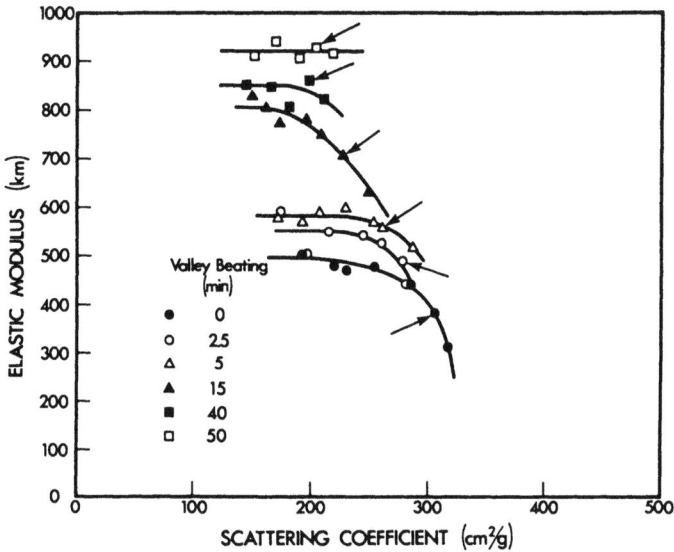

Fig. 3. Same type of plot as Figure 2 for a bleached kraft pulp of Southern U.S. pine.

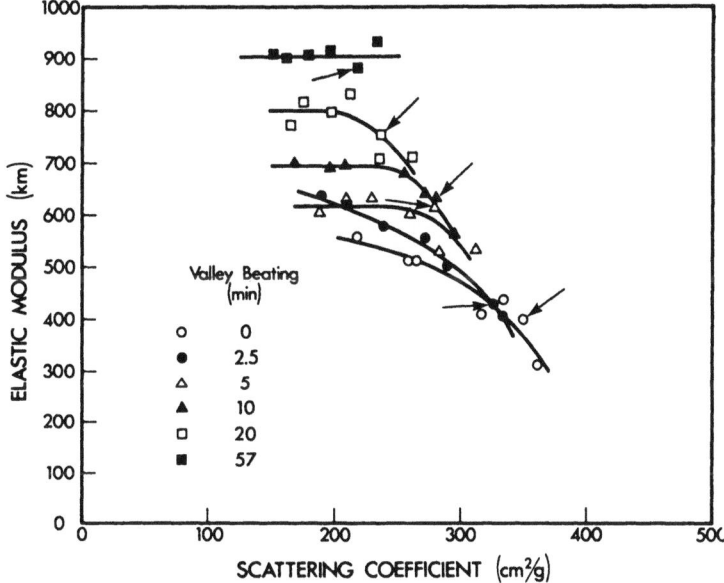

Fig. 4. Same type of plot as Figure 2 for a bleached kraft pulp of Western Canadian Douglas fir.

for different pulps and treatments. An important implication of
this is that there are at least two mechanisms that control the
elastic modulus of paper. One controls the increase in modulus
with wet-pressing at low pressing pressures; the other controls
the levelling-off modulus that is reached at high pressing pressures.

5. MECHANISMS FOR THE INCREASE IN MODULUS WITH BONDING

In his original paper Cox[1] pointed out that, in practical
cases such as short-fibre reinforced composites, the strain in a
fibre cannot be considered uniform and equal to the strain in a
homogeneous matrix. Since fibres are of finite length, stress
must be transferred from each fibre to its neighbours in the regions of the fibre ends. He showed that the stress distribution
in a fibre follows a pattern as shown in Figure 5. The stress is

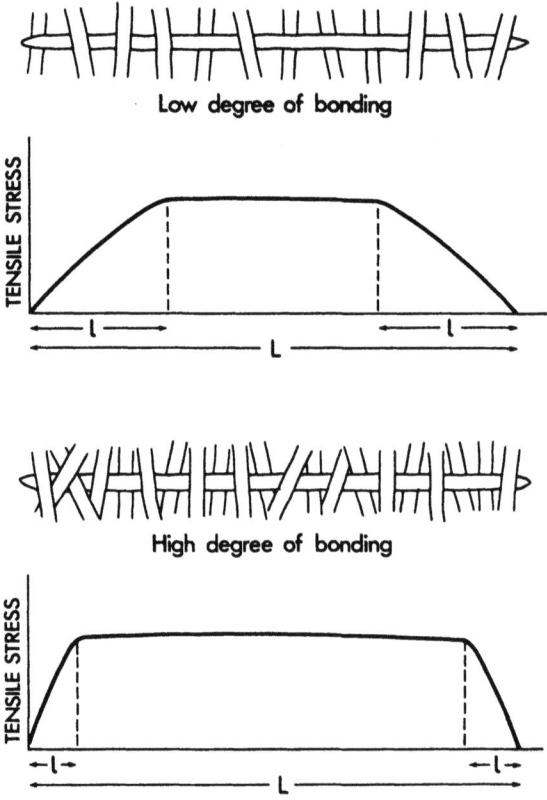

Fig. 5. The tensile stress distribution along the length, L, of a
fibre at low and high degrees of bonding.

maximum at the centre and diminishes to zero at the ends, the fall-off in stress being dependent on the shear modulus of the matrix. This causes the elastic modulus to fall short of the theoretical by an amount which depends on the ratio of the length, ℓ, over which the stress is transferred, to the fibre length, L.

An analogous situation exists in paper. A typical fibre in paper is bonded to a number of fibres crossing it and therefore, can be considered embedded in a matrix of other fibres. When paper is strained, the distribution of stress along the length of this fibre is similar to that considered by Cox[1].

If we consider sheets of different degrees of bonding, it is expected that the transfer of stress would occur over a shorter length, ℓ, the higher the bonding degree (Figure 5). The increase in modulus with bonding may thus be seen as a more efficient utilization of the entire fibre length, by reducing the length of the fibre used for stress transfer.

6. MECHANISMS FOR DIFFERENCE IN LEVELLING-OFF MODULUS

The plateau that is reached at high wet-pressing pressures varies markedly from pulp to pulp. It is highest for beaten dried chemical pulps or unbeaten never-dried chemical pulps, and is lowest for mechanical pulps and unbeaten dried chemical pulps. The reason for these differences becomes clear upon microscopic examination (Figure 6) of sheets of a dried pulp, unbeaten and beaten, that have been wet-pressed to identical scattering coefficients. Although the sheets have similar degrees of bonding, the beaten sheets have much higher elastic modulus.

The scanning electron micrographs of Figure 6 show that the fibres in the unbeaten sheets are clearly more crimped and kinked than those in the beaten sheets. They are also more frequently curled out of the plane of the sheet. Two mechanisms seem to be responsible for this appearance. Firstly, the beating action itself can swell fibres and remove crimps and kinks to some extent. Secondly, and probably more importantly, the mechanism proposed by Giertz[3] seems to be operating. At fibre crossings, the highly swollen fibres become bonded and during drying the transverse shrinkage stresses of one fibre form microcompressions in the crossing fibre as proposed by Page and Tydeman[4]. According to Giertz[3], this produces tension in the unbonded regions of the fibres which causes tightening of the sheet structure. If adequate, this can remove the out-of-plane curl and crimps in the fibres leaving them ready to take further stress. If the shrinkage stresses are not adequate, either because they are not high enough, or because the fibres are too curled and crimped for them to be fully straight-

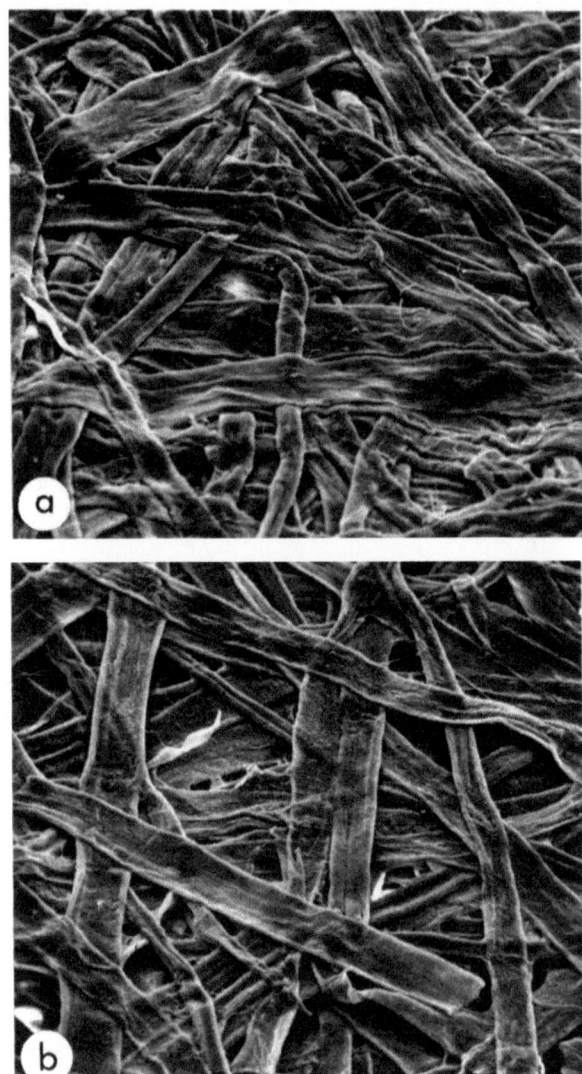

Fig. 6. The structure and physical properties of sheets made under different conditions from a commercial, dried, bleached softwood kraft pulp:

	Valley Beating Time (min)	Wet-Pressing Pressure	Scattering Coefficient (cm^2/g)	Elastic Modulus (km)
(a)	0	6.89 MPa	310	471
(b)	32	138 kPa	310	786

ened, some slackness in the structure remains.

It is, therefore, reasonable to conclude that the levelling-off modulus reached by a pulp at high wet-pressing pressures depends on the degree of residual curl and crimps in the fibres.

7. EFFECT OF CURLATION ON MODULUS

If the modulus depends on the extent of curl and crimps in the fibres, it would be expected that the introduction of crimps into otherwise straight fibres would reduce the modulus. This was confirmed from the experiment of Figure 7, which shows the modulus-scattering coefficient plot for a laboratory-made, never-dried, softwood unbleached kraft pulp with and without curlation. Curlation[5] which introduces severe kinks and crimps into the fibres, was carried out by beating the pulp at 20% consistency for four hours in a Hobart kitchen mixer. Sheets at different degrees of bonding were made by pressing them at various wet-pressing pressures. The curlated pulp has a lower modulus than the unbeaten pulp over the whole pressing range and levels off at a value 35% lower.

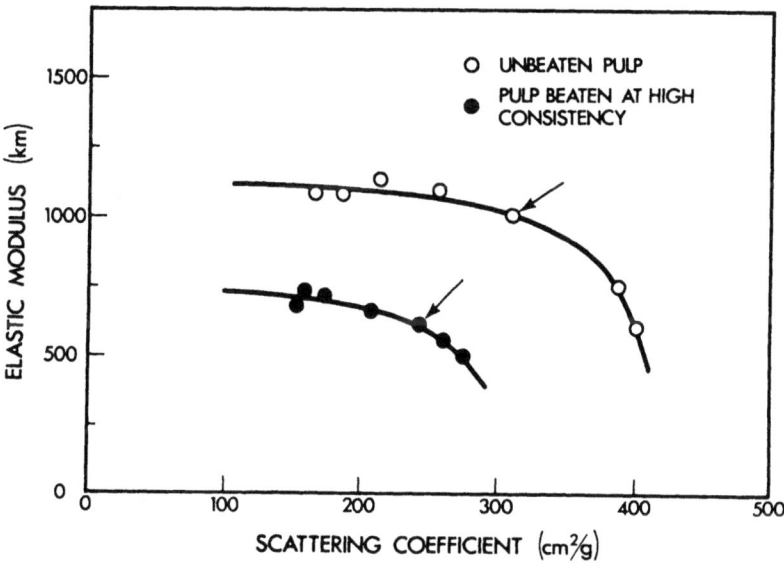

Fig. 7. Plot of elastic modulus against scattering coefficient for sheets made at different wet-pressing pressures from a softwood unbleached kraft pulp, unbeaten, and beaten at high consistency.

8. EFFECT OF LATENCY REMOVAL ON MODULUS

If the introduction of kinks and crimps in fibres reduces sheet modulus, their removal should increase modulus. The removal of latency from mechanical pulps is an example. During manufacture, mechanical pulp fibres emerge from the refining zone severely curled, kinked and crimped. If cooled, they set in their distorted form. Sheets of such a pulp have low strength properties while greater strength lies latent in the fibres. The fibres are straightened and the properties are restored by reheating and stirring the pulp at low consistency[6].

Handsheets were made at different wet-pressing pressures from a laboratory-made softwood refiner mechanical pulp with and without latency removed. The results on elastic modulus and scattering coefficient presented in Figure 8 show that latency removal increases the modulus over the entire range of pressures.

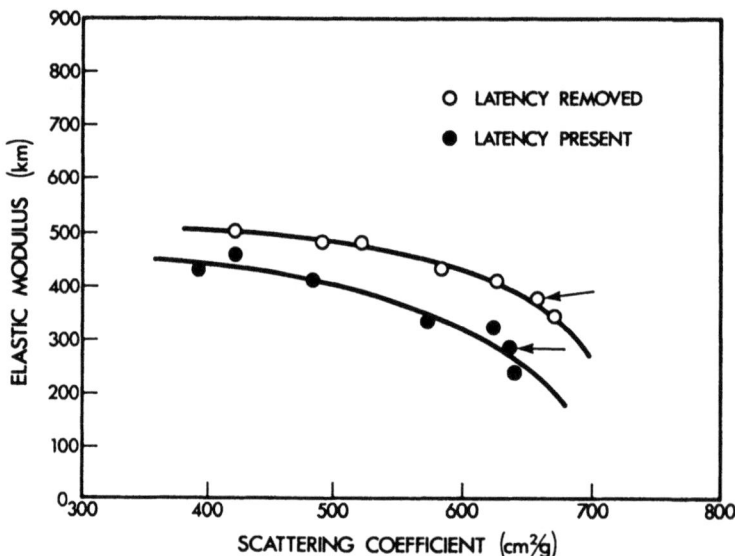

Fig. 8. Plot of elastic modulus against scattering coefficient for sheets made at different wet-pressing pressures from a softwood refiner mechanical pulp, with and without latency removed.

9. EFFECT OF DRYING RESTRAINT ON MODULUS

It was pointed out earlier that during drying, the structure of the sheet is tightened by the shrinkage in the fibres. Crimps are thus partly removed and the elastic modulus of the sheet increases. This only occurs, however, if the sheet is restrained from shrinkage. If shrinkage is allowed, no such tightening of the structure is permitted and the modulus is low. This is indicated in Figure 9. Sheets at different wet-pressing pressures were prepared from a laboratory-made, never-dried, softwood unbleached kraft pulp. The pulp was slightly beaten to enhance fibre shrinkage during drying. The sheets were dried without restraint on Teflon plates and were allowed to shrink freely. Sheets dried fully restrained were also prepared. Results for the two sets of sheets clearly show the effect of fibre shrinkage during drying on the sheet modulus. The modulus for the sheets dried without restraint is considerably lower.

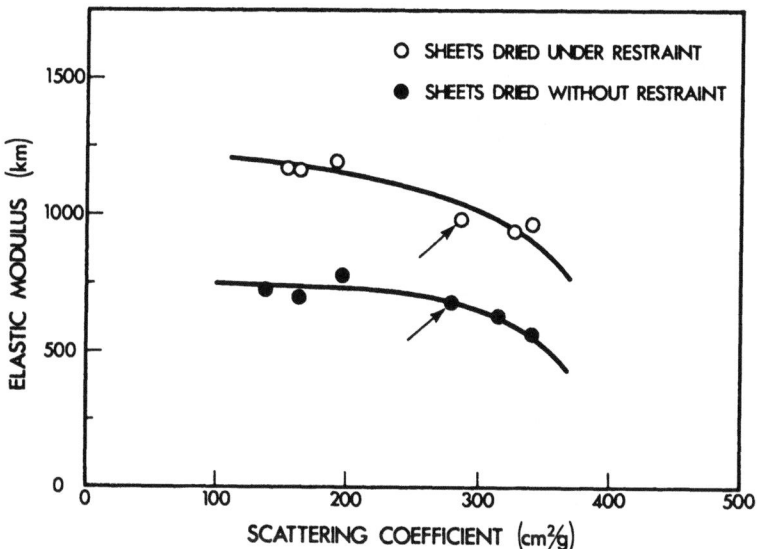

Fig. 9. Plot of elastic modulus against scattering coefficient for sheets made at different wet-pressing pressures, and dried with and without restraint, from a softwood unbleached kraft pulp, beaten 1000 revolutions in the PFI mill.

10. DISCUSSION

The theory presented here puts together three mechanisms which, it is believed, are entirely responsible for the elastic modulus of paper. The effect of these mechanisms is best summarised by reference to Figure 10 with the following commentary:

1. The elastic modulus of the fibres. For well bonded sheets of long straight fibres the modulus of the sheet is, experimentally, of the order of one-third of the modulus of the fibres as predicted by Cox[1]. The modulus of sheets in condition A in Figure 10 is controlled by this factor.
2. The degree of bonding. Because of the transfer of load from

the fibre to the network near the ends of the fibre, the modulus falls short of the theoretical value of one-third as in condition B of Figure 10. The extent to which it falls short depends on the degree of bonding.
3. The presence of curl, kinks, crimps or microcompressions. These lower the modulus of the sheet and are particularly important for sheets of unbeaten dried chemical pulps, for mechanical pulps with latency present, and for curlated chemical pulps. Such sheets are represented by condition C in Figure 10.

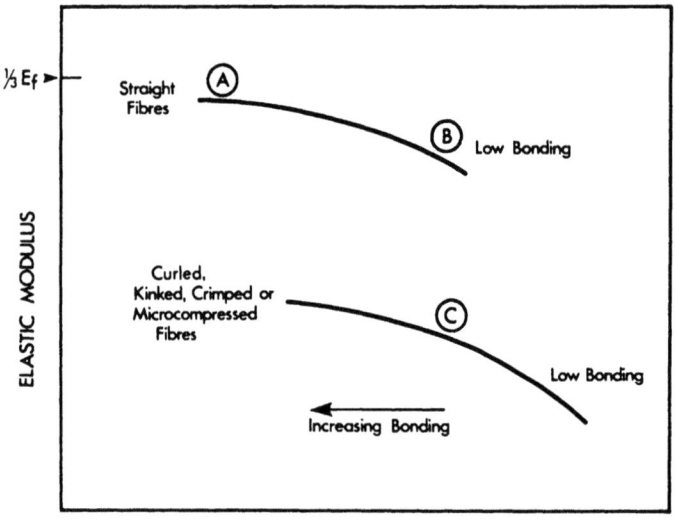

Fig. 10. A model diagram showing the effect of fibre properties and papermaking conditions on the elastic modulus of paper.

None of these concepts are individually new, and the present authors owe much to their proponents. However no previous authors have demonstrated that all three mechanisms are necessary for a description of the modulus of paper over the normal range of conditions found in practical papermaking.

Cox[1], who was responsible for the relation between fibre and sheet modulus, and for the effect of fibre ends, had no knowledge of the importance of curl and crimp.

Kallmes et al.[7] drew attention to the effect of curl in fibres and followed the lead of Giertz[3] and Lobben[8] in expressing the importance of shrinkage in tightening the structure. In their enthusiasm to promote this hypothesis however these authors ignored the effect of fibre ends and indeed specifically denied that relative bonded area, per se, has any effect on modulus.

Recently, Hollmark, Andersson and Perkins[9], in discussing the mechanical properties of low basis weight, low density sheets incorporated the modulus of the fibres and the effect of fibre ends. They also pointed out the difficulty of explaining the low modulus of freely dried sheets without introducing the concept of curl. Their supporting data are, however, very sparse.

It seems that all three mechanisms are important in controlling the modulus of paper in the range of conditions usually encountered in papermaking. Moreover, depending on the circumstances, different factors become dominant.

For example, for the pulps of Figure 2, 3 and 4, prepared at the standard wet-pressing pressure of 345 kPa, it is evident that the low elastic modulus of the unbeaten sheets is caused both by curl, crimps and kinks, and by inadequate bonding. Beating has two effects which are of equal importance. The curl, crimps and kinks are removed, and the bonding level is raised to the point where the effect of load transfer is small.

The low modulus of mechanical pulp sheets at the standard wet-pressing pressure is also limited by both effects (Figure 8). By contrast, the curlated pulp of Figure 7 is adequately bonded and the main cause of low modulus is an induced curl and crimp.

In this way the theory can be utilized, in practice, to determine the cause of low modulus, and the steps necessary to modify the situation are made evident.

11. CONCLUSION

There is a strong qualitative case for the views that the modulus of paper is controlled by fibre modulus, by the effect of load transfer from fibre to fibre, and by the presence of curl and kinks in fibre segments.

In subsequent reports, this case will be strengthened by the derivation of mathematical equations for modulus that are supported from extensive experimental data.

12. EXPERIMENTAL

Standard procedures of the Technical Section of the CPPA were generally followed for specimen preparation and testing. The latency of the mechanical pulps was removed by a standard hot disintegration treatment[10]. Except for the wet-pressing pressure which was varied, handsheets of 60 g/m^2 basis weight were made following the standard procedure. Wet-pressing was carried out in a laboratory press capable of reaching 14 MPa pressure on the sheet. The wet-pressing pressures generally ranged from 68.9 kPa to 12.4 MPa.

The scattering coefficient was determined from the reflectance measurements made on sheets at 681 nm wavelength with the Zeiss Elrepho photometer. The elastic modulus was determined from load-elongation curves obtained on 10 cm long, 1.5 cm wide specimens with a floor model Instron universal testing instrument. All measurements were made at 22°C temperature and 50% relative humidity.

REFERENCES

1. Cox, H.L., Brit. J. Appl. Phys. 3, 72 (1952).
2. Page, D.H., El-Hosseiny, F., Winkler, K. and Lancaster, A.P.S., Tappi 60(4): 114 (1977).
3. Giertz, H.W., Proceedings of EUCEPA/European Tappi Conference, Venice, 1964.
4. Page, D.H., Tydemen, P.A., In "Formation and Structure of Paper", (F. Bolam, Ed.), Tech. Sect. B.P. & B.M.A., London, 1962, p. 397.
5. Hill, M.S., Edwards, J. and Beath, L.R., Tappi 33(1): 36 (1950).
6. Beath, L.R., Neill, M.T. and Masse, F.H., Pulp Paper Mag. Can. 67(10): T423 (1966).
7. Kallmes, O.J., Bernier, G. and Perez, M., Paper Tech. Ind. 18(7): 222 (1977), 18(8): 243 (1977), 18(9): 283 (1977), and 18(10): 328 (1977).
8. Lobben, T.H., Norsk Skogind. 29(12): 311 (1975), and 30(3): 43 (1976).
9. Hollmark, H., Andersson, H. and Perkins, R.W., Tappi 61(9): 69 (1978).
10. Skeet, C.W. and Allan, R.S., Pulp Paper Mag. Can. 69(4): T222 (1968).

AN INTRODUCTION IN THE MECHANICS OF FIBER-REINFORCED COMPOSITES

P.S. Theocaris and G.C. Papanicolaou

Department of Theoretical and Applied Mechanics,
The National Technical University of Athens, Greece

ABSTRACT. The effect of the boundary interphase on the thermomechanical behaviour of Composites reinforced with short fibers is presented. The lectures include the following sections: 1. Introduction; 2. Classification of Problems in the Mechanics of Composite Materials; 3. Theories on the Longitudinal stress distribution along the fiber-matrix interface; 4. The Concept of the Boundary Interphase; 5. The problem of crack propagation in composite materials.

1. INTRODUCTION

A composite material is defined to be any material consisting of two or more distinct *constituents* or *phases* that are insoluble in one another. The main types of reinforcement are *particles, discontinuous fibers, continuous fibers* (or filaments) and *flakes*. By the term *"particulate composites"* we are referred to composites reinforced by particles having all dimensions of the same order of magnitude. Particulate composites are produced from a polymeric matrix into which a suitable metal powder has been dispersed and exhibit highly improved mechanical properties, better electrical and thermal conductivity than either phase, lower thermal expansivity, and improved dimensional stability and behaviour at elevated temperatures. Machine parts, structural components, etc., may be manufactured from such materials.

Epoxy resins are suitable matrices for this class of materials. This is due not only to their general properties such as linear mechanical behaviour, transparency, etc., but also to the possibility of modification of their mechanical and optical properties in a very wide range by using suitable modifiers. Their rheological behaviour [1] as well as their dynamic properties have been extensively investigated [2].

On the other hand, fiber reinforcement has the following advantages [3]: a) the composite becomes very strong; b) the matrix can be reinforced in the required direction by the design which ensures the maximum utilization of the fiber properties. However, a disadvantage of this kind of reinforcement is that the fibers are able to transmit the load only in the directions of their axis and there is no strengthening effect in the direction perpendicular to the axis, and in some cases even weakening may occur. The reinforcing fibers used must have the following properties: high melting point, low specific gravity, high strength over the whole range of working temperatures, minimum solubility in the matrix, high chemical stability, absence of phase transformations over the whole range of working temperatures, and nontoxicity during production and service.

The role of the matrix is to protect the filler from corrosive action of the enviroment and to ensure interactions between the fibers by mechanical and other effects. The mechanical properties of fiber composites are very dependent on the mutual position of the fibers in the monolithic materials. In the case of composites reinforced with fibers the deformation of the matrix is used to transfer stress by means of shear tractions at the fiber-matrix interface to the embedded high-strength fibers. On the other hand, fibers retard the propagation of cracks and thus produce a material of high strength.

2. CLASSIFICATION OF PROBLEMS IN THE MECHANICS OF COMPOSITE MATERIALS

The characterization of a composite system usually requires a great number of parameters [4]. Referring specifically to a two-phase system, by which a filler is dispersed in a matrix, the physical behaviour of the system depends on the geometrical, topological and physical properties of the dispersed phase. Problems of interactions at the interfaces between the different constituents are clearly of great importance in the design and fabrication of composite materials. One class of problems of this kind concerns the mechanisms by which load is transferred from the matrix to the fiber. Clearly, if the composite is to be effective the major proportion of the load must be supported by the fiber rather than by the matrix. However the load is usually applied mainly to the matrix, and so must be transferred to the fiber through adhesion or friction at the matrix-interface. This gives rise to complex stress and strain distributions, with length scale of the order of the fiber diameter, in both the fiber and the matrix.

In most theoretical models describing the mechanical behaviour of composite systems this adhesion is considered as perfect, i.e. the interface can ensure continuity of stresses and displacements. However, such a condition is hardly fulfilled in real composites.

In reality, around an inclusion embedded in a matrix a rather
complex situation develops, consisting of areas of imperfect
bonding, mechanical stresses due to shrinkage [5], high stress gradients or even stress singularities due to the geometry of the
inclusion [6], voids, microcracks etc. Moreover, the interaction
of the surface of the filler with the matrix material is usually
something much more complicated than a simple mechanical effect.
The presence of the filler actually restricts the segmental and
molecular mobility of the polymeric matrix, as adsorption interaction in polymer surface layers into filler particles occurs [7].
It is obvious that under these conditions, the quality of adhesion
can hardly be quantified and a more thorough investigation is
necessary.

Another large set of problems is associated with the relation
of the properties of the composite to the individual properties of
the fiber and the matrix. The problem of the prediction of elastic
moduli of macroscopically isotropic composites has been treated by
bounding techniques, using variational principles of the theory of
elasticity. Methods suitable for arbitrary phase geometry have been
given by Paul [9] and Hashin and Shtrikman [8] and for specified
(spherical inclusions) phase geometry by Hashin [10].

Finally, another set of problems concerns the macroscopic behaviour of the composite as a whole under specified loading.

In the present lecture we shall pay special attention to the
problem of interaction at the interfaces between the different
constituents of composites with linearly oriented fibrous fillers.
As a first step we shall present the existing theories concerning
this problem.

3. THEORIES ON THE LONGITUDINAL STRESS DISTRIBUTION ALONG THE FIBER-MATRIX INTERFACE

One of the first considerations has been developed by Cox [11]. He
considered an elastic fiber of length ℓ embedded in an elastic
matrix under a general strain ε. He made the assumptions of perfect
bonding between the two phases, as well as of equal lateral contraction of the fiber and the matrix. From the last assumption follows that there is no load transfer through the ends of the fiber.

He also defined the load transfer from the matrix to the fiber
as,

$$\frac{dP}{dx} = c(u_f - u_m) \qquad (1)$$

where P is the load in the fiber, u_f and u_m the axial displacements
in the fiber and matrix respectively and x the distance from the
end of the fiber, while c is a constant. From equation (1) follows
that:

$$\frac{d^2P}{dx^2} = c\left(\frac{du_f}{dx} + \frac{du_m}{dx}\right) \tag{2}$$

Moreover, the strain in the fiber can be expressed in terms of the difference in moduli $E=E_f-E_m$.

$$\frac{du_f}{dx} = \frac{P}{AE} \tag{3}$$

$$\frac{du_m}{dx} = \varepsilon \tag{4}$$

where A is the cross sectional area of the fiber.
Substituting in eqn.(2),

$$\frac{d^2P}{dx^2} = c\left(\frac{P}{AE} + \varepsilon\right) \tag{5}$$

By solving the differential equation (5) we obtain an expression of the type:

$$P = EA\varepsilon + c_1 \sinh\eta x + c_2 \cosh\eta x \tag{6}$$

where:

$$\eta = \left(\frac{c}{AE}\right)^{\frac{1}{2}} \tag{7}$$

and c_1, c_2 constants.
Next, we can evaluate the constants c_1 and c_2 from the following conditions:

$P = 0$ at $x = 0$ and $x = \ell$, then:

$$P = (E_f-E_m)A\varepsilon\left[1 - \frac{\cosh\eta(\frac{\ell}{2}-x)}{\cosh\eta\frac{\ell}{2}}\right] \text{ for } 0<x<\ell/2 \tag{8}$$

The final result from the above analysis is:

$$\sigma_f = \frac{P}{A} = \frac{(E_f-E_m)\sigma_c}{E_m}\left[1 - \frac{\cosh\eta(\frac{\ell}{2}-x)}{\cosh\eta\frac{\ell}{2}}\right] \tag{9}$$

The tensile stresses in the fibers when $x = \frac{\ell_c}{2}$ have the maximum value

$$\sigma_{fmax} = \varepsilon E_f \text{ at } x = \frac{\ell_c}{2} \tag{10}$$

where ℓ_c is the "*critical length*" of the fiber.

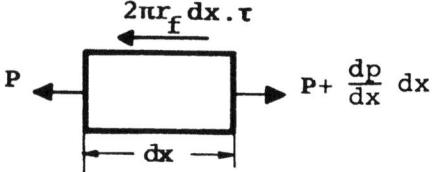

Fig.1. Forces acting on an element of a fiber

If we consider the equilibrium of the forces acting on an element of a fiber (Fig.1), then we can evaluate the shearing stresses in the fiber-matrix interface:

$$\frac{dP}{dx}dx = 2\pi r_f dx\, \tau \tag{11}$$

where

$$\tau = \frac{1}{2\pi r_f}\frac{dP}{dx} \tag{12}$$

From (8) and (12) we have the final result for the shearing stresses:

$$\tau = \frac{(E_f - E_m)A\,\varepsilon}{2\pi r_f}\,\frac{k\sinh\eta(\frac{\ell}{2}-x)}{\cosh\eta\frac{\ell}{2}} \tag{13}$$

The distibution of both the tensile and shearing stresses is shown in Fig 2.

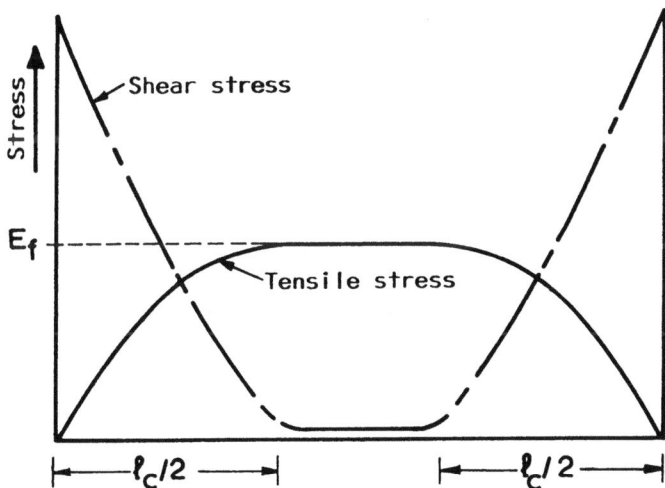

Fig.2. Distribution of tensile and shearing stresses

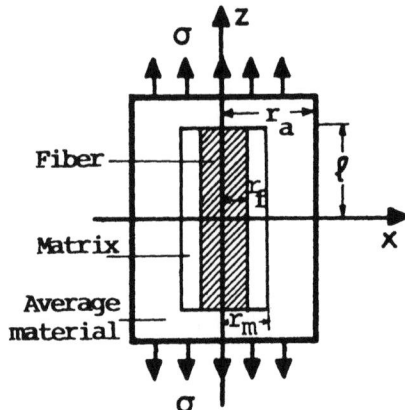

Fig.3. The model used by Rosen

Rosen [12] analyzed the shear-stress field interaction and force transfer between parallel fibers in a composite loaded in tension along the fibers in an approximate but useful manner. The model used is shown in Fig.3 and consists of a fiber surrounded by a matrix which in turn is imbedded within a composite material exhibiting "average composite properties". The fiber is assumed to carry only extension and the matrix to transmit only shear stresses. No stress is transmitted axially from the fiber end to the average material. Shear stresses in the average material are considered to decay in a negligible distance from the inclusion interface. For equilibrium, it is valid that:

$$\tau + (r_f/2) d\sigma_f/dz = 0 \qquad (14)$$

where τ is the shear stress on the cylindrical surface of the fiber and σ_f is the tensile stress in the fiber (Fig.4).
For equilibrium in the axial direction:

$$\sigma_f (r_f/r_a)^2 + \sigma_a [(r_a^2 - r_m^2)/r_a^2] = \sigma \qquad (15)$$

where σ_a = axial stress in average material, and σ = applied axial stress.

In this analysis the axial strains in the matrix are considered negligible (E_m=0), however, the shear strain γ_m is determined from the displacements on the outer surface of the fiber and the inner surface on the average material:

$$u_a - u_f = (r_m - r_f) \gamma_m \qquad (16)$$

Differentiating Eq.(16) twice and using the stress-strain relations

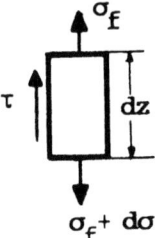

Fig.4. Stresses in a fiber element

yields:

$$\frac{1}{E_a}\frac{d\sigma_a}{dz} - \frac{1}{E_f}\frac{d\sigma_f}{dz} = \frac{r_m - r_f}{G_m}\frac{d^2\tau}{dz^2} \tag{17}$$

Next, differentiating Eq.(15) and substituting the result into (14) yields:

$$\frac{d^2\tau}{dz^2} - \eta^2\tau = 0 \tag{18}$$

where:

$$\eta^2 = \frac{2G_m}{E_f(r_m - r_f)r_f}\left[1 + \frac{E_f}{E_a}\left(\frac{r_f^2}{r_a^2 - r_m^2}\right)\right]$$

The solution of Eq.18 is of the form:

$$\tau = A\sinh\eta z + B\cosh\eta z$$

The constants A and B can be evaluated from the boundary conditions:

$$\tau(0) = 0$$

$$\sigma_f(\ell) = 0$$

Therefore

$$B = 0$$

$$A = \frac{G_m \sigma r_a^2}{\eta E_a(r_m - r_f)(r_a^2 - r_m^2)\cosh\eta\ell}$$

Hence:

$$\tau = \frac{G_m \sigma r_a^2 \sinh\eta z}{\eta E_a (r_m - r_f)(r_a^2 - r_f^2)\cosh\eta\ell} \tag{19}$$

From Eqs.(14) and (19) it can be deduced that:

$$\sigma_f = -\frac{\sigma r_a^2 E_f}{[E_a(r_a^2 - r_m^2) \pm E_f r_f^2]} \left(\frac{\cosh\eta z}{\cosh\eta\ell} - 1\right) \tag{20}$$

If $r_a \gg r_m$, then:

$$\eta^2 \approx 2G_m / r_f E_f (r_m - r_f) \tag{21}$$

and:

$$\sigma_f = -\frac{\sigma E_f}{E_a}\left[\frac{\cosh\eta z}{\cosh\eta\ell} - 1\right] \tag{22}$$

The maximum axial stress is:

$$\sigma_{fmax} = \sigma_f(0)|_{\ell\to\infty} = \sigma E_f / E_a \tag{23}$$

The stress at a point at distance δ from a fiber end is:

$$\sigma_f(\ell-\delta) = -\frac{\sigma E_f}{E_a}\left[\frac{\cosh\eta(\ell-\delta)}{\cosh\eta\ell} - 1\right] \tag{24}$$

The ratio of the stress at a point at distance δ from the fiber ends to the stress in the center of a continuous fiber is given by:

$$\Phi = \frac{\sigma_f(\ell-\delta)}{\sigma_f(0)|_{\ell\to\infty}} = 1 - \cosh\eta\delta + \tanh\eta\ell \cdot \sinh\eta\delta \tag{25}$$

For large ℓ:

$$\tanh\eta\ell = 1$$

Therefore:

$$\Phi = 1 - \cosh\eta\delta + (\cosh^2\eta\delta - 1)^{\frac{1}{2}} \tag{26}$$

from which:

$$\cosh\eta\delta = [1 + (1-\Phi)^2]/2(1-\Phi) \tag{27}$$

and:

$$\frac{\delta}{\delta_f} = \frac{1}{2}\left[(v_f^{-\frac{1}{2}}-1)\frac{E_f}{G_m}\right]^{\frac{1}{2}} \cosh^{-1}\left[\frac{1+(1-\Phi)^2}{2(1-\Phi)}\right] \tag{28}$$

At this point of the analysis it is useful to define an *"ineffective fiber length"*, by the condition that the elastic stress ratio, σ_f/σ_{fmax} is equal to Φ. Thus, Eq.(28) gives this length δ at which the stress in the fiber, measured from the end, has reached a value $\Phi\,\sigma_{fmax}$ ($\Phi<1$). Rosen points out that the choice of a value of Φ is somewhat arbitrary and suggests that $\Phi=0.9$ appears suitable and useful. The *"ineffective fiber length ratio"* δ/d_f indicates the length over which force is transferred from the matrix, to the fiber. On the other hand, *"effective length"* is that portion of the fiber in which the average axial stress is greater than 90% of the stress which would exist for infinite fibers.

A more general case is obtained when the load is applied to both fiber and matrix. This has been evaluated by Dow [13]. The model used in this theory was similar to that considered by Cox, except that no matrix was present at the end of the fiber. He also assumed a perfect bonding between the fiber and the matrix as well as straight lines before deformation remain straight after deformation. According to his theory, the following expressions are valid:

$$\sigma_f = \left(\frac{Peff}{A_f + \frac{A_m E_m}{E_f}}\right) \cdot \left(1 - \frac{\cosh\frac{\lambda z}{d_f}}{\cosh\frac{\lambda \ell}{d_f}}\right) \tag{29}$$

$$\tau = \frac{\lambda}{4}\left(\frac{Peff}{A_f + \frac{A_m E_m}{E_f}}\right) \cdot \frac{\sinh\frac{\lambda z}{d_f}}{\cosh\frac{\lambda \ell}{d_f}} \tag{30}$$

where:

$$Peff = P_m - \frac{A_m E_m P_f}{A_f E_f} \tag{31}$$

is the load producing shear deformation

and:

$$\lambda^2 = 4 \cdot W = \text{constant}$$

where:

$$W = \frac{2\sqrt{2}\left[\frac{G_f}{E_f}\right] \cdot \left[1 + \frac{A_f E_f}{A_m E_m}\right]}{(\sqrt{2}-1) + \frac{G_f}{G_m}\left\{\left(\frac{A_m}{A_f} + 2\right)^{\frac{1}{2}} - 2^{\frac{1}{2}}\right\}} = \text{const.} \qquad (32)$$

It is evident that in all theoretical models describing the thermomechanical, as well the mechanical properties of composite systems, a perfect adhesion between the main phases is considered. However, there is a region between fiber and matrix which contains both areas of adsorption interaction in polymer surface layers into fibers, as well as an area of mechanical imperfections. This is what we call *"interphase"*. We shall now consider the effect of this interphase on the thermomechanical behaviour of composites reinforced with unidirectional short fibers.

4. THE CONCEPT OF THE BOUNDARY INTERPHASE [14,15]

To aid our analysis a model similar to Rosen's model is considered (Fig 5). In this model, the fiber is surrounded by the interphase and this in turn is surrounded by the matrix. The difference from Rosen's model is that in our model we consider the interphase material and this is a better approximation to the real situation that exists in composite materials. On the other hand we make the following assumptions:

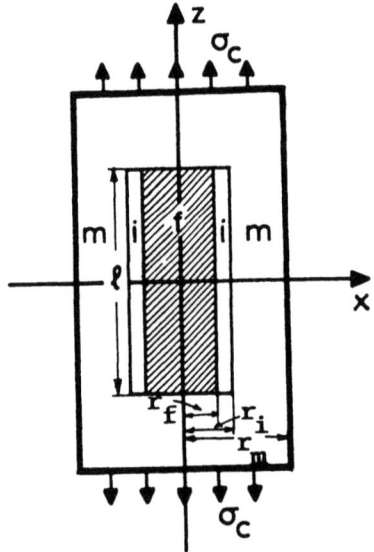

Fig.5. Basic model considered in the theory.

(i) Perfect bonding exists at all interfaces.
(ii) Fiber and matrix material only carry tensile stresses.
(iii) Interphase only carries shear stresses.
(iv) The interphase is homogeneous and isotropic material.
(v) Both matrix and fibers are homogeneous isotropic materials.
(vi) If perfect adhesion between the two main phases of the composite exists, it is expected that the thickness of the interphase vanishes, and the interphase layer tends to an interface.

Next, we consider that the model presented in Fig 5 is subject to a uniform temperature rise ΔT and on the same time to an external stress σ_c parallel to the direction of the fibers. The equilibrium of forces for an element of the model length dz is expressed by:

$$\frac{dN_f}{dz} = 2\pi r_f \frac{(u_f - u_m)}{(r_i - r_f)} \cdot G_i \tag{33}$$

where N_f the tensile force in the fiber, u the displacement, r the radius, and G is the shear modulus. Indices m, f and i denote matrix, fiber and interphase respectively.

According to Hooke's law it is valid that:

$$N_f = \frac{du_f}{dz} \cdot E_f \cdot \pi r_f^2 \tag{34}$$

or:

$$\frac{dN_f}{dz} = \frac{d^2 u_f}{dz^2} \cdot E_f \cdot \pi r_f^2 \tag{35}$$

From Eqns. (34) and (35) we derive the following differential equation:

$$\frac{d^2 u_f}{dz^2} - \eta^2 (u_f - u_m) = 0 \tag{36}$$

where:

$$\eta^2 = \frac{2G_i}{E_f \cdot r_f (r_i - r_f)} \tag{37}$$

By solving the differential equation (36), we obtain the following expression:

$$u_f = u_m \pm A \sinh \eta z + B \cosh \eta z \tag{38}$$

In order to determine the constants A and B we make use of the following boundary conditions:

For $z = \pm \ell/2 \Rightarrow \dfrac{du_f}{dz} = \varepsilon_f^T$ \hfill (38.1)

where ε_f^T is the thermal strain developed in the fiber.

Whence:

$$A = \dfrac{1}{\eta \cosh\dfrac{\eta\ell}{2}} \left[\varepsilon_f^T - \dfrac{du_m}{dz}\right] \quad (39.1)$$

$$B = 0 \quad (39.2)$$

From Eqns. (38) and (39) we can write that:

$$\dfrac{du_f}{dz} = \dfrac{du_m}{dz} + A\eta\cosh\eta z \quad (40)$$

and from Hooke's law:

$$\sigma_f = E_f \dfrac{du_f}{dz} \quad (41)$$

From Eqns. (39.1), (40) and (41) we derive the following expression for the tensile stress in the fiber:

$$\sigma_f = E_f \cdot \left[\dfrac{du_m}{dz}\left(1 - \dfrac{\cosh\eta z}{\cosh\dfrac{\eta\ell}{2}}\right) + \varepsilon_f^T \dfrac{\cosh\eta z}{\cosh\dfrac{\eta\ell}{2}}\right] \quad (42)$$

The variation of shear stress (τ) along the interphase is obtained by considering the equilibrium of the forces acting on an element of the fiber, that is:

$$\tau = -\dfrac{1}{2\pi r_f} \dfrac{dN_f}{dz} = -\dfrac{G_1}{(r_i - r_f)} A \sinh\eta z \quad (43)$$

or:

$$\tau = -\dfrac{G_1}{r_i - r_f} \cdot \dfrac{\sinh\eta z}{\eta\cosh\dfrac{\eta\ell}{2}} \cdot \left[\varepsilon_f^T - \dfrac{du_m}{dz}\right] \quad (44)$$

In polymeric composites the modulus of elasticity of the matrix depends on the temperature. This variation for the case of an epoxy resin is given in Fig.6.

Taking into account this variation, we can consider that the strain of the matrix consists of two terms, where the first term depends on the mechanical loading, as well as on the temperature indirectly and the second one is thermal strain of the matrix:

$$\varepsilon_m = \dfrac{\sigma_m}{E_m(T)} + \varepsilon_m(T) \quad (45)$$

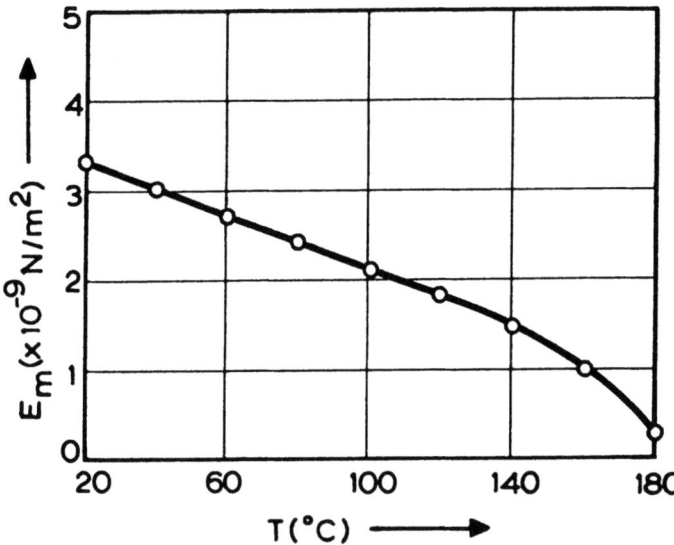

Fig.6. Modulus of elasticity of the epoxy-matrix plotted against temperature.

Differentiating Eqn.(45) with respect to T we obtain:

$$\frac{du_m}{dz} = \frac{\sigma_m}{E_m(T)}\left[1 - \frac{1}{E_m(T)} \cdot \frac{dE_m(T)}{dT} \cdot \Delta T\right] + \alpha_m \Delta T \tag{46}$$

where α_m expresses the linear thermal-expansion coefficient of the matrix. Equilibrium of forces in the direction of the applied uniaxial tensile stress σ_c gives:

$$\sigma_m v_m + \sigma_f v_f = \sigma_c \tag{47}$$

From Eqns (42), (46) and (47) we derive the equation:

$$\sigma_m = \frac{\sigma_c - E_f v_f \left[\alpha_m \Delta T\left(1 - \frac{\cosh\eta z}{\cosh\eta\ell/2}\right) + \epsilon_f^T\left(\frac{\cosh\eta z}{\cosh\eta\ell/2}\right)\right]}{v_m + v_f \frac{E_f}{E_m(T)}\left(1 - \frac{1}{E_m(T)} \frac{dE_m(T)}{dT} \Delta T\right)\left(1 - \frac{\cosh\eta z}{\cosh\eta\ell/2}\right)} \tag{48}$$

where v_m is the real volume fraction of the fiber; i.e. it is valid that:

$$v_m = 1-v_f-v_i \tag{49}$$

For a given composite it is possible to calculate the values of v_i, α_i, G_i, r_i and η by means of a theory which has been developed in previous publications [14,15].

Next, in order to verify the results of this theory we shall present some applications with a real composite consisting of an epoxy matrix reinforced with glass-fibers of finite length. The properties of the constituent materials are given in table I. The mechanical and thermal behaviour of the matrix were also investigated previously [16-20].

TABLE I

Properties of the constituent materials.

	E-glass	Matrix	Interphase.
Young's modulus	69.9×10^9	3.2×10^9	45.8×10^9
Poisson's ratio	0.20	0.35	0.29
Thermal expansion coefficient, $°C^{-1}$			
$\alpha_1 (T<T_g)$	5×10^{-6}	52.5×10^{-6}	2.4×10^{-6}
$\alpha_2 (T>T_g)$	-	109.0×10^{-6}	

From Fig.6 we can see that as the temperature of the matrix is increasing, we have a decrease of E_m. This variation of E_m is accompanied by an increase of the strain in the matrix, as the composite is loaded by an external load, while in the same time is heated (Fig 7).

The tensile stress in the fiber is increasing with temperature (Fig 8), while in the matrix has an opposite dependence on the temperature (Fig 9). On the other hand, the tensile stress in the fiber builds up to a maximum value, which remains constant through the central portions of the fiber and then decays rapidly at the ends of the fiber (Fig.10). Conversely, the tensile stress in the matrix has an opposite variation along the fiber length (Fig 11). A similar variation appears also for the strain developed in the matrix (Fig 12). Finally, the shear stress at the interphase peaks to a maximum value near the fiber ends and then decays rapidly toward the center of the fiber (Fig 13).

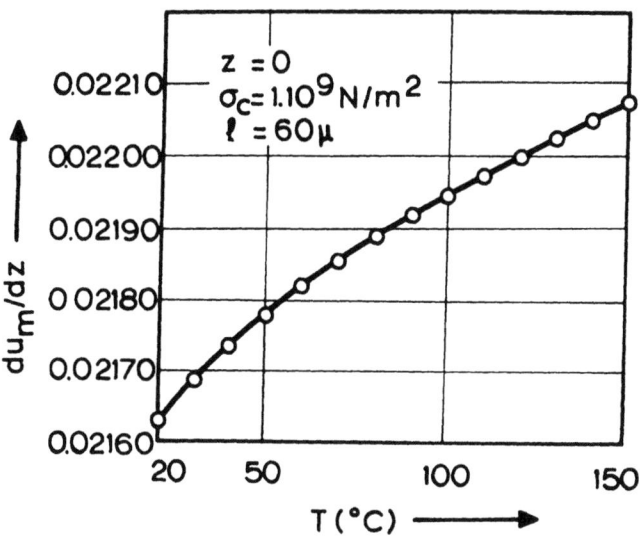

Fig.7. Longitudinal strain in the matrix plotted against temperature.

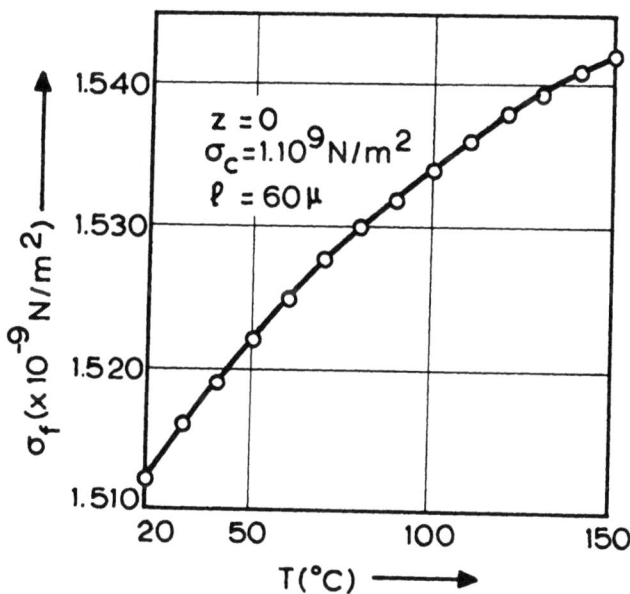

Fig.8. Tensile stress in the fiber plotted against temperature.

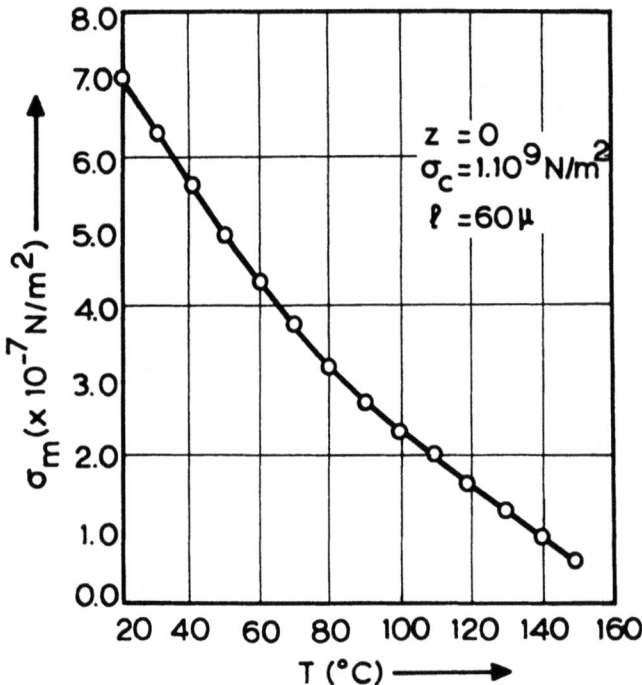

Fig.9. Tensile stress in the matrix plotted against temperature.

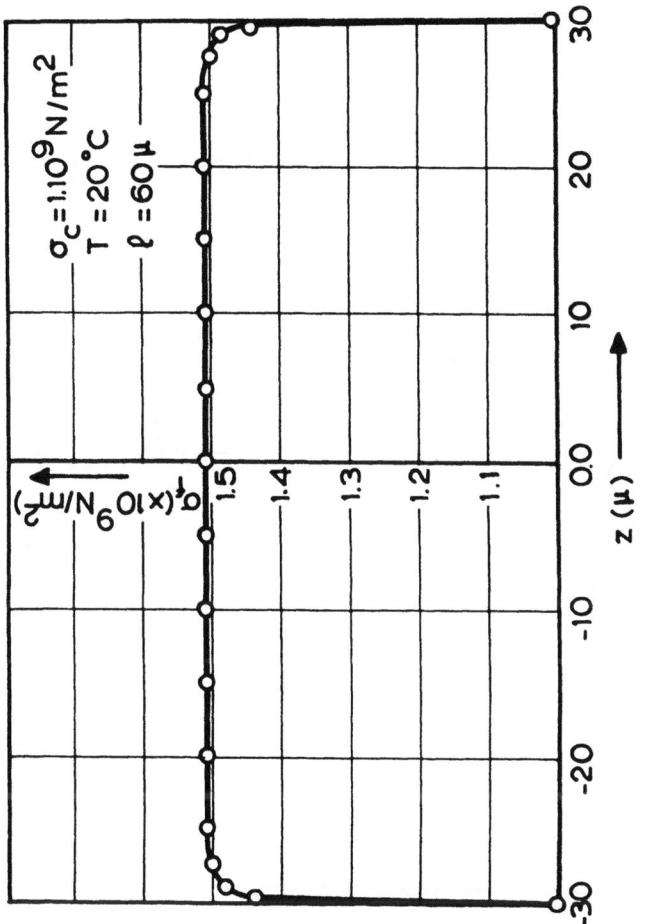

Fig.10. Distirbution of the tensile stress in the fiber.

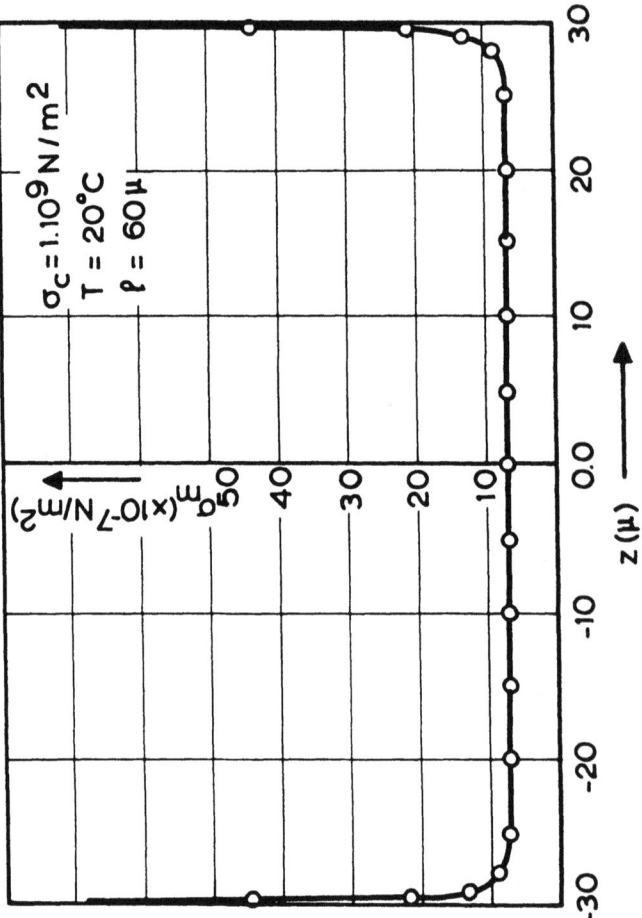

Fig.11. Distribution of the tensile stress in the matrix along the fiber length.

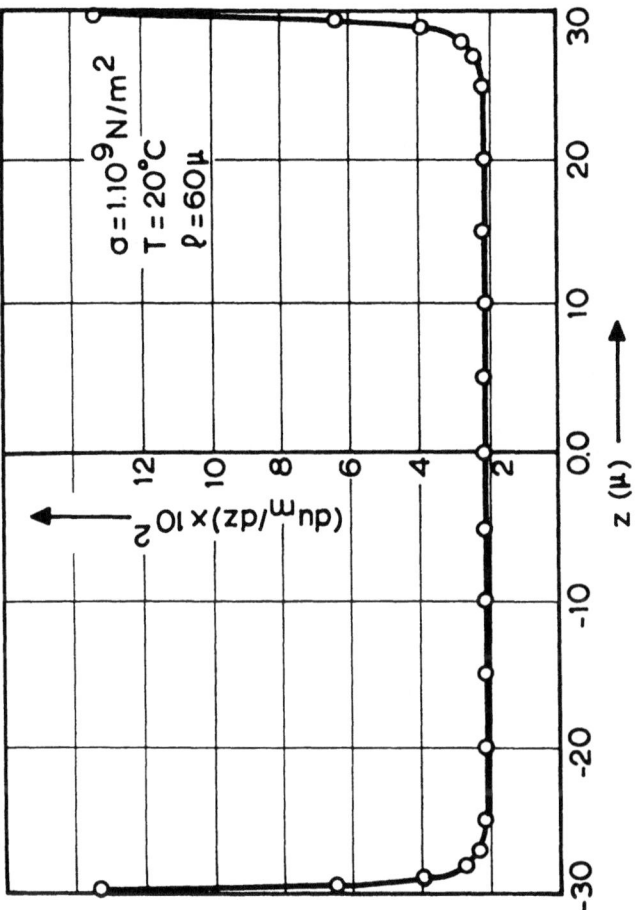

Fig.12. Distribution of the tensile strain in the matrix along the fiber length.

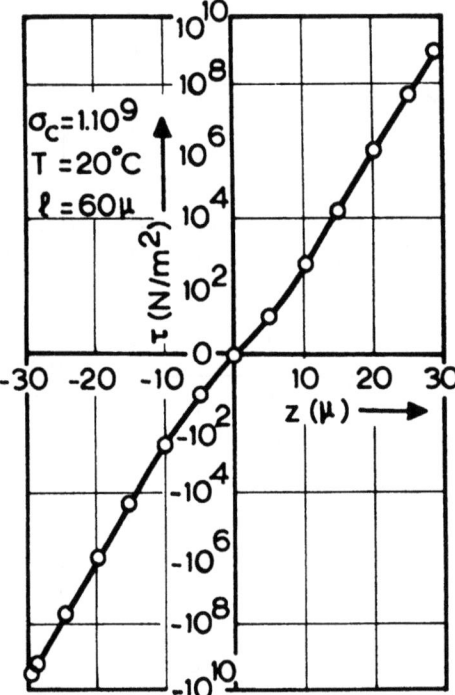

Fig.13. Distribution of the shear stress in the interphase material along the fiber length.

5. THE PROBLEM OF CRACK PROPAGATION IN COMPOSITE MATERIALS

The study of fracture mechanisms in composite materials is a difficult task. By virtue of the specific nature of the structure of composite materials (presence of components markedly differing in physical and mechanical properties) their complete strengthening depends on many factors. The main ones are the combination of the elastic characteristics E,G and ν of the inclusion and the matrix, the size, orientation, distribution, and volume fraction of fiber in the matrix, fabrication processes in which incomplete resin impregnation leaves elongated air voids between the fibers, poor wetting and bonding of the matrix to the fibers which allow an early separation of phases at low stresses, early fiber fractures causing high shear stresses at the broken fiber ends. These stresses induce fiber-matrix separation, moisture-induced fiber-matrix bond degradation, or any combination of these factors.

Moreover, the presence of "hard" inclusions generates stress concentrations, localized primarily at the interface between inclusion and matrix. This effect is related with Poisson's ratio differences between matrix and inclusion. In addition, we can consider the existence of a boundary interphase as a weak surface surrounding each one of the inclusions, and consisting from relatively weak matrix material containing microcracks, voids and other singularities and which under the presence of local stress concentrations can cause complex modes of internal nonelastic behaviour, such as plasticity and fracture, as well as debonding. The resulting failure zones are large in size and disturb the stress distribution at large distances from the crack tip.

When a fiber fracture occurs, there are several possibilities for the subsequent behaviour of the composite. First, if the bond between matrix and fibers is not sufficiently strong, the high interface shear stresses may produce the interface failure which could propagate along the length of the fiber reducing the fiber effectiveness over a substantial fiber length. Lamination may propagate without limit and bring about fracture of the whole composite. Second, if the bond between fiber and matrix is sufficiently strong so that it does not break immediately after formation of internal fracture in the fibers, the crack formed may propagate over the matrix in a direction perpendicular to the fiber orientation in accordance with Griffith's mechanism. Griffith assumed that brittle bodies are fractured because of spontaneous propagation of microcracks and their conversion into macrocracks. The cracks propagate spontaneously when the work of the external forces exceeds the energy required for forming a new surface. Finally, there is another possibility of fracture behaviour. According to this behaviour during axial-tension of composites two mechanisms of stress transmission exist in the reinforced fibers. The first one is an adhesional mechanism acting before the bond between the phases is broken, and the second one is a frictional mechanism after cleavage of this bond.

In order to clarify the effect of the boundary interphase on the crack propagation we shall give some experimental examples. The first experimental work deals with the study of crack propagation in two phase epoxy resin specimens by the method of high speed photography along with the optical method of caustics [21]. In this study it was proved that the interphase plays the role of a "barrier" to the crack propagation. Indeed the crack propagates with a certain maximum velocity in the first phase of the composite and then stops momentarily when it reaches the interphase, thus attaining later, in the second phase, a new maximum velocity.

The maximum velocity and the stress intensity factor in the second phase of the composite specimens are strongly depended on the material characteristics of the first (notched) phase and are also highly influenced by the crack arrest process itself. The crack propagation and the stress field concentrations at the crack tip in the first phase of the composite specimens is mainly independent of the material characteristics of the second phase. When the phase I (slitted) is composed out of a more ductile material than phase II, then an increased stress intensification in phase II of the composite occurs, which results in increased values of the crack propagation velocity and the stress intensity factor in phase II. This increased stress intensification results also in a diminishing of the crack arrest time at the interface and in increased values of the shear stresses at this area. The increased shear stresses at the interphase create local deviations of the propagating crack from the vertical direction as it intersects the interphase and even in crack bifurcations at this area.

The opposite effects were determined for the case of specimens with increased brittleness of phase I relative to phase II. Fig. 14 shows a typical case for the path of the crack through the composite specimen. (Phase I not plasticized, phase II 40% plasticized). For the inverse arrangement when the crack starts from the ductile material and enters later the more brittle one, the crack deviates from its original direction, as it enters phase II of the composite, thus forming an angle of about $60°-70°$ with the interface. After entering the second phase and at relatively short distance from the interface it takes again the initial vertical to the interface direction (Fig 15).

Finally, Fig 16 shows the fracture process of a composite specimen (phase I with 30% of plasticizer, phase II with a pure epoxy polymer). In this case where the bond between the two phases is not perfect, the unique crack splits into two branches starting at the interface and propagates along it. The two branches, after meeting some points of a stronger interface bondage, stop to propagate along the interface, they enter in phase II and progressively they are oriented in a direction normal to the interface direction.

We shall now analyse the fracture of a composite system which

Fig.14. Shape of crack for ph.I 20% pl., ph.II 40% pl.

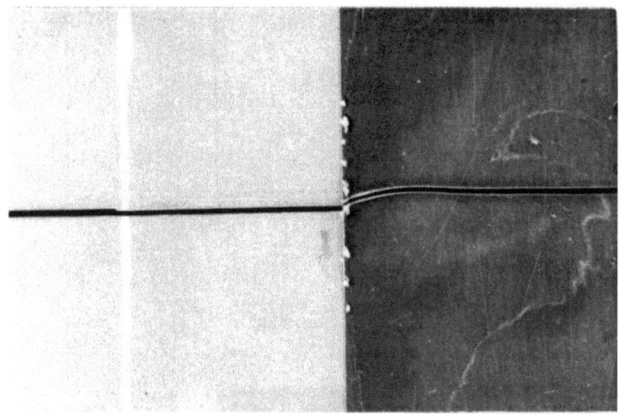

Fig.15. Shape of crack for ph. I 10% pl., ph.II 0% pl.

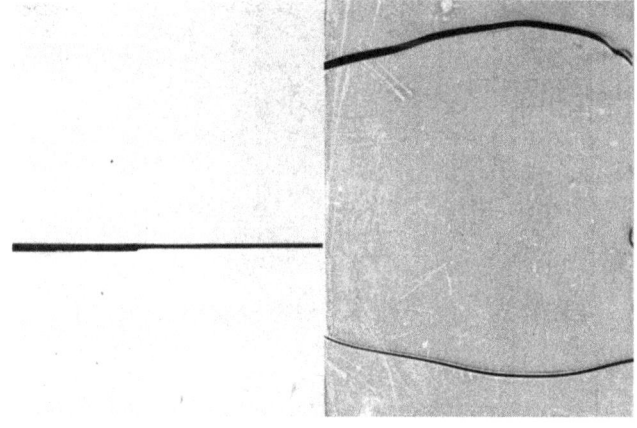

Fig.16. Shape of crack for ph.I 30% pl., ph.II 0% pl.

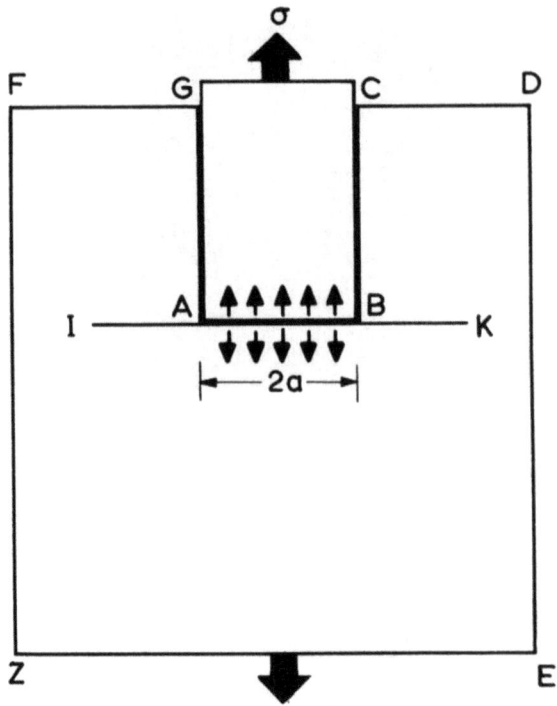

Fig.17a. Composite system subject in tension

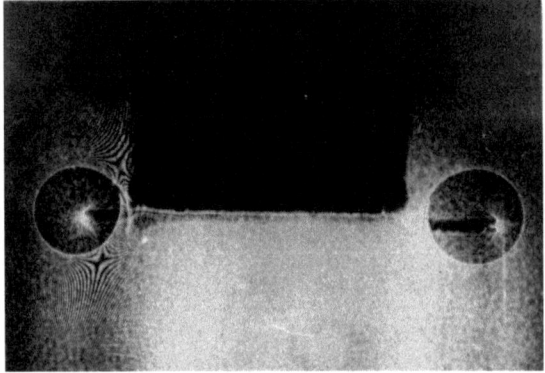

Fig.17b. Caustics created at the tip of the cracks

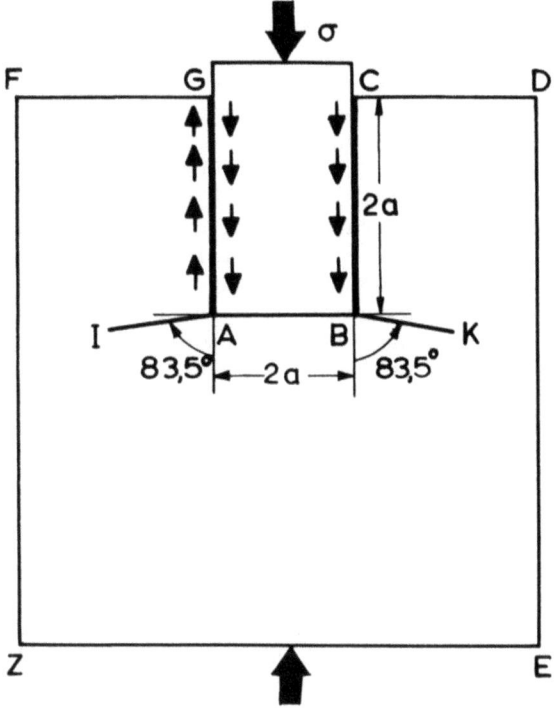

Fig.18a. Composite system subject in compression

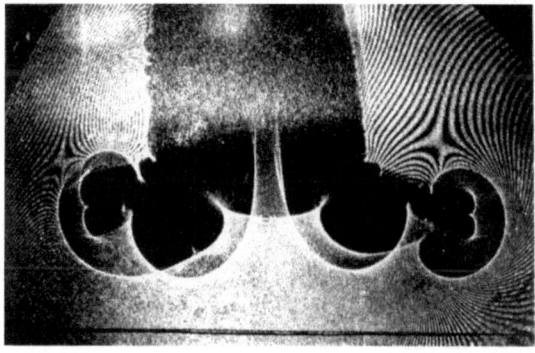

Fig.18b. Caustics created at the tip of the cracks

differ from the case discussed above [22].

In this case a rectangular inclusion is connected by adhesion with a plastic matrix and the whole system is subject in tension (Fig 17a). If the bond between matrix and inclusion is sufficiently strong, the shearing stresses lead to a first separation of the matrix from the inclusion across the boundary line AB and then this initial separation plays the role of an initial crack, which propagates in a direction perpendicular to the stress direction in accordance with Griffith's mechanism. In Fig 17b we can see the caustics created at the tip of the crack during crack propagation in a composite specimen having as matrix material plexiglass and reinforced with an aluminum inclusion of rectangular cross section. The two phases were connected by adhesion with an epoxy resin. These caustics yield all the necessary information for the stress singularity at the crack tip, as well as the value of the complex stress intensity factor there [23]. On the other hand, if the same composite specimen is subjected to compression (Fig 18a) then the crack propagates in a direction forming an angle of $83,5°$ to the stress direction. Fig 18b shows the caustics created at the tip of the crack during crack propagation.

REFERENCES

1. P.S. Theocaris, "Rheologic Behaviour of Epoxy Resins in their Transition Region", Journal of Applied Polymer Science, Vol.8, No.1, pp.399-412, (1964).
2. T. Hirai and D.E. Kline, "Effects of Heat Treatment on Dynamic Mechanical Properties of Nonstoichiometric, Amine-Cured Epoxy Resins" J. Applied Polymer Science, Vol.17, p.31, (1973).
3. I.N. Frantsevich and D.M. Karpinos, *Fibrous Composites*, Translated from Russian, Israel Program for Scientific Translations, Jerusalem, 1972.
4. L. Holliday and J. Robinson, "The thermal expansion of composites used on polymers" Journal of Material Science, Vol.8, pp.301-311, (1973).
5. P.S. Theocaris and S.A. Paipetis, "Constrained Zones at Singular Points of Inclusion Contours" Intern. Jnl. Mech. Sci., Vol.18, No.11-12, pp.581-587, (1976).
6. P.S. Theocaris and S.A. Paipetis, "The State of Stress Around Inhomogeneities by the Method of Caustics", Fiber Science and Technology, Vol.9, No.1, pp.19-39, (1976).
7. Yu. S. Lipatov, V.F. Babich, and V.F. Rosovizky, "Effect of Filler on the Relaxation Time Spectra of Filled Polymers" Journal of Applied Polymer Science, Vol.20, pp.1787-1794, (1976).
8. Z. Hashin and S. Shtrikman, "Note on a Variational Approach to the Theory of Composite Elastic Materials" Journal of the Franklin Institute, Vol.271, p.336, (1961).
9. B. Paul, "Prediction of Elastic Constants of Multiphase Materials" Trans. AIME, Vol.218, p.36, (1960).
10. Z. Hashin, "The Elastic Moduli of Heterogeneous Materials" Jour-

nal of Applied Mechanics, Vol.29, Trans ASME, Vol.84, Series E, p.143, (1962).
11. H.L. Cox, "The elasticity and Strength of Paper and Other Fibrous Materials" Brit. Journal. Applied Phys., Vol.3, p.72, (1952).
12. B.W. Rosen, "Mechanics of Composite Strengthening" Fiber Composite Materials, American Society for Metals, Metals Park, Ohio, pp.37-75, (1965).
13. N.F. Dow, "Study of Stresses Near a Discontinuity in a Filament-Reinforced Composite Metal" General Electric Company Report TIS R63SD61, (1963).
14. G.C. Papanicolaou, S.A. Paipetis, and P.S. Theocaris, "The Concept of the Boundary Interphase in Composite Mechanics" Colloid and Polymer Science, Volume 256, No.7, p.625 (1978).
15. G.C. Papanicolaou and P.S. Theocaris, "Thermal Properties and Volume Fraction of the Boundary interphase in Metal-Filled Epoxies" Colloid and Polymer Science, Vol.257, (1979).
16. G.C. Papanicolaou, S.A. Paipetis and P.S. Theocaris, "Cross-linking Studies in Plasticized Epoxies by Means of Dynamic Measurements" Journal of Applied Polymer Science, Vol.20, p.903, (1976).
17. P.S. Theocaris, S.A. Paipetis and G.C. Papanicolaou, "Indentation Studies in Plasticized Epoxy Polymers" Journal of Appl. Polymer Science, Vol.22, p.1417, (1978).
18. G.C. Papanicolaou, S.A. Paipetis and P.S. Theocaris, "Thermal Properties of Metal-Filled Epoxies" Journal of Applied Polymer Science, Vol.21, p.689, (1977).
19. S.A. Paipetis, G.C. Papanicolaou and P.S. Theocaris, "Dynamic Properties of Metal-Filled Epoxies" Fiber Science and Technology, Vol.8, p.221, (1975).
20. P.S. Theocaris, S.A. Paipetis and G.C. Papanicolaou, "Indentation Studies in Aluminum-Filled Epoxies" Journal of Applied Polymer Science, Vol.22, p.2245, (1978).
21. P.S. Theocaris and J. Milios, "Dynamic Crack Propagation in Composites" To be published in Engineering Fracture Mechanics"
22. J.N. Prassianakis, "General Methods for the Determination of Complex Stress Intensity Factors in Plane Elasticity Problems" Ph.D. Thesis at the Nat. Techn. University of Athens, 1979.
23. P.S. Theocaris, D. Raftopoulos, F. Katsamanis, "Static and Dynamic Stress Intensity Factors bt the Method of Transmitted Caustics", Jnl. Engrg. Mat. and Techn., Trans. ASME, Series H, Vol.99, No.2, pp.105-109, (1977).

MECHANICAL PROPERTIES OF COATED FABRICS

J. Skelton

FRL, An Albany International Company
Dedham, MA 02026

INTRODUCTION

Many of the woven fabrics that are produced for industrial applications are used as substrates in coated fabrics. The fabric in this application functions principally as a load bearing member and as an anchoring site for the coating component. These functions imply a set of requirements that place certain restrictions on the nature of the fabric structure and organization, and these restrictions in turn have an effect on the range of mechanical properties that can be usefully obtained. From a theoretical point of view, the restrictions make possible some significant simplifications in the analytical treatment of the fabric mechanical properties: this paper outlines these simplifications and presents a discussion of some of their consequences.

SPECIAL FEATURES OF SUBSTRATE FABRICS

The original work on cloth geometry was based on the intersection properties of cylindrical, radially undeformable, yet axially flexible yarns [1]. Subsequently, efforts have been made to take into account such factors as yarn flattening, finite bending stiffness, stress relaxation and a variety of weave constructions [2,3,4,5]. As is inevitable with such refinements, each increment of attention to analytical detail leads to an increase in the complexity of the results, with a consequent diminution in their direct usefulness, and it becomes increasingly difficult to relate the mathematics to the underlying physical principles. This is particularly true for a treatment of the stress-strain behavior of woven fabrics under biaxial load, which is a complicated subject for analysis with even the most simple of models [6]. Since many coated fabrics are used under conditions of biaxial loading, it is worthwhile to examine the special features of these fabrics to see what structural assumptions are justified and what simplifications can be made to the analysis.

As a general rule the woven substrate fabric will be required to be as efficient as possible in terms of load bearing capacity and tensile stiffness. This means that low twist, continuous filament yarns will be used. The tearing strength is an important parameter of coated fabrics, since the presence of coating reduces yarn mobility and can decrease the tearing strength of the substrate to marginally acceptable levels. In order to retain as much mobility as possible in the coated state, the substrate fabrics will usually be designed to minimize the number of yarn intersections per unit area, and accordingly, it is common to find large yarns woven in multi-end basketweave constructions. The consequence of these choices is that the aspect ratios of the cross sections of the yarns or yarn bundles is large and the amplitude of the crimp waveform is low. This type of construction minimizes the deviation of the individual filaments from their ideally straight configuration and gives maximum tensile stiffness and strength and tear resistance, all desirable characteristics. The fact that the absolute levels of crimp in the threads are low implies that the differences between the warp and filling tensile responses will be small, which generally is useful since it increases the design versatility of the fabric. Additional benefits of a secondary nature can accrue from this type of construction: the bending stiffness of the fabric is minimized, since the moment of area of the cross-section is low and the amount of coating material needed for coverage is also kept to a minimum, since the interstitial spaces and voids are reduced in size.

SIMPLE MODEL OF SUBSTRATE FABRIC

The load-elongation behavior of a simplified model fabric under uniaxial load will be considered first in order to illustrate the assumptions and the method of analysis, and the extension to biaxial stress will be presented in outline. The center lines of the threads are assumed to lie in circular arcs both before and after deformation. This is not a physically realizable assumption for a system that is deformed by the action of forces, but it does not appear to lead to any major difficulties, and the analytical simplication that is introduced by this means is so enormous that it is certainly justified at the present level of approximation. Examination of cross-sections of fabrics show that the circular arc geometry is a very reasonable model in most cases, and it has been used as the basis for a much more detailed study of biaxial behavior [7].

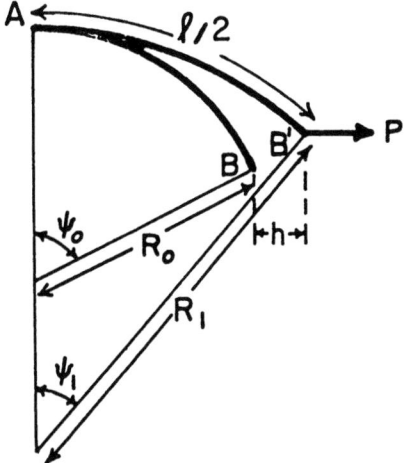

Fig. 1. Circular arc geometry - nomenclature.

The deformation caused by bending may be found from consideration of Fig. 1. The arc AB, of length $\ell/2$, represents the yarn center line, one-half of one of the component circular arcs being shown. The yarn, with flexural rigidity G, is acted upon by the force P so that it moves into the position AB'. The values of R_0, Ψ_0 are associated with the initial configuration and the values R_1, Ψ_1 with the strained configuration. The effects of any transverse forces at fabric crossover points are neglected, since no tensions are applied to the crossing threads under uniaxial load conditions.

The mean moment of the force P acting on the yarn is given by:

$$M = (PR_1/2)(1 - \cos\Psi_1) = (PR_1/2)[1-(1-\Psi_1^2/2)] \text{ approximately}$$

$$= P\ell^2/16R_1.$$

(1)

This mean bending moment produces a change in mean curvature given by M/G so we may write:

$$P\ell^2/16R_1G = (1/R_0) - (1/R_1) = (R_1 - R_0)/R_1R_0,$$

and manipulation leads to:

$$R_1 = R_0[1 + (P\ell^2/16G)].$$

(2)

The distance h is given by:

$$h = R_1 \sin\Psi_1 - R_0 \sin\Psi_0,$$

(3)

and if filament length is taken to be invariant, we may write

$$\ell/2 = R_1\Psi_1 = R_0\Psi_0. \qquad (4)$$

Writing $\sin\Psi = \Psi - \Psi^3/6$ and substituting equations (4) and (2) in equation (3) yields after manipulation the approximate relationship between the total deformation 2h caused by bending and the force P:

$$2h = (\ell\Psi_0^2/6)([\{1 + (P\ell^2/16G)\}^2 - 1] / \{1 + (P\ell^2/16G)\}^2]. \qquad (5)$$

The arc length ℓ can be related to the spacing, X_0, of the crossing threads in the unstressed fabric and the crimp, c, through the expression:

$$\ell = X_0(1 + c)$$

and the axial strain due to bending ε_B can be written:

$$\varepsilon_B = 2h/X_0 = c\left[\frac{(\{1 + (1 + c)^2 (PX_0^2/16G)\}^2 - 1)}{\{1 + (1 + c)^2 (PX_0^2/16G)\}^2}\right] \qquad (6)$$

Equation (6) is much more useful for the present purpose than the exact expression for the deformation involving elliptic integrals [8]; it may be shown that the maximum error in ε_B is approximately 10 percent over the range of parameters considered, and it is usually much smaller than this.

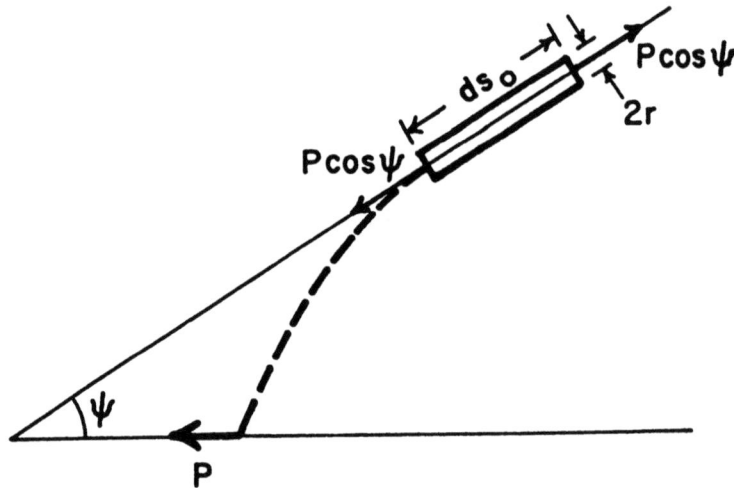

Fig. 2. Element of strained, bent filament.

In order to remove the restriction imposed by the assumption of unextensible filaments, it is necessary to consider the tensile deformation of a small element of a bent yarn. Fig. 2 shows the forces acting on such an element with cross-sectional area a.

The axial extension, d, of the element is given by:

$$d = (P \cos\Psi \, ds_0)/a^2 E, \tag{7}$$

where E is the Young's modulus of the filament material.

The extension of the element in the direction of the force P is given by $d \cos\Psi$, and the total extension of the filament in this direction is:

$$\int_0^\ell (P \cos^2\Psi \, ds_0)/(aE) = \frac{P}{aE} \int_0^\ell \cos^2\Psi \, ds_0. \tag{8}$$

$$= \frac{P\ell}{aE} [1 - \Psi_0^2/3] \text{ approximately} \tag{9}$$

$$= \frac{Pr^2 \ell}{4G} \cdot [1 - \Psi_0^2/3] \tag{10}$$

where r is the radius of the equivalent circular yarn, and G is its flexural rigidy. The elongation, ε_T, due to axial extension is:

$$\varepsilon_T = \frac{PX_0^2}{16G} \cdot 2(r/X_0)^2 \, (2 + c) \text{ approximately.}$$

The total fractional elongation ε of the yarn is given by the sum of the bending and the tensile components. For small crimp amplitudes the arc length of the loaded yarns can be approximated by the separation of the crossing yarns, X_0, and the expression for total elongation becomes:

$$\varepsilon = c[\{(1 + K)^2 - 1\}/(1 + K)^2] + 4K(r/X_0)^2 \tag{11}$$

where $K = PX_0^2/16G$.

This simple expression for the total elongation is quite useful for estimating purposes since it contains only readily available geometrical parameters of the fabric, and the yarn flexural rigidity G. The estimation of this latter factor is the major practical difficulty in this analysis since there is always some uncertainty as to the state of restraint of the individual filaments in the yarn. This matter is taken up more fully in a subsequent section.

Fig. 3. Model for biaxial stress-strain behavior.

The analysis can be extended, with some further, but often acceptable loss of accuracy, to the case of biaxial loading. The physical model to be considered in this case is shown schematically in Fig. 3, in which loads P and P' are applied to the ends of two sets of centrally interconnected leaf springs. This model may be treated in a manner very similar to the uniaxial case, but since the load is applied to both sets of yarns, it is necessary to include the effects of transverse loading. This is done by adding a load T at the free end of the element shown in Fig. 1 in a direction perpendicular to the force P.

An equation analogous to equation (1) can be written for each bent arc:

$$(PR_1/2)(1 - \cos\Psi_1) - (TR_1 \sin\Psi_1)/2 = G[1/R_0 - 1/R_1] \qquad (12)$$

and the horizontal deflection, h, and the vertical deflection, v, of the free ends of the arcs are given, after simplification, by:

$$h = \ell/12 [\Psi_0^2 - \Psi_1^2] \qquad (13)$$

$$v = \ell/4 [\Psi_0 - \Psi_1] \qquad (14)$$

and the ratio h/v is:

$$h/v = (\Psi_0 + \Psi_1)/3, \qquad (15)$$

for each set of threads. Since the vertical deflection, h, for the two sets of threads must be the same:

$$v = 3h/(\Psi_0 + \Psi_1) = 3h'(\Psi_0' + \Psi_1') \qquad (16)$$

and hence

$$h/h' = -\frac{(\Psi_0 + \Psi_1)}{(\Psi_0' + \Psi_1')}. \qquad (17)$$

This relationship between the horizontal deflections, and, hence, for an initially square fabric, for the bending elongations, is sometimes useful if the initial and final crimp angles are known. It is generally more useful to express the ratio in terms of the initial crimp angles only, and this can be done by use of the relationship embodied in equation (12). This leads to:

$$-(h/h_1) = \frac{\left[\Psi_0 + \dfrac{C - D + B\Psi_0' + BF\Psi_0}{A + BF}\right]}{\Psi_0' + \left[\dfrac{A[C - D + B\Psi_0' + BF\Psi_0]}{A + BF} - C + D\right]} \tag{18}$$

where

$A = P/8 + 2G/\ell^2$

$B = P'/8 + 2G'/\ell'^2$

$C = 2\Psi_0 G/P^2$

$D = 2\Psi_0' G'/\ell'^2$

$F = \ell/\ell'$

and

$\ell = (1 + c)X_0; \quad \ell' = (1 + c')X_0'.$

Equation 18, which describes a simplified model, treated in an approximate manner, is already algebraically complicated, and points out the difficulty of obtaining usable closed-form solutions for the deformations resulting from biaxial loading. The expression has been checked using experimental data found for a range of coated fabrics intended for air-inflated structures, and the agreement appears to be adequate for the determination of trends, but it is not good enough for detailed predictive purposes. There are two major complications: uncertainty in the appropriate value of the flexural rigidity; and the nature of the interyarn forces, which are in reality distributed rather than localized as is assumed in the model, and hence have a different mean moment. Simple approaches to the resolution of these difficulties are discussed below.

The flexural rigidity of a yarn in which there is a varying degree of association between the component filaments has been studied by several workers [9,10,11]. The two limiting situations, in which there is either complete freedom of relative motion of the filaments, or a complete lack of freedom, are relatively easy to deal with, and the ratio of the stiffness in the two cases can be shown to be approximately the same magnitude as the number of filaments in the yarn. Intermediate states, in which the freedom of movement is modified by frictional restraints or by the shear properties of an embedding matrix are much more difficult to analyze, and the results can generally not be put into a simple form [12,13]. Measurements and calculations on a range of coated fabrics have indicated that the freedom of motion is inhibited to a very considerable extent, in these structures and the "no freedom" model is usually appropriate.

The calculations of the theoretical flexural rigidity of a flattened yarn in fabric with complete restraint can be carried out simply and reasonably accurately using a model for the yarn structure in which the filaments are arranged in a rectangular array of a rows of filaments with b filament per row. This model is exact for a very low twist yarn with the minimum number of layers of filaments (b = 2), and is reasonably good over the practical range of flattened cross sections found in coated fabrics. The ratio of the stiffness G, G', of such a yarn with and without restraints is given by:

$$G'/G \sim 2(2a^2 + 1)/3, \tag{19}$$

and this ratio lies between 1 and 2 orders of magnitude for yarns with up to eight layers of filaments.

The nature and effect of the distribution of the normal forces between the two crossing yarns has been studied in a preliminary way by Freeston and Schoppee [7]. If it is assumed that the normal force is attributable to the tension in the filaments, and that the magnitude of the force at any point along the interface between the two is proportional to the number of filaments active, and hence to the thickness of the crossing yarn at that point, then it can be shown that a parabolic distribution of normal force is reasonable for the circular arc geometry. This distribution is analytically tractable, and can be used to make estimates of the effects of the interyarn forces on the biaxial tensile behavior.

CONCLUSIONS

Coated fabrics often have geometrical configurations that permit simplified theoretical treatments of their mechanical behavior, and the simplifications give good agreement between theory and experiment for the case of uniaxial load conditions. Biaxial loading introduces complications in the simplified model and the derived expressions relating load and elongation in the two directions are less easy to use, and much less accurate than in the uniaxial case. The discrepancies are related to the difficulties of estimating the flexural rigidity of constrained yarns, and the effects of distributed normal forces at yarn crossover points.

REFERENCES

[1] F. T. Peirce, J. Text. Inst. 1937, 45, 745.

[2] A. Kemp, J. Text. Inst., 1958, 49, T44.

[3] G. M. Abbott, P. Grosberg and G. A. V. Leaf, J. Text. Inst. 1973, 64, 346.

[4] B. Olofsson, J. Text. Inst. 1964, 55, T541.

[5] L. I. Weiner, Textile Fabric Design Tables, Technomic Publishing Co., Stamford, CT.

[6] W. D. Freeston, M. M. Platt and M. M. Schoppee, Text. Res. J., 1967, 37, 948.

[7] W. D. Freeston and M. M. Schoppee, U. S. Army Natick Lab. Tech. Report 73-25-GP., 1971.

[8] J. Skelton, J. Text. Inst., 1967, 58, 533.

[9] D. N. E. Cooper, J. Text. Inst., 1960, 51, T317.

[10] M. M. Platt, W. G. Klein and W. J. Hamburger, Text. Res. J. 1959, 29, 611.

[11] J. D. Owen, J. Text. Inst., 1968, 59, T313.

[]2] F. Tabaddor, Fiber Science and Technology, 1974, 7, 89.

[13] R. D. Adams and A. S. Weinstein, J. Eng. Mat. Tech., Paper 75 Mat. K.

THE ROLE OF TEXTILES IN PNEUMATIC TIRES

S. K. Clark

Department of Applied Mechanics and Engineering
Science, University of Michigan, Ann Arbor, Michigan

ABSTRACT. The role of textile cords in pneumatic tires is discussed from the point of view of the mechanics of the tire as applied to loads induced in the textiles. Consideration is given first to the composite properties of the rubber and textile assembly. The influence of the geometry of the composite is shown, and the classical ways of examining loads induced in the textile cords are described. A review is given of measurement methods for determining loads or strains in textile cords. Data taken from such measurements is given for both bias and radial tire constructions.

1. INTRODUCTION

Textile cords may be viewed as the primary load carrying element in a pneumatic tire, or may be viewed as one element in a cord-rubber composite. In either event, the textile cord operates in an environment made up of the following: (a) fluctuating loads, mostly tensile but on occasion compressive, (b) temperatures as high as 125%C, (c) moisture, and (d) surface deformations induced by being bonded tightly to a rubber matrix along its outer surface.

While this environment is hostile, it should be recognized that the textile cord used in a pneumatic tire is part of a complex elastomeric composite made up of three major elements. These elements are: (a) the rubber matrix, usually of quite low modulus and high extensibility, (b) the reinforcing cord, more usually of much higher modulus and lower extensibility than the matrix, and (c) the adhesive film which bonds the cord to the matrix.

A schematic representation of this is given in Fig. 1A.

For reasons of manufacturing economies, the typical construction practice used in the tire industry involves calendering sheets of rubber around an array of parallel cords to form a flat, essentially two-dimensional highly anisotropic composite sheet made up of a rubber matrix surrounding parallel textile cords. These cords commonly have substantial twist and often are multiplyed, that is, made up of two or three oppositely twisted yarns. These sheets are then assembled into various tire configurations designed to utilize their properties to best advantage. Examples of widely used designs are given in Fig. 1B, where it is shown that the typical bias or cross-ply design utilizes two or more, usually an even number, of plies laid at alternate diagonal angles to one another. Fig. 1B also shows the typical radial tire construction which is widely used in Europe and becoming dominant in the United States. Here, the carcass is represented by a single ply structure involving radially oriented cords while the tread area is reinforced by a belt structure of relatively small angle with respect to the tire center line. This tends to make a very stiff longitudinal reinforcement for the tread area, while at the same time providing flexibility for the vertical deflection needed in passenger car service.

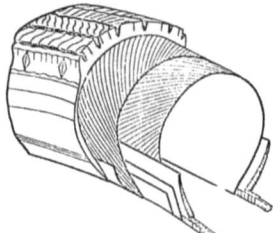

Figure 1B. Cross-bias tire

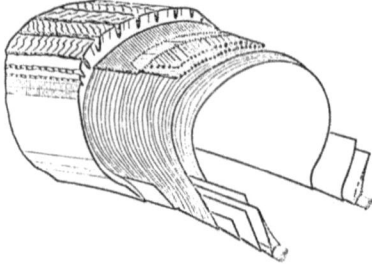

Figure 1B. Radial ply rigid breaker tire

2. CONSTITUENT MATERIAL PROPERTIES

2.1 Rubber Properties

For relatively small strains rubber may be treated as a homogeneous, isotropic material with two independent elastic constants. In the structural analysis of cord-rubber composites which follows, it is convenient to employ the Young's modulus E_r and the Poisson's ratio ν_r of rubber as the two independent constants. They are

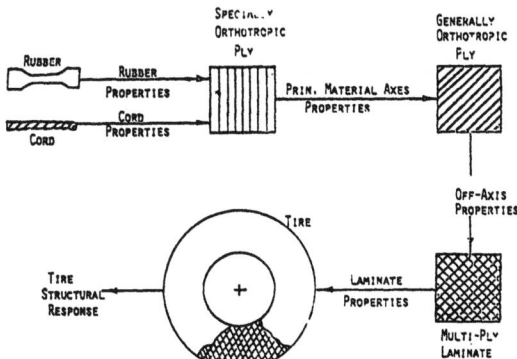

Fig. 1. Flowchart showing objectives of cord-rubber composite analysis.

related to the shear modulus by the expression

$$G_r = \frac{E_r}{2(1+\nu_r)}$$

Young's modulus, also commonly referred to as the modulus of elasticity, is the initial slope of the rubber stress-strain curve. It is usually determined from a uniaxial tension test but can also be determined from compression, bending or torsion experiments. Young's modulus may be as low as 100 psi for some nonreinforced (unfilled) elastomers to as high as 100,000 psi for highly vulcanized (high sulfur) compounds such as ebonite.

The Young's modulus or initial slope of the stress-strain curve of a rubber specimen is affected by physical testing and chemical vulcanization parameters, including (a) the rate of deformation and temperature of the test (that is, the viscoelastic nature of the rubber), (b) the cyclic loading history of the specimen (that is, appreciable stress-softening occurs in filled systems), (c) the compounding ingredients mixed with the rubber, such as carbon black, sulfur, and oil, and (d) the state of cure of the rubber. When rubber is deformed, little if any volume change occurs compared to that which occurs in other materials, so that:

$\nu_r \to \tfrac{1}{2}$, $K_r \to \infty$, and $E_r \to 3G_r$.

In other words, only one independent elastic constant exists for incompressible isotropic materials.

Typical values of rubber elastic constants are given in Table 1.

TABLE 1 Typical Values of Elastic Constants for
Rubber Used in Calendared Plies of Tires

Rubber Skim Stock	Young's Modulus E_r, psi	Poisson's Ratio ν_r
Textile Body Ply	800	0.49
Textile Tread Ply	3000	0.49
Steel Tread Ply	2000	0.49

2.2 Cord Elastic Constants

The hundreds of continuous, oriented, polymeric filaments that constitute the typical organic plied and cabled cord used in tires should individually be considered as being transversely isotropic with five independent elastic constants: (a) an extensional Young's modulus, (b) an extensional Poisson's ratio, (c) a transverse Young's modulus, (d) a transverse Poisson's ratio, and (e) a torsional (shear) modulus. For such filaments, isotropy exists in planes perpendicular to the direction of fiber drawing as discussed by Ward [1]. It is apparent then that twisted tire cord, which also has a certain amount of void content between its filaments and plies, cannot be considered as a homogeneous, isotropic material like rubber. Brewer [2] in his stress analysis of bias ply aircraft tires, approximated the material properties of twisted cord as being transversely isotropic, which is strictly true only for a single fiber without twist--i.e., monofilament. However, to be consistent with approximations commonly employed, we use only the extensional Young's modulus and Poisson's ratio as well as the shear modulus of the cord; we will neglect the transverse Young's modulus and Poisson's ratio of the cord for which no theoretical or experimental data are available. It appears that the transverse Young's modulus and Poisson's ratio of the cord do not strongly affect many of the properties of two-dimensional cord-rubber composite systems as used in tires.

In addition to the complications introduced by viscoelastic effects (strain rate and temperature dependent material properties), tire cord elastic constants are also significantly affected by the amount of ply yarn and cable twist employed. For example, to a first approximation, the Young's modulus E_c of twisted singles yarn in tension decreases with increasing twist according to the expression [4]

$$E_c = \frac{E_f}{1+4\pi^2 R^2 T^2}$$

translated directly on a one-for-one basis into the elastic and
strength characteristics of tire structures. The reason for this
is that in almost all cases textile cords are not used as direct
zero-angle tension members in tire construction, but rather are
used in the form of laminates made up of parallel arrays of cords
running at some angle to the principal stresses imposed on them.
They are often in the form of symmetric laminates such as illustrated in Fig. 2. Fig. 2a shows the common type of composite
textile-rubber structure commonly used in bias tires, which are
widely manufactured for automotive service up until a few years
ago, and are still used for many truck and aircraft tires. This
type of structure has a number of advantages to it since there is
a considerable body of technology which has been built up on its
use in various specialized tires. In Fig. 2a the principal stress
directions are often taken as being the bisectors of the cord
angles. However, these are only the simplest stress states since
the more complex ones involving cornering, camber, and road irregularities may be in any general direction.

In Fig. 2b is illustrated the type of structure commonly used
in belt areas of radial tires which have now nearly taken over the
passenger car tire design area and are becoming more and more popular for trucks. This type of structure utilizes three layers of
textiles imbedded in rubber and in that sense is a redundant net
structure. Clearly it is considerably stiffer in both principal
directions than the bias ply structure. This represents one of
the inherent advantages of the radial construction over the bias,
in that the tread belt region is stiffer and hence is less subject
to slip due to elastic forces while in contact with the roadway.
This design reduces the wear of tires with radial construction

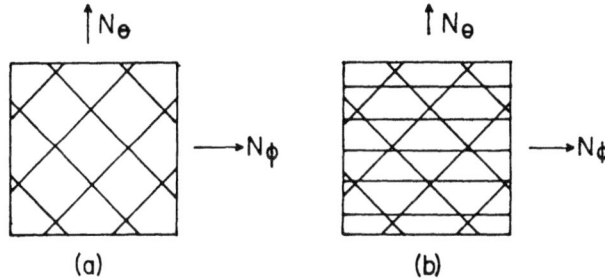

Fig. 2. Bias and radial tire cord geometry.

substantially over tires with normal bias construction.

Tests on the stress-strain curves of textiles normally show substantial nonlinearities in their behavior. Often this is considerably modified when tests are done on composites where the textile cords run at an angle to the direction of loading. The rubber is relatively elastic in the small strain range, and this tends to cause the composite structure to act more like a linearly elastic solid than the cord itself. Fig. 3 shows a typical stress-strain curve for a tubular specimen using rayon textile reinforcement in a rubber matrix. One may see that for relatively small strains the representation of the material characteristics by a linearly elastic solid is an adequate one. In spite of fairly large deformations, most pneumatic tires do not operate with carcass strains much in excess of 10%.

In tire service it is possible for the textile to come into compression, a state which is difficult to attain outside of encapsulation in a matrix. Experiments run some years ago by the writer with heavy cylinders loaded in compression showed that the textile carried considerable load in compression. It is known from experimental studies which will be quoted later that a similar thing can occur in actual tire structures. Fig. 4 illustrates the type of compression specimen used while Figs. 5 and 6 represent stress-strain curves in both tension and compression for rayon and nylon tubular specimen of this design. It is interesting to note that the modulus of the entire specimen in compression is not zero, but that the textile cords carry load. In this particular specimen

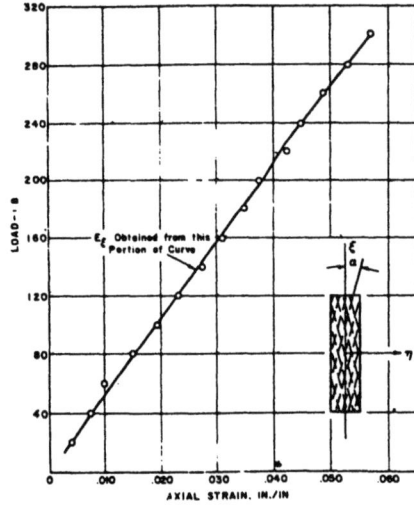

Fig. 3. Typical load-axial strain curve for cylindrical tubes jected to axial tension.

where E_f is the filament modulus, R the yarn radius, and T the twist; the Poisson's ratio ν_c of two ply cord in compression decreases with increasing twist according to the relation

$$\nu_c = (4\pi^2 R^2 T^2)^{-1} \text{ [3]}.$$

In some cases, twisting fiber into tire cord can result in as much as a one-third decrease in the extensional Young's modulus for typical belt ply cords and a one-half decrease in the extensional Young's modulus for typical body ply cords. However, twisting is needed in order to provide adequate cord fatigue life under tire service conditions.

Some idea of the range of values for the Young's modulus of typical belt and body ply cords can be gained from inspection of Table 2.

The stress-strain properties, including Young's moduli, of a variety of belt cord materials are contained in the data of Draves, et al. [5]. The shear moduli of organic tire cord constructions are taken to be approximately 700 psi, though measurements to confirm this value are lacking; measurements on steel cord (5x1x0.010 in) are more easily conducted and indicate that G_c is approximately 1×10^6 psi [6]. Poisson's ratios of tire cords are often in excess of 0.5 due to twist; the higher the twist level, the larger is ν_c.

3. COMPOSITE MATERIAL CHARACTERISTICS

The material characteristics of textile cords are not usually

TABLE 2 Typical Values of Young's Modulus for Tire Cord

Cord Construction	Young's Modulus E_c, psi
Belt Ply	
5x1x0.010 in. Steel	15.9×10^6
1500/2 Kevlar	3.6×10^6
1650/3 Rayon[a]	1.6×10^6
Body Ply	
1000/2 Polyester	575,000
840/2 Nylon	500,000

[a] This is a relatively low twist rayon (with a high Young's modulus) compared to the higher twist level rayon used in the body plies of tires.

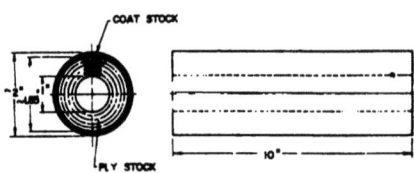

Fig. 4. Typical specimen geometry, with cord specimen shown. Rubber specimens are of identical outside dimensions.

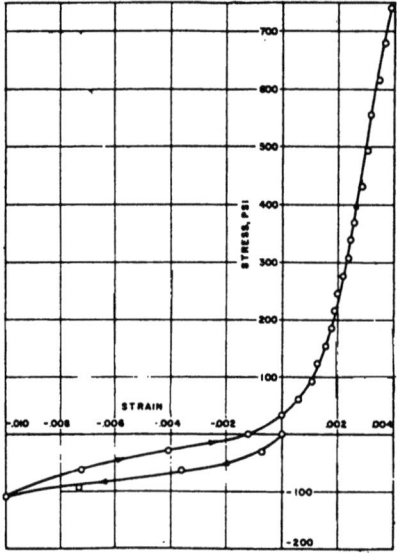

Fig. 5. Stress-strain curve for rayon cord specimen.

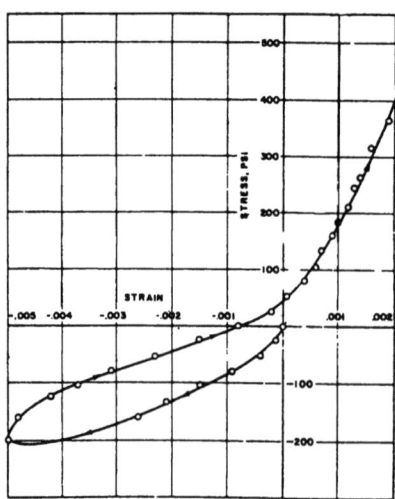

Fig. 6. Stress-strain curve for nylon cord specimen.

the cords ran parallel to the axis of the cylinder.

Such compression loadings are believed to be the source of many textile failures, and good tire design practice attempts to avoid cords in compression whenever possible.

In summary, then, while the textile cords themselves are important in defining tire stiffness and tire structural strength, the cord properties do not themselves directly translate into tire structural properties but are modified by the presence of the elastomer, and further modified by the geometry of the composite. For that reason a considerable amount of effort has been expended in composite theory applied to tire structures.

4. ANALYSIS OF CORD LOADS IN TIRES

The loads normally encountered by textile cords in pneumatic tire service must be kept to a small fraction of ultimate tensile load since the key to good tire design is long fatigue life. The cord loads are fluctuating with each cycle of tire revolution, and many millions of cycles are normally encountered during the life of the tire.

This means that the textile cord first must be securely bonded to the elastomeric matrix surrounding it, and this requires careful and balanced adhesive systems. Much care is given to these in the tire industry. Secondly, resistance to fatigue requires that each of the individual textile cords be isolated from rubbing or chafing contact with any others. Fabric design must avoid square wovens and instead concentrate on linear arrays where each textile cord lies parallel to its neighbor. This array is then calendared into a rubber sheet and the entire assembly is then a single ply of so-called tire fabric. This is illustrated in Fig. 1.

The tire designer may then use tire fabric in various geometric layups or arrays in order to achieve the particular load carrying objectives of a given tire design.

Calculation of the loads in typical tire cords is an extremely difficult and complex process, and one which is not yet completely settled at this time. The reason for this is the complicated nature of the tire itself and its loading in service. Often it has been considered simplest to break the tire loads down into several constituent parts, such as: (a) inflation alone, (b) vertical load, (c) steering forces, (d) road irregularities and bumps, (e) camber, (f) speed, and (g) torque. Of this complex array of forces, we have reasonable information only for the inflation part of the load at this time. The other forces provide inputs to textile cord loads which are basically unknown in our present state of knowledge.

4.1 Inflation Loads

Due to the axisymmetric nature of inflation, its effect on textile cord loads is more easily determined than are other influences. Two fundamental static equilibrium conditions always govern the state of membrane forces in a pneumatic tire. These are given by first Fig. 7, where an axisymmetric section of the toroidal shell is shown under the action of inflation pressures. By equilibrating the horizontal components of the total forces represented by Fig. 7, one may solve for the membrane force component in the meridianal direction for any position around the meridian of the tire as given by Eq. 4.1.

$$p\pi(r^2-r_0^2) = 2\pi r N_\phi \sin\phi$$
$$N_\phi = \frac{p(r^2-r_0^2)}{2r\sin\phi} \qquad (4.1)$$

Similarly, each small element of the curved shell surface must be in equilibrium with the inflation pressure and this gives a relationship between the curvatures of the element and the membrane loads which it must sustain in Eq. 4.2. This is illustrated in Fig. 8.

$$\frac{N_\theta}{r_2} + \frac{N_\phi}{r_1} = p \qquad (4.2)$$

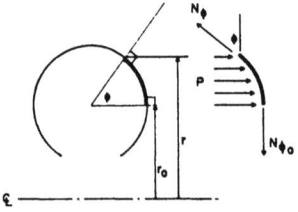

Fig. 7. Equilibrium of an annulus.

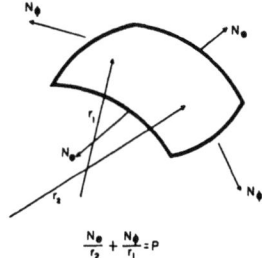

Fig. 8. Equilibrium of a curved section.

While equations 4.1 and 4.2 are quite useful in assessing the state of cord tension in simple tire construction such as the two-ply bias tire, they do contain the seeds of several highly misleading elements. First of all, the tire does not retain a constant geometry during inflation so that it is very difficult to determine with any exactness the radius r_0 shown in Fig. 7. This radius changes during the inflation process, as do the radii r_1 and r_2 in Fig. 8. Therefore, one cannot measure these ahead of time on an uninflated tire and hope to predict the influence of the inflation pressure on cord stress.

Lauterbach and Ames have shown that if one uses the final inflated goemetry of the tire then the use of Eqs. 4.1 and 4.2 yields results for membrane forces which are in relatively close agreement with measurement.

Another fundamental difficulty with Eqs. 4.1 and 4.2 is that for tires having multiple plies one must assume equal distribution of membrane forces between plies or perform a complex analysis in order to assign the various plies their particular membrane force. This becomes a particularly severe problem in aircraft or off-the-road tires where a large number of plies are the rule. We do not have a simple method for distributing a membrane force among the various plies of a large multi-ply tire.

Further difficulty with Eqs. 4.1 and 4.2 is that they are not adequate to describe the state of membrane force in regions of greatest interest to the tire designer, namely, the bead region, where a multiple set of reinforcements is located, and the tread region of the tire which sometimes now is formed in a belted configuration.

4.2 Net Theory

From the point of view of analysis, the simplest tires to consider are two-ply bias construction tires exemplified by relatively small automobile passenger car tires in use up until a few years ago. For these tires, it was common to assume that the two plies each carried the same cord loads as shown in Fig. 9.

Under this assumption a net, which neglects completely the role of the matrix, assumes that the cords are attached at their cross-over points. It requires that there be a definite ratio between the total membrane forces carried in the two perpendicular directions, given in Fig. 9 as the directions θ and ϕ. This limits severely the loads which can be carried by a net without angle change, since the stresses generated by the inflation of the tire as given by Eqs. 4.1 and 4.2 are quite independent of the net directions. It is this basic difference which caused so much uncertainty for so many years concerning the exact cord loads even during the inflation of a simple two-ply bias tire, and further which led to the concept of pantographing or angular change of the net structure in order to accommodate the differing stress levels. Basically Eqs. 4.1 and 4.2 are not in agreement with Fig. 9. These

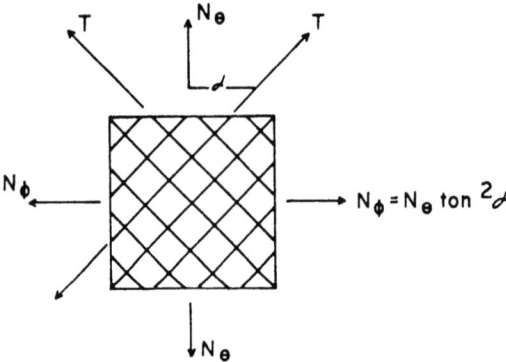

Fig. 9. Cord forces in a net.

equations predict a ratio of membrane stresses N_θ and N_ϕ which is different from the ratio predicted by Fig. 9. It is only in rare instances that the ratios given by these two different expressions will coincide. It is the difference between the ratios of membrane forces N_θ and N_ϕ as expressed by Eqs. 4.1 and 4.2, and Fig. 9 that gives rise to interply shear stresses in the rubber matrix between the cord.

4.3 Composite Theory

An alternate method of approaching the problem in textile loads in tire structures is to consider the structure as a composite material. This requires the use of concepts originally developed for plywood and for rigid composites such as used in aerospace applications. These rigid composites generally do not change geometry in structural applications, while the pneumatic tire normally undergoes reasonably large deformations and can exhibit changes of cord angle and structural configuration. For this reason the application of composite theory to tire structures is substantially more difficult than to conventional composites, but there has been considerable work done in this area because of the usefulness of composite theory in describing the elastic stiffness characteristics of composites which in turn are needed in modern finite element analysis programs. The role of composites in pneumatic tires has been recently reviewed by Walter [7] and some of the subsequent material presented here is taken from his work. Most of this is directed toward determination of the elastic constants of cord-rubber composites as opposed to the cord loads themselves, but nevertheless is useful as a prerequisite to stress analysis work since the redundant structure requires the elastic constants prior to determination of cord forces.

Walter shows that the elastic constants of a cord-rubber composite laminate in the principal directions, as shown in Fig. 10,

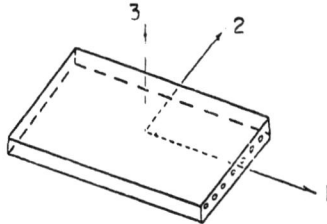

Fig. 10. Unidirectional calendared ply of cord and rubber showing natural or principal material axes.

can be approximated by Eqs. 4.3 - 4.7, where ζ_1 and ζ_2 are factors depending on cord geometry and spacing, and v_c is the volume fraction of the cord in calendared ply given by Eq. 4.8.

$$E_1 = E_c v_c + E_r(1-v_c) \qquad (4.3)$$

$$\nu_{12} = \nu_c v_c + \nu_r(1-v_c) \qquad (4.4)$$

$$E_2 = \frac{E_r[E_c(1+\zeta_1 v_c)+\zeta_1 E_r(1-v_c)]}{E_c(1-v_c)+\zeta_1 E_r(1+v_c/\zeta_1)} \qquad (4.5)$$

$$G_{12} = \frac{G_r[G_c(1+\zeta_2 v_c)+\zeta_2 G_r(1-v_c)]}{G_c(1-v_c)+\zeta_2 G_r(1+v_c/\zeta_2)} \qquad (4.6)$$

$$\nu_{21} = \nu_{12} E_2 / E_1 \qquad (4.7)$$

$$v_c = \frac{\pi R^2 (epi)}{t} \qquad (4.8)$$

with R = cord radius, t = calendared ply thickness, and epi = cord end count (ends per inch). Note that for a planar network of parallel cords as used in the plies of tires, cord volume fraction is identical to the ratio of cord cross-sectional area to the total area of cord and rubber.

One precaution in the application of the Halpin-Tsai equations to cord-rubber composites is that the initial or small strain modulus of many textile tire cords is significantly different in the vulcanized tire from that in greige, untreated condition on which cord tests are likely to be conducted, due to sensitivity to both heat and moisture. Whenever any doubt exists concerning the suitability of these equations for predicting the elastic constants of a one-ply system, it is advisable to make direct experimental measurements on the ply.

Typical comparisons of Eqs. 4.3 through 4.7 with experimental data are given in Figs. 11 and 12.

As an example of the use of these relationships, consider a typical 1650/3 rayon rubber ply used in pairs in the carcass of a typical radial tire with an uncured textile end count of 22 ends per inch. In the cured tire, the end count varies continuously from a maximum at the bead of 22 to a minimum of the crown of 14. Using the sidewall as a typical area where the cured end count is approximately 18 per inch, and the cured ply gage 0.048 inches, one may compute the final composite modulus using the individual rubber properties E_r = 800 psi, E_c = 740,000 psi, ν_r = 0.49, ν_c = 0.66, v_c = 0.34, ζ_1 = 2, ζ_2 = 1, we obtain for the composite sheet

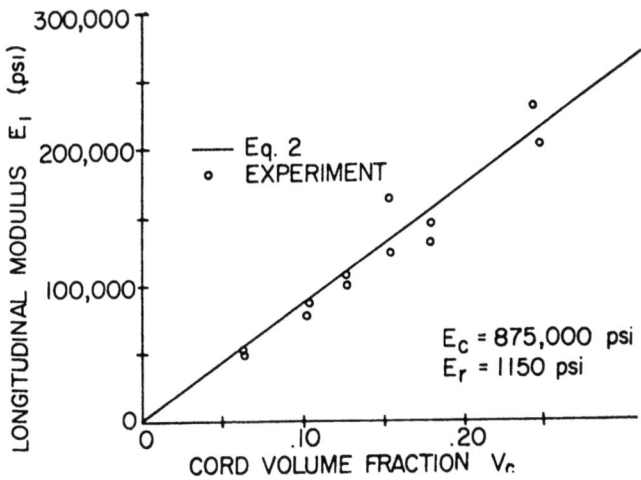

Fig. 11. Comparison between theoretically predicted and experimentally measured values of longitudinal modulus E_1 as a function of volume fraction v_c for 1500/2 Kevlar-rubber ply.

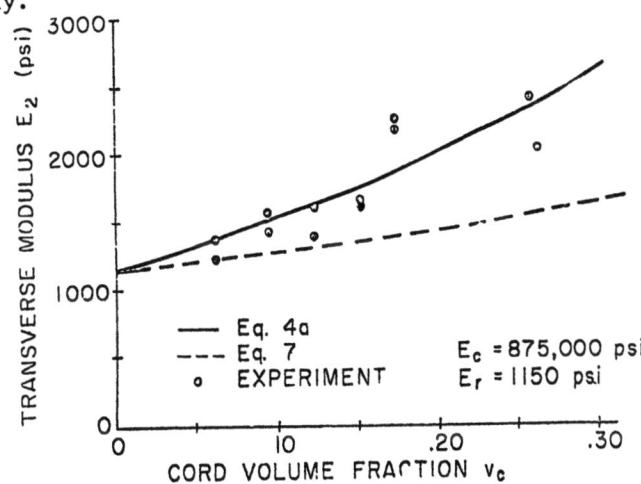

Fig. 12. Comparison between theoretically predicted and experimentally measured values of transverse modulus E_2 as a function of volume fraction v_c for 1500/2 Kevlar-rubber ply.

$E_1 = 252 \times 10^3$ psi

$E_2 = 2.04 \times 10^3$ psi

$G_{12} = .363 \times 10^3$ psi

$\nu_{12} = 0.547 \qquad \nu_{21} = 0.004$

With such material properties known from calculation, radial tire sidewall strains can be obtained from Hooke's law if the stresses can be calculated for some particular loading condition.

The preceding developments dealing with composite properties gives the tire designer some physical insight into the effects on mechanical properties of combining two dissimilar materials, cord and rubber, into one single composite. For example the individual steel monofilaments used in the belt cords of a radial tire initially possess Young's modulus of approximately 28×10^6 psi. When five of these filaments are twisted into a 5x1x0.010 inch cord construction, the modulus decreases to $15\text{-}18 \times 10^6$ psi. When this steel cord is combined with rubber with a modulus of 800 to 3000 psi, the composite modulus in the direction of reinforcement is further reduced to 1 to 3×10^6 psi depending on end count and ply thickness. This value is achieved only if the composite is loaded in tension since compressive Young's moduli for steel cord-rubber composites may be three orders of magnitude lower.

Typical values of composite sheet moduli parallel and at right angles to the reinforcing cord directions are given in Table 3.

5. MEASUREMENT OF CORD LOADS IN TIRES

5.1 Measurement Methods

The measurement of cord loads has been one of the most important fields of activity for the experimental stress analyst in the tire industry. A large number of methods have been proposed for such measurements. Among the earliest of these were various techniques for measurement of cord loads using grid or elongation marks. For example, marks on cords have been used to measure the outer ply cord strains in the sidewall of an inflated and statically

TABLE 3 Longitudinal and Transverse Moduli Comparison for Rigid and Flexible Composites

Filamentary Composite System	Longitudinal Young's Modulus E_1, psi	Transverse Young's Modulus E_2, psi	Degree of Anisotropy E_1/E_2
Glass-Epoxy	7,500,000	2,600,000	2.9
Graphite-Epoxy	30,000,000	750,000	40
Nylon-Rubber	163,000	2,000	80
Rayon-Rubber	253,000	2,000	125
Steel-Rubber	2,540,000	3,000	850

loaded tire. Such studies establish the fact that in inflated and
deflected bias tires, cords in the contact zone lose the tension
initially imposed by inflation pressure all along their length.
Similar techniques have been applied to measure the strains in the
innermost ply of an inflated and loaded tire, in this case by the
use of small knots as bench marks in the cord. In some cases it
has been possible to use x-ray photography to more clearly define
cord strains. Loughborough [8] used fine steel wires wrapped
tightly around the cords as markers. The markers were put on the
same cord and the change in length between markers due to inflation
and deflection loads gave a measure of the strain. Weickert [9]
treated one strand of the twisted cord with a metal salt and used
the shadowed cord in x-ray photographs as the marker. He also used
small diameter steel balls as markers to detect carcass rubber
strains between the second and third ply for a 4-ply nylon tire.
All of these methods are limited by the fact that they can only be
used on the static nonrolling tire, and require careful interpretation of the results.

Miniature force transducers using resistance foil strain gages
have been developed by Patterson [10], Clark and Dodge [11], and
Walter [12] which permit the direct measurement of the cord loads
in a tire under operating conditions. These devices are placed in
series with the cord and are imbedded in the tire during building.
In service, they provide a reproducible and easily monitored electrical signal which is an accurate measure of tire cord load. These
transducers are much smaller than either the clip gage, the rubberwire gage, or the liquid metal gages previously discussed.

The force transducer used by Clark and Dodge is shown in Fig.
13; it is a 0.50 inch long thin-walled beryllium-copper tube with
tire cord bonded through it. Extensive cord force measurements
made at the crown, shoulder, and sidewall in the innermost ply of

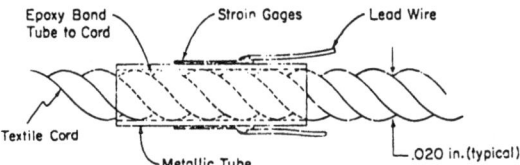

Fig. 13. Tubular cord force transducer.

two and four-ply tires at different loads and pressures have been reported using this device.

The basic geometry of one of two types of load transducers used by Patterson is shown in Fig. 14. It is an aluminum alloy billet of rectangular cross section which averages the force of two adjacent cords in a tire.

A thorough review of a number of methods for measurement of textile strain is given in The Proceedings of the Conference "Stresses and Strains in Textile Structures" held at the Shirley Institute in 1973 [13].

The numerical values of cord force obtained in different tire constructions are dependent entirely on the type of tire structure and its loading. It is not possible to assign single numerical values to such forces because of the wide variation of loading conditions and structures. Nevertheless, an effort will be made to give some typical examples of cord forces which have been measured in pneumatic tires.

The magnitude of cord forces due to inflation alone has been measured by a number of workers using various types of gages. One set of data obtained using tubular metallic elements bonded over the cords and carrying resistance strain gages is given in Figs. 15 and 16. These measurements show that on a typical bias-ply

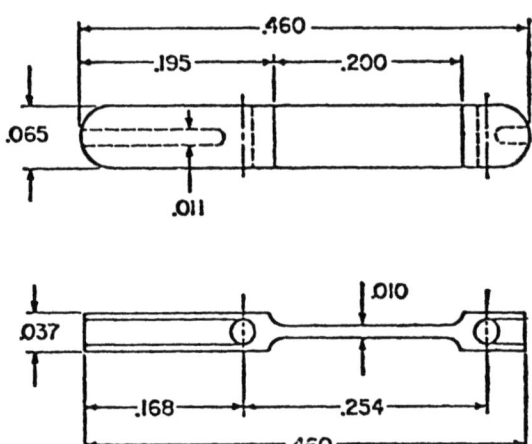

Fig. 14. Cord force transducer used by Patterson (lead wires and strain gages not shown; dimensions in inches).

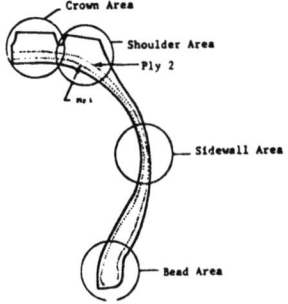

Fig. 15. General location designations.

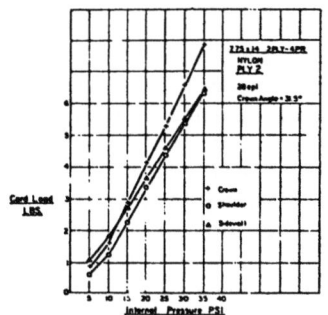

Fig. 16. Cord load vs. internal pressure.

tire, cord loads induced by direct inflation are nearly linearly proportional to inflation pressure, but in general are small compared to the total strength of the individual cord. For example in the data shown in Fig. 16 the average strength of the individual cord is of the order of 30 to 40 pounds, so that normal inflation pressures utilize only about 10% of cord strength. Similar data taken on aircraft tires of much thicker wall construction is given in Fig. 17. While there is some scatter associated with this

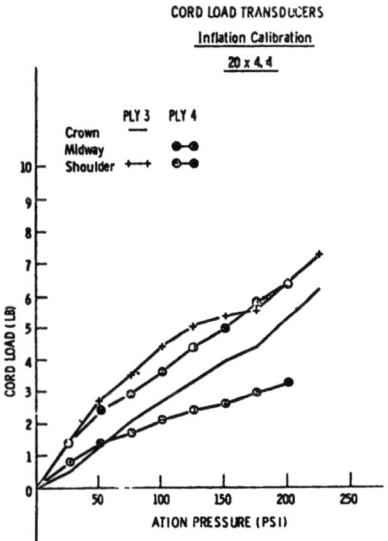

Fig. 17. Cord load vs. inflation pressure, plies 3 and 4.

latter data, the general conclusions seem to be nearly the same.

The next simplest type of cord load induced in the tire is due to the load carried by it. As the tire rolls slowly, various positions in the tire exhibit cord loads which fluctuate fairly widely. Typical fluctuations are shown in Fig. 18. Generally the load induced in textile cords in the crown of the tire is significantly lessened as the tire cord rolls through the tire contact patch. On the other hand in the shoulder and sidewall, for bias-ply constructions, the loads are first increased and then decreased as the tire rolls through contact. In both cases a substantial fluctuation of cord load is present and the cord may in some cases actually go into compression. This is illustrated in Fig. 19 where it is seen that in a simple bias-ply tire the lower curve shows

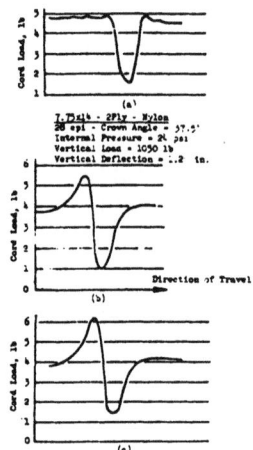

Fig. 18. Typical cord load cycle for bias-ply tire. (a) crown region; (b) shoulder region; and (c) sidewall region.

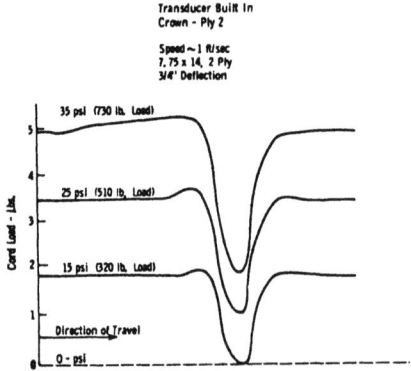

Fig. 19. Cord loads under constant vertical deflection--varying internal pressure.

that the cord tension goes to zero at an inflation pressure of 15 psi and load of 320 pounds. Were the load larger, negative cord load or cord compression would be obtained.

Typical cord loads induced in radial tires are shown in Figs. 20 and 21. Notice that in all cases the process of the tire rolling through the contact patch results in a fluctuation in the tire cord load about that tensile value induced by inflation. This could be generalized in the form shown in Fig. 22.

Steering induces additional loads into the textiles of the typical pneumatic tire. This is illustrated from the work of J. D. Walter shown in Fig. 23 where the cord loads induced in a particular tire are shown for slip angles other than zero. It is seen that relatively small amounts of steer cause very large increases in the textile cord loads, although the numbers involved are still small compared with the absolute strength of the cords. Important points on the fluctuating cord load diagram are shown. In turn, this may be further generalized to the form shown in Fig. 24 where the range of upper and lower values of tire cord fluctuation are plotted against various operating parameters of

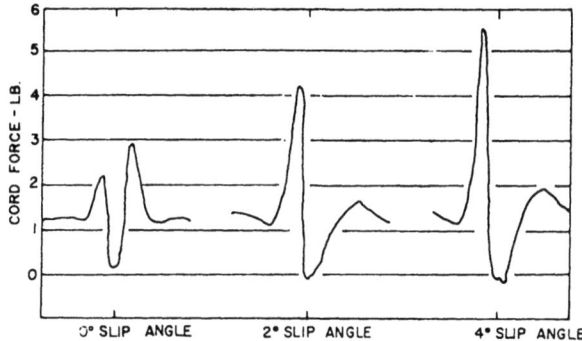

Fig. 20. Cord loads as influenced by steer.

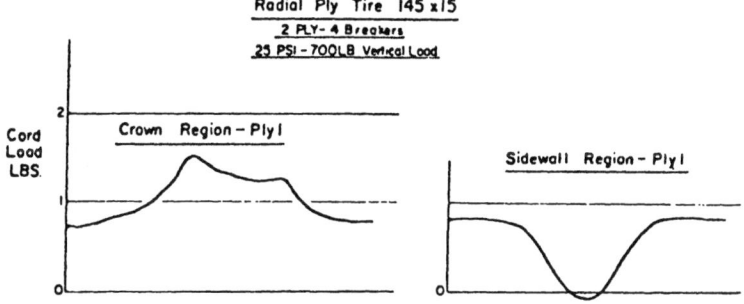

Fig. 21. Typical cord load cycles for carcass of radial ply tire.

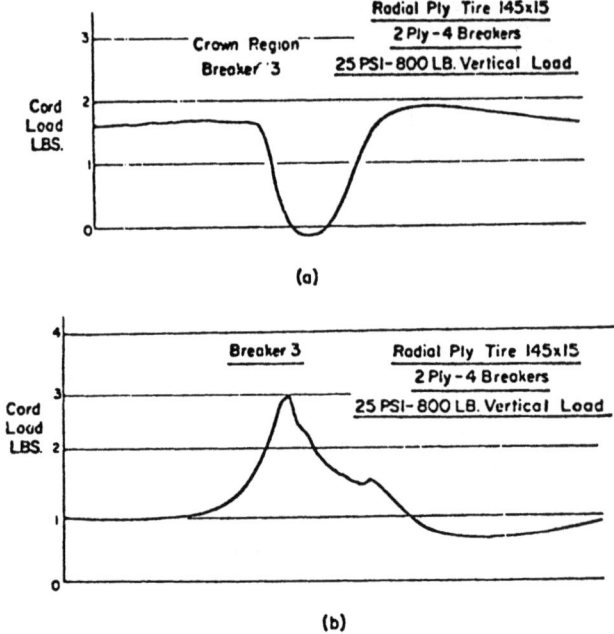

Fig. 22. Typical cord load cycles for belt of radial ply tire; (a) belt 3-crown, and (b) belt 3-edge.

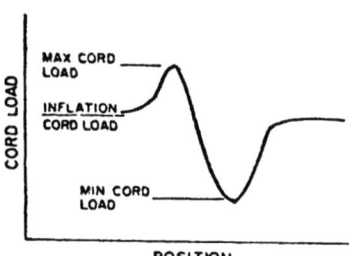

Fig. 23. Basic characteristics of cord load fluctuation in rolling tire.

the tire, i.e., its inflation pressure and vertical load. Of course this represents only the cord load at one single point in the textile cord and such plots should be obtained for various locations throughout the tire. Some typical data condensed in this fashion for a bias-belted tire are shown in Fig. 25 which shows that for several different inflation pressures, compressive cord loads can be introduced into the tire in the shoulder area.

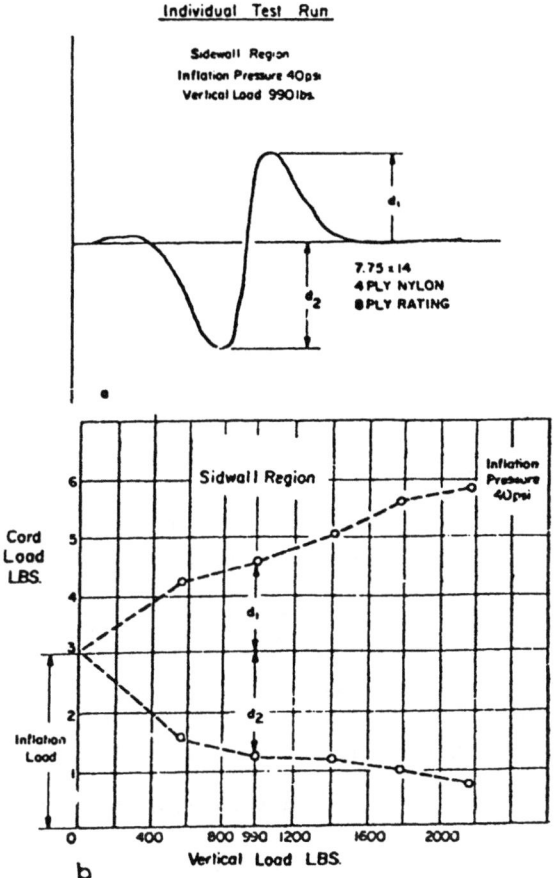

Fig. 24. Typical data reduction; (a) typical load fluctuation for one complete rotation, and (b) condensed data from a number of curves.

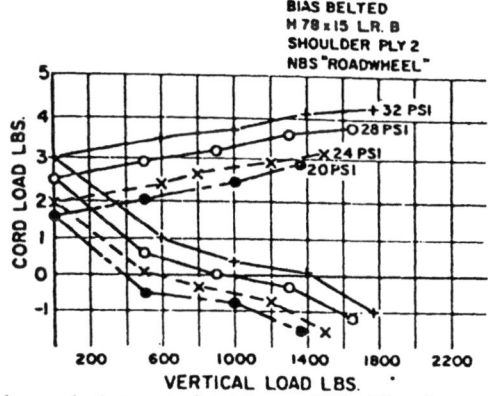

Fig. 25. Condensed data reduction, H78x15 tire, load range B, ply 2, bias-belted construction, shoulder location.

REFERENCES

1. I. M. Ward, <u>Mechanical Properties of Solid Polymers</u>, 226-229, Wiley-Interscience, New York, 1971.
2. H. K. Brewer, Stresses and Deformations in Multi-Ply Aircraft Tires Subject to Inflation Pressure Loading, 1970, AFFDL-TR-70-62.
3. J. O. Wood and G. B. Redmond, Tyre-Cord Behaviour Under Compressive Stresses, <u>Journal of the Textile Institute</u>, <u>56</u>, T191-T204, 1965.
4. J. W. S. Hearle, P. Grosberg, and S. Backer, <u>Structural Mechanics of Fibers, Yarns, and Fabrics</u>, <u>1</u>, Wiley-Interscience, New York, 1969.
5. C. Z. Draves, Z. S. Lee, and L. Skolnik, Survey of Cord Candidates for Radial Tire Belts, <u>Rubber World</u> (4), <u>164</u>, 41-47, 1971.
6. D. J. Lehmicke, Firestone Tire & Rubber Company Research Report, 1976, unpublished.
7. J. D. Walter, Elastic Properties of Cord Reinforced Rubber, unpublished manuscript to appear in <u>Mechanics of Pneumatic Tires</u>.
8. D. L. Loughborough, J. M. Davies, and G. E. Monfore, The Measurements of Strains in Tires, <u>Canadian Journal of Research</u>, <u>28</u>, 490, 1950.
9. B. Weickert, Methods for Measuring Static Strains in Automobile Tires, <u>Textile Research Journal</u>, <u>32</u>, 705, 1962.
10. R. G. Patterson, The Measurement of Cord Tensions in Tires, <u>Rubber Chemistry and Technology</u>, <u>42</u>, 812-822, 1969.
11. S. K. Clark and R. N. Dodge, A Load Transducer for Tire Cord, SAE Paper No. 690521, 1969.
12. J. D. Walter and G. L. Hall, Cord Load Characteristics in Bias and Belted-Bias Tires, SAE Paper No. 690522, 1969.
13. Stresses and Strains in Textile Structures, Shirley Institute Conference, September 12-13, 1973, Shirley Institute Publication S-10.

MECHANICS OF FABRICS IN TENSION STRUCTURES

Rainer Blum

Institut für Mechanik (Bauwesen)
Universität Stuttgart (TH)

ABSTRACT. Two Problems are treated here: First the nonlinear equations for the deformation of a prestressed hyperelastic membrane under any load are deduced from the principle of the minimum of the potential energy. Solving methods are discussed. Second the relations between stress and shape of surfaces are discussed. The equation for minimal surfaces are set up and their importance is mentioned.

1. Introduction

Tension structures like air halls are characterized by a membrane that forms, without bending stiffness and under tension stress, a roof and receives all loads, such as wind, snow, etc. The difference between this and normal structures is, that here without an external load an initial stress is necessary in order to secure functional capability. In pneumatic structures this initial stress is introduced by the inner pressure and in tent-like membrane structures by cables and pylons. Under external loads, such structures undergo large deformations, resulting in non-linear calculations. Using the theory of membrane structures, two problem areas can be distinguished: Since the conditions of equilibrium for a membrane are statically determined internally, one can find a statically possible equilibrium configuration under surface and boundary loads, independant of any constitutive equation. This one problemarea treats these statically possible surfaces. Examples of such surfaces are minimal surfaces, which

show a hydrostatic stress state at every point. Pneumatically stressed surfaces with a
hydrostatic stress state at every point are, to be sure, not minimal surfaces in the mathematical sense, but will also often be counted in this group. Wether these statically possible surfaces can be produced of a definite material with a given constitutive equation and how the initial state must look so that the statically possible form is achieved under the boundary and surface loads, requires additional examination.
The solution of this problem of cutting patterns is very important, since without an exact cutting pattern undesired wrinkles can form in the membrane. Not as yet can this problem be considered conclusively solved.
One can also perform the usual deformation calculation, which calculates the deformation of the membrane's initial configuration under external forces. Special difficulties now arise in the areas, in which one expects wrinkles. Here one abandons the range of validity of the membrane theory.
In order to make the structure of the things dealt with here as clear as possible, a covariant presentation will be foregone and we will work with a Cartesian coordinate system. The Einstein summation convention is used throughout. Greek indices assume the values 1 and 2, Latin indices run from 1 to 3. Vectors are indicated by letters with a line over them, second rank tensors by underlined letters. An exeption are base vectors, which are indicated by small gothic letters.
The first one dealt with is the deformation of a two-dimensional plane membrane, with or without initial stress. For that purpose, the geometry of deformation will first be given.

2. DEFORMATION GEOMETRY OF TWO-DIMENSIONAL PLANE SURFACES

In a three-dimensional Euclidean space a Cartesian coordinate system X_k, $k = 1,2,3$ with the base vectors $\mathfrak{n}_k$ is choosen, so that the two-dimensional membrane to describe in the undeformed configuration lies in the X_1-X_2 plane. A material point P of the membrane is described through the assertion of its position, viz. its radius vector $\bar{R}$:

$$\bar{R} = X_\alpha \mathfrak{n}_\alpha \tag{2.1}$$

in the reference configuration. The vector $d\bar{R}$:

$$d\bar{R} = dX_\alpha \mathfrak{n}_\alpha , \tag{2.2}$$

which connects two infinitesimal neighboring material points P and Q, has the length dS with

$$dS^2 = \delta_{\alpha\beta}\, dX_\alpha\, dX_\beta = dX_\alpha\, dX_\alpha. \qquad (2.3)$$

The surface element $d\bar{O}$ of the membrane, which is spanned by two vectors $d\bar{R}_1 = dX'_\alpha\, \mathcal{N}_\alpha$ and $d\bar{R}_2 = dX^2_\beta\, \mathcal{N}_\beta$ is defined by:

$$d\bar{O} = d\bar{R}_1 \times d\bar{R}_2 = E_{\alpha\beta}\, dX'_\alpha\, dX^2_\beta\, \mathcal{N}_3 \quad \text{with}$$

$$E_{\alpha\beta} = (\mathcal{N}_\alpha \times \mathcal{N}_\beta)\, \mathcal{N}_3. \qquad (2.4)$$

The $E_{\alpha\beta}$ are the components of the anti-symmetric permutation tensor with the components:

$$E_{\alpha\beta} = \begin{pmatrix} 0 & 1 \\ -1 & 0 \end{pmatrix}. \qquad (2.5)$$

After deformation the material point P of the membrane with the radius vector $\bar{R}$ in the reference configuration is in the position $\bar{r} = \bar{r}(X_1, X_2)$:

$$\bar{r} = \bar{R} + \bar{U}(X_1, X_2). \qquad (2.6)$$

$\bar{U}$ is the displacement vector of P. In components one obtains with $\bar{r} = x_k\, \mathcal{N}_k$ and $\bar{U} = \bar{U}_k\, \mathcal{N}_k$:

$$x_k = \delta_{\alpha k}\, X_\alpha + U_k. \qquad (2.7)$$

Displacements out of the X_1-X_2 plane are also possible. The material vector $d\bar{r}$, which results from $d\bar{R}$ of (2.2) through deformation, yields:

$$d\bar{r} = dx_k\, \mathcal{N}_k = \mathcal{J}_\alpha\, dX_\alpha \quad \text{with}$$

$$\mathcal{J}_\alpha = F_{k\alpha}\, \mathcal{N}_k, \quad F_{k\alpha} = \frac{\partial x_k}{\partial X_\alpha} = \delta_{\alpha k} + \frac{\partial U_k}{\partial X_\alpha}. \qquad (2.8)$$

The vectors $\mathcal{J}_\alpha$ can be considered as the metric vectors of the deformed surface. For the length ds of $d\bar{r}$ the following relation is valid:

$$ds^2 = g_{\alpha\beta}\, dX_\alpha\, dX_\beta \quad \text{with}$$

$$g_{\alpha\beta} = \mathcal{J}_\alpha\, \mathcal{J}_\beta = F_{k\alpha}\, F_{k\beta}. \qquad (2.9)$$

The $g_{\alpha\beta}$ are the components of the metric tensor of the deformed surface. The components of the Green deformation tensor $\varepsilon_{\alpha\beta}$ are defined by:

$$\varepsilon_{\alpha\beta} = 1/2 \, (g_{\alpha\beta} - \delta_{\alpha\beta}) = 1/2 \, (F_{k\alpha} F_{k\beta} - \delta_{\alpha\beta}) \quad (2.10)$$

$$= 1/2 \, (\frac{\partial U_\alpha}{\partial X_\beta} + \frac{\partial U_\beta}{\partial X_\alpha} + \frac{\partial U_k}{\partial X_\alpha} \frac{\partial U_k}{\partial X_\beta}) \, .$$

The $\varepsilon_{\alpha\beta}$ depend only quadratically on the displacement U_3 out of the X_1-X_2 plane. This means, statically, the equilibrium in the $\mathit{1\!\!\!/}_3$-direction can only be set up on the deformed element (theory of second order). From the metric vectors g_α, which span the tangential plane on the deformed membrane, the normal unit vector $\mathit{1\!\!\!/} = \mathit{g}_3$ of the surface is constructed by:

$$\mathit{g}_3 = \frac{\mathit{g}_1 \times \mathit{g}_2}{|\mathit{g}_1 \times \mathit{g}_2|} = \frac{1}{\sqrt{g}} \, \mathit{g}_1 \times \mathit{g}_2 \quad \text{with}$$

$$g = |\mathit{g}_1 \times \mathit{g}_2|^2 = |g_{\alpha\beta}| \, . \quad (2.11)$$

In general, g_3 is a broken rational function of the $F_{k\alpha}$ of (2.8). By quadratic approximation, the following results for g_3:

$$\mathit{g}_3 = (1 - 1/2 \, \frac{\partial U_3}{\partial X_\lambda} \frac{\partial U_3}{\partial X_\lambda}) \mathit{1\!\!\!/}_3 + (\frac{\partial U_\lambda}{\partial X_K} - \delta_{\lambda K}) \frac{\partial U_3}{\partial X_\lambda} \mathit{1\!\!\!/}_K$$

$$= F_{k3} \mathit{1\!\!\!/}_k \quad \text{with} \quad (2.12)$$

$$F_{\ell 3} = \frac{\partial U_3}{\partial X_\lambda} (\frac{\partial U_\lambda}{\partial X_K} - \delta_{\lambda K}), \, F_{33} = 1 - 1/2 \, \frac{\partial U_3}{\partial X_\lambda} \frac{\partial U_3}{\partial X_\lambda} \, .$$

(2.12) and (2.8) can be combined, leading to

$$\mathit{g}_k = F_{mk} \mathit{1\!\!\!/}_m \quad (2.13)$$

with the inversion

$$\mathit{1\!\!\!/}_m = \overset{-1}{F}_{km} \mathit{g}_k \, . \quad (2.14)$$

The F_{mk} are the components of the deformation gradient $\underline{F}$ of the membrane deformation.
For the surface element $d\bar{o}$, resulting from $d\bar{O}$ of (2.4) one obtains:

$$d\bar{o} = \mathit{g}_\alpha \times \mathit{g}_\beta \, dx^1_\alpha \, dx^2_\beta \quad \Rightarrow \quad \text{with (2.11)}$$

$$= \sqrt{g} \, E_{\alpha\beta} \, dx^1_\alpha \, dx^2_\beta \, \mathit{g}_3 \, . \quad (2.15)$$

Therewith, the relation between $dO = |d\bar{O}|$ and $do = |d\bar{o}|$

follows:

$$do = \sqrt{g}\ dO \qquad (2.16)$$

If one wants to calculate the deformation of a prestressed membrane, it must be studied how the total deformation is composed from the initial deformation and the additional one. For that purpose, the different configurations must be indicated. In the stress-free initial configuration the material membrane point P should have the position $\overset{\circ}{\bar{R}}$:

$$\overset{\circ}{\bar{R}} = \overset{\circ}{X}_\alpha\ \mathbf{1}_\alpha. \qquad (2.17)$$

In the prestressed configuration the same point has the position $\bar{R}$:

$$\bar{R} = \overset{\circ}{\bar{R}} + \overset{\circ}{\bar{U}} = X_\alpha\ \mathbf{1}_\alpha \quad \text{with}$$
$$X_\alpha = \overset{\circ}{X}_\alpha + \overset{\circ}{U}_\alpha. \qquad (2.18)$$

$\overset{\circ}{\bar{U}} = \overset{\circ}{U}_\alpha\ \mathbf{1}_\alpha$ is the displacement vector of the initial configuration into the prestressed configuration. Finally, P comes to the end position $\bar{r}$:

$$\bar{r} = \overset{\circ}{\bar{R}} + \overset{G}{\bar{U}} = \bar{R} + \bar{U} = x_k\ \mathbf{1}_k \quad \text{with}$$
$$x_k = \delta_{k\alpha}\overset{\circ}{X}_\alpha + \overset{G}{U}_k = \delta_{k\alpha}X_\alpha + U_k. \qquad (2.19)$$

$\overset{G}{\bar{U}}$ is the vector of the total displacement, $\bar{U}$ is the vector of the displacement of the prestressed configuration into the final configuration. It is true, naturally:

$$\overset{G}{\bar{U}} = \overset{\circ}{\bar{U}} + \bar{U}. \qquad (2.20)$$

For the relation of the coordinate differentials $d\overset{\circ}{X}_\alpha$, dX_α and dx_k one obtains:

$$dX_\alpha = \overset{\circ}{F}_{\alpha\beta}\ d\overset{\circ}{X}_\beta,$$
$$dx_k = \overset{G}{F}_{k\beta}\ d\overset{\circ}{X}_\beta = F_{k\alpha}\ dX_\alpha \qquad (2.21)$$

with the abbreviations:

$$\overset{\circ}{F}_{\alpha\beta} = \partial X_\alpha/\partial \overset{\circ}{X}_\beta = \delta_{\alpha\beta} + \partial \overset{\circ}{U}_\alpha/\partial \overset{\circ}{X}_\beta$$
$$\overset{G}{F}_{k\beta} = \partial x_k/\partial \overset{\circ}{X}_\beta = \delta_{k\beta} + \partial \overset{G}{U}_k/\partial \overset{\circ}{X}_\beta \qquad (2.22)$$
$$F_{k\alpha} = \partial x_k/\partial X_\alpha = \delta_{k\alpha} + \partial U_k/\partial X_\alpha$$

From (2.21) one gets the relation:

$$\overset{G}{F}_{k\beta} = F_{k\lambda} \overset{\circ}{F}_{\lambda\beta} . \qquad (2.23)$$

For the deformation tensor $\overset{\circ}{\varepsilon}_{\alpha\beta}$ between the initial configuration and the prestressed one one obtains:

$$\overset{\circ}{\varepsilon}_{\alpha\beta} = 1/2 \, (\overset{\circ}{F}_{\mu\alpha} \overset{\circ}{F}_{\mu\beta} - \delta_{\alpha\beta}) . \qquad (2.24)$$

On the other hand one obtains for the tensor of the total deformation $\overset{G}{\varepsilon}_{\alpha\beta}$:

$$\overset{G}{\varepsilon}_{\alpha\beta} = 1/2 \, (\overset{G}{F}_{k\alpha} \overset{G}{F}_{k\beta} - \delta_{\alpha\beta}) , \qquad (2.25)$$

whereas the tensor $\varepsilon_{\alpha\beta}$ which describes the deformation between the pre-stressed configuration and the final one yields:

$$\varepsilon_{\alpha\beta} = 1/2 \, (F_{k\alpha} F_{k\beta} - \delta_{\alpha\beta}) . \qquad (2.26)$$

Placing (2.23) in (2,24) and considering (2.26), one obtains for the composition of $\overset{G}{\varepsilon}_{\alpha\beta}$ from $\varepsilon_{\alpha\beta}$ and $\overset{\circ}{\varepsilon}_{\alpha\beta}$:

$$\overset{G}{\varepsilon}_{\alpha\beta} = \overset{\circ}{\varepsilon}_{\alpha\beta} + \overset{\circ}{F}_{\kappa\alpha} \varepsilon_{\kappa\lambda} \overset{\circ}{F}_{\lambda\beta} . \qquad (2.27)$$

3. STATICS OF PRE-STRESSED PLANE SURFACES

In this chapter, the equations of equilibrium of a plane pre-stressed membrane under tangential and normal loading shall be derived. We consider the membrane as a two-dimensional hyperelastic continuum without any bending stiffness. In order to obtain the equilibrium conditions we shall start from the principle of the total potential energy. Thereby, it is assumed that all forces possess a potential. Then for equilibrium is valid:

$$\delta\Phi = 0 \qquad (3.1)$$

and the condition for the stability of the equilibtium is:

$$\delta^2\Phi > 0 . \qquad (3.2)$$

If π is the potential energy of the internal forces and $\overset{\circ}{w}$ that one of the external forces per surface element $d\overset{\circ}{O}$ of the undeformed membrane, one can state for Φ up to boundary terms:

$$\Phi = \iint (\pi + \overset{\circ}{w}) \, d\overset{\circ}{O} \qquad (3.3)$$

π can be considered as a function of the components $\overset{G}{\varepsilon}_{\alpha\beta}$ of the Green deformation tensor of (2.25):

$$\pi = \pi(\overset{G}{\varepsilon}_{\alpha\beta}) = \pi(\overset{\circ}{\varepsilon}_{\alpha\beta} + \overset{\circ}{F}_{\kappa\alpha}\, \varepsilon_{\kappa\lambda}\, \overset{\circ}{F}_{\lambda\beta}) \,. \tag{3.4}$$

The structure of π suggests a development in a power series:

$$\pi(\overset{G}{\varepsilon}_{\alpha\beta}) = \pi(\overset{\circ}{\varepsilon}_{\alpha\beta}) + \frac{\partial \pi}{\partial \overset{\circ}{\varepsilon}_{\alpha\beta}} \overset{\circ}{F}_{\kappa\alpha}\, \varepsilon_{\alpha\beta}\overset{\circ}{F}_{\lambda\beta} \tag{3.5}$$

$$+ 1/2\, \frac{\partial^2 \pi}{\partial \overset{\circ}{\varepsilon}_{\alpha\beta}\partial \overset{\circ}{\varepsilon}_{\gamma\delta}} \overset{\circ}{F}_{\kappa\alpha}\varepsilon_{\kappa\lambda}\, \overset{\circ}{F}_{\lambda\beta}\, \overset{\circ}{F}_{\mu\gamma}\, \varepsilon_{\mu\nu}\, \overset{\circ}{F}_{\nu\delta} + \cdots$$

which will be broken off after the quadratic term. In

$$\frac{\partial \pi}{\partial \overset{\circ}{\varepsilon}_{\alpha\beta}} = \overset{\circ}{K}_{\alpha\beta} \tag{3.6}$$

one recognizes the Piola-Kirchhoff stress tensor of second kind of the pre-stressed configuration.
The Euler stress tensor $\overset{\circ}{n}_{\mu\nu}$ of the pre-stress state is defined by:

$$\overset{\circ}{n}_{\mu\nu} = \frac{1}{\sqrt{\overset{\circ}{g}}}\, \overset{\circ}{F}_{\mu\alpha}\, \overset{\circ}{K}_{\alpha\beta}\, \overset{\circ}{F}_{\nu\beta}\,. \tag{3.7}$$

Therewith and with the abbreviation:

$$\overset{\circ}{n}_{\mu\nu\kappa\lambda} = \frac{\partial^2 \pi}{\partial \overset{\circ}{\varepsilon}_{\alpha\beta}\partial \overset{\circ}{\varepsilon}_{\gamma\delta}} \overset{\circ}{F}_{\kappa\alpha}\, \overset{\circ}{F}_{\lambda\beta}\, \overset{\circ}{F}_{\mu\gamma}\, \overset{\circ}{F}_{\nu\delta}\, \frac{1}{\sqrt{\overset{\circ}{g}}} \tag{3.8}$$

one obtains for π up to the inessential constant $\pi(\overset{\circ}{\varepsilon}_{\alpha\beta})$:

$$\pi(\overset{G}{\varepsilon}_{\alpha\beta}) = \sqrt{\overset{\circ}{g}}\, \overset{\circ}{n}_{\mu\nu}\varepsilon_{\mu\nu} + 1/2\, \sqrt{\overset{\circ}{g}}\, \overset{\circ}{n}_{\kappa\lambda\mu\nu}\, \varepsilon_{\kappa\lambda}\, \varepsilon_{\mu\nu} + \cdots \tag{3.9}$$

The $\overset{\circ}{n}_{\kappa\lambda\mu\nu}$ can be considered as the components of the tangential stiffness of the membrane material. In the case of non-linear constitutive equation they are dependent on the pre-stress configuration. With (3.9) one obtains for the total potential energy Φ:

$$\Phi = \iint (\overset{\circ}{n}_{\mu\nu}\varepsilon_{\mu\nu} + 1/2\, \overset{\circ}{n}_{\kappa\lambda\mu\nu}\, \varepsilon_{\kappa\lambda}\varepsilon_{\mu\nu} + \overset{\circ}{w}/\sqrt{\overset{\circ}{g}}\,)\, \sqrt{\overset{\circ}{g}}\, d\overset{\circ}{O} \tag{3.10}$$

$\sqrt{\overset{\circ}{g}}\, d\overset{\circ}{O}$ is in accordance with (2.16) nothing other than the surface element $d\overset{\circ}{O}$ of the pre-stress configuration and $\overset{\circ}{w}/\sqrt{\overset{\circ}{g}}$ is the potential of the external forces per predeformed surface element $d\overset{\circ}{O}$:

$$w = \overset{\circ}{w}/\sqrt{g} \;, \tag{3.11}$$

so that Φ results in:

$$\Phi = \iint (\overset{\circ}{n}_{\mu\nu}\,\varepsilon_{\mu\nu} + 1/2\, \overset{\circ}{n}_{\kappa\lambda\mu\nu}\,\varepsilon_{\kappa\lambda}\,\varepsilon_{\mu\nu} + w)\, d0\;. \tag{3.12}$$

The potential of the external forces should be calculated for two kinds of loads:
1) for gravitational loads and
2) for pressure loads.

The potential energy w_G for gravitational loads per unit of surface, with t as the load per unit surface and $\mathbb{1}_G$ as the direction of gravity results in:

$$w_G = -\,t\,\bar{U}\,\mathbb{1}_G\,. \tag{3.13}$$

The load per unit surface, $\bar{p}$ of the pressure is indicated with the scalar pressure p and with the normal unit vector $\mathfrak{y}_3$ of the deformed surface:

$$\bar{p} = p\,\mathfrak{y}_3\;. \tag{3.14}$$

In a displacement $d\bar{U}$ the pressure does the work dA:

$$dA = \iint p\,\mathfrak{y}_3\,d\bar{U}\,do = \iint p\,\mathfrak{y}_3\,d\bar{U}\,\sqrt{g}\,d0\;. \tag{3.15}$$

The total work in a displacement $\bar{U}$ is obtained through the integration from $\bar{U} = 0$ until the final displacement:

$$A = \iint (p\int \mathfrak{y}_3\,\sqrt{g}\,d\bar{U})\,d0\;, \tag{3.16}$$

wherefrom one deduces the potential w_p per unit surface to be:

$$w_p = -p\int \mathfrak{y}_3\,\sqrt{g}\,d\bar{U} = -p\int F_{m3}\,\sqrt{g}\,dU_m\;. \tag{3.17}$$

Conclusively, one can assert the total potential energy Φ up to boundary terms:

$$\Phi = \iint (\overset{\circ}{n}_{\mu\nu}\,\varepsilon_{\mu\nu} + 1/2\,\overset{\circ}{n}_{\kappa\lambda\mu\nu}\,\varepsilon_{\kappa\lambda}\,\varepsilon_{\mu\nu} - t_m\,U_m$$
$$- p\int F_{m3}\,\sqrt{g}\,dU_m)\,d0 \tag{3.18}$$

with $t\,\mathbb{1}_g = t_m\,\mathbb{1}_m$.

To simplify things we limit ourselves here to the boundary conditions $\bar{U} = 0$. Then eventual boundary terms disappear and (2.18) presents the final value of Φ. In equilibrium the first variation $\delta\Phi$ must vanish. In accordance with the laws of variational calculus one obtains:

$$\delta\Phi = -\iint \left[\frac{\partial}{\partial x_\nu} (\overset{\circ}{n}_{\mu\nu} + \overset{\circ}{n}_{\mu\nu\kappa\lambda} \varepsilon_{\kappa\lambda}) F_{k\mu} + t_k + p\, F_{k3} \sqrt{g} \right] \delta U_k \, do. \tag{3.19}$$

This expression must disappear for all δU_k compatible with the boundary conditions $\bar{U} = 0$, wherefrom the equilibrium conditions are:

$$\frac{\partial}{\partial x_\nu} \left[(\overset{\circ}{n}_{\mu\nu} + \overset{\circ}{n}_{\mu\nu\kappa\lambda} \varepsilon_{\kappa\lambda}) F_{k\mu} \right] + t_k + p\, F_{k3} \sqrt{g} = 0. \tag{3.20}$$

This is a system of quasi-linear partial differential equations for the determination of the displacement components U_k of a membrane with the initial Euler stress tensor $\overset{\circ}{n}_{\mu\nu}$ and the tangential stiffness tensor $\overset{\circ}{n}_{\mu\nu\kappa\lambda}$ under pressure p and gravitational loads t_k. The equation from (3.20) with k = 3 is discussed further. It reads:

$$\frac{\partial}{\partial x_\nu} \left[(\overset{\circ}{n}_{\mu\nu} + \overset{\circ}{n}_{\mu\nu\kappa\lambda} \varepsilon_{\kappa\lambda}) F_{3\mu} \right] + t_3 + p\, F_{33} \sqrt{g} = 0. \tag{3.21}$$

Therefrom, after differentiation and under consideration of the equilibrium conditions of the pre-stress configuration:

$$\frac{\partial \overset{\circ}{n}_{\mu\nu}}{\partial x_\nu} = 0 \tag{3.22}$$

one obtains with the assumption that the $\overset{\circ}{n}_{\mu\nu\kappa\lambda}$ are homogeneous throughout the membrane:

$$(n_{\mu\nu} + n_{\mu\nu\kappa\lambda} \varepsilon_{\kappa\lambda} + n_{\lambda\nu\kappa\mu} \frac{\partial U_3}{\partial x_\kappa} \frac{\partial U_3}{\partial x_\lambda}) \frac{\partial^2 U_3}{\partial x_\mu \partial x_\nu} +$$
$$+ n_{\mu\nu\kappa\lambda} \frac{\partial U_3}{\partial x_\mu} \frac{\partial U_\rho}{\partial x_\kappa} \frac{\partial^2 U_\rho}{\partial x_\lambda \partial x_\nu} + t_3 + p\, F_{33} \sqrt{g} = 0 \tag{3.23}$$

Under the assumption, in which the U_ρ are known, this equation for U_3 is elliptic when:

$$\left| \overset{\circ}{n}_{\mu\nu} + \overset{\circ}{n}_{\mu\nu\kappa\lambda} \varepsilon_{\kappa\lambda} + \overset{\circ}{n}_{\lambda\nu\kappa\mu} \frac{\partial U_3}{\partial x_\kappa} \frac{\partial U_3}{\partial x_\lambda} \right| > 0 \tag{3.24}$$

and hyperbolic when:

$$\left| \overset{\circ}{n}_{\mu\nu} + \overset{\circ}{n}_{\mu\nu\kappa\lambda} \varepsilon_{\kappa\lambda} + \overset{\circ}{n}_{\lambda\nu\kappa\mu} \frac{\partial U_3}{\partial x_\kappa} \frac{\partial U_3}{\partial x_\lambda} \right| < 0 . \tag{3.25}$$

Here the type of the differential equation depends on the solution itself and can, for example, change in the course of a iterative solution process. In general, under tension stress one expects the elliptical case to occur. Later, it will be shown that for the hyperbolic case one infringes the stability condition (3.2). When

using approximate methods for solving (3.23), it can occur through numerical inaccuracy that on the border between both cases the elliptic case goes over to the hyperbolic case. Then, all hitherto known solving methods diverge.
If one neglects the effect of the displacement U_ρ on the U_3 displacement, then with

$$\varepsilon_{\kappa\lambda} = \frac{\partial U_3}{\partial X_\kappa} \frac{\partial U_3}{\partial X_\lambda} \cdot 1/2 \qquad (3.26)$$

for this case one obtains:

$$\left[\overset{\circ}{n}_{\mu\nu} + (1/2\, \overset{\circ}{n}_{\mu\nu\kappa\lambda} + \overset{\circ}{n}_{\lambda\nu\kappa\mu}) \frac{\partial U_3}{\partial X_\kappa} \frac{\partial U_3}{\partial X_\lambda} \right] \frac{\partial^2 U_3}{\partial X_\kappa \partial X_\lambda} +$$
$$+ t_3 + p\, F_{33}\, \sqrt{g} = 0. \qquad (3.27)$$

This equation is always elliptic when $|n_{\mu\nu}| > 0$. That means, that with the approximation (3.27) one cannot describe instability. All approximations of (3.27) converge better than approximations of the total system.
By linearizing one obtains the classical membrane equation:

$$n_{\mu\nu} \frac{\partial^2 U_3}{\partial X_\mu \partial X_\nu} + t_3 + p = 0. \qquad (3.28)$$

Here, the deflexion U_3 is independent of the material and will be determined by the pre-stress alone.
If the second variation of Φ is performed, one obtains from (3.18):

$$\delta^2 \Phi = \iint \left\{ [(\overset{\circ}{n}_{\mu\nu} + \overset{\circ}{n}_{\mu\nu\kappa\lambda}\, \varepsilon_{\kappa\lambda})\, \delta_{kl} + F_{1\kappa}\, F_{k\lambda}\, \overset{\circ}{n}_{\mu\nu\kappa\lambda}] \delta F_{k\mu} \delta F_{1\nu} \right.$$
$$\left. - p\, (\frac{\partial F_{13}}{\partial F_{k\alpha}} \sqrt{g} + F_{13} \frac{\sqrt{g}}{\partial F_{k\alpha}})\, \delta F_{k\alpha}\, \delta U_1 \right\} dO. \qquad (3.29)$$

It is sufficient for stability, that for arbitrary variations δU_1 and $\delta F_{k\alpha}$ is valid:

$$\delta^2 \Phi > 0. \qquad (3.30)$$

It should be examined, that only δU_3 is different from zero. Then the Legendre necessary condition for $\delta^2 \Phi > 0$ reads:

$$\left| n_{\mu\nu} + n_{\mu\nu\kappa\lambda}\, \varepsilon_{\kappa\lambda} + n_{\kappa\nu\lambda\mu}\, F_{3\kappa}\, F_{3\lambda} \right| > 0. \qquad (3.31)$$

That is exactly the condition for the ellipticity (3.24) of the differential equation for U_3 with assumed U_ρ. With this is shown that (3,23) becoming hyperbolic is equivalent to the loss of stability. Then wrinkles form

in the membrane. It can be shown, that the form of the wrinkles cannot be described by membrane theory. In this case bending terms must be introduced into the theory.
In solving the nonlinear membrane equations, various methods are at one's disposal. One can try to solve the variation problem Φ = minimum directly with statement functions, which satisfy the geometric boundary conditions. The resulting nonlinear algebraic equations can be solved by a Newton-Raphson method. The method of finite elements is based on the same principle, however makes use of local statement functions.
Further, one can try to solve the differential equations iterative through use of finite difference method. It is often proved favorable to perform the iteration of U_3 apart from the iteration of U_α. A possible iteration scheme would appear as follows: First the differential equation for U_3 alone without U_α-terms is approximated until numerical stability. Then, with the so-obtained U_3-solution, the equation for U_1 and U_2 would be calculated up to numerical stability. The so-obtained approximate for U_1 and U_2 would be used again in order to obtain a better approximation for U_3 etc.
Above all, difficulties occur when the determinant (3.24) is too small. In this case, all methods converge badly. If the problem is so posed that the determinant increases in the course of iteration, one can employ the following method: Start with an essentially too high initial stress and, with every iteration step, take back some of this initial stress, until the desired size is achieved. Natur-ally , it is sometimes also advisable to gradually increase the load.
The following diagram shows the results of the calculation of a quadratic rubber membrane under pressure load. Here we have the displacement U_3 in the middle of the membrane as a function of the pressure. The size of the membrane was 20×20 cm.
Curve 1 shows the displacement with a hydrostatic pre-stress of 0,153 kp/cm.
Curve 2 shows the displacement without pre-stress.
Curve 3 shows the result of linear calculation, with the same pre-stress as in curve 1.
The result of a calculation of U_3 alone with neglecting the influence of U_α cannot be distinguished from curve 1 in this scale.

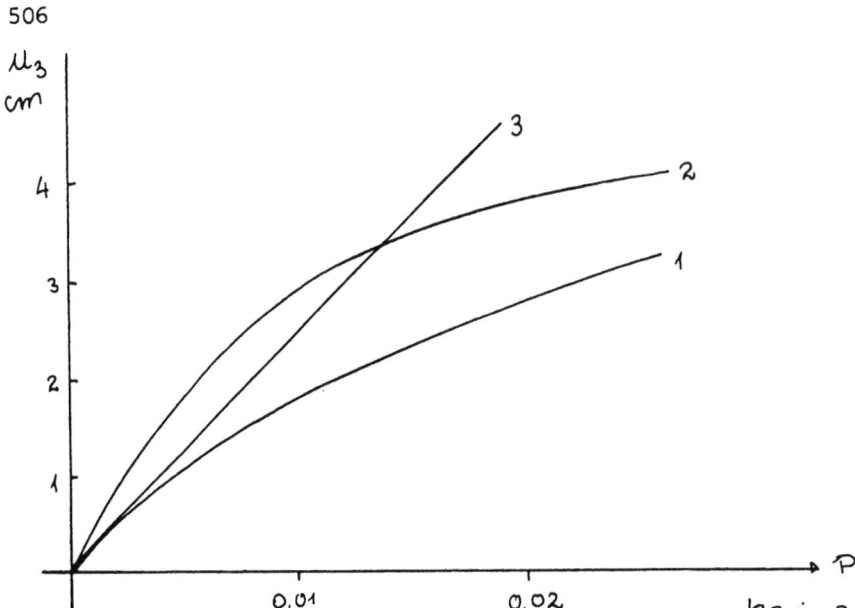

The theory above is valid for a membrane with known hyperelastic constitutive equation and with negligible bending stiffness. When treating a membrane made of fabric, one has to choose the appropriate tangential stiffnesses $\overset{\circ}{n}_{\mu\nu\kappa\lambda}$. If one assumes the fabric to have an orthotropic linear constitutive equation and when the warp-direction is parallel to the X_1-axis and the weft-direction is parallel to the X_2-axis, the following components $\overset{\circ}{n}_{\mu\nu\kappa\lambda}$ do not disappear:

$$\overset{\circ}{n}_{1111}, \overset{\circ}{n}_{2222}$$

$$\overset{\circ}{n}_{1122} = \overset{\circ}{n}_{2211} \tag{3.32}$$

$$\overset{\circ}{n}_{1212} = \overset{\circ}{n}_{2121} = \overset{\circ}{n}_{2112} = \overset{\circ}{n}_{1221}.$$

$\overset{\circ}{n}_{1111}$ and $\overset{\circ}{n}_{2222}$ are the E-moduli in the warp and the weft direction, $\overset{\circ}{n}_{1122}$ describes the transverse contraction and $\overset{\circ}{n}_{1212}$ is the shear modulus.

4. GEOMETRY OF CURVED SURFACES

This chapter will deal with the essential geometric properties for the treatment of statically possible stress states in curved surfaces. We restrict ourselves here to surfaces, whose radius vector is represented by:

$$\bar{r} = x_k \, \mathbb{1}_k = x_\alpha \, \mathbb{1}_\alpha + x_3(x_1, x_2) \, \mathbb{1}_3 . \tag{4.1}$$

At every surface point q with radius vector $\bar{r}$, a point P with radius vector $\bar{r}_p$ will be constructed through the projection of q into the x_1-x_2 plane:

$$\bar{r}_p = x_\alpha \mathbf{1}_\alpha . \tag{4.2}$$

For the vector $d\bar{r}$, which connects neighboring surface points, one obtains from (4.1):

$d\bar{r} = \mathcal{Y}_\alpha \, dx_\alpha$ with

$$\mathcal{Y}_\alpha = P_{k\alpha} \mathbf{1}_k \, , \quad P_{k\alpha} = \delta_{k\alpha} + \frac{\partial x_3}{\partial x_\alpha} \delta_{3k} = \frac{\partial x_k}{\partial x_\alpha} . \tag{4.3}$$

The $\mathcal{Y}_\alpha$ span the tangential plane on the surface. The metric tensor on the surface is given by:

$$g_{\alpha\beta} = \mathcal{Y}_\alpha \mathcal{Y}_\beta = P_{k\alpha} P_{k\beta} = \delta_{\alpha\beta} + \frac{\partial x_3}{\partial x_\alpha} \frac{\partial x_3}{\partial x_\alpha} . \tag{4.4}$$

The determinant $|g_{\alpha\beta}| = g$ is calculated by:

$$g = 1 + \frac{\partial x_3}{\partial x_\alpha} \frac{\partial x_3}{\partial x_\alpha} . \tag{4.5}$$

For the inverse $\bar{g}_{\alpha\beta}^{-1}$ of $g_{\alpha\beta}$ one obtains:

$$\bar{g}_{\alpha\beta}^{-1} = \frac{1}{\sqrt{g}} \begin{pmatrix} g_{22} & -g_{12} \\ -g_{12} & g_{11} \end{pmatrix} . \tag{4.6}$$

From the tangent vectors $\mathcal{Y}_\alpha$, the normal unit vector $\mathcal{Y}_3$ is constructed by:

$$\mathcal{Y}_3 = \frac{1}{\sqrt{g}} \mathcal{Y}_1 \times \mathcal{Y}_2 = \frac{1}{\sqrt{g}} (\mathbf{1}_3 - \frac{\partial x_3}{\partial x_\alpha} \mathbf{1}_\alpha) . \tag{4.7}$$

The components of the curvature tensor yield:

$$b_{\alpha\beta} = - \frac{\partial \mathcal{Y}_3}{\partial x_\alpha} \mathcal{Y}_\beta = \frac{1}{\sqrt{g}} \frac{\partial^2 x_3}{\partial x_\alpha \partial x_\beta} . \tag{4.8}$$

The surface should be cut into two parts by a curve with the radius vector $\bar{r} = \bar{r}(s) = x_k(s) \mathbf{1}_k$. The appertaining projection curve is given by $\bar{r}_p = {}^k \bar{r}_p (s) = x_\alpha(s) \mathbf{1}_\alpha$. The vector $d\bar{r} = \mathcal{Y}_\alpha \, dx_\alpha$ is tangential to the curve. An intersecting element $d\bar{s}$ to $d\bar{r}$ is formed, which should lie in the tangential plane to the surface and should be perpendicular to $d\bar{r}$, having the length $ds = |d\bar{r}|$. The statement:

$$d\bar{s} = d\bar{r} \times \mathcal{Y}_3 = \mathcal{Y}_\alpha \times \mathcal{Y}_3 \, dx_\alpha \tag{4.9}$$

satisfies all conditions. Since $d\bar{s}$ lies in the tangential plane on the surface, one can decompose:

$$d\bar{s} = ds'_\alpha \, \vartheta_\alpha \quad \text{with}$$

$$ds'_\alpha = g_{\alpha\beta} \sqrt{g} \; E_{\beta\mu} \, dx_\mu \; . \tag{4.10}$$

The dot at ds'_α should indicate, that it was calculated with the base-vectors ϑ_α. By projection, the intersecting element $d\bar{s}_p$, which results from $d\bar{s}$, is:

$$d\bar{s}_p = d\bar{r}_p \times \mathit{1}_3 = ds_{p\alpha} \, \mathit{1}_\alpha, \quad ds_{p\alpha} = E_{\alpha\beta} \, dx_\beta \; . \tag{4.11}$$

Therewith, it follows that the relation between ds_α and $ds_{p\alpha}$ is:

$$ds_\alpha = \overset{-1}{g}_{\alpha\beta} \sqrt{g} \; ds_{p\beta} \; . \tag{4.12}$$

5. STATICS OF CURVED SURFACES

In this chapter, the stress distribution in surfaces and the relation with geometry will be discussed. First, stresses should be defined. Through an intersecting element $d\bar{s}$, defined in (3.9), of an intersecting curve, a force $d\bar{k}$ is transferred. One can now consider this force as a function of $d\bar{s}$ and, therewith, has defined the Euler stress tensor $\underline{n}$:

$$d\bar{k} = \underline{n} \; d\bar{s} \; . \tag{5.1}$$

Another possibility exists therein: To consider $d\bar{k}$ as a function of the projected intersecting element $d\bar{s}_p$:

$$d\bar{k} = \underline{L} \; d\bar{s}_p = \bar{L}_\alpha \, ds_{p\alpha} \; . \tag{5.2}$$

$\underline{L}$ is a kind of Piola-Kirchhoff stress tensor, $\bar{L}_\alpha$ are the pertinent stress vectors. Since no forces in normal direction are transferred in the membrane, the following result is true:

$$\bar{L}_\alpha = L'_{\alpha\beta} \, \vartheta_\beta \; . \tag{5.3}$$

One can also decompose $\bar{L}_\alpha$ according to the base vectors $\mathit{1}_k$:

$$\bar{L}_\alpha = L_{\alpha k} \, \mathit{1}_k \; . \tag{5.4}$$

From the equivalence of the two decompositions it follows through multiplication with $\mathit{1}_m$ and with (4.3):

$$L'_{\alpha\beta} \, P_{m\beta} = L_{\alpha m} \; . \tag{5.5}$$

For $m = \mu$ one obtains with $P_{\mu\beta} = \delta_{\mu\beta}$:

$$L^{\bullet}_{\alpha\beta} = L_{\alpha\beta} , \qquad (5.6)$$

for $m = 3$ it follows:

$$L_{\alpha 3} = L^{\bullet}_{\alpha\beta} P_{3\beta} = L_{\alpha\beta} \frac{\partial x_3}{\partial x_\beta} . \qquad (5.7)$$

If one cuts out a part of the membrane along a closed curve c with the projection curve c_ρ, applies the external forces at the boundaries, and establishes the equilibrium conditions, then one obtains with the surface load $\bar{p} = p_k \mathbb{1}_k$ per projected surface element do_p:

$$\oint L_{\alpha k} \, ds_{p\alpha} + \iint p_k \, do_p = 0 . \qquad (5.8)$$

Therefrom, with the help of the Gauss theorem and regarding the free choice of the cut out part the following results:

$$\frac{\partial L_{\alpha k}}{\partial x_\alpha} + p_k = 0 . \qquad (5.9)$$

For $k = \kappa$ one obtains:

$$\frac{\partial L_{\alpha \kappa}}{\partial x_\alpha} + p_\kappa = 0 , \qquad (5.10)$$

for $k = 3$ it follows with (5.7) and (5.10):

$$\frac{\partial L_{\alpha 3}}{\partial x_\alpha} + P_3 = L_{\alpha\beta} \frac{\partial^2 x_3}{\partial x_\alpha \partial x_\beta} + P_3 - P_\beta \frac{\partial x_3}{\partial x_\beta} = 0. \qquad (5.11)$$

For the momentum equilibrium of the cut out part, referring to the origin of the coordinate system, one obtains:

$$\oint \bar{r} \times d\bar{k} + \iint \bar{r} \times \bar{p} \, do_p = 0 , \qquad (5.12)$$

wherefrom with (5.12) it follows:

$$P_{k\alpha} L_{\alpha l} = P_{l\alpha} L_{\alpha k} . \qquad (5.13)$$

With $P_{\alpha\beta} = \delta_{\alpha\beta}$ it follows for $k = \kappa$, $l = \lambda$:

$$L_{\lambda\alpha} = L_{\alpha\lambda} , \qquad (5.14)$$

and for $k = 3$, $l = \lambda$:

$$L_{\lambda 3} = P_{3\alpha} L_{\alpha\lambda} = L_{\alpha\lambda} \frac{\partial x_3}{\partial x_\lambda} , \qquad (5.15)$$

viz. the condition (5.7). We have now three equilibrium conditions (5.10) and (5.11) for the symmetrical stress tensor $L_{\alpha\beta}$. The stress problem in the membrane is also statically determined internally and can be solved, with consideration of static boundary conditions, without consideration of constitutive equations. The physical components of the stress tensor should be calculated from the $L_{\alpha m}$. The normal stress N_s, acting on the intersecting element $d\bar{s}$ can be considered as the component of $d\bar{k}$, in relation to ds acting in the direction of $d\bar{s}$, the shear stress T_s is the corresponding component in the direction of $d\bar{r}$. If one now selects an intersecting element $d\bar{s}_1$, so that the projected intersecting element $d\bar{s}_{p1}$ shows in the η_1-direction,

$$d\bar{s}_{p1} = dx_2\, \eta_1 , \quad d\bar{s}_1 = \overset{-1}{g}_{1\alpha} \sqrt{g}\, dx_2\, \mathcal{J}_\alpha , \quad d\bar{r} = \mathcal{J}_2\, dx_2 \qquad (5.16)$$

then one can state for $d\bar{k}$:

$$d\bar{k} = L_{1m}\, dx_2\, \eta_m = (N_1\, \overset{-1}{g}_{1\alpha} \sqrt{g}\, \mathcal{J}_\alpha + T_1\, \mathcal{J}_2)\, dx_2 ,$$

wherefrom it immediately follows:

$$L_{11} = N_1\, \overset{-1}{g}_{11} \sqrt{g} , \quad L_{12} = N_1\, \overset{-1}{g}_{12} \sqrt{g} + T_1 . \qquad (5.17)$$

With the choice of a second intersecting element $d\bar{s}_2$ with $d\bar{s}_{p2} = dx_1\, \eta_2$ one obtains similarly:

$$L_{22} = N_2\, \overset{-1}{g}_{22} \sqrt{g} , \quad L_{21} = N_2\, \overset{-1}{g}_{12} \sqrt{g} + T_2 . \qquad (5.18)$$

Should a hydrostatic stress state exist, then the same normal stress N affects every intersection. In this case, $\underline{L}$ has the following components:

$$L_{11} = \overset{-1}{g}_{11} \sqrt{g}\, N , \quad L_{12} = L_{21} = \overset{-1}{g}_{12} \sqrt{g}\, N ,$$
$$L_{22} = \overset{-1}{g}_{22} \sqrt{g}\, N \qquad (5.19)$$

or generally:

$$L_{\alpha\beta} = \overset{-1}{g}_{\alpha\beta} \sqrt{g}\, N . \qquad (5.20)$$

The equation of equilibrium (5.11) with such a state of stress given by (5.20) reads:

$$\overset{-1}{g}_{\alpha\beta} \sqrt{g}\, N\, \frac{\partial^2 x_3}{\partial x_\alpha\, \partial x_\beta} = 0 \qquad (5.21)$$

in absence of forces p. One can show, that the conditions

(5.10) reduce to the same equation. The equation of a surface, on which a stress-state (5.20) is statically possible, must then satisfy equation (5.20), which reads in full after multiplication with $\sqrt{g}$:

$$\left[1 + \left(\frac{\partial x_3}{\partial x_2}\right)^2\right] \frac{\partial^2 x_3}{\partial x_1^2} - 2 \frac{\partial x_3}{\partial x_1} \frac{\partial x_3}{\partial x_2} \frac{\partial^2 x_3}{\partial x_1 \partial x_2} +$$

$$\left[1 + \left(\frac{\partial x_3}{\partial x_1}\right)^2\right] \frac{\partial^2 x_3}{\partial x_2^2} = 0 . \qquad (5.22)$$

Thereby, a boundary curve $x_3 = x_3(x_1(s), x_2(s))$ can be asserted. (5.22) is the known minimal surface equation, which a surface $x_3 = x_3(x_1, x_2)$ with an alleged boundary curve must satisfy, so that the area of the surface will become minimal. The variation principle for the minimal surface reads:

$$\delta \iint \sqrt{g} \, dx_1 \, dx_2 = 0 \qquad (5.23)$$

and can be used for approximate solution of (5.22). The minimal principle for a surface with a hydrostatic stress in every point under a pressure load p has the form:

$$\delta \iint (\sqrt{g} + p \, x_3) \, dx_1 \, dx_2 = 0 . \qquad (5.24)$$

The problems handled here have attracted many scientists and there exist many examples in the literature. The minimal surfaces can be produced with soap bubbles. The form ot these soap bubbles is measured and serves for finding the form of the structure to be built. Still yet the cutting patterns are taken from the calculated and measured soap bubble forms. Thereby the elasticity and the deformation of the real membrane is neglected ore introduced pragmatically. Since the structures to be built in most cases are made of fabrics and since these fabrics have in most cases a little shear stiffness, the form measured on soap bubbles will be reached approximately. If the material will have a larger shear stiffness, the cutting pattern have to be very exact ore wrinkles will occur.

CONCLUDING REMARKS
The equations for a pre-stressed membrane without any bending stiffness have been derived here. It has been shown, that under circumstances the differential equation for the normal displacements U_3 can change the character from a elliptic one to a hyperbolic one. In this case we will have instability. It cannot be shown here but it is true, that this condition of instability

is equivalent to the fact that one of the mean stresses of the Euler stress tensor is negative and one is positive. Therewith and with the fact, that the form of a wrinkle has a positive Gaussian curvature it can be concluded, that wrinkles cannot be described by a pure membrane theory. One has to add bending terms. Namely, in a region with positive Gaussian curvature the two mean stresses must have opposite sign, asuming a membrane theory alone. And this is a contradiction to the equilibrium to be stable.
Further, the general equations for the relations between the stress distribution and the geometry of a surface have been derived for a pure membrane theory. The equations of minimal surfaces have been derived and their importance for cutting patterns have been mentionned. The problem of cutting patterns is, to my knowledge not yet solved.

LITERATURE

Blaschke, W. : Vorlesungen über Differentialgeometrie
    Berlin 1945

Blum, R. : Theorie vorgespannter Membranen, to appear

Blum, R.: Beitrag zur nichtlinearen Elastizitätstheorie in covarianter Schreibweise mit Anwendung auf die Schalentheorie
    Dissertation Stuttgart 1970

Flügge, W. , Chou, S.C.: Large eformation of Very Thin Shells- The Inverse Problem
    J. of A lied Mechanics, 1972

Krauss, W.: Membrantheorie großer Verformungen in kovarianter Darstellung
    Dissertation Stuttgart 1978

Leonhard, J.W.: State of the Art in Inflatable Shell Research
    J. Eng. Mech. Div. 1974

Oden, J.T.,Sato, T.: Finite Strains and Displacements of Elastic Membranes by the Finite Element Method   Int.J.Solids structures 1967

Otto, F.,Trostel,R.:Zugbeanspruchte Konstruktionen
    Frankfurt Berlin 1962

BEHAVIOUR OF GEOTEXTILES

J. P. Giroud

Director, Geotextiles and Geomembranes Group,
Woodward-Clyde Consultants, Chicago, USA.

ABSTRACT. Geotextiles are the textiles used in geotechnical engineering. This presentation includes the following sections: 1. Functions of geotextiles; 2. Relationship between functions and properties of geotextiles; 3. Design examples.

1. FUNCTIONS OF GEOTEXTILES

Geotextiles are used in all the branches of geotechnical engineering: earthworks (earthfills, retaining structures, landslide stabilization, erosion prevention, erosion control); hydraulic works (earth dams, canals, bank and shore protection, coastal works); pollution control (pond linings, solid waste disposals, groundwater recharge); traffic structures (highway embankments, paved and unpaved roads, parking lots, railroad structures); drainage (soil stabilization, control of groundwater, agriculture); etc.

In a given application, a geotextile can perform one or several functions. Sixteen functions have been identified. They are presented below with a definition and an example for each.

Drain: the geotextile is _placed in_ a soil of low permeability through which water is seeping slowly; its function is to gather the water and to convey it towards the outlet.

An example is a chimney drain in an earth dam (Fig.1a).

Waterproof film (or waterproof membrane): the geotextile is

impregnated and/or coated* with an impermeable material, such as asphalt or plastic; its function is to stop liquids and gas.

. An example is a pond lining (Fig. 1b).

Solid filter: the geotextile is placed between the fine soil which has to be drained and the coarse soil which constitutes the drain; its function is to allow water to go through it but to prevent movement of fine soil particles.

Two situations must be distinguished:

. One-way steady flow: an example is a trench drain (Fig.1c);

. Two-ways dynamic flow: an example is a bank protection with rocks (Fig. 1d).

Liquid filter: the geotextile is placed across a flow of liquid carrying fine particles in suspension; its function is to stop the fine particles while allowing the water to go through it.

. An example is a settling pond for groundwater recharge (Fig. 1e).

Support: the geotextile is placed between a waterproof film and a material containing void spaces; its function is to prevent bursting of the film over the voids.

. An example is the lining of an old cracked canal pavement (Fig. 1f).

Separator: the geotextile is placed between two soils which have a tendency to mix when they are squeezed together by the applied loads; its function is to keep apart these two soils.

. An example is a railroad track structure (Fig. 1g).

Surfacing: the geotextile is placed on the ground; its function is to provide a flat and clean surface for traffic.

. An example is a helicopter pad (Fig. 1h).

Curtain: the geotextile is hung alongside an earth or rock mass; its function is to prevent the passage in one way of light and/or in the other way of falling rocks.

. An example is a rock fall net (Fig. 2a).

*This function is different because here the Geotextile is modified.

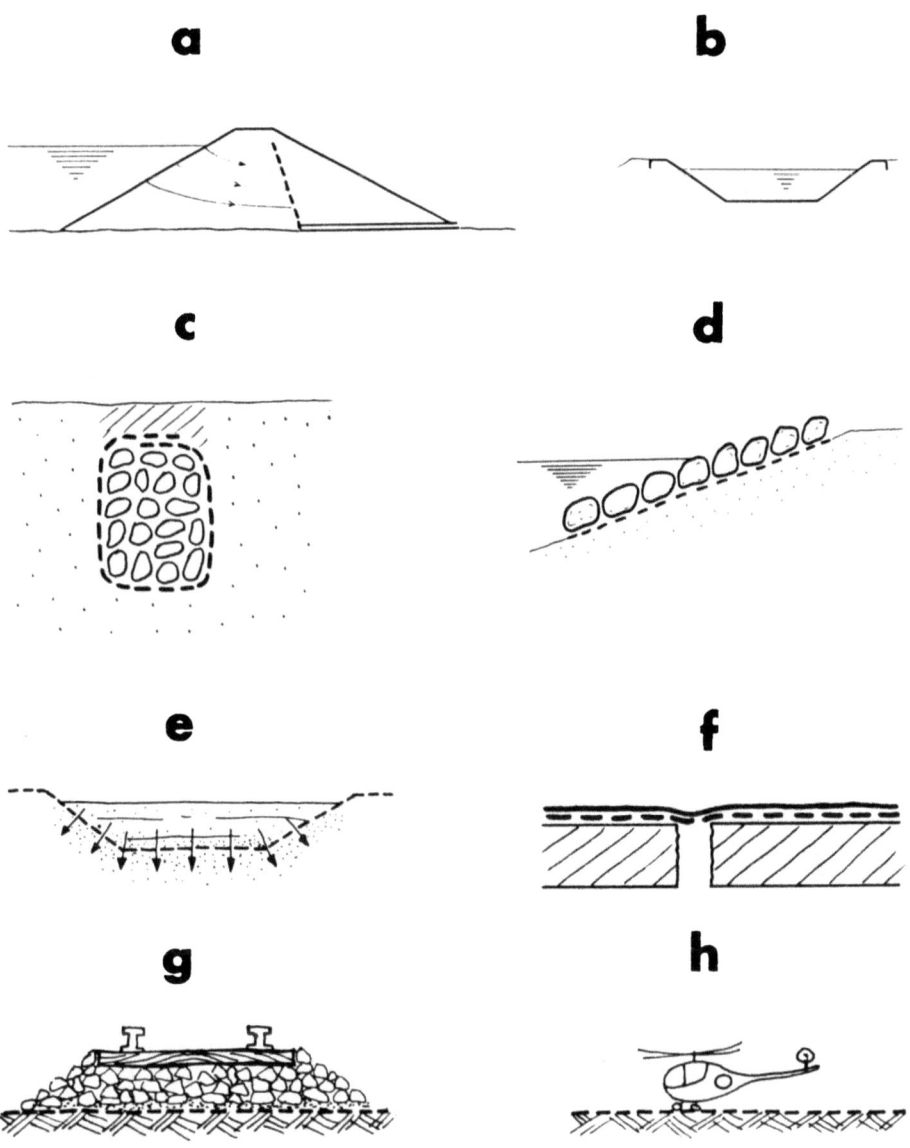

Fig. 1. Drain; (b) Waterproof membrane; (c) Solid filter steady flow; (d) Solid filter dynamic flow; (e) Liquid filter; (f) Support; (g) Separator; (h) Surfacing.

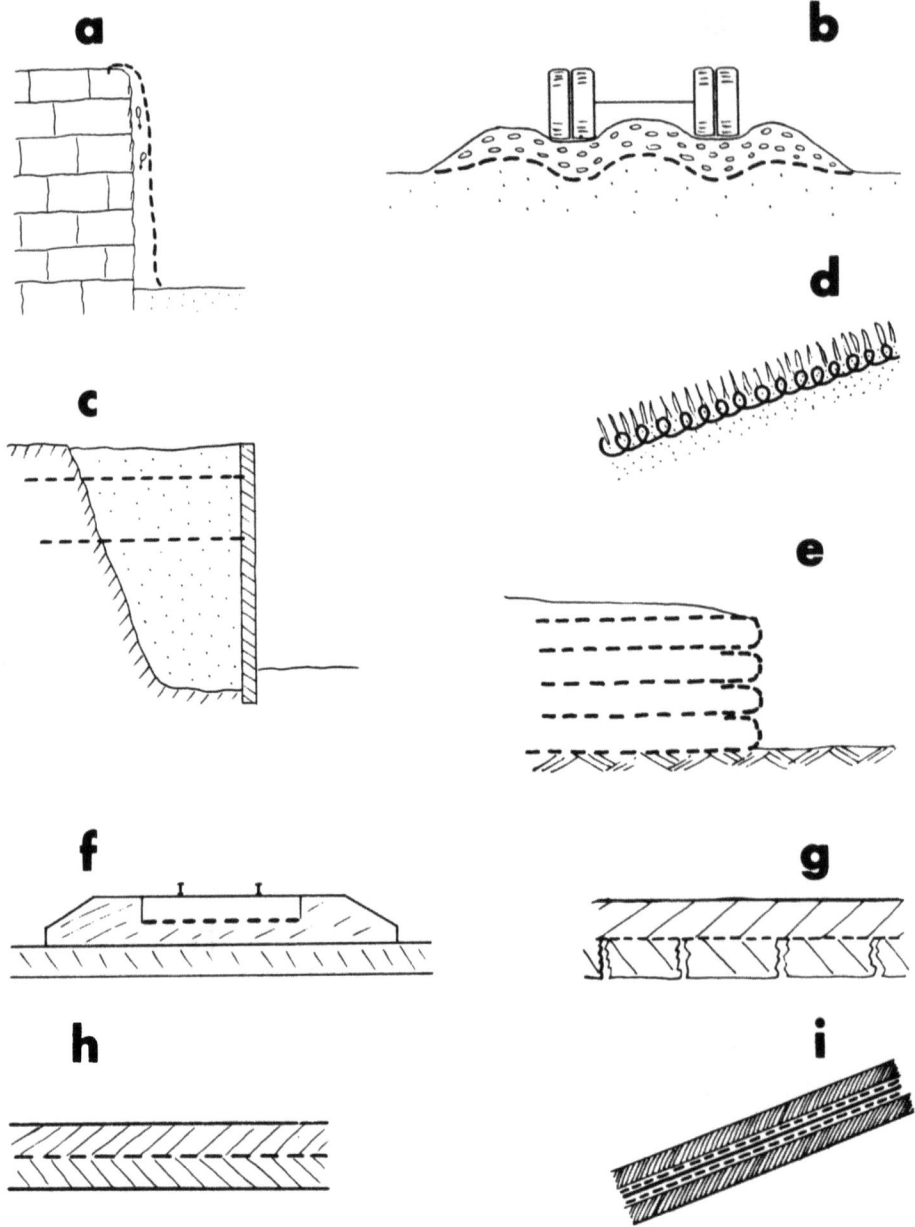

Fig. 2. (a) Curtain; (b) Mechanical membrane; (c) Tie; (d) Binder; (c) Reinforcement; (f) Absorber; (g) Crack barrier; (h) Bond; (i) Lubricator.

Membrane (or mechanical membrane): the geotextile is <u>placed between</u> two materials having different pressures; its function is to contain the material having the highest pressure.

> . An example is the use of a geotextile in an unpaved road to prevent upheaval of the soft subgrade soil between the wheels (Fig. 2b).

Tie: the geotextile is <u>attached to</u> two earth, rock or concrete masses which have a tendency to move apart; its function is to keep them together.

> . An example is a tie-back wall (Fig. 2c).

Binder: the geotextile is <u>placed on</u> a soil the particles of which have a tendency to move apart when they are submitted to small mechanical actions; its function is to keep them together.

> . An example is erosion prevention of a slope (Fig. 2d).

Reinforcement: the geotextile is <u>placed in</u> a soil which is not able to withstand tensile loads applied on it; its function is to carry the tensile loads.

> . An example is a multilayer soil-geotextile reinforced mass (Fig. 2e).

Absorber: the geotextile is <u>placed on</u> a solid mass submitted to shocks and vibration from outside; its function is to reduce the intensity of shock and vibration transmitted to it.

> . An example is the interposition of a geotextile between railroad ties and a bridge deck (Fig. 2f).

Crack barrier: the geotextile is <u>placed between</u> two materials which have a tendency to exhibit cracks; its function is to prevent cracks in one material to extend to the other one.

> . An example is the prevention of reflective cracking in road pavements (Fig. 2g).

Bond: the geotextile is <u>placed between</u> two materials which must not have any differential movement; its function is to increase interlocking (adhesion and friction) between these two materials.

> . An example is a geotextile placed in a pavement (Fig. 2h).

Lubricator: the geotextile is <u>placed between</u> two materials which have to move along each other; its function is to reduce interlocking (adhesion and friction) between these two materials.

An example is a multilayer concrete-geotextile-geomembrane pavement for a canal lining where differential movements are expected (Fig. 2i).

2. RELATIONSHIP BETWEEN FUNCTIONS AND PROPERTIES OF GEOTEXTILES

2.1 Hydraulic Functions

The first four functions described previously are the hydraulic functions. Their relationship with the physical properties of geotextiles is presented in Fig. 3.

The hydraulic behaviour of a geotextile is influenced by its mechanical behaviour. Therefore the role of tensile stresses and compressive stresses must be considered.

To the best of our knowledge, the influence of tensile stresses on the hydraulic behaviour of a geotextile has not been evaluated by means of tests. However, it appears that, in the case of a woven geotextile, tensile stresses increase the space between threads. This results in an increase of permeability to both liquids and solids. This also may be true in the case of a heatbonded nonwoven. It seems more difficult to predict the behaviour of a needlepunched nonwoven. With regard to this type of geotextile, tensile stresses have two opposite effects: they increase the length and decrease the thickness of the geotextile. Therefore, it is impossible to predict whether the spaces between filaments of a needlepunched nonwoven increase or decrease when it is submitted to tension. Since, in many cases, the geotextile filters are submitted to tensions, it is strongly recommended that tests be conducted to evaluate the effect of tensile stresses on the filtration characteristics of a geotextile.

The influence of compressive stresses on permeability is very likely to be unimportant for woven and heatbonded nonwoven geotextiles which exhibit very low compressibility. Therefore, only the case of needlepunched nonwoven geotextiles is considered. The coefficient of permeability to liquids is defined by Darcy's formula:

$$Q/A = ki \qquad (1)$$

where:

Q: discharge of water (m^3/s);
A: area of the cross section of the flow (m^2);
k: coefficient of permeability (m/s);
i: hydraulic gradient (dimensionless).

The hydraulic gradient is:

$$i = \frac{\Delta p}{\rho_w g x} \qquad (2)$$

where:

Δp: hydraulic head loss (N/m^2);
ρ_w: mass per unit volume of the liquid (kg/m^3)
g : gravity $(9.81 \ m/s^2)$;
x : length of the flow path (m).

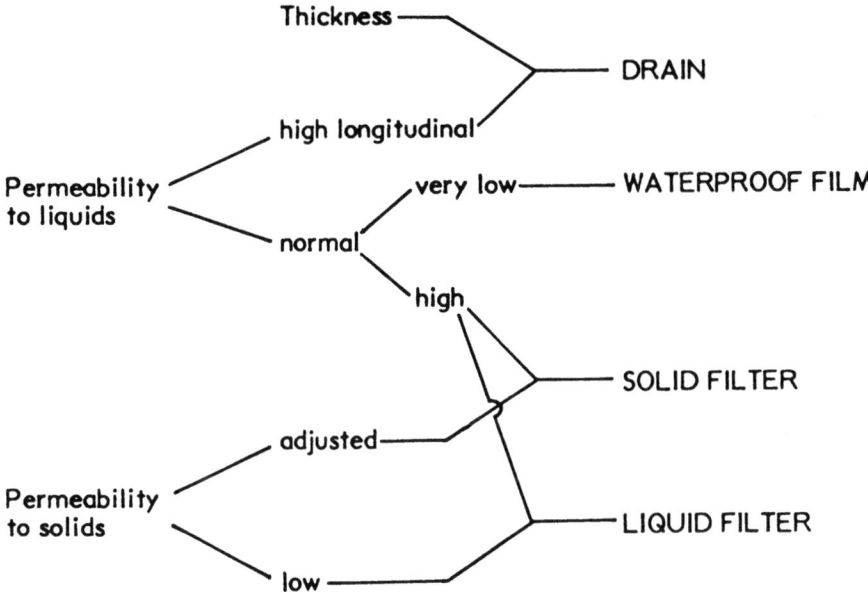

Fig. 3. Relationship between properties and hydraulic functions of geotextiles.

Two cases of flow must be considered: normal flow (Fig. 4a) and longitudinal flow (Fig. 4b). Normal flow occurs when the geotextiles acts as a filter while longitudinal flow occurs when the geotextile acts as a drain.

In the case of normal flow:

$$A = LB$$
$$i = \Delta p / \rho_w g h$$

Hence:

$$Q/LB = (k_N/h)(\Delta p/\rho_w g) \qquad (3)$$

where:

k_N: coefficient of normal permeability (m/s).

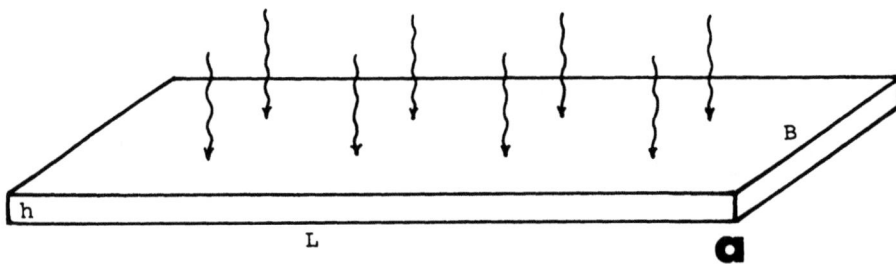

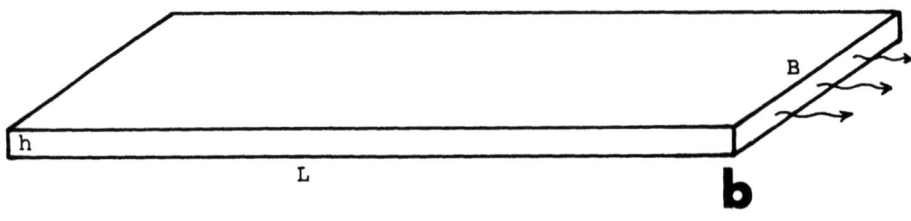

Fig. 4. Normal and longitudinal flows.

In the case of longitudinal flow:

$$A = Bh$$
$$i = \Delta p / \rho_w g L$$

Hence

$$Q/B = (k_L h)(\Delta p/\rho_w gL) \qquad (4)$$

where:

k_L: coefficient of longitudinal permeability (m/s).

It appears that a normal flow is governed by k_N/h and a longitudinal flow by $k_L h$. The following nomenclature is proposed:

k_N/h: permissivity of the geotextile;
k_L/h: transmissivity of the geotextile

Tests have been performed at the University of Grenoble (France) on a needlepunched nonwoven geotextile. The influence of a compressive stress on the thickness and the permeability to water of the geotextile has been measured. Typical results are presented in the upper part of Fig. 5.

For the considered geotextile, k_N and k_L have close values: therefore only an average value for k is presented. Numerical values for permissivity and transmissivity are presented in the lower part of Fig. 5. They have been deduced from the values of k and h. It may be seen that the longitudinal flow is more affected by a compressive stress than a normal flow. As discussed previously, normal flow occurs when the geotextile acts as a filter (Fig. 1c, d, e) while longitudinal flow occurs when the geotextile acts as a drain (Fig. 1a).

No tests have been performed to evaluate the influence of a compressive stress on the permeability of a geotextile to solids. However, in the case of needlepunched nonwovens, a simple formula can be proposed:

$$\frac{O + d}{O' + D} = \sqrt{\frac{h}{h'}} \qquad (5)$$

where:

O : average space between filaments corresponding to thickness h (mm) for the considered geotextile;

O': average space between filaments corresponding to thickness h' (mm) for the same geotextile;

d : diameter of filaments (mm).

The demonstration of this formula stems from the fact that all the geometrical models used for describing the structure of a nonwoven geotextile lead to the same type of expression for the spaces between filaments:

$$O = d (\lambda \sqrt{h\rho_f/m} - 1) \qquad (6)$$

where:

  m : mass per unit area of the geotextile (kg/m^2);
  ρ_f: mass per unit volume of the filaments (kg/m^3);
  x^f: dimensionless coefficient depending on the geometrical model.

2.2 Mechanical functions

Relationship between properties of geotextiles and their twelve mechanical functions is presented in Fig. 6. It appears that stresses exerted on the geotextiles fall in two classes: the distributed stresses and the concentrated stresses. As shown in Fig. 7, there is a correspondence between distributed and concentrated stresses and the following pairs can be associated: compression-puncturing, tension-tearing, distortion-snag, membrane-bursting and bending folding.

Actually, these stresses are not equally important: (1) because of the high flexibility of geotextiles, stresses due to bending-folding are negligible; (2) in the case of membrane-bursting, the compressive stress is the cause of the phenomenon, but its direct effect on the geotextile is usually negligible; therefore, the membrane-bursting effect results in a quasi-pure tension of the geotextile; (3) it is unlikely to exert shear without compression.

Therefore, the discussion can be restricted to three types of stresses: compression, tension and triaxial.

As aforementioned, tests have been performed to evaluate the compressibility of needlepunched nonwoven geotextiles (Fig. 5). But this is not sufficient to have a complete knowledge of the compressive behaviour of geotextiles. For example, tests under repeated loads would be necessary to establish the utilization of geotextiles in such functions as absorber or crack-barrier (Fig. 6).

There is much more to say about tensile behaviour. Many various tests are used (Fig.8). Some are standard tests currently used in the textile industry. Others are rather new tests developed by civil engineers. A classification is necessary and a tentative one is proposed in Fig. 9. It is noteworthy that

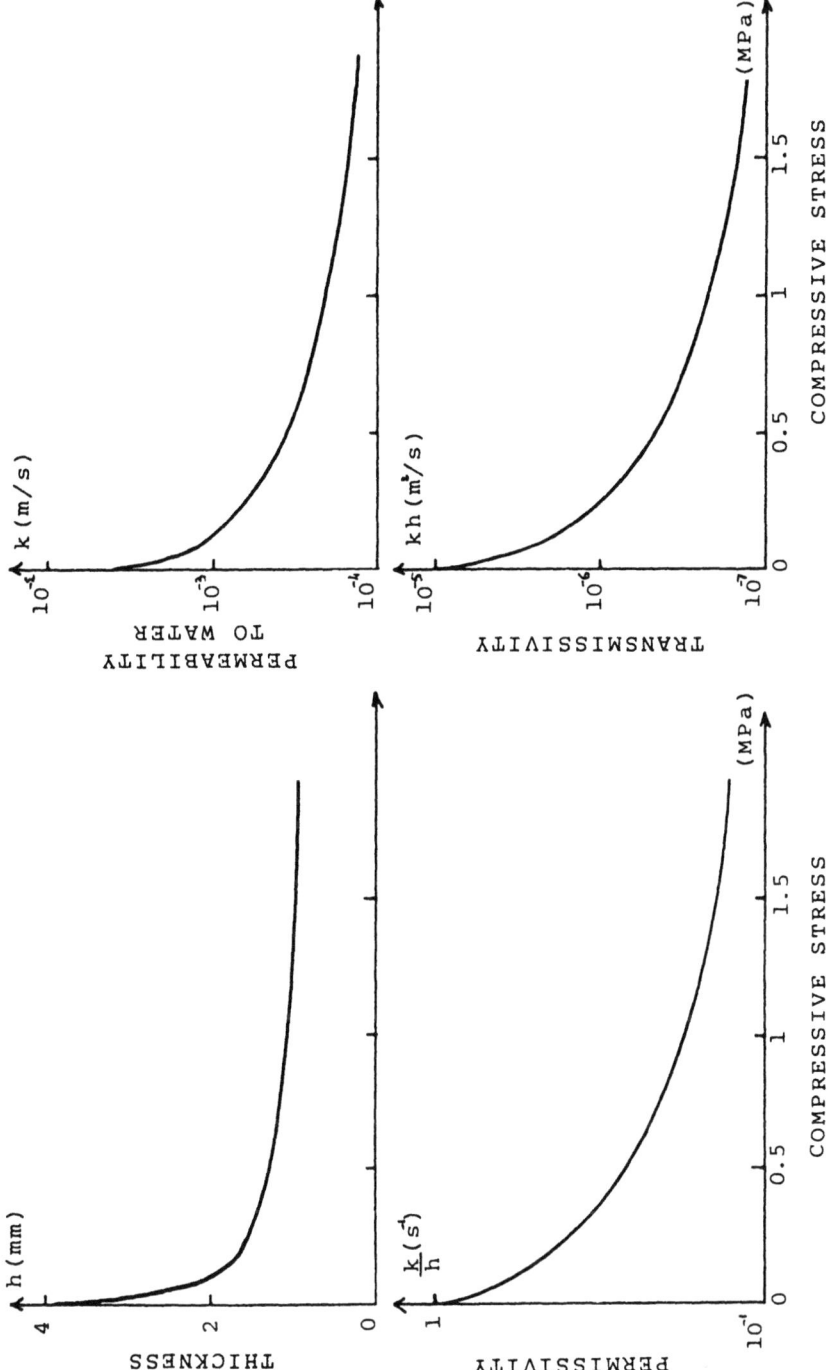

Fig.5. Thickness, permeability, permissivity and transmissivity of a needlepunched geotextile.

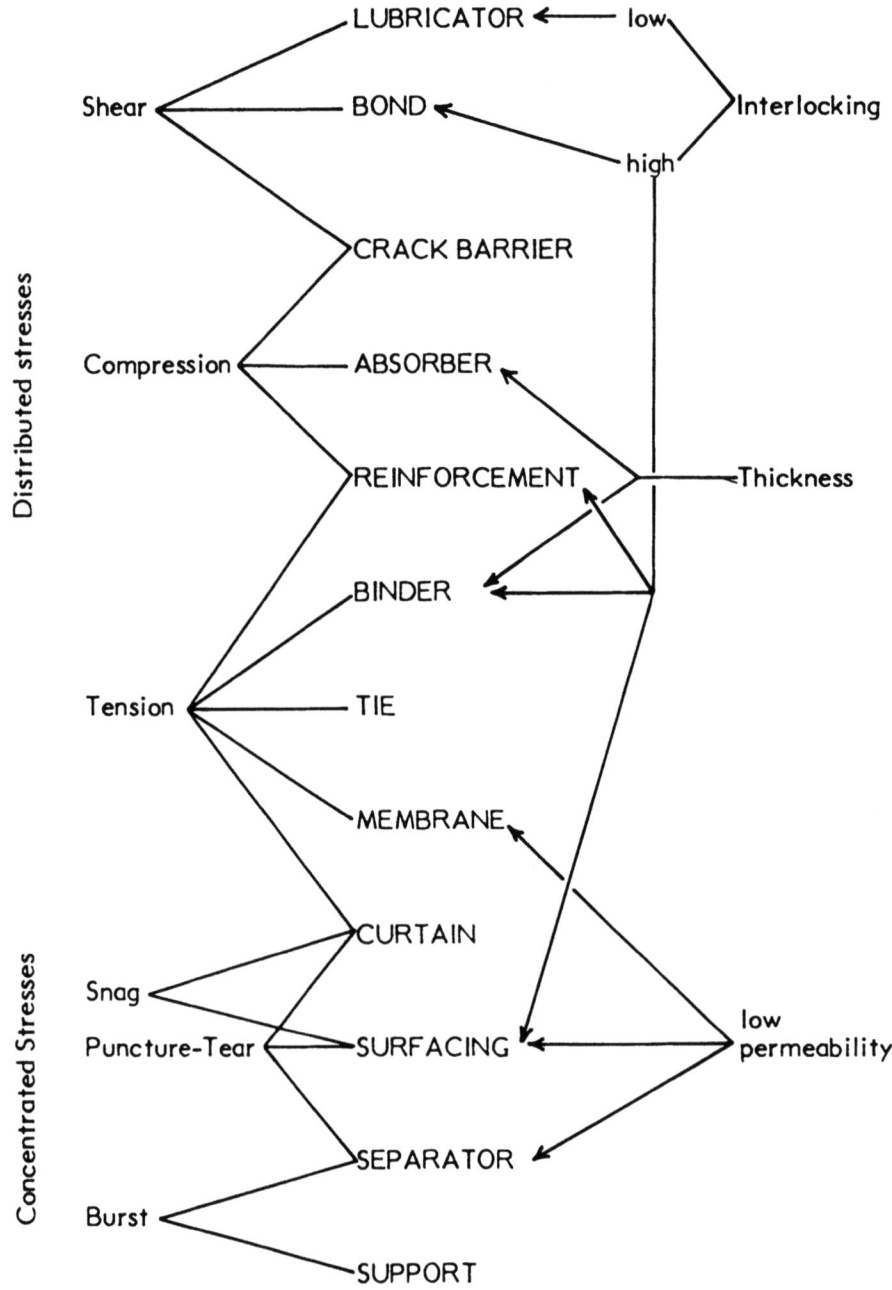

Fig. 6. Relationship between properties and mechanical functions of geotextiles.

the most classical tensile tests from the textile industry, the
strip test and the grab test, are both unidirectional. However,
in field situations, compound stresses are more likely to occur
than elementary stresses. As shown in Fig. 7, this results in
bi- or tridirectional situations. Therefore bi- and tridirectional
tests must be developed to meet the needs of geotechnics.

Attempts to develop tridirectional tests on geotextiles have been
made in some geotechnical laboratories. The main result of these
tests is that the tensile strength of a geotextile is increased
if a compressive stress is exerted simultaneously with the
tensile stress.

3. DESIGN EXAMPLES

The following examples are related to an actual earth dam. They
illustrate the manner in which the aforementioned properties
are considered in the design of a civil engineering structure.

3.1 Hydraulic design

A geotextile can be used as a chimney drain in earth dam as shown
in Fig. 10a. To evaluate the requirements needed for this
application, the following procedure is suggested.

The first step is to evaluate the discharge of water through the
earth which constitutes the upstream portion of the dam, between
the reservoir and the chimney drain. The exact method would
consist of drawing a flow net. This method is quite tedious and
is beyond the scope of this paper. A simpler calculation can be
made if the horizontal permeability of the earth is much greater
than its vertical permeability (this is very often the case
because of the construction of dams in successive layers). In
this case, a good approximation consists of assuming that the
flow of water is horizontal. Therefore it is sufficient to
know the coefficient of horizontal permeability of the earth.
Assuming that its value is $k_E = 10^{-8}$ m/s, the discharge through
the dam, dQ, per unit width, B, at depth z, can be expressed
according to Darcy's formula (Eq.1):

$$\frac{dQ}{B} = k_E \frac{z\,dz}{L} = k_E \frac{z\,dz}{3(9+z)} \tag{7}$$

where:

L: length of the flow path (Fig. 10a).

Hence, the discharge in the chimney drain at depth z, is:

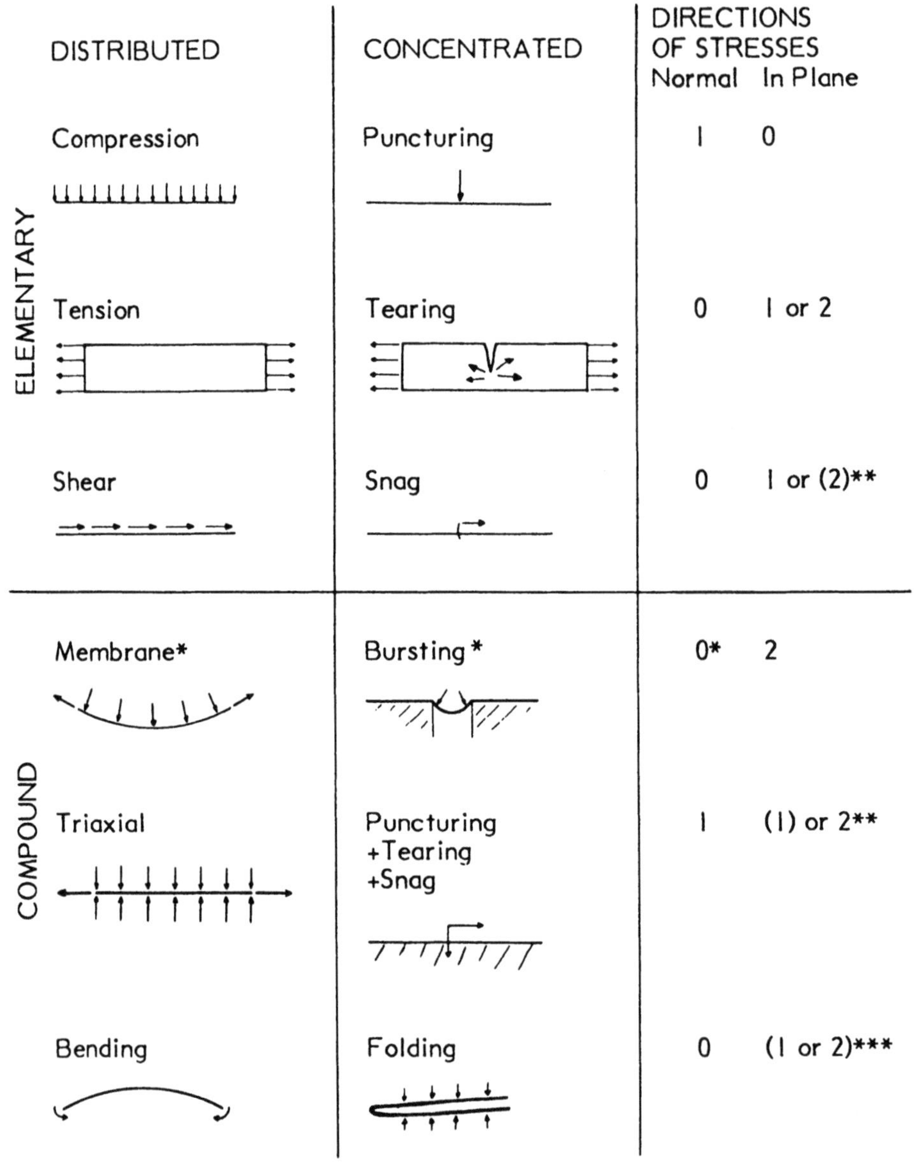

	DISTRIBUTED	CONCENTRATED	DIRECTIONS OF STRESSES Normal	In Plane
ELEMENTARY	Compression	Puncturing	1	0
	Tension	Tearing	0	1 or 2
	Shear	Snag	0	1 or (2)**
COMPOUND	Membrane*	Bursting*	0*	2
	Triaxial	Puncturing +Tearing +Snag	1	(1) or 2**
	Bending	Folding	0	(1 or 2)***

Fig. 7. Classification of stresses exerted on a geotextile. Notes: (*) In the case of membrane - bursting, the applied compressive stress is "transformed" into a tensile stress, this is why 0 is mentioned in the normal direction: (**) the figure in parenthesis is more unlikely to occur than the other; (***) Bending-folding is not important since geotextiles are usually very flexible (except when they are impregnated with bitumen).

$$\frac{Q}{B} = \frac{k_E}{3} (z - 9 \log \frac{z+9}{9}) \qquad (8)$$

The second step consists of evaluating the required transmissivity for the geotextile from the above calculated discharge. Again the Darcy's formula must be used (Eq.1):

$$\frac{Q}{B} = kh \sin\beta \qquad (9)$$

where:

- k : coefficient of permeability of the geotextile (m/s);
- h : thickness of the geotextile (m);
- kh : transmissivity of the geotextile (m^2/s);
- $\sin\beta$: hydraulic gradient in the geotextile.

If the geotextile were laid perfectly flat, β would be the slope of the chimney drain (β = 60°, Fig. 10a). But the geotextile is likely to have an accordion shape. Therefore it is more conservative to take:

$$\alpha = 45° \qquad \sin\alpha = \sqrt{2}/2$$

The required value for the transmissivity of the geotextile is calculated from Eq. 8 and Eq. 9.

$$kh > \frac{10 k_E \sqrt{2}}{3} (z - 9 \log \frac{z+9}{9}) \qquad (10)$$

(where 10 is a factor of safety)

The numerical values of kh for different values of depth z are tabulated in Fig. 10b. Also tabulated is the earth pressure at depth z, expressed by:

$$p = \rho g (z + z_0) \qquad (11)$$

where:

- ρ : mass per unit volume of the earth (assumed value: 2000 kg/m^3);
- g : gravity (9.81 m/s^2);
- z_0 : depth of earth above the water level (Fig.10a) (z_0=5m).

In Fig. 10c, the required transmissivity thus obtained is compared with the transmissivity of the geotextile presented in Fig. 5. It is shown in Fig. 10c that the use of this geotextile as a chimney drain is possible only in the upper 21m of the dam.

3.2 Mechanical design

If the foundation of the considered dam consists of a cracked rock, (Fig. 11a) a geotextile appears to be an interesting solution to bridge the cracks. The procedure for selecting the geotextile is described below.

The geotextile, in this application, is submitted to a high compressive stress. As shown in Fig. 7, there is a risk of puncturing and of bursting. The resistance to puncturing can be evaluated only by tests while the resistance to bursting can be determined by a theoretical analysis.

If a crack of infinite length and of width B (Fig. 11b) is considered and if the pressure, p, exerted by the earth is assumed to be normal to the geotextile, the shape of the geotextile is circular (Fig. 11c). From this assumption, it can be shown that:

$$t = \rho B f(\varepsilon) \qquad (12)$$

where:

t : tension of the geotextile (N/m);
p : earth pressure (N/m^2);
B : width of the crack (m);
ε : elongation of the geotextile.

The function $f(\varepsilon)$ is defined by two equations:

$$f(\varepsilon) = \frac{1}{4} \left(\frac{B}{2d} + \frac{2d}{B} \right) \qquad (13)$$

$$1 + \varepsilon = \frac{1}{2} \left(\frac{B}{2d} + \frac{2d}{B} \right) \text{Arc sin} \frac{2}{\frac{B}{2d} + \frac{2d}{B}} \qquad (14)$$

$f(\varepsilon)$ is tabulated in Fig. 11d.

The case of a circular hole is much more complicated than the case of a crack of infinite length. The reason is that the shape of the cross section of the geotextile is no longer circular. However, a reasonable approximation is obtained by assuming a spherical shape. The expression thus obtained for the tension is:

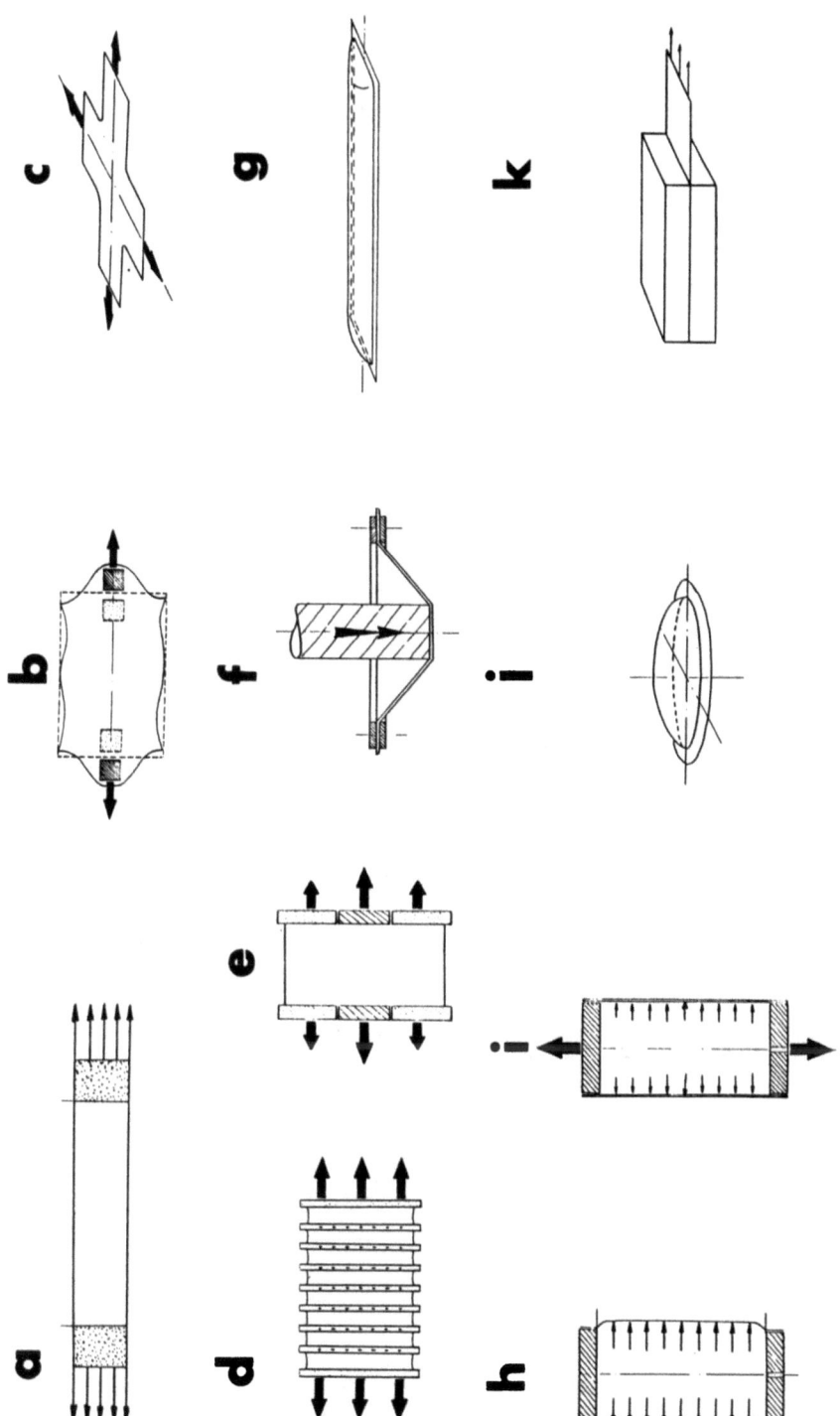

Fig. 8. Illustrations of various types of tensile tests.   Note: Letters a through k refer to Fig. 9.

$$t = \frac{pBf(\varepsilon)}{2} \tag{15}$$

For practical applications, two different situations must be considered:

(1) Tensile strength, t, and elongation at failure, ε of the geotextile are known from a "bidimensional test" (see Fig. 9). Then it is possible to determine the bursting pressure over a crack of infinite length and width B:

$$p = \frac{t}{B f(\varepsilon)} \tag{16}$$

or over a circular hole of diameter B:

$$p = \frac{2t}{B f(\varepsilon)} \tag{17}$$

Moreover, for geotextiles having the same structure, the tensile strength is proportional to the mass per unit area:

$$t = mT \tag{18}$$

where:

    m : mass per unit area (kg/m^2);
    T : unit tensile strength (($N/m)/(kg/m^2) = m^2/s^2$)).

For example, for typical needlepunched nonwoven geotextiles made of continuous polyester filaments, a value for T on the order of 80,000 m^2/s^2 is reasonable.

(2) The bursting strength, po, of the geotextile is known from a bursting test performed with standard equipment (see Fig. 8j).

Then it is possible to determine the bursting pressure over a crack of infinite length and width B:

$$p = \frac{P_o B_o}{2B} \tag{19}$$

or over a circular hole of diameter B:

$$p = \frac{P_o B_o}{B} \tag{20}$$

where:

    Bo : diameter of the hole in the apparatus used for the bursting test (Bo = 30.5 mm for ASTM

Direction of stress	Distribution of stress and strain in plane of the geotextile		Variation of the three principal stresses			Variation of the three principal strains			Ability to measure anisotropy	Test	
			σ_1	σ_2	σ_3	ϵ_1	ϵ_2	ϵ_3			
Unidirectional	uniform		> 0	0	0	> 0	< 0	< 0	Yes	Strip	(a)
	nonuniform		> 0	?	0	> 0	?	?	Yes	Grab	(b)
Bidirectional	uniform	general case	> 0	> 0	0	> 0	> 0	< 0	Yes	Cross	(c)
		no lateral strain	> 0	> 0	0	> 0	0	< 0	Yes	Constrained strip	(d)
										Widestrip	(e)
	nonuniform		> 0	> 0	0	> 0	> 0	< 0	No	Punching	(f)
Membrane	uniform	general case	> 0	> 0	(< 0)	> 0	0	< 0	Yes	?	
										Trough test	(g)
		no lateral strain	> 0	> 0	(< 0)	> 0	0	< 0	Yes	Sleeve	(h,i)
	nonuniform		> 0	> 0	(< 0)	> 0	> 0	< 0	No	Burst	(j)
Tridirectional	uniform	general case	> 0	> 0	< 0	> 0	> 0	< 0	Yes	Cross with normal stress	
		no lateral strain	> 0	> 0	< 0	> 0	0	< 0	Yes	Constrained strip with normal stress Wide strip	
	nonuniform		> 0	> 0	< 0	> 0	> 0	< 0	Yes	Pull out test	(k)

Fig. 9. Classification of various types of tensile tests

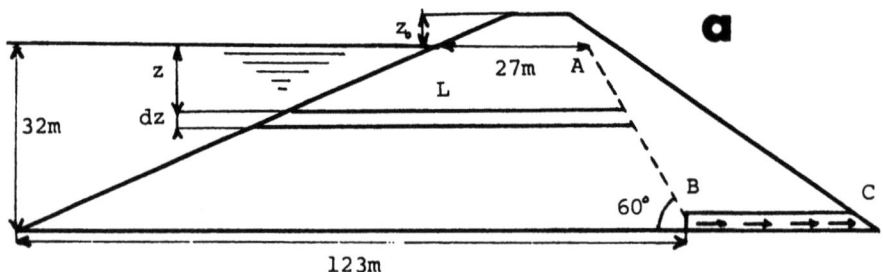

Depth Fig. 10a z (m)	Earth Pressure Eq. 11 p (kPa)	Required Transmissivity Eq. 10 kh (m^2/s)
0	98	0
1	118	2.44×10^{-9}
10	294	1.54×10^{-7}
15	392	2.91×10^{-7}
20	490	4.46×10^{-7}
25	589	6.15×10^{-7}
30	687	7.92×10^{-7}
32	726	8.65×10^{-7}

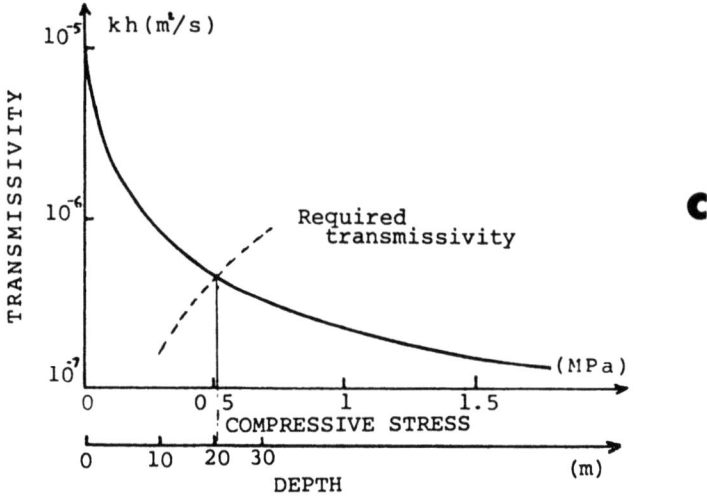

Fig. 10. (a) Cross section of the dam; AB is the location selected for the chimney drain; BC is the outlet pipe; (b) Earth pressure and required values for the transmissivity of the geotextile; (c) Transmissivity of the considered geotextile (from Fig. 5.) compared to the required transmissivity (from Fig. 10b.).

Mullen test; Bo = 35.7 mm for AFNOR).

Eq. 19 and 20 have been developed from Eq. 16 and 17 assuming that, for a given geotextile, failure occurs at the same elongation in a burst test and in a bidirectional tensile test. Preliminary laboratory tests indicate that this assumption is reasonable.

Example problem. The following properties of a geotextile are known:

mass per unit area, $m = 0.27$ kg/m^2
bursting strength (ASTM), $P_o = 2.7 \ 10^6$ N/m^2
"bidimensional" tensile strength, $t = 21000$ N/m
"bidimensional" elongation at failure $\varepsilon = 35\%$

The width of the cracks is 0.1m. Is this geotextile suitable?

There are two possibilities to make this evaluation. The first possibility is to use Eq. 16 and Fig. 11d:

$$p = \frac{21000}{0.1 \times 0.52} = 404 \text{ kN/m}^2$$

The second possibility is to use Eq. 19:

$$p = \frac{2.7 \ 10^6 \times 0.0305}{2 \times 0.1} = 412 \text{ kN/m}^2$$

The two results are close. This demonstrates that bursting and "bidimensional" tests conducted on this geotextile were very consistent. By comparing the value of p thus obtained with the value of the earth pressure at the base of the dam, 726 kN/m^2, as tabulated in Fig. 10b, it is clear that the considered geotextile is not sufficient. If the same type of geotextile is to be used, the mass per unit area should be higher, according to Eq. 18. The minimum mass per unit area to be selected is:

$$m = \frac{0.27 \times 726}{412} = 0.48 \text{ kg/m}^2$$

Actually, a higher value should be used in order to have a factor of safety.

Because of the scope of this paper, rather simple examples have been presented. The selected cases correspond to situations in the field which can easily be related to simple laboratory tests. In general, more complicated kinematics have to be considered.

4. CONCLUSION

So far, in most of the cases, geotextiles have been used with little or no design. This can be accepted for many minor and/or temporary structures. But such an approach hampers the development of the use of geotextiles for major structures. This paper presents an attempt at a rational approach to design major structures with geotextiles. It is hoped that this will help to foster the use of geotextiles in major structures.

The author is indebted to W. G. Salomone and L. Noiray for many valuable comments and help for the edition of this paper.

DISCUSSION I

YARNS, CORDS AND ROPES

Leader   : Professor S. Backer
Rapporteur : Dr. H. Behery

On the Use of Energy Methods

The energy method is a powerful tool for finding approximate solutions to complex textile mechanical problems (Huang). However, for some problems it has its limitations, examples of which are as follows:

(i)   The accuracy of the energy method usually depends on the assumptions made.

(ii) The energy method often gives an overall picture of the deformation, while the force method provides detailed information of it.

(iii) Inaccuracy may also result when using the energy method if either separation of fibres (or filaments), local buckling, or contact deformation are involved.

Some of these are avoided in work by Huang and Funk [1] on the extension of continuous filament yarns. In this work they use the yarn element model previously adopted by Hearle, but the assumption of uniform contraction with radial direction and equal lateral forces acting in the transverse direction of the yarn element is relaxed. In order to include the possibility of fibre separation the yarn is divided into two regions. In the central region it is assumed that there is no separation of fibres, and in the region near the yarn surface there is a separation of fibres in the longitudinal direction. Differential equations are set up for these two regions and continuity conditions in displacement and stress are used at the interface. It is found that the assumption used by Hearle concerning the uniform radial contraction and uniform lateral pressure on the yarn element is only valid in the central region where there is no fibre separation. The findings suggest that (a) it is necessary to accept the idea of separation of fibres in a continuous filament yarn subjected to extension or to use a yarn element model different from Hearle's, and (b) the fibre separation in a yarn subjected to extension can be detected by the force method, and not by the energy method.

It was then remarked (Levinson) that there is no single energy method for elastic systems, but there are various methods based upon several different principles. The canonical principle is the Hu-Washizu principle enunciated in the 1950's. In this principle all the fields - stress, strain and displacement - may be varied independently to yield the equilibrium, stress-displacement and constitutive equations as well as all the boundary conditions. This principle is stated in terms of the strain energy density. If one imposes certain of the field equations as constraints on admissible trial fields, one obtains the potential energy principle from the canonical principle and the complementary principle from Reissner's principle. Castigliano's theorem is a special case of the complementary energy principle for a system of a finite number of discrete applied forces. It may be easily extended to nonlinear systems. The potential energy principles may also be extended to nonlinear systems, but the general extension of the strict complementary energy principle to the nonlinear case is an open question.

For exact formulations, all the principles lead to the same boundary value problems. On the other hand, the choice of an energy principle to serve as the basis of an approximate method to solve a particular problem should be governed by what the problem solver best understands about the system being studied. If, for example, the displacement field can be approximated with some confidence, then the use of a potential energy field is indicated.

With regard to the applications of an energy method to continuum models for fabrics, Levinson pointed out that it may not allow for one-sided constraints, such as tension, being inadmissible. A possible alternative method is based on a field far removed from textile mechanics, namely, the phenomenological modelling of soil-structure interaction. Levinson, Bharatha and others view soils as "quasi-continua" based on the fact that full continuum interaction does not occur between soil elements just as it does not between fabric components. They have developed a general theory of elastic foundations which is now being

studied in detail for the very simple linear isotropic case. The
theory allows one to specify the degree of material continuity
in the foundation. More continuity means more complicated
equations. The approach used to derive the appropriate field
equations is an equilibrium one. The boundary conditions are
not, however, so easily obtained.

It is instructive to recall the development of classical plate
theory, where equilibrium methods led to the correct differential
equation. However, the proper specification of the boundary
conditions from equilibrium considerations eluded Lagrange, Navier
and Poisson during the first half of the nineteenth century and
in 1850 Kirchhoff found them by means of the potential energy
principle. Thompson (Lord Kelvin) and Tait explained the meaning
of the subtle free edge condition in mechanical terms 29 years
later. The obvious conclusion, from the above illustrations is
that one should know one's problem and use the most expedient
approach to solve it – even use different approaches for the
various aspects of a single problem.

Experimental Data on Yarns and Ropes (Stevens)

In order to illustrate the effect of structure on the dependence of
loading history, data was presented on the load-extension recovery
characteristics of nylon fibres, yarns and ropes. The replacement
of natural fibre by nylon in a twisted rope, in order to achieve
better strength/weight ratios, produced undesirable effects such
as higher extensibility, large torque under load, and kinking
after a sudden retraction. An 8-end braided rope of twisted
members can be made which is free of torque but still rather
extensible; 48% to break. Fig. 1 shows load-extension recovery
cycles made at low speed. After the first cycle only about 40% of
the work done to stretch the rope is recovered and the second
loading curve is much steeper. As a result of further loading
cycles, the rope extends further due to creep, which varies
somewhat logarithmically with the number of cycles. The complete
load extension recovery curve at the 120th cycle is shown,
indicating that the hysteresis is then much less. The same trend
can be observed with a braid-over-braid structure of the same
calibre, which has only 30% extension to break.

Fig. 2 [3] shows a comparison of the fibre, the yarn, a tubular
woven webbing, and two sizes of braided ropes, all woven from
nylon (ICI, Type 114), on a specific tension basis against strain.
Since it is possible to determine factors for the effective linear
density of each rope to that of the yarn and also of the increased
stress factor due to the geometry of a yarn in the weave, it should
be possible to assess the residual factors in strain due to the
distribution of the weave and further compression of the fibres.

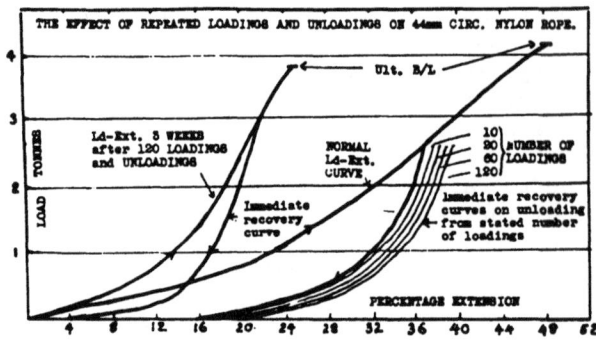

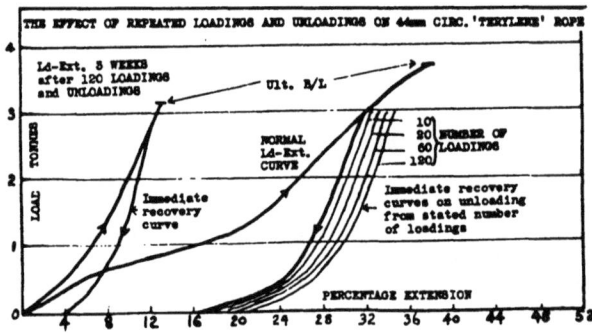

Fig. 1

Yarn Strength - Blended Yarns

The mechanics of twisted structures have been developed to a high
degree on the basis of fairly simple models (Backer). The single
helix model of the single yarn and the double helical model of
the plied yarn can be applied with success to analytical predictions
of selected yarn properties, based on knowledge of fibre properties.
The most successful analyses are concerned with predictions of
tensile elastic moduli or with formulation of expressions for
generalized stiffness, relating structural resistance to tensile
displacements, or to twisting displacement or to bending
curvatures or, in some cases, to combinations of these. But
predictions of structural strength efficiency leave a good deal
to be desired.

On occasion, the assumption of complete structural failure
immediately upon rupture of the first fibre in the system, permits
reasonable estimates of the tensile strength of twisted structures,

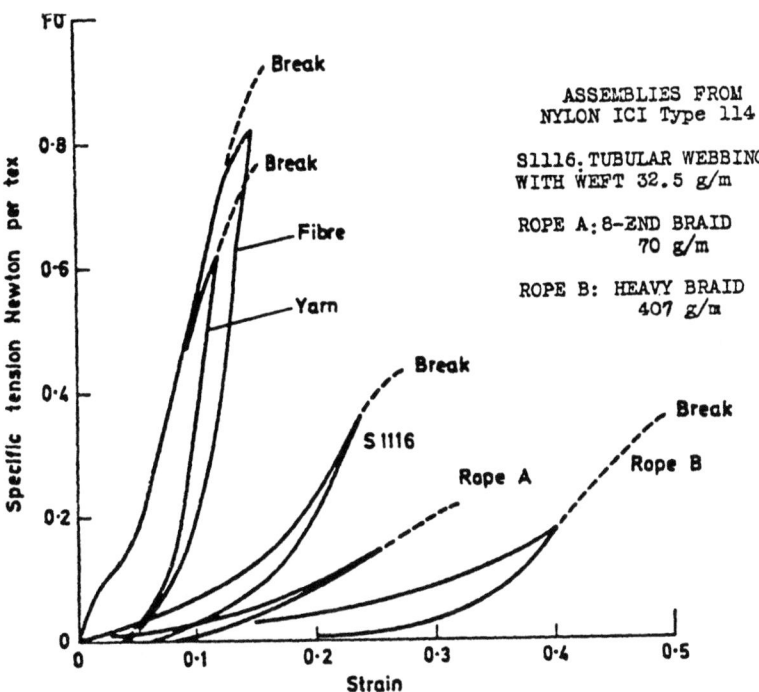

Fig.2 Load vs. Extension for Nylon fibre, yarn and ropes

and in a limited number of cases reasonable predictions of structural elongation to failure. But the success of such a simple failure model depends on there being a high level of fibre uniformity in properties, in geometry, and in structural placement. When nonuniformity is encountered, the initial fibre failure and the propagation of the rupture across the structure become stochastic problems, which can be dealt with by using the methods developed by Phoenix for parallel fibre bundles with varying amounts of interaction. The role of fibre interactions during rupture is accentuated if there is a bi-modal distribution of fibre properties. A typical case is that of the polyester/cotton blend.

Experiments were described (Backer [4,5]) on a model blended twisted strand consisting of 90 yarns of continuous filament polyester and staple cotton. The blends varied from cotton/polyester percent ratios of 10/90 to 67/33 with a few specimens at 2/98. The twist mutiples varied from about 1 to 2 to 3 (in the cotton system). The placement patterns consisted essentially of uniform distribution, of cotton core/polyester sheath and cotton sheath/polyester core and migration of the components was kept to a minimum.

The results provide support for some predictions based on assumptions of uniform properties. In particular of critical segment lengths as a function of strand twist and strand extension, and the critical blend ratios and strand twists at which the strand converts from a level of polyester elongation-to-rupture to that of cotton elongation-to-rupture. They also show that it is possible to use a tracer method for determining strain distribution in twisted strands, and provide visual evidence of rupture propagation from an initial component failure and identification of the conditions under which rupture propagates to strand failure. Finally, they provide a data base for further stochastic studies of strength efficiency, based on fibre-bundle theory.

Another important property which, for certain proportions of the constituents in a blended yarn, has not been adequately taken into account, in explaining the loss of mechanical properties, is the inter-fibre friction (Goswami). A recent study [6] shows that the breaking energy of cotton/polyester yarns, when the polyester content is 60%, could primarily be attributed to the surface interactions (inter-fibre friction). The first step is to determine the difference between the theoretically calculated value of breaking energy of a blended yarn and those obtained by experimental observations. The theoretical value can be obtained by a slightly modified form of the relationship obtained by application of the law of mixtures.

$$(BE)_b = \frac{1}{2} \left\{ f_c (BE)_c + f_p (BE)_p (E) \right\}$$

where (BE) denotes breaking energy, f fractional component, with suffices b, c, p corresponding to blend, cotton and polyester respectively, and (E) is an interaction term which accounts for the fact that the blended yarn breaks above the cotton breaking extension E_c and below the breaking extension of polyester E_p.

Fig. 3 shows theoretical and experimental breaking energies as functions of blend ratio. ΔBE is the difference between the two.

The shear friction behaviour of the fibres is then determined, in order to separate the effects of friction between similar components from those between dissimilar components. This is achieved by writing

$$H_e = f_c H_c + f_p H_p + H_{cp}$$

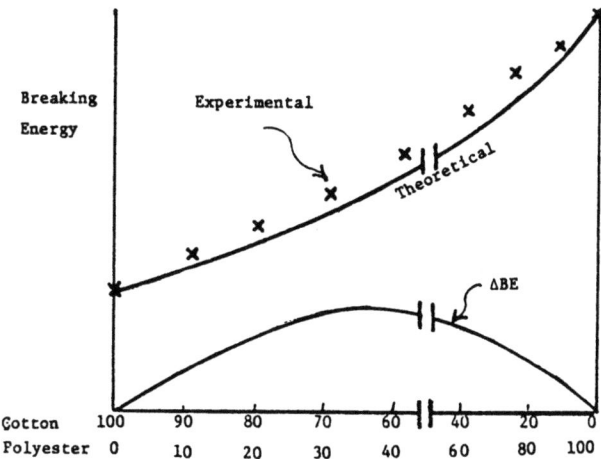

Fig.3 Breaking energy of cotton/polyester blends.

where H denotes shear friction with suffices as above and also e, cp corresponding to experimentally obtained values for the blend, and between the two components, respectively. Thus H_{cp} is a function of the blend ratio. The difference ΔH between experimental shear friction and theoretical, obtained by the above rule, varies with blend ratio in the same way as ΔBE. This indicates that fibre surface interactions play a very important role in determining the loss in breaking energy of blended yarns. However, the influence of fibre migration, fibre packing and the clustering of fibres in the yarn cross-section cannot be ignored.

Lateral compression of yarns (Carnaby)

A brief outline was given of an extension to the author's energy method for analysing the mechanics of bulky yarns. In this new development successive axial zones were treated in such a way that they could interchange their radial ordering in a manner simulating the dynamics of yarn formation. This enabled one to derive the migration pattern and strain distribution in a freshly spun yarn theoretically.

It was stated that any attempt to model lateral compression of a yarn should include terms to account for steric interference of the helices. In fact the curvatures associated with packing could be expected to swamp the curvatures calculated from deformed helix shapes after a very small amount of compression. This steric interference could be modelled using an expressing for bending energy derived from the local specific volume distribution in the yarn. However as the annular symmetry was no longer valid, the grid used to model this packing variability should not consist of concentric cylinders but should also be subdivided in the circumferential direction.

REFERENCES
1. N.C. Huang and G.E. Funk, Textile Res. J., 45, 14 (1975).
2. G.W.H. Stevens, AIAA, paper 75 - 1364
3. G.W.H. Stevens, Aero, Res. Council, CP1327 (1975)
4. C.J. Monego and S. Backer, Textile Res.J., 38, 762. (1968).
5. C.J. Monego and S. Backer, Textile Res. J., (in press).
6. K.E. Duckett, B.C. Goswami and H.H. Ramey, Textile Res.J., 49, 262.(1979).

DISCUSSION II

NONWOVENS

Leader   : Dr. Mackeprang
Rapporteur : Dr. A. Newton

This session consisted largely of two contributions by Popper and Dent. Both contributions consisted of modifications of the theoretical treatment outlined by Hearle in his second lecture. Popper's contribution concerned the role of fibre buckling in nonwovens during uniaxial extension, and its effect on the Poisson's ratio of the fabric.

Nonwoven Mechanics - Effects of Fibre Buckling (Popper)

Classical nonwoven theory provides a means for computing load deformation response and Poisson contraction from the properties and geometrical arrangements of the fibres. The theory has several deficiencies however - most notable of which is the failure to include bonding level. This omission creates unusually large discrepancies between the measured and computer values of Poisson's ratio. Measured values are often greater then 3, but the predictions are usually less than 0.5.

We found it is possible to significantly improve the classical theory by specifying the mechanism by which the fabric resists lateral contraction during load. Based on observations, we found that this resistance comes primarily from the <u>compressive force in buckled fibres</u> and <u>not</u> from an axial compressive force as previously assumed. By making this seemingly small change, the theory can include many additional factors such as: fabric weight, fibre denier, and fibre cross-section.

The modification to the theory is made by the following steps:

1. Modify the fibre force-elongation curve to include a limiting compressive load equal to the buckling load. As shown in Fig. 1 this level occurs when the strain reaches a value dependent on the ratio of free span between bonds to fibre diameter.

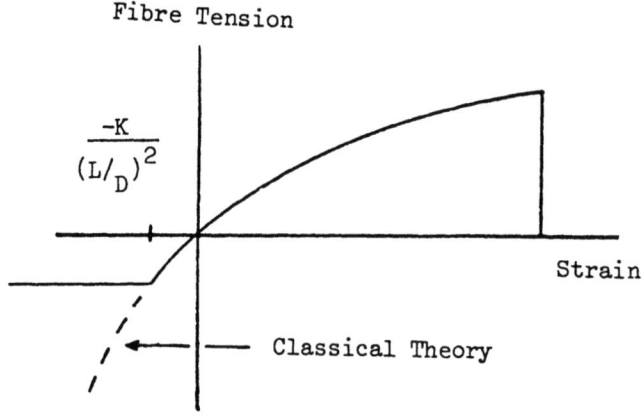

Fig. 1.

2. Compute the free span between fibre crossings from the statistical geometry of the system (details not given). The frequency of bonds along a fibre can then be found by multiplying the crossing frequency by the bonding fraction (see Fig. 2).

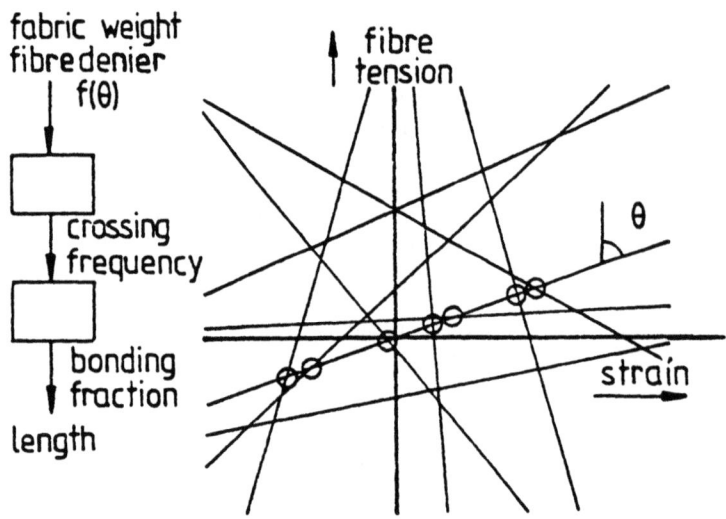

Fig. 2

3. Specify fabric weight, angle distribution, fibre properties etc. and solve the classical nonwoven equations with the revised fibre property relation by iterating to make the cross-directional force equal zero. In our analysis we included large strains, fibre failure, and reorientation of fibres due to strain. (Note that the free span between bonds may vary with fibre angle).

Typical results are shown on Figs. 3 - 5 for a specific fabric bonded by varying amounts. Fig. 3 shows the relatively small effect of bond level on the force-elongation plot and Fig. 4 shows the enormous effect on Poissons ratio. At low strain levels, the ratio apprachs the classical theory. It increases as more fibres buckle, but decreases at very high strains due to fibre reorientation. Fig. 5 gives the stress results in normalized form relative to an equal weight parallel array. The "inefficiency" due to fibre buckling increases with strain.

An alternate method of computing this effect in nonwovens is to eliminate the classical assumption of affine deformations. This can be done by Monte Carlo simulation of random arrays with arbitrary bonding and computation of force-elongation response. Results of this simulation are shown in Fig. 6 for a nonwoven stretched to varying levels.

In answer to questions addressed to Dr. Popper the following emerged:

1. The treatment outlined by Popper used Euler Buckling equations and not slender body theory.

2. The theory outlined by Hearle could be used by including a fibre stress-strain relationship which limits the compressive stress at negative strains. However, appropriate statistical geometry relations would have to be added.

3. The term "Poisson's Ratio" as used in Popper's and other workers' publications was really the ratio of lateral to longitudinal extension, and was not related to the value of 0.3 - 0.5 which was obtained in most solid materials. Values in nonwovens may be considerably higher (e.g. greater than 4), and can also be negative in the thickness direction (in other words, the fabric expands in thickness during extension). The Poisson's ratio of a nonwoven is structurally determined and varies considerably with the orientation distribution of the constituent fibres. The type of bond is not critical. The effect of fibre diameter can be predicted. For larger diameters and fixed fabric weight: fibre rigidity increases; length between bonds increases (lowering the buckling load); and the Poisson Ratio increases.

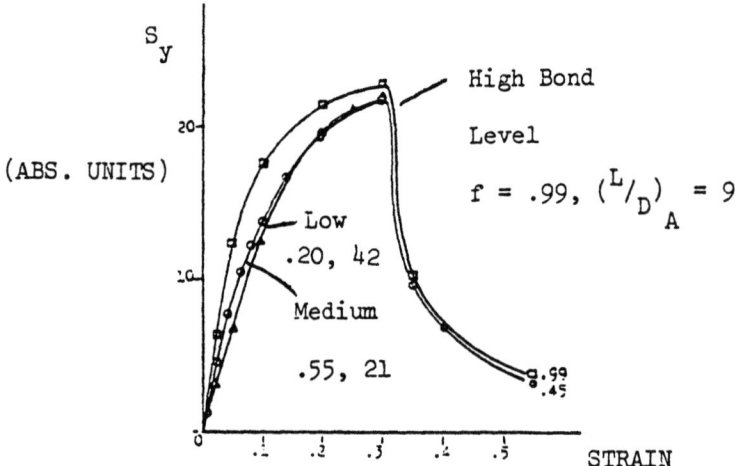

Fig. 3. Computed Stress-Strain relation for varying Bond levels.

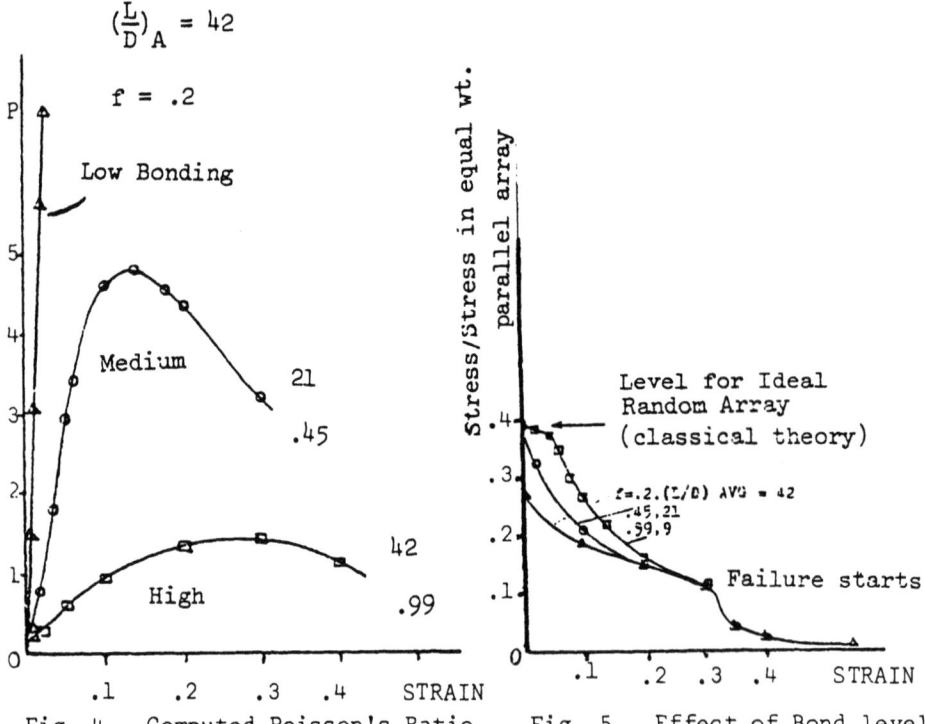

Fig. 4. Computed Poisson's Ratio function for varying Bond levels

Fig. 5. Effect of Bond level on normalized stress ratio.

547

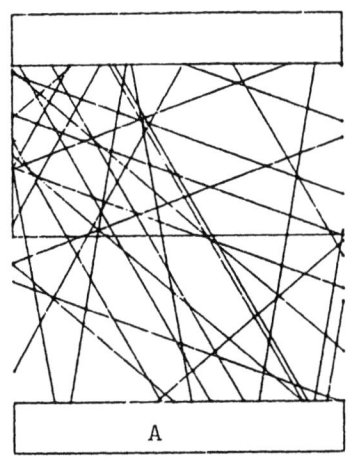

A

INITIAL GEOMETRY
JAWS AT EDGES 2 AND 4. EDGES 1 AND 3 FREE
TENSILE STRAIN=5.00 %
SCALE FACTOR=.15 UNITS PER INCH OF CHART
PERCENT OF INTERSECTIONS WHICH ARE BONDED= 50.00 %

B

TENSILE TEST - FREE EDGES
JAWS AT EDGES 2 AND 4. EDGES 1 AND 3 FREE
TENSILE STRAIN=5.00 % TENSILE STRESS=31.40 LBS./IN.

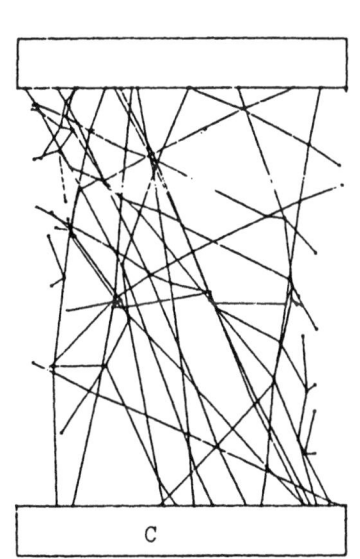

C

INITIAL GEOMETRY
JAWS AT EDGES 2 AND 4. EDGES 1 AND 3 FREE
TENSILE STRAIN= 25.00 %
SCALE FACTOR=.15 UNITS PER INCH OF CHART
PERCENT OF INTERSECTIONS WHICH ARE BONDED= 50.00 %

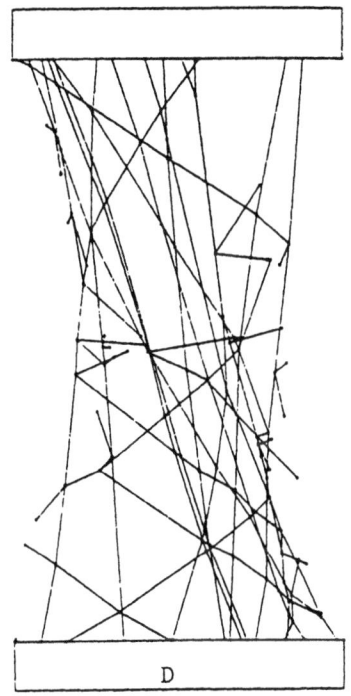

D

INITIAL GEOMETRY
JAWS AT EDGES 2 AND 4. EDGES 1 AND 3 FREE
TENSILE STRAIN= 75.00 %
SCALE FACTOR=.15 UNITS PER INCH OF CHART
PERCENT OF INTERSECTIONS WHICH ARE BONDED= 50.00 %

Fig 6. Simulation of Nonwoven Arrays in Uniaxial test

In the theory described by Popper the buckling length between bonds was not experimentally verified and the effect of interaction between fibres in compression and extension was not taken into account.

Straight Filament Web Modulus (Dent)

For small web strains $\varepsilon_F \to 0$, the filament strains ε_f, are given by

$$\varepsilon_f \approx \varepsilon_F (\cos^2\theta - \sigma \sin^2\theta)$$

where θ is the filament angle to the test direction and σ is the "apparent" Poisson's ratio. It is thus seen that for filaments where $\tan\theta \gtrsim 1/\sqrt{\sigma}$, ε_f becomes negative and such filaments buckle.

It can then be shown that sources of energy other than filament extension must be included in the analysis. Inclusion of filament bending and bond rotation gives:

Normalized web modulus = $\pi F/EA\varepsilon_F$ (or $E_F \rho f/E$)

where F is the web load at small web strains ε_F, E is the filament modulus, A the filament cross sectional area, and ρ_f is the filament density.

The full derivation is somewhat long and will be published shortly.

DISCUSSION III

EFFECTS OF VARIABILITY; STRENGTH CONVERSION

Leader   : Dr. P. Popper.
Rapporteur : Miss M. Arponen

The discussion was opened with the comment (Popper) that fibre strength variability often reduces strength efficiency markedly. This effect can overshadow the effect of fibre orientation. In some cases, such as twisted continuous filament yarns, variability may even reverse the usual relation between strength and orientation (Fig. 1).

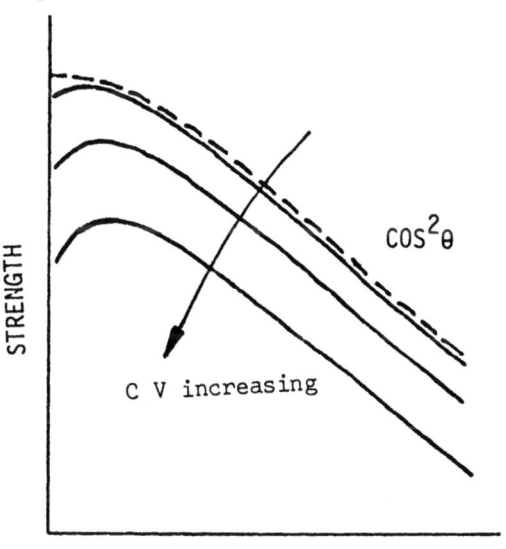

Fig.1 Effect of variability on strength conversion

Alternative Approaches to Strenth Conversion

(A) Chain-of-bundles Theory (Phoenix)

The statistical chain-of-bundles model for the tensile strength of yarns, ropes and cables can be illustrated by applying the Weibull distribution to a nine fibre bundle model, with a shape factor $\rho = 10$. The strength efficiency of a nine fibre chain is about 75% when compared with the median fibre strength. For a nine fibre bundle the equal-load-sharing rule results in 70% and the local-load-sharing rule in 68% efficiency. As the bundle size increases this figure falls to 25% asymptotically.

In order to illustrate the size effect, the median strength of a one-million-nine-fibre bundle is compared with that of one million single fibres in a chain. Under equal load sharing the efficiency is about 50% and under local load sharing about 42%. But one million single fibres in a chain have an efficiency of only about 25% (Fig. 2).

The model was developed for filament yarns. Variations in the number of elements per bundle introduce an increased amount of variability which can be modelled e.g. by the binomial distribution. In staple yarns, the random fibre strength and strength reductions can be analysed as functions of the helical angle [1]. It is suggested (Carnaby) that end slippages could also be considered in the model.

(B) Weak Link Theory (Dent)

If a linear assembly has n unit cells in which the breaking strength is a random variable of mean $\bar{x}_1$ and s.d.σ_1, the breaking strength of the assembly has a mean minimum $\bar{x}_n$ which, in general, is given by

$$\frac{\bar{x}_1 - \bar{x}_n}{\sigma_1} = f(n).$$

Since $\bar{x}_n/\bar{x}_1$ is a measure of strength efficiency, $f(n)$ is a scaled measure of how this falls with increasing n.

For any continuous distribution of the unit cell breaking strength (referred to as asymetrical) $f(n) \leq \frac{n-1}{\sqrt{(2n-1)}}$

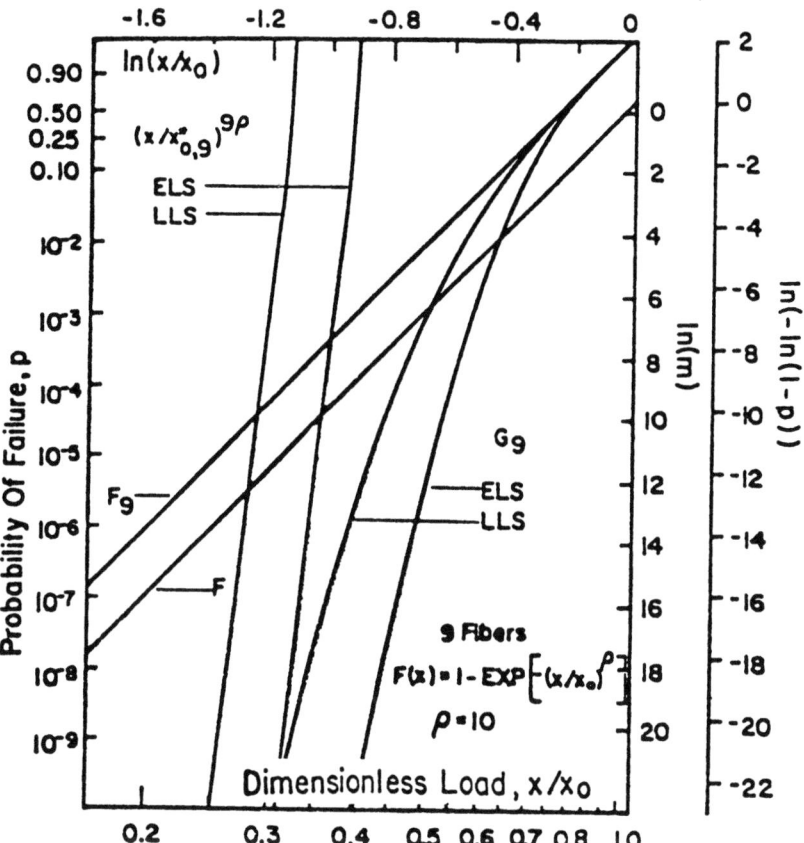

Fig. 2. Probability distribution for the strength of a nine fibre bundle under equal (ELS) and local (LLS) load sharing

If the distribution is symmetrical this bound is reduced approximately to $\frac{1}{2}\sqrt{(n+1)}$ for $n > 5$. For a discontinuous distribution (asymmetrical based on sample values, the bound is greater

$$f(n) \lesssim \sqrt{(n-1)}$$

Particular examples of continuous distributions are

(a) Uniform $f(n) = \sqrt{3} \left(\frac{n-1}{n+1}\right)$

(b) Gaussian $f(n) \simeq 4.21 \, (1 - n^{-1/5})$; exact values tabulated as $w(n)$ [2].

(c) Weibull $f(n) = (1 - n^{-1/\rho})/\sqrt{\dfrac{\Gamma(1+2/\rho)}{\Gamma^2(1+1/\rho)} - 1}$

where ρ is the shape parameter.

The way in which strength efficiency falls with increasing n is shown in Fig. 3. Similar relations exist for the coefficient of

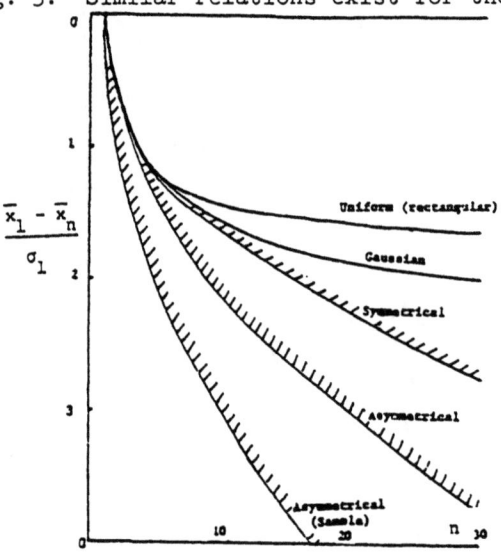

Fig.3 Weak link effect for various distributions

variation as a function of that for the breaking strength for a unit cell.

If the breaking strengths in adjacent unit cells are not independent, but are correlated in some way, the theory is modified [3].

$$\frac{\bar{x}_1 - \bar{x}_n}{\sigma_1} = f(n)D(n)$$

where $D^2(n) = (1 + \frac{1}{n}) - \frac{2}{n} \sum_{i=0}^{n} (1 - \frac{i}{n}) r_i$

and r_i is the ith autocorrelation coefficient.

For a yarn, the effects of different levels of long-term influence can be seen by putting

$$r_i = (1 - \frac{i}{n})^\gamma$$

where γ represents the long-term association within an individual filament in the yarn or alternatively within a yarn in a bundle. In the latter case it would depend upon yarn twist and staple length. γ increases as long-term influences decrease ($\gamma = \infty$ corresponds to the random case first considered). Fig. 4 shows the effect of changing γ for a Gaussian distribution of breaking strength.

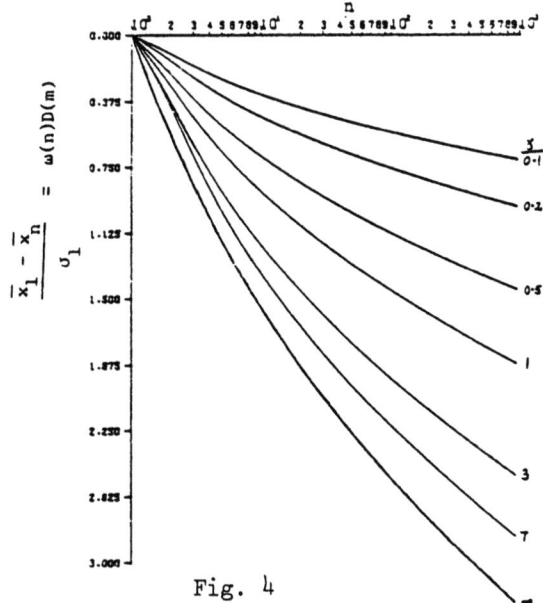

Fig. 4

Both arguments can be extended to deal with an nxm matrix of unit cells, representing, for example, a twisted yarn of m fibres or a non-woven. There are corresponding functions, f(n,m) depending on the unit cell distribution and D(n,m) given by

$$D^2(n,m) = 1 + \frac{1}{nm} - \frac{2}{mn} \sum_{i=0}^{n} \sum_{j=0}^{m} (1 - \frac{i}{n})^\alpha (1 - \frac{j}{m})^\beta$$

where α and β depend upon the bonding in a non-woven, the twist in a yarn, etc. The case for which $\alpha=\beta=2$ and the basic distribution is Gaussian, when f(n,m) is written w(n,m), is illustrated in Fig. 5.

It is suggested (Popper) that the autocorrelation functions eliminate the problem of infinite fibre strength at zero unit cell length. The same problem can be dealt with (Phoenix) by using a double Weibull distribution in which one shape factor is much greater than the other.

The Role of Variability in Rupture Propagation in Blended Yarns (Backer).

For the case of high cotton blend percentages the first rupture is observed to occur in the central component, the break then spreads rapidly towards the yarn surface in a narrow zone. Upon approaching the outer layer of the yarn, the rupture zone broadens significantly

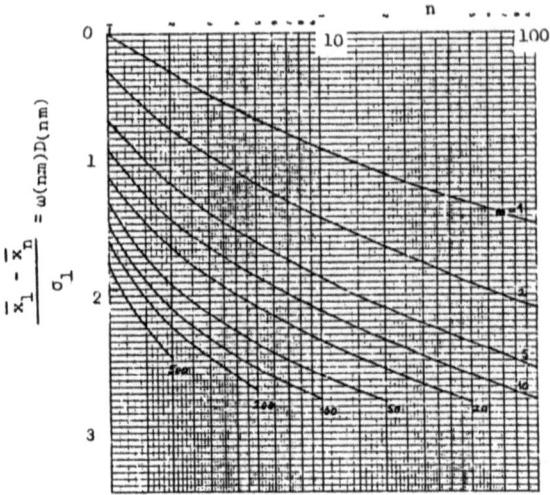

Fig. 5

reflecting the combined effect of lowered lateral pressure and of cotton component strength variability.

When a fibre breaks the load it bore is locally redistributed on its neighbours as shown in Fig. 6(a). The tension build up on them has the shape of a "tent". The distribution of weak spots in the cotton yarn is reflected in the observed strength variation with gauge length. This variation can be represented by a "slack tent" whose local height indicates the highly probable strength dependence on gauge length. If this "slack tent" is centred at the location of a break in an adjacent yarn then its intersection with the friction "tension-buildup tent" dictates the zone in which the next break will occur. Fig. 6(b) illustrates the case in which the tension build-up is high compared with the maximum variation with gauge length. The friction tension-buildup tent pierces the slack tent in a narrow zone near the first failure. In this case the rupture will propagate in a narrow zone. Fig. 6(c) illustrates the case in which the apex of the friction tension - buildup tent does not rise above the slack tent. For the case with a high friction-pressure product no rupture will propagate. For the low friction-pressure product the flat friction tension-buildup tent intersects the slack tent at a distance from the narrow zone and the rupture propagates, wandering from the narrow zone. Such were the experimental observations of a blended model study.

A similar effect is observable in the stretch-breaking process, but in this case it is complicated by the process dynamics, which result in the possibility of several cycles of loading on a particular element of tow before it is ruptured.

Variability in Experimental Testing (Stevens)

Because of variability in experimental testing, there is a need to design experiments to cover properly the factors involved. Work on textile material for aerospace applications had resulted in a good picture of the design of experiments and statistical methods for the study of the complex instructions in applied textiles [4].

Concerning the contentious matter of the variability in experimental method in relation to the variability of the material, in strength tests on a nylon ribbon the scatter at low rate of strain was 8%, but at a strain rate of 10/sec the scatter was 20%. The mean was however, slightly higher. Can the fibre bundle theory be used to predict anything about the effect of rate of strain on breaking strengths? Also, in its application to the effect of length upon strength, can it give any guide to the relation between the best length of specimen to be used in the laboratory and the length in the designed system? For example, a 100 mm length of parachute cord would be tested in the laboratory although the cord length used in the parachute would be 10 m or more.

The prediction of the strength of long fibrous structures from tests on short sections, using the weakest-link theory can be dangerous (Phoenix). Fortunately, the theoretical results of the chain-of-bundles model suggest that such an extrapolation, using it, may yield conservative estimates.

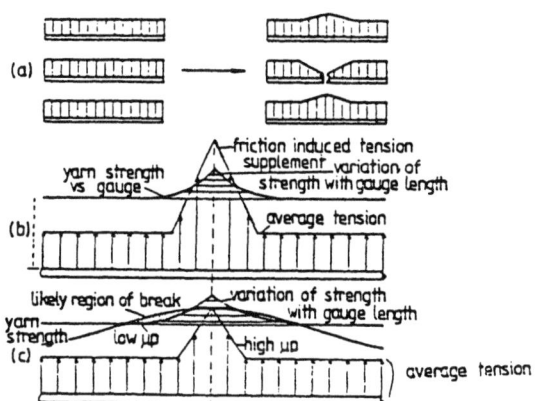

Fig.6. Load sharing in Blended Yarns

REFERENCES

1. L.Phoenix, <u>Textile Res. J.</u>, 49,407. (1979)
2. L.H.C. Tippett, <u>Biometrika</u>, 17, 364,
3. J.L. Spencer-Smith, <u>J.Text.Inst.</u>, 38, P257.(1947)
4. J. Swallow et al. <u>Royal Aircraft Establishment</u> TR 68070 and 76031

DISCUSSION IV

ONE DIMENSIONAL STRANDS; TEXTURED YARNS

Leader   : Mr J.J. Thwaites
Rapporteur : Dr J.I. Curiskis

As a preliminary to the discussion the following table was
presented (Hearle), illustrating the range of methods available
and the state of understanding of problems of the relations between
the set of forces and the set of deformations of fibres and yarns.

Force \ Deformation	Extension	Curvature 1	Curvature 2	Twist
Tension	*	?	?	*
Bending moment 1	?	*	?	?
Bending moment 2	?	?	*	?
Torque	*	?	?	*

* fairly well understood

? further study needed

Problems relating to the extension and twist of one-dimensional strands are fairly well understood, and those associated with the remaining diagonal elements of the table can be tackled by the methods described in the lectures. However, the remaining problems require further attention.

Methods of Formulation for 3-dimensional Problems

The relative merits of Cartesian (body) co-ordinates and Euler angles (global co-ordinates) were discussed. Even the method of de Jong and Postle uses the latter (Thwaites). A general view, held even by some who have used Euler angles, is that the cartesian system is better from a computational standpoint.

A possible problem with Euler angles (Leech) is that the matrix relating their rates of change to the angles themselves can become singular. Of the 9 direction cosines used in the cartesian formulation only 3 are independent, which suggests the possibility of using Hamilton's unimodular quaternions [1], the rate-of-change matrix for which is never singular.

The value of the kinetic analogy is small (Konopasek), because the elastica problem is largely computational rather than one of formulation. There is, however (Leech), the possibility of borrowing some of the solution techniques used in dynamics; in particular the Jacobian elliptic functions. There are 12 of these (sn, cn, dn, etc) which satisfy the differential equations.

$$\dot{\alpha}_i = \lambda_i \alpha_j \alpha_k \qquad i, j, k = 1, 2, 3$$

where the λ_i are of the same form as the bending/torsion rigidity terms in the Euler equations of bending. They also include oscillatory and divergent functions.

Problems involving Initial Planar Crimp

The prediction of the tensile deformation of a filament crimped into the form of a plane wave can be described (Lloyd [2]) in terms of dimensionless entities: crimp level, slenderness, i.e. ratio of thickness to wave length, and shape function, i.e. normalized curvature. The use of dimensionless variables in the equation for this extensivle planar elastica facilitates the making of 'intelligent guesses' for solution curves which can then be improved iteratively. It is particularly valuable to use the set of normalized shape functions devised by Konopasek [see below] for the initial shape.

In the solution it is possible by using log-log plots, to look for normalized curves which separate the extension and bending effects. Good agreement has been obtained with experimental data for tendon, which consists largely of aligned collagen fibrils with planar crimp. Some scepticism was expressed about 'drawing asymptotic tangents' (de Jong) and attention was drawn to the classical (non-computer) work on this problem by Olofsson [3].

It is important to be able to describe the initial shape of an elastica simply (Konopasek). This is best done by expressing the original curvature and torsion in terms of the arc length. For the planar case, despite the association of sinusoidal shapes with waves, they are best avoided. There are, however, two different cubic-exponential functions which define similar families of periodic functions. Depending upon the variation of a single parameter α, the range is from alternating semi-circular arcs ($\alpha = -\infty$) via sine-like functions ($\alpha = 0$) to saw-tooth functions ($\alpha = \infty$). The initial curvature k_0 (s, α) is given over the first quarter-period (s = 0, 1), in the first variant by

$$k_0 = \begin{cases} (u - 1)e^{\alpha u} + 1 & \infty < \alpha \leq 0 \\ u e^{\alpha(u - 1)} & 0 \leq \alpha < \infty \end{cases}$$

where $2u = 3s - s^3$

and in the second variant by

$$k_0 = \begin{cases} \frac{1}{2}(3u^2 - u^3) & \text{where } u = se^{\alpha(1 - s^2)} & -\infty < \alpha \leq 0 \\ \frac{1}{2}(3u - u^3) & \text{where } u = se^{\alpha(s^2 - 1)} & 0 \leq \alpha < \infty \end{cases}$$

These are shown in Fig. 1.

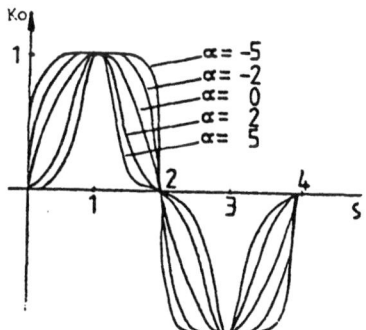

Fig.1 Cubic exponential functions

Approximate Solutions: Textured Filaments (Hearle & Thwaites)

The shape of filaments textured by the twist-heatset-detwist (false twist) method, under tensile loading, is of importance in determining the performance of the textured yarns which comprise them. Because the filament shape when heat set is not known precisely, various representative simple, heat-set geometries can be considered, for example, a regular helix. Computation of the detwisted shape by exact methods even for these simple cases is extremely difficult and costly in terms of computer time [4]. It is therefore sensible to use such computation only as an infrequent check on the information produced by approximate methods.

The first such problem to be considered was the formation of snarls in a filament heat-set in a straight, but twisted configuration [5, 6]. The method is to construct a potential energy function in terms of the two (independent) variables unsnarled length and helix angle in the snarl. A solution is obtained by minimization. The key assumption is that the strain energy is confined to the straight, unsnarled, length and to the regular part of the snarl. The contributions from the tip of the snarl and the transition zone are neglected (A in Fig. 2).

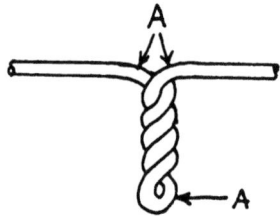

Fig.2 Snarl in a straight, twisted filament

A similar technique can be applied to the detwisted shape assumed by a filament set in a helical configuration. It is observed in practice that this consists of two almost helical segments of opposite sense, together with a coupling member. The key assumption here is that the strain energy of this coupling member can be neglected, i.e. that it is replaced by a 'hinge' (Fig. 3). The problem is then one of determining the helix angle, radius and number of turns in each of the helical segments. This is a constrained optimization problem - there are five constraints. A minimum energy state can be found by a search [7, 8], which can be costly, or by using the Lagrange multiplier technique [9] in which case the problem is reduced to solving a single non-linear algebraic equation. A large amount of valuable information results, concerning the state of stress as well as the geometrical shape. One can also readily determine in what state of detwist/

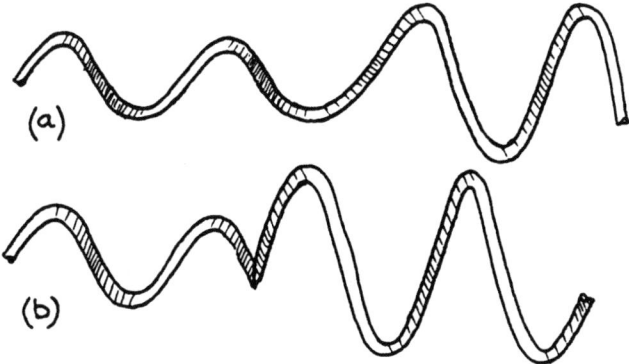

Fig.3 Detwisted helix (a) as observed (b) model

relaxation the detwisted geometry changes from this form to one including snarls or one which is a simple helix (Fig. 4).

Extensions of the above method can be applied to other configurations which approximate to the heat-set shape of a filament. The most obvious is to take account of migration in the twisted yarn by assuming this shape to be either partly helical and partly straight or partly helical with larger and smaller radii [8, 10]

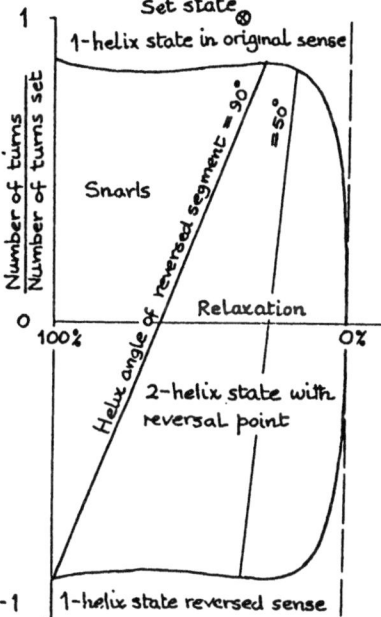

Fig.4 The state of a detwisted helical filament

REFERENCES

1. E. Leimanis and N. Minorsky, Dynamics and Non-linear Mechanics Vol.2, Surveys in Applied Mathematics, Wiley, New York, 1958.
2. C.P. Buckley, D.W.Lloyd, M.Konopasek, 'On the deformation of slender filaments with planar crimp' (to be published).
3. B. Olofsson, J.Test.Inst., 55, T541.(1964).
4. M. Konopasek and K. Bruggisser, Applied Polymer Symposia, 33, 203.
5. J.W.S. Hearle, J.Text.Inst., 57, T491.(1966).
6. J.W.S. Hearle and A.E. Yegin, J.Text.Inst., 63, 477 and 64, 601.
7. J.J. Mertens, in Bulk Strength and Texture, Textile Institute, Manchester 1966.
8. A.E. Yegin, Ph.D. Thesis, UMIST 1969.
9. J.J. Thwaites, in Symposium on the use of computer methods in calculating the properties of textiles for design purposes, UMIST, 1971.
10. M.J. Denton, J.Text.Inst., 59, 550.(1968).

DISCUSSION V

WOVEN AND KNITTED FABRICS.

Leader   : Professor P. Grosberg
Rapporteur : Dr. Frank Ko.

In his opening remarks Professor Grosberg indicated that the
session would be concerned with the analysis of woven and knitted
fabrics in one plane which includes the lectures given by Leaf,
Grosberg, Hepworth, Moghe, de Jong, and part of Skelton's lecture.
The analytical approaches presented in these lectures are quite
similar, and they seem to deal with the improvement of the accuracy
of the various models. The inclusion of friction does make the
model more realistic. In general, most people have used simple
models for a very complex problem.

Virtual work method and limit theory (Leech)

The application of force and energy methods to textiles has been
discussed; however there are other technques which can be applied.
If, for example, the collapse load of a structure is required,
then the limit theories applied to rigid perfectly plastic
materials can be used. These theories can be applied using
admissible or compatible, fields to find upper and lowerbounds to
these collapse loads. Then by scanning those admissible fields,
one can search for greatest lowerbounds and least upper bounds and
thus bracket these loads. A further use of limit theories is to
search for limit displacements (jammed or locked configurations)
by applying for plastic perfectly rigid materials, dual principles
to the limit load theories. This latter application expanded by
Prager (1,2) should be considered for the jamming analysis in
knits and certain weaves.

The energy method of Hearle and Shanahan, as applied to for example to the Peirce geometry, was developed from the explicit expression of constraint. However sometimes the equations of constraint are not explicit, and in this case the constraint equations can be handled using Lagrange multipliers. In this approach, the principle of virtual work, the Lagrange multiplers are the forces of constraint. For example if the constraint to be used expresses mathematically the contact of warp and weft then the Lagrange multiplier is that force of contact.

REFERENCES

1. W. Prager. Trans.Soc.Rheology $\underline{1}$, 169 (1957)
2. W. Prager. Qu.Appl.Math. $\underline{27}$, 128 (1969).

The accuracy of theoretical models (De Jong)

The problem of deciding which theoretical model of fabric mechanics predicts more closely the mechanical properties of fabrics is a difficult one. The energy method, described in the lecture, enables one to span the range of models between Peirce's geometrical model of the woven fabric and his mechanistic or force determined model. When one compares actual experimental results with the predictions of the energy method, the yarn load-extension characteristics need to be determined as well as the way in which the yarn rounds up during the tensioning, or in some cases, compresses. Furthermore either realistic values have to be assigned to the curvature constraint of the yarn inside the fabric, or to the yarn compressibility function. Experimental methods to determine yarn crimp also give variable results for staple yarns. The actual conditions of test method may also affect the fabric tensile characteristics, such as when the 'uniaxial' test is more accurately represented by a strip biaxial test. These conditions are known to reduce fabric extensibility results greatly (1). Useful trends in the mechanical properties of woven fabrics can be predicted however.

When the model for the plain knit fabric is compared with actual results, the trends are again confirmed provided realistic values of yarn compression are fed into the programme, such as to give reasonable inter-yarn distances. However, a variety of other factors such as fabric take-down tension and setting conditions are known to alter the shape of the loop and the fabric dimensions. The theoretical trends for the effect of increasing the yarn compression index whilst keeping the minimum inter-yarn distance fixed, is effectively to give the yarn in the structure more room to move, thereby producing greater fabric extensibility. The effect of actual changes in yarn compression and inter-yarn distance have only a small effect on the fabric load extension modulus (2).

The freedom of the yarn to move within the structure is specified in the models either by constraints on the yarn curvature or by specifying high compression factors in combination with a low compression index. In real fabrics, this is specified by a 'tightness' factor for knitted fabrics or cover factor for woven fabrics. This leads to the concept of a universal tightness factor for fabrics relative to a compression index, which results from the energy and analysis as: (3)

$$\frac{CL^{3-a}}{B} r_o^a$$

where C and B refer to the yarn compression and bending rigidities, r_o is a yarn diameter, a the compression index and L is the yarn loop length in the structural repeat.

The factor (Cr_o^a/B) is a yarn property. The effect of fabric tightness alone can therefore be investigated by varying L. As the value of L decreases, the tightness factor $(Cr_o^a/B) L^{3-a}$ increases (since $10 < a < 20$ for practical fabrics) and hence the compression term in the energy to be minimised becomes more important and tends to swamp the contributions of the bending and tension energy to the geometry of the structure. It is for this reason that geometrical models have been found more acceptable in woven fabrics (L~0.05 cm) than knitted fabrics (L~0.5 cm).

Perhaps better and more accurate descriptions of yarn compression behaviour can be found than was used in the present analysis. However, it is clear that the concept of a tightness factor is very important in determining the behaviour of the structure of a fabric.

Comparison of the effect of making yarns from different fabrics or filaments with different structure and packing fractions is then also possible.

REFERENCES

1. S de Jong and R. Postle, "Modified Equations for the Energy Analysis of the Plain Weave", letter to the Editor, J.I.I., in publication.
2. S. de Jong and R. Postle, J.Textile Inst. 68, 307. (1977)
3. S. de Jong and R. Postle, Textile Res.J. 1978, 48, 127.(1978)

Interdependence of energy terms (Carnaby)

I believe that there is a major problem with independence of some yarn energy terms, especially compression and bending. I have found that local pressures such as clamps affect the bending rigidity further along the yarn in some yarn buckling experiments that I have been doing. This distance may exceed 1 cm. I would point out that such a distance is in excess of the yarn crossover distance in fabrics and therefore, of major importance. In fact measurements of bending rigidity on free yarn could be out by a factor of several times for a length of yarn trapped between a cross over in a fabric. To be quite rigorous, and probably to get reasonable accuracy, it is necessary to consider this dependence during the course of fabric deformation.

Frictional Elastica (Konopasek)

A planar elastica model may be used for investigation of large cylindrical deflection of sheet materials. Examples are the analysis of heavy sheet with finite bending rigidity by Shanahan, Lloyd and Konopasek, or analysis of tubelike fuel tank (heavy and extensible coated fabric, finite bending rigidity, hydrostatic pressure) by Olson and Konopasek.

Accounting for friction-elastic moment-curvature relationships is the new step in making the elastica model of fabric bending more realistic. This problem was treated in the past by Kawabata and later, more fully, by Bruggisser (M.Sc.Thesis, Georgia Tech. 1976). He analysed friction-elastic horizontal cantilever, ring and heavy cantilever with changing fixation angle, under both loading and deloading conditions.

The phenomenon generally neglected in the analysis of the geometry and mechanics of woven fabrics is the non-uniformity of the shape of normal yarn cross-section along the weave repeat. This is primarily due to compliance and mutual accommodation of warp/weft yarn surfaces especially under extensive pressures in some finishing processes and in use of the fabric.

I built a model of closely packed and locked plain woven fabric in which the mass of fibres of a unit cell of fabric fills completely the volume of an orthogonal parallelepiped. This may be considered a limiting-case woven structure made of perfectly compliable plastic material as opposed to another limiting case, Peirce elastica model assuming non-deformable circular cross-section of a yarn with finite bending rigidity. All geometrical models and mathematical models (elastica or other) with deformable yarn cross-section would fall between these two limits.

A concentration of stress and wear may be expected at and around the edges of the closely packed yarns at the surface of the fabrics. This is similar to the situation at the edges of the strands of a well worn rope: the cross-section of a two-ply rope consisting initially of two near-ellipses changes gradually to circular ones consisting of two almost half circles, with those sharp edges where the wear disintegration starts.

Structural Modifications on Woven Fabrics and their Effects on Bending Behaviour. (Skelton).

If a constant curvautre is imposed on a fabric which is assumed to have a circular arc geometry, it can be shown that the amplitude of the crimp waveform, measured as the difference between the radii of the inscribed and escribed circles, is increased. The increase is larger for large crimp amplitudes and for high values of imposed curvatures, and can lead to a reduction in the inter-yarn forces in a fabric that is bent to high curvatures. It also necessitates a redistribution of yarn material between the inner and outer arcs, a redistribution that is only of minor consequence if a single cycle of bonding is considered, but which can have importance if the loading is continued for a large number of cycles as in a flex fatigue tests. When the imposed curvature is removed the fabric attempts to regain its original configuration, but is opposed by the frictional forces and a small fraction of the amplitude increase is returned. On the next cycle of bending, the amplitude increase is slightly larger, as a consequence of the functional relationship described above, and the returned fraction is also larger. Accordingly the amplitude of the crimp waveform, and hence the thickness of the fabric can be seen to increase, slowly at first and then quite rapidly, as the number of flex cycles increase. Thickness increases of up to 5 times the initial thickness have been observed, and the increase appears to be limited only by the onset of jamming that accompanies the associated shortening of the fabric structure.

Set fabrics (Williams)

A series of treated cotton fabric varying in equal increments over a wide range of recovery properties, were examined for distortions introduced by laundering. Analysis of the patterns and their relationship to various fabric properties demonstrates that acceptable apparel fabric performance (smooth appearance) requires two conditions to be met; stabilization against internal distortional forces, and the ability to recover from externally applied distortions. (1,2,3). Essentially all modern apparel fabrics meet these conditions, regardless of fibre type, yarn type, construction weave, etc.

These conditions require the fibres and yarns to be set in a relaxed minimum energy condition in their actual configuration in the fabric. It is suggested that theoretical mechanical analyses of fabrics be based on detailed models consistent with this condition, including crimp configuration of yarns, collapsed (e.g. typically, elliptical) yarn cross-section, fibre/filament migration in the yarn, etc. all in a relaxed, equilibrium, minimum energy state.

REFERENCES

1. Textile Research J. 27, 129, (1957)
2. Amer.Dyestuff Reptr., 48, 37, (1959).
3. Ibid, 48, No.13, 27, (1959).

DISCUSSION VI

COMPLEX FABRIC DEFORMATIONS, BUCKLING AND DRAPE

Leader   : Dr R.W. Dent
Rapporteur : Dr R.W. Rennell

Complex Fabric Deformation   (Lloyd)

The types of deformation possible are:

1. Planar deformation
2. The "tension membrane"
3. Plate and shell problems
4. Buckling
5. Post-buckling behaviour

Of these the "tension membrane" and planar deformation can be solved. The buckling problem can be solved in so far that the conditions for the onset of buckling and the modes of buckling can be obtained. Plate and shell problems and post-buckling in textiles are as yet unsoluble but are of the greatest interest to the general problem of drape.

The models for material behaviour that can be included in the general analysis can be listed in order of increasing realism: small displacements; small strain, large displacements; large strain, large displacements. However as the realism of the marerial behaviour increases so do the problems of mathematical modeling.

The models possible for material behaviour may also be tested in increasing order of realism: linear elasticity; non-linear elasticity; viscoelasticity.

The main difficulties associated with the mathematical modelling of complex fabric deformation are:

1. The mechanical properties. The difficulties of formulation and measurement of the relevant properties.
2. The methods of analysis. Analytical solutions are generally impracticable, and this of course implies numerical methods using computers.
3. Continuum models. Here there are restrictions imposed because of the limit on the size of the elements that can be considered.
4. Coarse net models of drape.
5. A large strain, large deflection theory is needed for a satisfactory model of drape using continuum mechanics.

There are probably two "lines of least resistance" that might be pursued for realistic modelling of the drape problem using combined mechanics.

1. An enhanced Torbe-type element which includes a bending and twisting deformation mode.

2. A geometric approach using a geometric description of the fabric surface. This would allow strains, curvatures, strain energy, etc. to be calculated.

Complex Fabric Deformations (Hearle)

Complex fabric deformations, involving double curvature, can be studied in various ways. Skelton has referred to the need for shear in the forced conformation of open weave fabric to the shape of bottles. Cusick [1] used the drapemeter, based on earlier FRL research, in which a circle of fabric drapes over a over a smaller circular plate in a number of nodes. In other studies at UMIST we have tried to develop laboratory tests for the bending of tubes of fabric, in order to simulate such practical situations as the buckling of a sleeve on the inside of the elbow joint.

More recently, we have realised that the fundamental problem to study is three-fold creasing in crow's-foot form, illustrated in figure 1(a). The toes A, B. C of the crow's foot are close to single curvature, while the central dome D is in double curvature. More complicated buckling patterns can probably be regarded as collections of three-fold units. In some cases, two three-fold creases may coalesce to give a four-fold crease as illustrated in figure 1(b). In principle higher orders might form by further coalescence but this does not appear to occur in practice. Lloyd tried several forms of experimental arrangement, and Amirbayat (see below) has now developed a three-point contractile system which promises to be satisfactory.

Cusick's [1] studies of drape indicated that good drape required

a combination of ease of bending and ease of shear. He also
investigated the relations between subjective and objective assessments of drape, along lines similar to Kawabata's study of fabric
hand.

The statement that double curvature requires shear is only a partial
truth. The essential feature is that double curvature over finite
areas can only be accommodated with area changes in the sheet of
material: this is the converse of the problem of forming a map of
the world on a plain sheet of paper. Where area changes are
relatively easy, through shear in woven fabrics, through various
modes of deformation of knitted fabrics, or through thickness
change in rubber, complex buckling occurs with smooth and sizeable
domes of moderate curvature at the centre of crow's feet. But
where area change is strongly resisted, as in paper, the zones
of double curvature collapse to singularities of infinitesimal
area, apparent as sharp points in the buckled material.

Finally a comment may be made on the shear (ε_3) mechanism in
fabrics. The area change $\delta A/A$ comes from the trellis action
shown in figure 2, leading to the reduction in the length y.
The general polynomial form would be:

$$\delta A/A = a_0 + a_1 \varepsilon_3 + a_2 \varepsilon_3^2 + \ldots$$

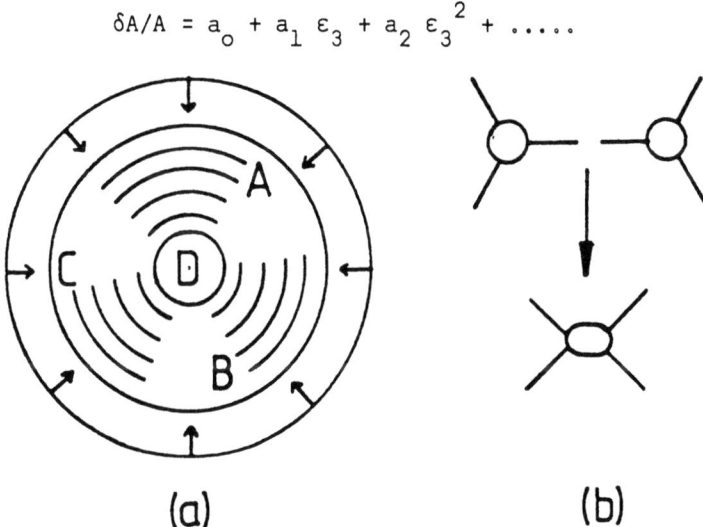

(a) (b)

Figure 1(a) Idealised indication of buckling due to uniform
contraction of the perimeter of a circular sheet, forming a
crow's-foot with toes A, B, C and central dome D.

(b) Two three-fold creases coalescing to give a four-fold
crease.

The first term a_0 is zero by definition of the zero of δA. In any symmetrical fabric the second term $a_1 \varepsilon_3$ must also be zero, since a finite term would imply that shear in opposite directions caused length (and hence area) changes in opposite directions, which would not be so. Consequently the simplest expression to consider is $\delta A/A = a_2 \varepsilon_3^2$: the problem is inherently non-linear, and could not be analysed by any treatment limited to linear strains. The second order quantity must have a large effect for levels of strain at which one would usually expect to be able to neglect the second-order terms.

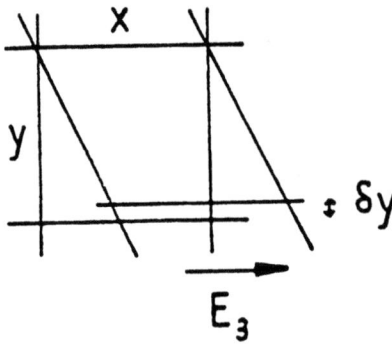

Figure 2. Trellis action in shear (ε_3) a woven fabric. Initial area $A = xy$. Area change $= \delta A = x\delta y$.

REFERENCE

1. J.W.S. Hearle, P. Grosberg and S. Backer, Structural Mechanics of Fibres, Yarns and Fabrics. Volume 1, Wiley-Interscience 1969, Chapter 12.

Study of Three Dimensional Deformation of Fabrics and Some Other Sheet Materials (Amirbayat)

Study of curvatures, which are main factors governing the bending strain energy, is the first step towards understanding of complex deformations. Our steps towards curvature measurements are briefly given below.

In order to deform fabrics, paper and sheet rubber into a familiar "y" shape buckling, many attempts have been made starting from a plane circular sample and moving three parts near the circumference towards the centre. Results for paper and sheet rubber were

reproducible, but for fabrics due to the significant effect of
the weight, the shapes obtained were not the same for the same
material. Using a simple set-up adopting a three-point lathe
chuck, a thin wooden back plate and a low pressure air stream
(to keep the fabric fluidized) was most satisfactory of all the
different attempts.

Flat samples of 14 cms diameter were connected to the chuck at
three points 10 cms apart and the points were moved towards the
centre to reduce the distance between pins to 8 cms. The samples
were fixed onto the back plate and taken for height measurements
Fig. 3.

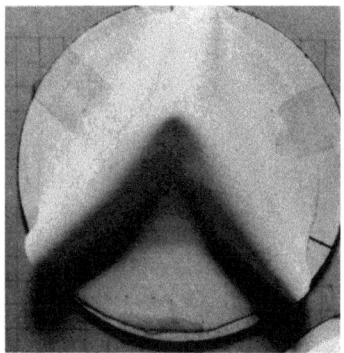 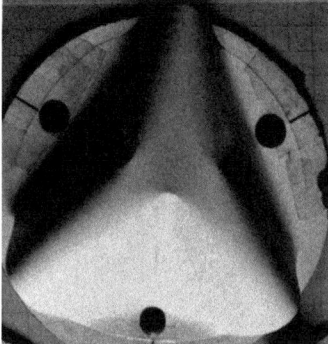

Fabric   Paper

Fig. 3.

The method applied for height measurements was close-range photo-
grammetry. Two parallax free photographs were taken from knwon
distances (between cameras, plane of cameras and the samples) while
a sufficient number of metal rods of known heights were located
on the back plate as control points.

Using two photographic plates obtained from this stage in a
stereocomparator, about 400 heights were measured on the surface
of each sample. These heights were used as data points for a
computer program called GPCP, to get heights at a regular grid
mesh of 1 x 1 mm. Fig. 4. shows the contour lines of fabric
and paper samples plotted from frame height measured by stereocomparator.

The following have been observed from deformation of different
materials:

1. Sheet rubber deforms three-dimensionally with almost no point

with single curvature.
The ability of rubber to produce double curvature at all points seems to be connected with the ease of area (and volume) change.

2. Paper collapses at any point forced to assume double curvature. Unlike the rubber, lack of readiness to change the area (especially under compression which results in delamination) is responsible for this behaviour.

3. Fabrics behave between the two extremes. More balanced fabrics show more 'papery' behaviour and unbalanced ones behave more rubberlike.

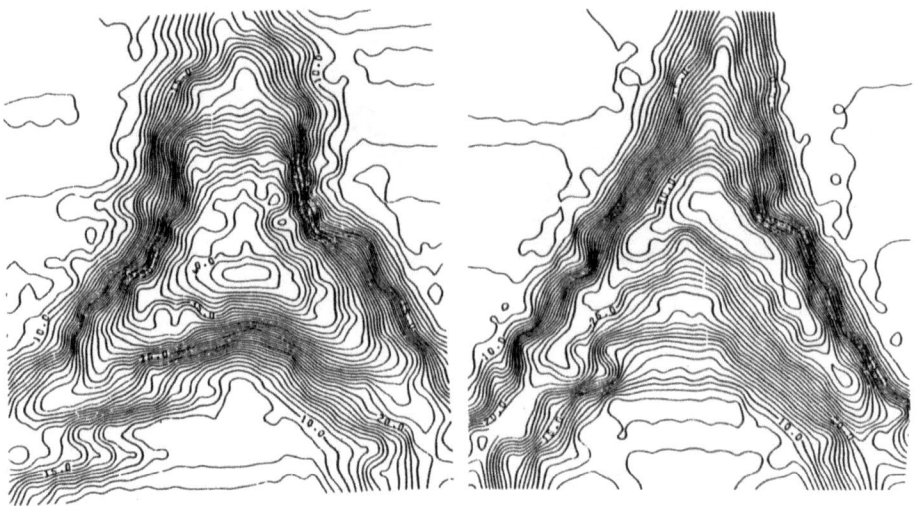

Fig. 4. Contour lines of paper and fabric samples

DISCUSSION VII

CHARACTERISATION OF STRUCTURE : GEOMETRIC AND TOPOLOGICAL

Leader   : Professor K. Slater
Rapporteur: Dr A. Newton

The Formation of Multi-Ply Nonwoven Fabrics (Newton)

The conventional squared paper representation of woven fabrics in which a square containing an 'X' represents a warp yarn lifted over a weft yarn and a blank square represents a weft yarn raised is an exact statement of the topological relationships of the yarns in a woven fabric. In multiply fabrics, the yarns separate automatically into different layers. It is possible to analyse a weave diagram to determine the nature of the fabric it represents. The analysis is based on the search for circuits, defined by J. Lourie, namely the arrangement $\begin{smallmatrix}XO\\OX\end{smallmatrix}$ in a weave which will cause yarns to interact and form a fabric. If this pattern does not occur in any two columns or rows of a weave pattern, then the two yarns will lie in different layers. A search for circuits in a stitched fabric weave representation results in a situation where if the stitch is correctly positioned, the number of circuits in the total pattern is higher than it would have been had the stitch been absent, but circuits do not exist between the stitching yarn and the adjacent yarns. If the stitch is incorrectly placed, in the weave representation, the stitched yarn will be found to have caused circuits to appear with the adjacent yarns.

An extension of these principles allows a complete analysis (by computer program) to be carried out on any weave representation which will determine the number of layers into which the resultant fabric will separate, and the location of the stitching points.

Structures and Topology (Konopasek)

1. Two-three- or, generally, n-axial woven fabrics may be defined as systems of two, three or n sets of quasi-parallel yarns interlaced or entangled in space. We have managed to generalise the concept of "elementary cell" and "repeat", as used in connection with planar projections of biaxial fabrics, for triaxial, four-axial and, generally, n-axial woven fabrics. The elementary cells must form tesselating patterns, i.e. they must fully cover the projection plane.

The number of unit cells or interlacing points in an elementary cell increases fast with increasing axiality. There are 1, 3, 8, 24.....interlacing points in a unit cell of biaxial, triaxial, four-axial and six-axial fabrics. Consequently there are always 2 different cells reflecting all the possible combinations of k interlacing points.

In the regular class there are two generic types of woven fabrics: one with the axiality 2, 4, 8, ...,2^n and another with the axiality 3, 6, 12, ...,3×2^n. The fabrics of the non-regular class with axiality different from the above may be generated by removing a certain number of sets of yarns from a fabric with higher axiality. If all the interlacing points in each cell are represented by real yarn, the fabric is called complete; otherwise it is incomplete. A repeat consists of a certain number of elementary cells. The repeats have to pack closely in a biaxial (tetragonal) grid.

2. There are three levels of models and modelling of textile structures: topological, geometrical and mechanical. Topological models take into account only positional relationships (i.e. entanglement or interlacing) of constituent yarns or their segments and neglect their geometrical or mechanical properties. Geometrical models take into account geometrical properties of constituent yarns under the conditions of defined topological constrains, and neglect the mechanical properties. Mechanical models take into account mechanical properties of yarns under the conditions of defined topological and geometrical constraints.

3. Any production process may be interpreted as a process of ordering the mass of material in a certain way. The purpose of most of the textile processes is to impose a certain order on the mass of the fibres and yarns or their segments. Every fibre assembly may be seen as a deterministic or probabilistic system of entangled and oriented slender bodies (fibres, yarns) and may be investigated at any or all of the above mentioned levels (topological, geometrical, mechanical).

4. Some of the concepts of the theory of knots or topology of

line structures may be meaningfully applied to modelling and
studying the topology of textile structures (see an article "The
Theory of Knots" by Lee Neuwirth in "Scientific American", June
1979, and W. Lietzman, "Visual Topology" published by Chatto &
Windus, London 1965). My framework for modelllng the topology of
textile structures may be briefly defined as follows:

(a) A planar projection of central axes of fibres or yarns
constituting a textile structure is considered to be an oriented
graph with defined vertices, edges and faces.

(b) The topology of the structure represented by a graph with
n numbered edges is defined by a n x 6 T-matrix with one row
for each edge. The six matrix elements in each row are: (1)
pointer to itself; (2) pointer to previous edge; (3) pointer to
next edge; (4) pointer to an edge branching from the same vertex;
(5) crossing function (+1 or -1 determining upper or lower position
of the edge at the originating vertex); (6) torsion function (+1
or -1 determining a Z- or S-twistedness at the originating vertex).

The following theorems follow from the definition of the T-matrix:

$$t_{t_{t,2},3} = t_{t_{i,3},2} = t_{t_{i,4},4} = i$$

$$r_{i,5} = -t_{t_{i,4},5} \qquad t_{i,6} = t_{t_{i,4},6}$$

These theorems are used in several algorithms developed for the
purpose of modelling the topology of textile structures and
simulating the processes of their modifications (for details
see M. Konopasek, "A Knot Exercise", Georgia Tech., Atlanta
1979).

Isotropy Groups in Woven Fabrics (Fitzgerald)

As is well known in linear elasticity, there are 32 crystal
classes. For classes in the plane, there are but four anisotropic
space and point groups, namely 360°, 180°, 120°, and 90° representing orthorhombic, triangular, and cubic symmetries. All higher
order structures appear isotropic.

In second order elasticity, however, we pick up $\frac{2\pi}{5}$ & $\frac{2\pi}{6} = 60°$
symmetries. Thus hexagonal symmetry is no longer transversely
isotropic. Further, the triangular point group becomes hexagonal,
so it is isotropic for linear elasticity, but not for 2nd order
elasticity.

DISCUSSION VIII

DYNAMIC RESPONSE

Leader   : Dr. R.G. Shephard
Rapporteur : Dr. C. Delides

Energy Absorption by Fibre Assemblies

Experiments on the response of single yarns to high-speed transverse impact were described (Williams) and the results discussed in relation to theoretical models.

The strain pattern and longitudinal and transverse stress wave forms appear to conform closely to those of the theoretical two-dimensional and experimental three-dimensional results described by Leech and Hearle. However, lower total energy absorption at break is observed in yarns of high modulus which are known to give superior high speed transverse impact resistance in fabric form.

This seemingly anomalous result led to discussion on how energy is absorbed. Are visco-elastic effects observable?

Within the speed- and time-scales concerned the yarns appear to behave elastically because the front of the kink wave is straight and appears to advance at a constant speed (Stevens). It is important to remember that the input is displacement, the stress adjusting itself to the deformed situation. If one cannot detect the slowing down of the kink then viscoelastic effects are insignificant. However, at higher speeds of impact, for nylon, aerodynamic drag on the yarn may show up as a slight curvature of the yarn.

It was also suggested that the broken fibre ends, when viewed in an electron microscope, would show evidence of melting.

A feature which can be overlooked (Stevens) is the effect of the finite diameter of a yarn. For longitudinal waves, simple theory predicts a sharp fronted stress/strain step. A real fibre with a real Poisson's ratio would need to contract. It cannot do so without approaching the new equilibrium diameter through a radial oscillation. The dilatation wave goes ahead and sets up radial stresses which cause the diameter to contract beyond its equilibrium value, causing a section of high longitudinal stress to occur (Fig. 1). This is one of the features in impact damage that explain why a fibre like nylon, which has a breaking strain of about 11% on a direct dynamic tensile test, breaks at speeds of normal impact that induce breaking strains of about 6% only.

Consider now the effects of the structure of the yarn. The first step is to consider a fibre in a shallow helix (Fig. 2). To a first approximation, a step-stress pulse travels along the fibre at the same sonic velocity and introduces a particle velocity. Behind the fibrewise wave-front there are radial forces, a centripetal tension component due to the curvature and a small centrifugal component due to the angular velocity of the moving particles. If the radius of the helix is a, the pitch of the helix b radians/unit length and the strain e, then it is possible to show [1] that the fibre performs an oscillation with a wavelength

$$2\pi/be^{\frac{1}{2}}(1-e)^{\frac{1}{2}}$$

However, this relates to an oscillation which crosses the axis. A fibre in a yarn cannot do this because it impacts against the other fibres. If the interyarn impact pressures are of an order much greater than the forces controlling the outer fibres then, in fact, these fibres would perform an oscillation with a wavelength half that calculated above.

In impact experiments done sometime ago [2] measurements were made of the strain along a yarn at a particular time after impact. Fluctuations in the strain along the yarn of about 30 mm wavelength were observed (Fig. 3). Although these fluctuations were said to be significant, no explanation was offered. If the data of 3 tpi and 2% strain are inserted into the above equation for wavelength, the value obtained is 60 mm, approximately twice that observed. Thus there is an indication that the observed strain fluctuations are due to transverse motion of many fibres in the yarn.

There is a case for repeating these experiments with yarns of different twists and also with yarns in transverse contact under known reaction. The transverse fluctuation in the incident strain step could cause uncertain friction, which may be, effectively, very low. This would be of significance in the propagation of waves in fabrics.

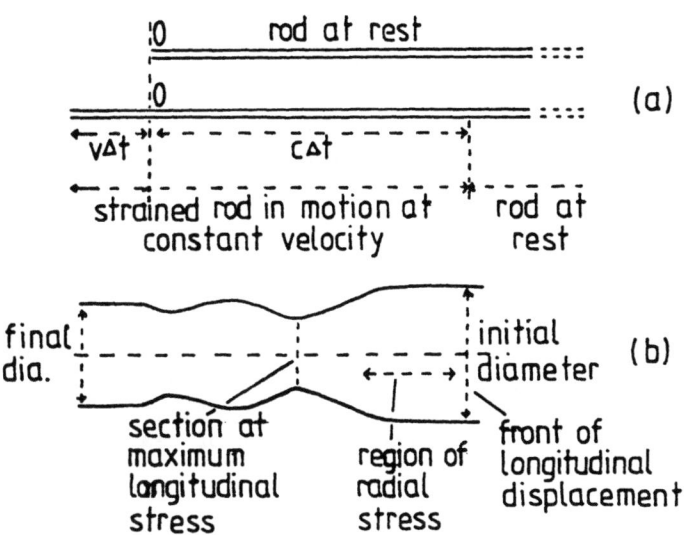

Fig. 1 Basic mechanics of a step wave

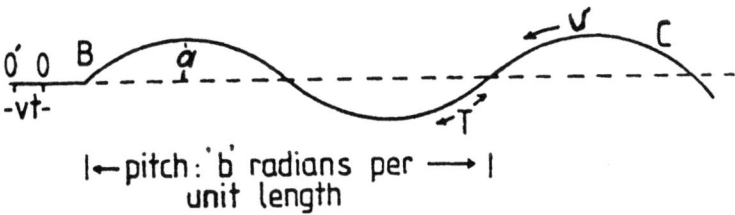

Fig. 2
Plan of shallow helix of radius a and pitch b.
Tension T produces centripetal force Tab^2.
Velocity v produces smaller centrifugal force mv^2ab^2
Equation of motion $\quad -\ddot{a} = c^2b^2e(1-e)a,$
Frequency given by $\quad 2\pi f = bce^{\frac{1}{2}}(1-e)^{\frac{1}{2}}$
Wavelength given by $\quad \Lambda = 2\pi/be^{\frac{1}{2}}(1-e)^{\frac{1}{2}}$

On the absorption of energy by fabrics, (Shephard) it is suggested that reflection and transmission coefficients at yarn crossovers are the key to the performance of textiles in absorbing energy. On the one hand high modulus yarns are used because the waves spread quickly along them and bring into play the maximum amount of yarn, while on the other, crossing yarns are necessary in order to spread energy to other parts of the textile. The existence of crossing yarns causes reflections which contribute to the maximum stress at the impact point and hence to failure there. By controlling the reflection and transmission coefficients (e.g. by friction) it may therefore be possible to optimise the energy absorbing capacity of a textile structure.

It was shown (Stevens) that the transmission to transverse yarns is non-linear, as follows.

Reflection of Waves at Joints in Nets (Stevens [1])

In normal impact of a kink wave on an elastic string, having initial small strain e_1 (as in practice, i.e. $> 5\%$), with impact velocity V, the strain e induced and the force of reaction F are given by

$$\left(\frac{V}{c}\right)^2 \simeq 2(e_1 + e)^{\frac{1}{2}} e \qquad (1)$$

$$F = 2(mT)^{\frac{1}{2}} V = 2mc(e_1 + e)^{\frac{1}{2}} V \qquad (2)$$

where m is the linear density and c the sonic velocity.

When a stress/strain pulse (T_o, e_o) with its associated particle velocity v_o reaches a joint in an orthogonal net, the joint and also the cross string are moved by partial reflection of the pulse. If the joint were fixed the pulse would be totally reflected and if the string broke it would retract with velocity $2v_o$. Thus there is an equilibrium equation for partial reflection.

$$T_o\left(2 - \frac{V}{v_o}\right) = F + mcV$$

where mcV is the transmitted tension. Substituting for F and T_o (= mcv_o) it follows that the direct reflection coefficient is given by

$$\frac{v}{v_o} = \frac{1}{1+x}$$

where $x^2 = e_1 + e$.

The branch transmission coefficient v/v_o or e/e_o is then found from equation (1). It is

$$\frac{e_o}{2x(1+x)^2}$$

which reduces to $(e_o/4)^{1/3}$ when $e_1 = o$.

Thus propagation in a net is non-linear even when the component strings are perfectly elastic.

Dynamic Elastic Moduli

The importance of knowing dynamic elastic moduli at all stress levels was emphasized (Stevens). For cords the longitudinal modulus increases with stress, specific modulus varying substantially linearly with specific stress. The weave has an effect which is more apparent at lower stress. In nylon cords, loose braids are softer than tight braids which are softer than tubular webbing with a weft. The relation of modulus to stress is the same for all three.

Concerning its measurement, the simple method of oscillating one end of a cord, the other end of which is fixed, at a frequency away from resonance, is adequate. But for a substantial specimen such as webbing, it should be possible to excite at the centre, both ends being fixed, without adding too much mass. It might then be possible to resolve the contributions to modulus from fibre properties alone and from the nature of the weave.

A method for fabrics was described (Blum), as follows.

Wave Propagation in a Coated Fabric (Blum)

Measurements of wave propagation in a coated fabric under tensile pre-stress give information, not only about the dynamic elastic properties of the material, but also about the state of stress. Consider the fabric to be an orthotropic, hyperelastic membrane without bending stiffness. The equations of motion are

$$n_{\mu\nu} \frac{\partial^2 u_3}{\partial x_\mu \partial x_\nu} = \rho \frac{\partial^2 u_3}{\partial t^2} \qquad (3)$$

$$E_{\nu\lambda x\mu} \frac{\partial^2 u_x}{\partial x_\lambda \partial x_\mu} = \rho \frac{\partial^2 u_\nu}{\partial t^2} \qquad (4)$$

$$\mu,\nu; \; x,\lambda = 1,2$$

where '1' and '2' represent the warp and weft directions respectively. $n_{\mu\nu}$ is the pre-stress and $E_{\nu\lambda x\mu}$ is the tangential E-modulus.

The characteristics of equation (3) with initial conditions $t = 0$, $x_1 = x_2 = 0$ are ellipses with semi-axes $\sqrt{(n_{11}/\rho)}\,t$ and $\sqrt{(n_{22}/\rho)}\,t$ where n_{11} and n_{22} are the mean tensile pre-stresses. Measurements show that these can thus be determined to an accuracy of 3%.

Equation (4) has two characteristics, one of which can be measured (Fig. 4). Measurements of the other, which can be seen, remain to be done. The first characteristic gives the dynamic elastic moduli in the warp and weft directions. In the former the difference between the dynamic and static moduli is not large, but for the weft direction it is substantial. This difference between warp and weft directions shows that the material exhibits anisotropic visco-elasticity owing to the difference in yarn geometry in the two directions.

Non-linear Visco-elasticity of Fibrous Materials (Ko [3, 4])

Polymeric materials such as textile fibres are composite structures. They should exhibit nonlinear viscoelastic behaviour. In order to understand the structure-property relationships of fibres and to generate design data for industrial products and surgical implants, the viscoelastic behaviour of a wide variety of materials including fibres, yarns, braids and biomaterials, has been measured.

It appears that, without exception, these fibrous structures have nonlinear stress-strain behaviour at all strain levels. Their hysteresis or damping characteristics are generally insensitive to strain-rate effect. Their stress relaxation and creep behaviour depend upon strain level and stress level respectively. This is characteristic nonlinear viscoelastic behaviour.

To summarise experimental observations, a Quasilinear Viscoelastic Model (QVM) has been used to relate stress history, $T(t)$, and strain history, $\lambda(t)$:

$$T(t) = T^\rho[\lambda(t)] + \int_0^t T^\rho[\lambda(\tau)]\dot{G}(t - \tau)d\tau$$

where the elastic response, $T^\rho[\lambda(t)]$, is expressed in terms of a spline function. The second term of the QVM reflects the history dependent behaviour, including stress relaxation $G(t)$, creep $V(t)$ and dynamic response $\mu(\omega)$. These history-dependent functions are expressed in terms of a continuous relaxation spectrum $S(\tau) = C/\tau$, for $\tau_1 < \tau < \tau_2$. Consequently, the history dependent behaviour of the materials can be summarised conveniently by the three parameters C, τ_1 and τ_2.

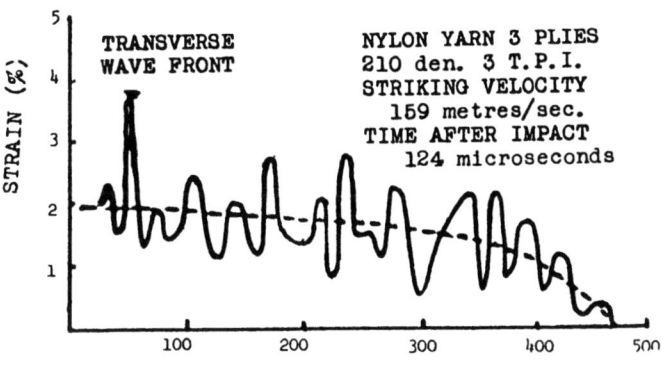

Fig.3 Strain versus position along the yarn (based on Fig. 7, ref. [2])

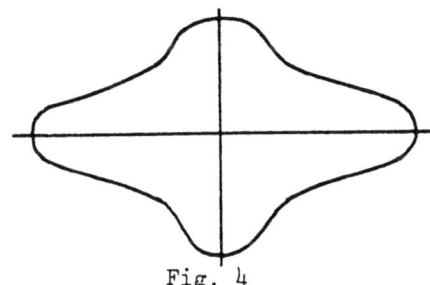

Fig. 4

REFERENCES:

1. G.W.H. Stevens, AIAA paper 79-0423.
2. D.R. Peterson et al., Text. Res. J., 30, 411 and 422.
3. Chen et al., Polymer Eng. and Sci., 16, 406.
4. F.K. Ko, Ph.D. Thesis, Georgia Institute of Technology, 1976.

DISCUSSION IX

RELATED MATERIAL; PAPER, RIGID COMPOSITES

Leader   Dr. A.R. Bunsell
Rapporteur Dr. O.G. Bilir

There were three fairly distinct contributions; on 'failure in fibre re-inforced composites', 'elasto-plastic analysis of short fibre composites' and 'wood and paper as flexible fibre assemblies'. These are summarised below. General discussion centred on the application of textile technology to the manufacture of composite materials.

Thus (Skelton) in many respects composites are the antithesis of traditional textiles because the properties of strength and stiffness are emphasized above all others. Hitherto composites have been made by hand-lay-up techniques. Now that they have proved themselves as primary structural materials, more cost-effective methods of production are being sought. The traditional textile industry, with its vast experience of materials handling, is becoming increasingly involved in the developing composites sector and may eventually be the main contributor to the commerical success of these high technology materials.

Examples of fabric use in composites (K_o) are as follows: fibre-glass tubular braids for helicopter blade reinforcement and knitted fibre-glass fabrics for aircraft body and interior panel reinforcement. Once the material is made the structures are achieved by traditional metal stamping techniques.

Failure in Fibre-reinforced Composits (Bunsell)

There are many similarities between textiles and composites as well as differences. For example the isolation of a fibre break in a twisted yarn, due to frictional forces produced by the attempted relative movement of the broken ends with respect to other intact fibres, is exactly analogous to what occurs in a composite. The shear of the matrix around a fibre break leads to a build up of load along the fibre, resulting in a reduced effective length and load transfer to neighbouring fibres. But the increasing transverse loads which occur in a yarn do not exist in a composite.

In many composites the fibres dominate the behaviour and control the failure of the material. The fibres must be broken to break the composite. This is not true of all composites and depends on the definition of what constitutes failure but it is obviously true for unidirectional composites. For example the creep of Kevlar-reinforced epoxy composite has been found to depend on the creep of the fibres.

There are two schools of thought concerning fracture in composites The dominant one uses the ideas of linear elastic fracture mechanics. This is basically a metallurgical approach in which the material is treated as a continuum and the nature of the reinforcement is ignored. Another approach is to consider the mechanisms of failure of the material using bundle theory.

Composites make up a family of materials and many types of failure are possible. In Asbestos-Cement, in which short fibres are randomly distributed in two dimensions in a brittle matrix, crack growth is dominated by the matrix with fibres inhibiting crack growth by bridging the cracks and microcracks. The theory of fracture mechanics works well for this composite and one can apply the model used by Dugdale to overcome difficulties of plastic deformation at crack tip in metals. Instead of an imaginary closing pressure, however, there is a real closing pressure due to the fibre bridges.

Boron-Aluminium, a metal matrix composite, breaks by different mechanisms depending upon the Aluminium alloy used. Cracks in the matrix always initiate from broken fibres. AL-1200 matrix cracks run parallel to the fibre direction, probably due to Poisson contraction normal to the fibres leading to tensile cracking. AL-6061 cracks run in a perpendicular direction to the fibres but pass around intact fibres. Stress concentrations at crack tips are insufficient to break fibres. In AL-2024, because the yield point is high, the load transfer between fibres is sufficient to break neighbouring fibres around a fibre break. This results in a series of fibre breaks without cracking of the matrix. In fatigue, the matrix cracks after breaking of the fibres. For Boron-Aluminium, fracture mechanics does not work.

In structures made of Carbon Fibre Reinforced Plastics (CFRP)
there are often millions of fibres so that individual fibre breaks
cannot be followed. However, they can be heard. Acoustic Emission
techniques indicate that unidirectional CFRP behaves like a fibre
bundle modified by the time-dependent properties of the matrix.

Experimental measurements, made by traditional techniques such as
strain gauges, indicate that CFRP is elastic but sometimes under
creep or fatigue conditions the material breaks. The Acoustic-
Emission technique shows that the material is not perfectly
elastic and that, under constant or cyclic loading, internal
damage continues. The rate of damage can be determined using
the technique and it is believed that the amount of damage necessary
for failure is independent of the type of loading.

If a fibre bundle model is considered, the probability of fibre
failure is given by a Weibull distribution and a curve of damage
as a function of stress, which is a function of this distribution,
can be drawn (Fig. 1a). A constant load on the bundle at
$\sigma_A < \sigma_{max}$ will break a certain number of fibres but the remaining
intact ones will hold the load. If the fibres are perfectly
elastic there will be no creep.

Similar diagrams can be drawn for unidirectional and, in certain
cases, cross-piled composites, but there is continued damage at
a constant load indicated by a vertical line AB (Fig. 1b). To
break the composite it is necessary for AB to intersect the upper
curve at C. The rate of damage accumulation is a logarithmic
function of time which is given by Acoustic-Emission measurements
so that minimum life times can be calculated. The failure
mechanism is that an increase of load transfer length, due to
viscoelastic behaviour of the matrix, leads to greater probability
of fibre failure.

Elasto-Plastic Analysis of Short-fibre Composites (Curiskis)

Although there has been much work on the analysis of continuous
filament reinforced composite materials, there have been only
two attempts to model composites reinforced by discontinuous
filaments. The first [1] is an axisymmetric finite element
analysis of a short-fibre bonded array, with the boundary conditions
imposed by an essentially trial-and-error approach. Such an
analysis excludes a finite amount of matrix material.

The effects of this matricular exclusion can be seen by means of
a three dimensional finite element analysis, but assuming that
the fibres are packed in a square-array [2]. For this case, the
boundary conditions on the nodal displacement pattern allow a

simple simulation of a uniaxial tensile test. The results (Fig.2) dramatically demonstrate the restraining influence of the excluded matrix (and of the fibre) in that higher loads can be supported than are predicted by the axisymmetric analysis. This difference is chiefly due to the high triaxial stress state (virtually hydrostatic) in the matrix region between colinear fibres. This can be seen from the internal stress distributions and the progression of the plastic zone boundaries with increasing applied load. There have been no experimental studies on comparable model composites.

Wood and Paper as Flexible Fibre Assemblies (Ansell)

The microstructure of the softwood cell (or tracheid) can be described as a hollow tube containing cellulose microfibrillar reinforcement. The matrix materials, hemi-cellulose (semi-crystalline) and lignin (amorphous) and the microfibrils in the thick S_2 layer are helically arranged at an angle of about $10°$. Not only is the arrangement of the crystalline reinforcement essential, but also the moisture content of wood, in order to obtain a material with high specific stiffness and high work of fracture. When the wood cell extends, micro-shear interactions take place between the microfibrils and the matrix as the helix extends. Stress-strain curves for wood having about 12% equilibrium moisture content are essentially straight lines to failure. Dry wood is stiffer but fails at much lower strains, resulting in a lower work of fracture.

When wood cells are delignified and made into paper, the arrangement of flatened cells is fixed when the paper has dried. The microfibrils now have a zig-zag orientation, and the cells are brittle, owing to their low moisture content and lack of lignin. In the flat assembly of cross-linked fibres, complex deformations at a cellular level are therefore not possible and paper behaves more like a polymer film than a woven fabric. In woven fabrics, complex deformations are possible, involving shear and bending of yarns and their constituent fibres. Skelton's approximate correlation between shearing and bending behaviour in fabrics is therefore not observed in paper undergoing complex compressive deformation, because paper is not a true flexible fibre assembly.

It is suggested that, owing to the complex shearing and bending of microfibrils which occurs in wood cells during loading, it is more reasonable to think of wood as a flexible fibre assembly than it is of paper.

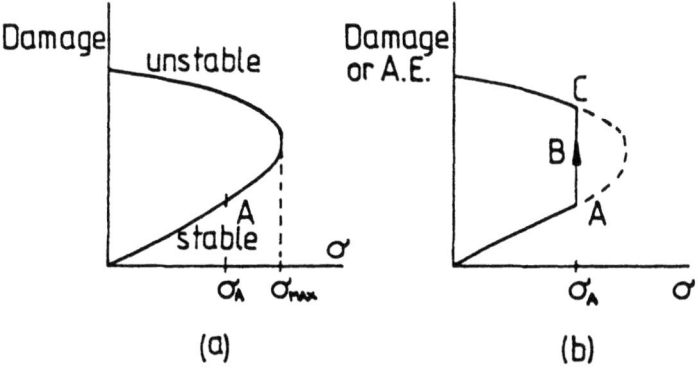

Fig.1 Damage as a function of applied load:
(a) in a fibre bundle (b) in a composite material

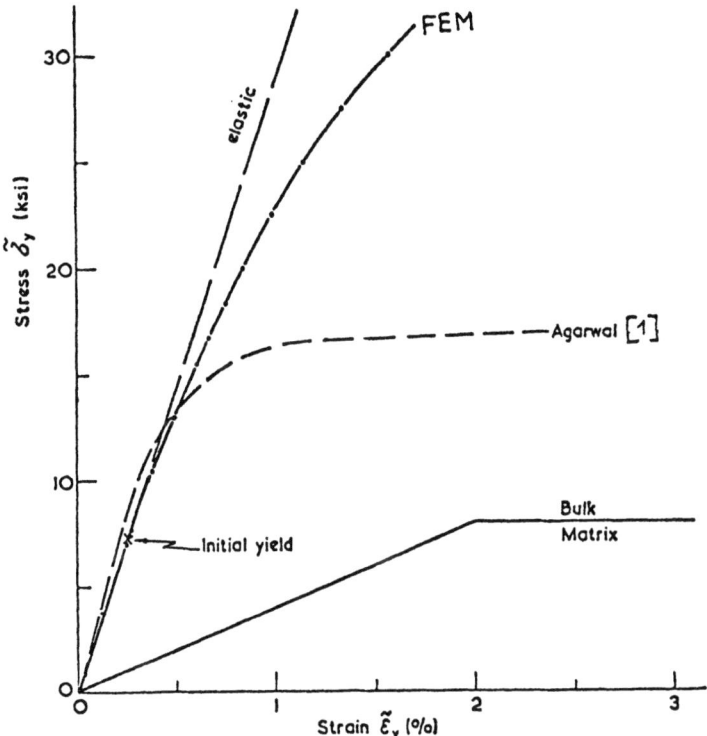

Fig. 2. Predicted composite stress-strain curve due to matrix yielding [2].

REFERENCES
1. B.D. Agarwal, J.M. Lifshitz and L.J. Broutman, Fibre Sci. Technol., 7, 45.
2. J.I. Curiskis, Ph.D. Thesis, Univ. of N.S.W. 1978.

DISCUSSION X

STRESS-CONCENTRATION, TEAR, WEAR, etc.

Leader   : Professor K. Slater
Rapporteur : Dr. R. Shishoo

General Discussion

Racah stated that the ability of a fabric to withstand crack propagation is related to its ability to distribute load concentrations on fairly large widths using its low shear stiffness and yarn slippage properties. In the trapezoidal tear test these qualities do not manifest themselves because the yarns ahead of and after the crack tip are not stressed due to the specimen geometry. This test should be replaced by a "fracture mechanics" type test in order to generate a suitable stress distribution around the tip of the crack.

Konopasek pointed out that the tongue-tear test seems to reflect only the specific tongue-tearing behaviour; and the trapezoidal test does not fully reflect practically important aspects of tear resistance. Until better machines and methods are available for testing tear resistance under biaxial tension, the use of the widely available bursting test can be made.

Shishoo pointed out the difficulty in measurement of
(a) tearing strength of a sheet material which has not a preferred direction of tear, e.g., nonwovens and some knitted fabrics,
(b) dynamic tearing strength of coated fabrics where the Elmendorf tester is not capable of producing tear.

Tearing (Stevens)

Stevens pointed out that the tear test had been illustrated with the untorn piece of material shown straight and at right angles to the torn tongues. In a practical situation, such as tearing a tongue to open a parachute pack, the pull may be in any orientation. More often than not, the tear force would distort the pack due to initial reaction at the speed at which the tear takes place and wrinkles at right angles to the tear line would appear. Is there any evidence on what these wrinkles would do? Would they stop the tear? Another method of pack-opening is to use a wedge-knife edge. Such means affects tearing behaviour if the fabric wrinkles crossways in front of the knife.

Barbage reported that when one is stitching together two fabric layers, using sewing speed of up to 7,000 stitches per minute, the sources of stress concentration are the areas of needle penetration as well as the thread itself. It has been observed that the strength of a sewing thread can undergo a 10 - 20% loss after a sewing operation.

Damage and failure modes of fabrics (Williams)

Most fabrics are tailored for their end use. There is always a compromise (Engineering Trade-off) in properties. Fabrics function well until "damaged" i.e., some critical performance characteristic drops below minimum requirements. The means by which they are "damaged" are more important than the final failure mode, as exemplified in Table I.

Table I

Use	Fabric	How Damaged	Failure Mode
Sheeting	Light wt. plain weave	Abrasion	Filling Tear
Boy's pants	Medium wt. Twill	Abrasion	Warp Tensile
Shirting	Light wt. plain	Abrasion	Frayed Edges
Fatigue Pants	Medium wt.	Abrasion	2-D Tear
Tentage	Very heavy twill	Exposure	Tensile
Men's Suiting	Medium wt. plain 1)55/45 PES/Wool 2)100% text.PES	Style Change Abrasion/ Extension	Discard Snagging Bagging
Parachute	Very light wt. plain	Uncertainty	Discard

The following two qualifications/explanations relative to the Table should be taken into account in order to avoid misinterpretations or misuse.

Due to the wide variety of conditions encountered, there are usually a number of different failure modes commonly found in a given fabric/end-use situation. The description in the Table of one typical failure mode for each situation was made to provide an example to support the discussion point, and does not imply that the failure mode given is the only or even the dominant one.

Multiple use of the designation "abrasion" is deliberately ambiguous. In each instance its meaning is different, due to the different balance of several contributing factors in the end uses, e.g., laundering of sheets, wear exposure of boys' pants etc. Due to the multiplicity and inadequacy of abrasion tests, the comments should not be considered to suggest applications of abrasion tests as useful life-predictors.

The point of the whole comment is simple, as a message to the Applied Mechanics/Theoretical/Analytical Scientists present: do not base your failure theories on models of fabric in the original as-made state. Identify the predominant mechanisms that are damaging the fabric, in the critical aspect of its end-use.

Crack Propagation in Paper (Seth)

The Elmendorf tear tester is used to determine tear resistance of paper. In paper sheets one tries to produce products having balanced tensile strength and tear resistance. Normally the paper sheet is characterised by plotting its tensile strength against tearing strength. With increased bonding, tensile strength increases but tearing strength decreases.

Pressroom breaks are low-tensile brittle failure caused by propagation of a pre-existing flaw. The mode of failure is tensile or opening mode, and for such failures Elmendorf tear is not a satisfactory criterion. Linear Elastic Fracture Mechanics (LEFM) is used to determine resistance to crack propagation in the opening mode. K_c varies with specimen width, but for large specimen width K_c^c and hence G_c as a material property could be obtained. Fracture resistance (R) of paper has also been determined by using quasistatic crack propagation technique of Gurney and Hunt. (Proc. Roy. Soc. 1967).

Excellent agreement between G_c and R for several papers has been obtained. It has been shown that fracture resistance increases with bonding. Dr. Seth pointed out that fracture resistance and Elmendorf tear are different mechanisms because of the different modes of crack propagation. Fracture resistance is now used to characterise printing-grade papers.

REFERENCES

1. R.S. Seth, D.H. Page, J.Mater. Sci., <u>9</u>, 1745, (1974)
2. "   "   "   "   TAPPI (1975)
3. "   "   "   "   TAPPI July 1979.

Wear in Parachute Cords (Stevens)

Mr. Stevens also drew attention to his work some ten years ago on ply-tear webbing, a textile assembly with two plys, bound by warp binders, which can be torn apart by breaking the binders sequentially. This is reported in Various R.A.E. Farnborough Reports and in NAS Tech. Memo X-58106, Nov. 1972. Webbings have been torn at up to 100 m/s separation speeds. The breaking energy of the binder is about 50% of the thermal energy of the fibres, nylon or polyester, from ambient conditions.

There was a problem of life and loss of strength in aircraft brake parachute cords. The ends of an aircraft break parachute undergo a dynamic loading at each use (stream). The cords have about 12 kN breaking load and are subjected to a peak load of 2.25 $\pm$ 0.3 kN on each stream. The cord strength is found to decrease with repeated use and after 50 streams a wholly nylon cord has lost about 30% of its strength. Thus there is a need to life these cords.* Recent work has shown that cords covered with either Nomex or Cotton deteriorate less, about 5% and 12% respectively, so such cords have a potentially longer life.

Currently cords are discarded after 50 streams, sometimes earlier if there is obvious visual deterioration. However, visual assessments of whether a cord is weakened or not are not very good. Tubular woven webbings, which looked good, were found to be weaker than detectably fluffy looking braids.

*i.e. to restrict the period of usage before descarding.

DISCUSSION XI

PNEUMATIC STRUCTURES AND TYRES

Leader   : Professor M. Levinson
Rapporteur: Professor B. Kaftanoglu

Measurement of Material Properties

Much of the discussion concerned the determination of the
properties of the materials used. Thus (Ansell) when designing
the shape of coated fabric panels, which are welded together to
form an airhouse shell, it is critical that stress concentrations,
buckling and over-compex detailing are avoided in the inflated
structure. For this reason accurate biaxial creep data are
required to characterise the fabric strain under pre-stress loads.
Preliminary work is in progress at Bath aimed at providing such
data for use by architects and building engineers.

Uniaxial (to failure) and biaxial (prestress) tensile characteristics
for a PVC-coated polyester fabric have been obtained, from which
it is clear that stressing the airhouse membrane above pre-stress
levels would result in irreversible residual strain, leading to
a slacker membrane. Weathering of the membrane would further
complicate the situation, so that analysis of the mechanics of
coated fabrics under pre-stress loads is far from easy.

Other contributors remarked on (a) the desirability of having a
biaxial tester in which the jaws could move independently and (b)
the importance of also testing the fabrics before coating.

Considerable interest was expressed in Dr. Blum's method of
measuring the dynamic elastic modulus of coated fabrics by
determining the velocity of propagation of a pulse. This velocity
(Stevens) would be a group velocity rather than a phase
velocity because one would expect frequency dispersion within the

material. It would be of interest to measure the effective
modulus also by vibrating specimens of known length. Experience
with nylon ropes is that the two measurements differ and that in
order to obtain the effective dilatation modulus, the figures
obtained by pulse propagation need to be increased. The problem
of measuring dynamic modulus in a complex visco-elastic fibre
assembly is not easy and there is much yet to be resolved.

Professor Clark's reference to the difficulty in determining the
transverse properties of cords, for later use in calculations
related to the basic composite ply structure in tyres was taken
up by Dr. Curiskis. The method of 'back calculation' from
measured composite properties using existing composite analyses
does not appear to have been pursured very far, but his experience
of estimating the constituent properties of keratin, when
idealized as a simple two-phase composite, suggests a possible
(and simpler) scheme of "back-calculation" of the transverse, and
indeed, of other, properties of cords or yarns, especially for
use in composite systems. Such a scheme would involve:

(a) making and testing simple model composites with known
(perhaps standardised)
(i) geometry   (ii) material properties,

(b) utilising existing, or other, micromechanics analyses to form
normalized "design" curves from which the yarn or cord
properties can be estimated. (Fig. 1)

Although problems exist as to the definition of the yarn/cord
cross-section, and therefore of volume fraction and of cord-matrix
interface, such an approach may find wide application in textile
mechanics for the determination of yarn/cord transverse mechanical
behaviour.

The Buckling of Membrane Structures

It was pointed out (Racah) that avoiding compressive stress is not
a good criterion. For example a cylindrical shell subjected to
internal pressure buckles at a critical level of that pressure,
even though the material is under biaxial tension throughout.
More complicated buckling (Fig. 2, Stevens) occurs in inflatable
air bags, used as shock-absorbers under heavy platforms when
dropped by parachute.

Concerning the form which buckling takes, as distinct from
whether or not it occurs, Professor Hearle suggested that bending
resistance is of importance. It is probably justifiable to
neglect it in the determination of the relation between forces

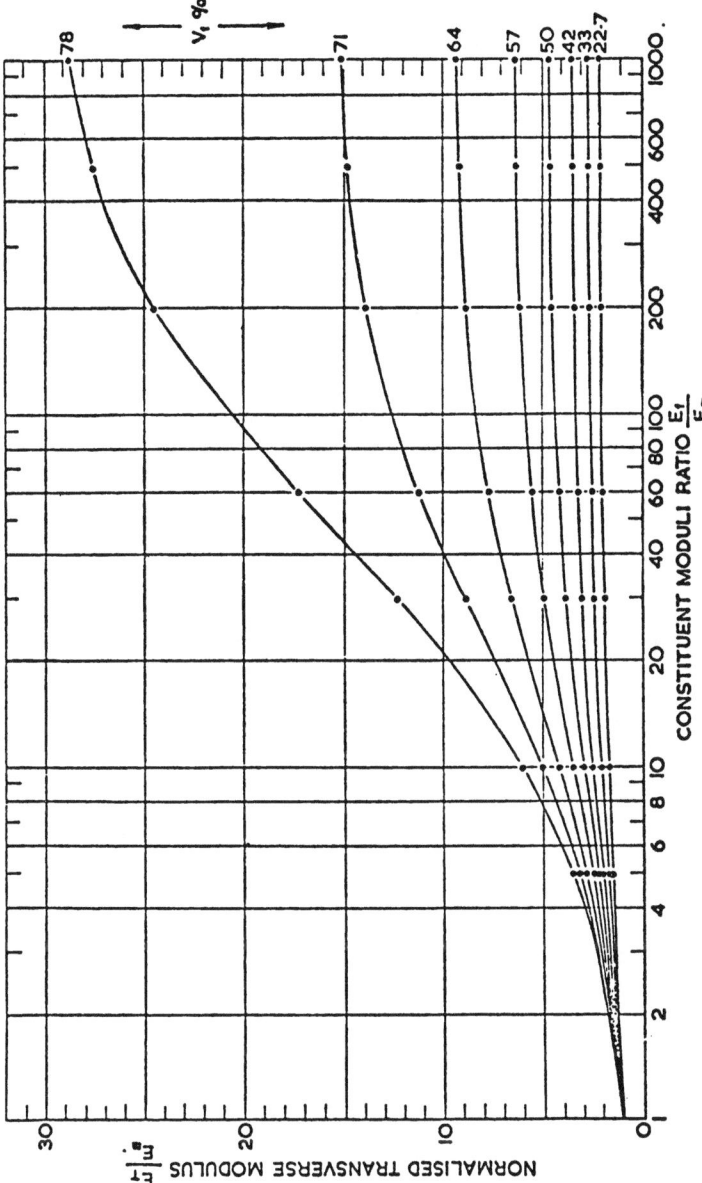

Fig. 1

Transverse modulus normalised with respect to matrix modulus [1]

and deformation in engineering fabrics. Nevertheless, bending could play an important role in determining the form of deformation. By way of analogy, one can consider a waterfall coming off a high plateau; the height variations of the plateau have negligible effect on the velocity of the water at the bottom of the fall, but are the dominant factors in locating the site of the waterfall.

Loss Characteristics of Textiles and Textile Composites (Clark)

The large number of vehicles in use in the United States require very large amounts of fuel, all in the form of liquid petroleum. Approximate numbers of vehicles are 150×10^6, approximate mileage 10^4 per year. Using typical values for pneumatic tyre rolling resistance, one can estimate that about 20% of vehicle fuel can be assigned to the rolling resistance of the tyres.

Tyre rolling resistance is almost wholly caused by internal hysteresis of the cord-rubber composite. It is a function of

(a) the load carried by the tyre,
(b) the tyre inflation pressure,
(c) the tyre composite loss characteristics.

The load and pressure effects are to a great extent determined by automobile design. The composite loss is the area of greatest unknown effect. At the present time, essentially all that is known is that some structural geometries, such as the radial tyre, are more efficient than others, such as the bias tyre.

There has been a great deal of work reported in both the textile and composite mechanics literature on the elastic properties of textiles and textile composites. What is badly needed now is more work on the loss characteristics of textiles and textile composites. Even a small decrease in tyre losses can result in immense fuel savings. For example, in the United States a 5% reduction in tyre rolling resistance in the entire fleet would result in a fuel saving of 20×10^6 barrels of gasoline per year.

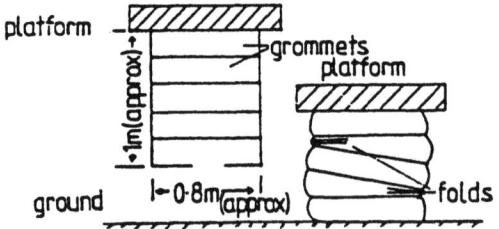

Fig. 2. Buckling of Shock-Absorbing Airbag in Vertical Descent

REFERENCE

1. J.I. Curiskis, Ph.D. Thesis, Univ. of N.S.W., 1978

DISCUSSION XII

GEOTEXTILES

Leader   : Professor K van Harten
Rapporteur : Messrs A. Nijhof and H. Dorr

Reasons for the Success of Geotextiles (Giroud)

The various applications presented illustrate some very practical reasons. For example:

    there are savings in materials and reductions in labour time,
    there is less wear of equipment,
    it is simpler to control the execution, and
    the structure has improved performance.

But geotextiles meet a more fundamental need in civil engineering. All, or nearly all, of our natural environment is three-dimensional, for example, the bodies of animals, masses of soil or rock and trees. There are examples of one-dimensional natural products in animal, vegetable and mineral fibres and in the way of two-dimensional objects, there are bark, leaves and the skins of animals. However if we consider the mineral kingdom on the scale that is of interest to engineers, there is practically nothing two-dimensional, except for schists, slates and mica. There is thus a considerable deficiency in Nature, and man has always applied himself to filling this gap with two-dimensional industrial products: leather, wooden planks, glass, tiles, paper, sheet metal, plastic sheets and panels, and all manner of textiles.

Until recently, in the construction field, these flat materials have been used as coverings. This is merely imitating the skins and barks of nature. The real revolution took place when a textile was used within a mass of soil, that is to say, actually inside a natural medium; Nature was no longer being imitated, it was being offended. This unusual application of a manufactured material that is virtually nonexistent in nature is profoundly original. We may, therefore, be certain that the use of flat materials such as synthetic sheeting and geotextiles in civil engineering is here to stay.

The Nature of Soil/Geotextile Interaction (Giroud)

One cannot offend Nature with impunity. Nature abhors discontinuities. Putting two different materials into contact, or the presence of a foreign body in a natural medium, creates discontinuities and Nature tends to eliminate them. This defensive reflex can assume various forms. For example, it can be simply the destruction of the foreign body (in this case the geotextile) by the chemical and biological action of the soil, just like the rejection of a graft by a living organism. But if the foreign body resists this destructive action, Nature produces a phenomenon which tends to remove the discontinuity. A typical example is filtration in which the discontinuity is due to contact between two different media.

Consider water flowing in succession through two granular media in contact with each other; first a medium of low permeability (silt for example), then a more permeable medium (gravel, for example). A natural phenomenon occurs (transport of particles), which tends to increase the permeability of the low-permeability medium (which loses particles by leaching) and to reduce the permeability of the high permeability medium (which clogs by trapping particles).

The leaching/clogging phenomenon can be avoided by using filtration criteria which determine the optimum grain size distribution of granular media in contact. These criteria represent only one kind of compatibility rule which must be observed when two bodies are in contact. This is particularly true when one of the bodies is an intrusion in the medium aurrounding it, such as a geotextile buried in the soil. The rules of compatibility for geotextiles are physicochemical, hydraulic and mechanical. They govern the changes ("ageing") in the geotextile within the soil, namely:

(a) physico-chemical ageing of the constituent material of the geotextile;
(b) hydraulic ageing, by clogging;
(c) mechanical ageing, by creep or fatigue.

Work Required for Future Progress (Giroud)

Clearly a better understanding of soil/geotextile interaction requires cooperation between soil and textile researchers. But this is only one example of the research required for the future use of geotextiles. The following is a tentative list of desirable improvements in knowledge concerning the geometry and mechanical properties of geotextiles.

Geometry:

(a) Methods for describing and quantifying the pore size distribution of nonwovens.
(b) Knowledge of the influence of compression and extension on the pore-size distribution of geotextiles.
(c) The relationship between the pore size distribution of a geotextile and the likelihood of granular particles passing through it.

Mechanical properties:

(d) The development of methods for performing biaxial tensile tests and triaxial tests (biaxial tension plus compression) on geotextiles.
(e) The experimental study of creep in uniaxial, biaxial and triaxial tests.
(f) Theoretical analysis of the bursting test, particularly the relationship between bursting strength and tensile strength and between bursting strength and size of the hole.
(g) Development of the mechanics of nonwovens, particularly the relationships between the mechanical behaviour of filaments and the fabric, and between the tensile strength as measured in various tests.
(h) Experimental and theoretical study of the compressive and distortive behaviour of geotextiles (distortion means shearing over the thickness of a geotextile), in particular the distortive behaviour of coated fabrics, in order to evaluate the risk of delamination.

Concerning (f) it is suggested (Kaftanoglu) that existing numerical solutions for the stresses and strains in bursting metallic membranes with large deformations and/or initial curvatures might be transformed in a manner suitable for geotextile problems.

The Testing of Geotextiles (Fitzgerald)

In the biaxial testing of fabrics, problems arise with square testpieces. The main one is the influence of the fixed clamping on the stresses and strains in the testpiece. This can be overcome

by using cross-like testpieces with fixed clamps on one axis and
movable clamps on the other (Fig. 1). A moiré-grid is put on the
fabric in the mid square in order to measure the deformation. In
the deformed state the pattern is like an eyeball, but the
interpretation of the measurements is complicated by the
viscoelastic properties of the material. For this reason a strain
hysteresis programme is used in conjunction with a digital
computer.

Other tests involve special arrangements of geotechnical fabrics
in combinations with soil, sand or gravel. For example, a fabric
laid between horizontal layers of gravel can be viewed in a
2-dimensional framing between transparent side plates. One can
see the shear planes caused by a load on a small part of the
surface, and observe the function of the fabric (Fig. 2). Besides
separation, and in several cases membrane function, the geotextile
fabric has a boundary effect; for instance it has been noted that
a different packing of the gravel occurs near the fabric.

In a similar 3-dimensional test, a cylinder of height equal to
diameter is partly filled with clay, which is then covered with
a geotextile fabric and a further layer of coarse material. (Fig.3)
Load cycling can be carried out on a small part of the surface on
the axis of the cylinder and results in a permanent set of the
fabric. This test illustrates a practical situation and the aim
is to correlate the results with effects noted in practice.

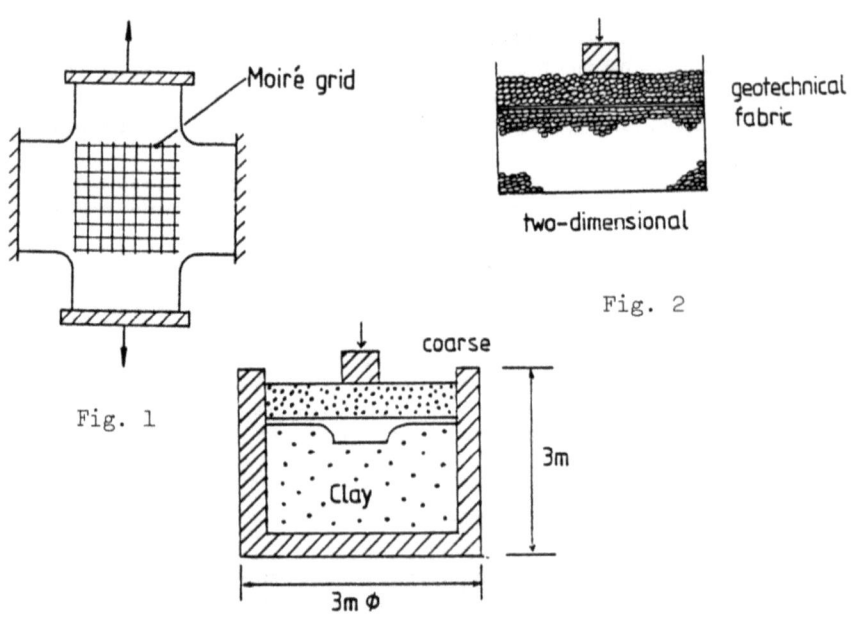

A Special Application: The Block Mattress (Dorr)

In the history of the Netherlands there have been many floods; this has been inevitable, because so much of the country, particularly in the south-west, is low-lying. Here the rivers Scheldt, Maas and Rhine flow through a delta to the sea, and centuries of interaction between the rivers, the wind and the tides have produced a region in which the coastline is contorted and the sea-water channels deep. As protection, dikes have been built along these channels, which stretch far inland, but again and again the defences have proved inadequate. The most recent disaster occurred in February 1953 when the number of deaths alone was 1,850 and the losses in livestock, buildings and agricultural land were immense. The 'Delta Plan' now being carried out, involves the shortening of the coastline and incorporates several dams.

On the sea-bed, soil conditions are bad and the current is so strong and turbulent that enormous amounts of sand are constantly in motion. There is thus a need to prevent erosion during the building of the dam in order to ensure the stability of the construction. Underneath and at both sides of the dam a large area of bottom-protection is necessary. In the past this was done with mats made of osierwood fascines which were sunk by ballasting with a large number of stones. The constructions were large and the labour costs high. Once it was appreciated that the filter properties of the fascines could be taken over by a woven fabric, other advantages became apparent. The cost of raw material and manufacturing are much less and above all, because of the greater strength, better and more efficient methods can be used to sink the mat.

The bottom protection required for the Delta Plan had to meet the following requirements:
- It must be permeable to water and impermeable to sand.
- It must not be affected by chemical and biological influences.
- The structure has to be capable of resisting heavy currents.
- The ballast layer must be fixed to the mats in such a way as to prevent shifting on steep slopes 40 m beneath the surace

This resulted in the development of the so called blockmattress made of polypropylene woven fabric with concrete blocks fixed on top, and in order to produce the required amount of 4.5×10^6 m^2, a special factory was built.

At the beginning of the production line six rolls of woven fabric are reeled off and immediately seamed together automatically, forming a strip 30 m wide. Then plastic pins are forced from beneath into the fabric in a prearranged pattern. These serve to anchor the concrete blocks that are to be cast on the mat at a

density of 200 kg/m^2. After the casting and hardening of the concrete, the mat is ready to be wound on a floating drum measuring 10 m in diameter and about 43 m in length.
When a completed mat of length 280 m is wound on the drum, it is towed to a pontoon, with the help of which the mat is unwound and laid on the bottom, which may be up to 40 m beneath the pontoon (Fig. 4).

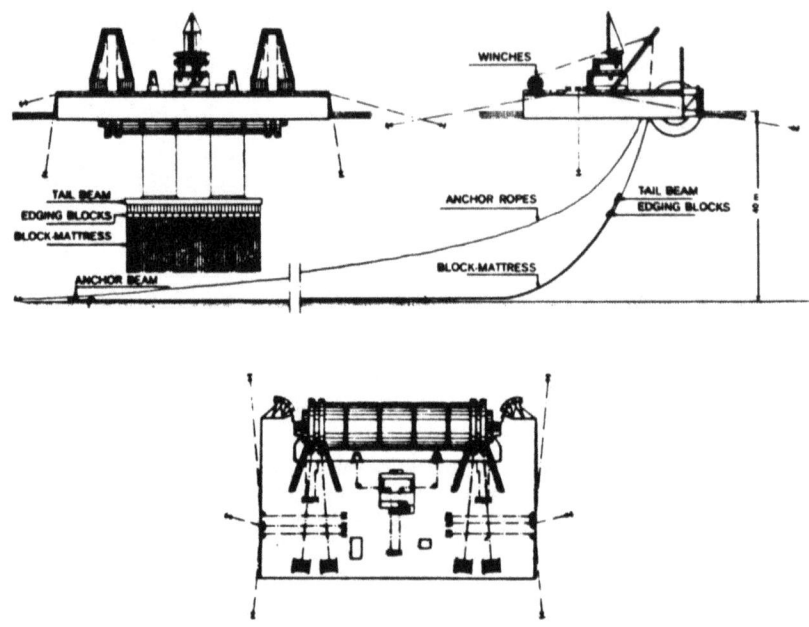

Fig. 4

The design of the fabric involved the achievement of high strength with low contraction, of adequate water permeability with fabric pores no greater than 300μm, and of long life. There were problems relating to seaming, to the interaction between pins, fabric and concrete blocks and to the end connections to the pontoon. There were also problems caused by a tight schedule and changing specifications and boundary conditions. These were solved by trial and error roughly as shown in the table. Tests were carried out on a large scale, with testpieces up to 1.0 m in width and, after definite designing, some tests on widths of 5.0 m. These were done to measure the influence of width on contraction and strength. The only problem not solved in time was the variation in filter properties and this was circumvented by laying a nonwoven on top of the woven fabric.

The fabric as designed is a poplin-like plain weave of polypropylene split fibre with the following dimensions: warp 1,900 tex;

weft 1,000 tex; thickness about 4 mm; mass per unit area 1.15 kg/m^2. It warp strength is 300 kN/m and its weft strength 70 kN/m. The predicted lifetime, based on oven tests, partly after extraction in boiling sea water or teepol, is 200 years.

Because of the manner in which the design was accomplished, much scientific work remains to be done for further development. This is largely in the field of the mechanics of fibre assemblies and includes the following:

- Problems of strength, in particular the phenomenon that strength decreases with increasing width.
- The mechanics of the fabric containing pins and the stress concentration caused by them.
- The way in which the load is carried at the end of the mat to the clamps on the pontoon.

In the general discussion it was remarked (Busching) that the use of geotextile fabrics in airport pavements was an economic factor. The necessary thickness of the top layer of improved soil can be reduced by putting a geotextile between it and the weak soil. He also mentioned the decreasing use of asbestos felt as a built-up roofing membrane and the increasing use of glass fibre assemblies, one-ply (elastomeric) systems and organic (paper) felt. The temperature of the insulation depends on the colour of the surface and the thickness of the insulation.

Concerning the use of geotextiles in different countries (Giroud), in the USA geotextiles are mainly used in railway construction, in the Netherlands and in northern West-Germany as part of coast-protection, in France partly (2/3) in earthworks and partly (1/3) in hydraulic works. The types of fibre and fabric in use are as follows: polyethylene, polyamide and polyester; 80% non-woven and 20% woven fabrics. The mass per unit area varies from 0.1 to 2 kg/m^2. A typical non-woven geotextile measures 0.3 kg/m^2 and costs about 1\$/m^2. In civil engineering, geotextiles generally account for a small part of the total project cost; quality is more important than low price.

Concerning the lifetime of geotextiles it was suggested (Giroud) that one builds for 50 to 100 years and no longer. But there is a lack of knowledge about corrosion etc. In Rotterdam harbour, polypropylene splitfibre woven fabrics, including additives, which had functioned as dike-like protections for more than 10 years, have been investigated and it is observed that the strength has not appreciably decreased over that period (van Harten).

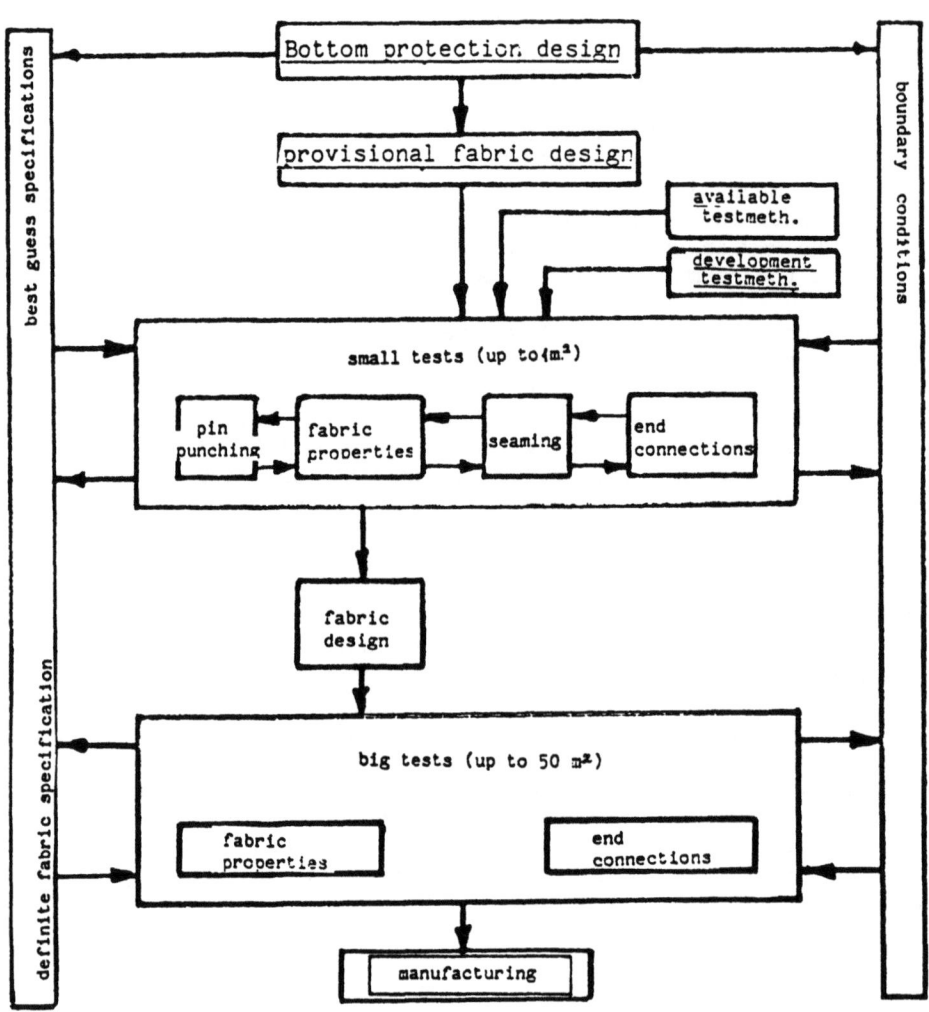

DISCUSSION XIII

APPLICATIONS OF MECHANICS IN RELATION TO CLOTHING

Leader   : Prof. S. Backer
Rapporteur : Dr B.C. Goswami

Clothing Manufacture (Backer)

Clothing alone accounted for a large portion of the total consumption of textile fabrics, followed by home furnishing and industrial end uses. He suggested that the consumption of textiles in the U.S. apparel market is quite stable; however, there is a strong possibility of growth in the home furnishing market in the near future and a long range upward trend in the industrial uses e.g. geotextiles.

He then discussed the purposes and the functions of clothing, the interaction between garment and the human body and the operations involved in the production of garments.
Table 1 shows the parameters involved in the design and production of clothing.

One of the most important functions of clothing is to provide 'comfort' to the human body. In other words clothing has a thermal function and it acts to provide a balance between body and environment i.e. it maintains the energy balance within limits between heating and cooling of the body. Clothing also has certain mechanical functions such as protection of a soldier's body from flying bullets and missiles (clothing for body armour). Clothing is essentially selected and adjusted to attain comfort and/or at least some protection from adverse environment and external conditions.

An additional function of clothing is 'Symbolic' in nature and that includes sexual and social status.

Table I

CLOTHING DESIGN AND PRODUCTION
    PROTECTIVE FUNCTION
        THERMAL MECHANICAL
    SYMBOLIC FUNCTION
        SOCIAL STATUS
        SEXUAL
    GARMENT – BODY INTERACTION
        NON RESTRAINING GARMENTS
        SEMI RESTRAINING
        BODY GARMENTS
    TAILORING 2D 3D
        DESIGN PATTERNS
        PATTERN MARKING
        CUTTING
        JOINING
            FUSING
            SEWING
        PRESSING
    MECHAINICS OF TAILORING 2D   3D
        SHAPING: VIA PATTERN
                   VIA PRESS SHAPING
                   VIA PRESTRESSING
                   VIA DIFFERENTIAL JOINING
    SEWING
        SHAPING
        SCULPTURING
        TACKING

In order to understand the physical and mechanical behaviour of clothing, it is important to analyse how garments designed for various functions interact with the body. The garment styles include those that provide minimum coverage of the body to the ones that cover almost the entire surface (100%) of the body. Garments can be designed that provide complete freedom of movement (no-restraint) to the body or they can be semi-restraining in nature such as loose shirts, jackets, etc. On the other hand, a third class of garments are those that put complete restraint or fully conform with the body movement e.g. hosiery and girdles etc. These types of garments can exert local strains that may go as high as 50% in the body. The garments that adhere to the body can either slip or stretch, can generate uncomfortably high levels of strains that will in turn cause high local stress levels in the body and may affect blood flow. It is, therefore, important to study the effect of yarn and fabric structure and the garment design on the mechanics of multiaxial deformation of fabrics.

The process of fabrication of garments (tailoring) involves; the transformation of a two-dimensional structure (planar sheet fabric) into a 3-dimensional form (garment) that involves the formation of extreme curvatures. The tailor achieves this by first designing a garment, then placing the pattern on the flat fabric sheet and marking it, and then he cuts it and joins the pieces either by sewing or fusing and finally pressing it. There have been some extraordinary developments in the field of cutting. One such development is the use of a laser beam which can cut through several layers of fabrics and still produce a clean edge.

The joining of the fabrics is achieved by sewing the two pieces with sewing threads. There are many problems associated with the performance of sewing threads during the sewing process, that can reach speeds of up to 140-165 km/hr. During the stitching process the sewing thread is subjected to complex kinematic and dynamic conditions e.g. tensile and bending stresses that take place at very high speeds, localized heating that may cause reduction of strength of thermoplastic sewing threads by as much as 60%. In addition, the mechanics of the thread structure and their properties in the seams after sewing include seam strength; stretch and pucker are also extremely important. Fusing of non-thermoplastic either by using an intermediate layer of a thermoplastic or of thermoplastic fabrics by direct heating, ultrasonics, microwaves and dielectrics also need attention of engineers and textile scientist. The process of sewing has received disproportionate amount of attention relative to the rest of the operation of clothing manufacture. It must be pointed out that sewing only accounts for approximately 1% of the total time in the construction of a garment.

The process of pressing is important in achieving the conformance of fabric to three-dimensional shape and retention of extreme curvatures. In addition to the most obvious purpose of pressing to improve the smoothness of fabric surface, it is also used to create creases, folds, and three dimensional shapes from flexible materials. In the shaping of garments, fabrics may be pre-stressed or, during sewing, one piece may be held back and the other let go producing differential shear in materials. In all three processes, fabric is an 'active' participant and there is a lot that needs to be done in the area of the mechanics of clothing formation if some further progress is to be made in mechanising and reducing the waste in the apparel industry. Professor Backer pointed out the published work of Joel Lindberg on the 'Mechanics of Tailoring' as well as some of the Japanese work on the application of analytical techniques to the design and fashion industries. He mentioned some of the activity, now in progress, in the U.S. that deals with the problems and the sagging futures (foreign competition) of the apparel industry.

Durable-press treatments (Williams)

Williams addressed the questions of cost and the conformance of post-cured durable-press treated fabrics, especially by the vapor phase system. He suggested that the problems of fabric conformance to 3-dimensional form and the retention of shape in garment impose certain performance limitations and improve performance; but have actually added more cost and made the problem more complex. Some attempts to reduce the cost and the complexities in the processes have been made through certain approaches such as, in-situ formation of fibres, spray bonding of fibres on a form and fabric moulding. So far, fabric moulding is the only process that has achieved any degree of success, but the major material and technological limitations continue to exist and there is still a great deal of work that needs to be done in the understanding of the mechanics of cutting and shaping of fabrics into garments.

Clothing Performance (Slater)

The ornamental, status (sexual aspects etc.) and modesty considerations of clothing are governed mainly by current fashion trends, but the protective role is dictated by the external environment. However, the design of fashion fabrics should not preclude the considerations of the mechanical properties of the material that are essential in achieving the aesthetic performance in garments.

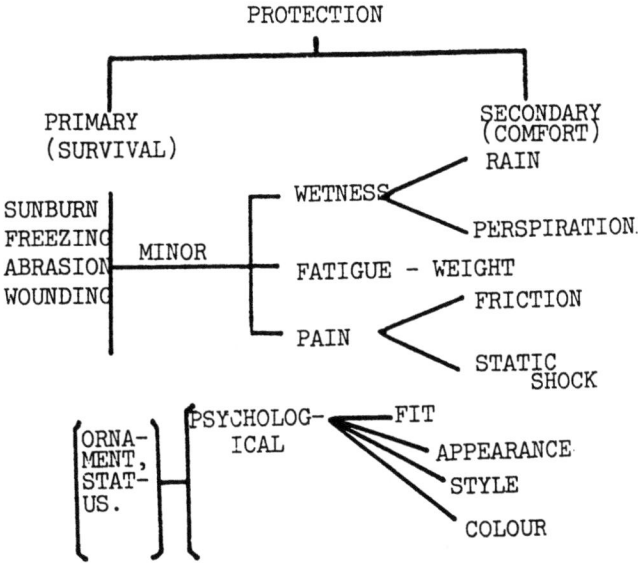

The protective aspect of clothing may be divided into two major categories i.e. (i) primary (survival) and, (ii) secondary (comfort). Primary protection from sunburn, cold, abrasion or wounding can literally save life, while minor problems in these areas can lead to severe discomfort. Comfort is also adversely affected by wetness (from rain or perspiration), pain (from excessive friction or static shock), fatigue (from undue fabric weight) and by psychological factors such as style, appearance, colour etc. The mechanical properties of fabrics must be adjusted to comply with all three requirements; this situation demands the balance of an extraordinary complex combination of factors, such as weight, bulk, compressibility, porosity to air, water or moisture vapor, thermal insulation, handle, drape and shear. Frequently an insoluble dilemma results, and a compromise solution must be found (e.g. the desirability of combining high thermal resistance with high moisture vapor permeability in Arctic conditions where blizzards or wind chill are encountered) - see the figure below:

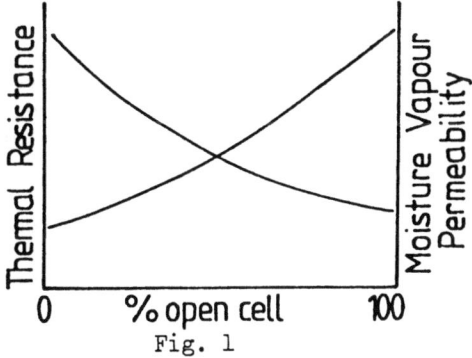

Fig. 1

A clothing system generally consists of a number of layers of garments (fabrics); thickness may vary because of seams etc., and the nature of garment fit to the body automatically produces differences in inter-layer spacing, pressure, temperature and relative humidity at different areas of the system. If, under certain circumstances, one of the components performs unsatisfactorily, then the entire assembly is made ineffective.

It is not always possible to engineer fabrics at the stage that will meet the requirements for the specific end-use performance. Consequently in designing fabrics the mechanical performance of materials should include a 'safety factor' that allows for the adverse effects of light, heat, moisture, perspiration, atmospheric contaminants, and faulty maintenance (including laundering, dry cleaning, bleaching, drying etc.) procedures.

Mechanics of Tailoring (Shishoo)

Shishoo discussed the mechanics of tailoring garments and pointed out the problems encountered in engineering clothing structures for specific end-uses. He first discussed the overall scheme of processing in which the fabric goes through various manufacturing steps and finally to the consumer as shown in Table II. He indicated that it is important to recognise the fact that the properties of fabrics are liable to change during lay-up and cutting because of hygral expansion especially for hygroscopic materials. The existence of high humidity in the laying and cutting rooms can later cause buckling and puckering of seams during the sewing operation. In the area of cutting some further work needs to be done in design and pattern grading as was mentioned earlier by Professor Backer. The present techniques used for cutting fabrics with blades has limitations as far as the number of layers of fabrics that can be laid. However, some new developments have taken place in this field where a very large number of fabric layers can be cut at the same time in a very short time. These developments include the use of laser and water jet. The use of laser causes melting and fusing in synthetic fibres and the water jet poses problems for woven and knitted fabrics made from natural fibres. Another major problem is that of fraying of fabric edges in the garment. This can be prevented to a certain extent by using overlock stitch. On the other hand, there is a great deal that requires attention from the point of view of yarn and fabric structure and the mechanics of edge sewing.

There are some major problems encountered in creating proper stiffness in such areas as the collar, placket and cuffs. In the older system, the stiffness in the placket was achieved by using

Table II

FLOW CHART

FOR

MANUFACTURE OF CLOTHING

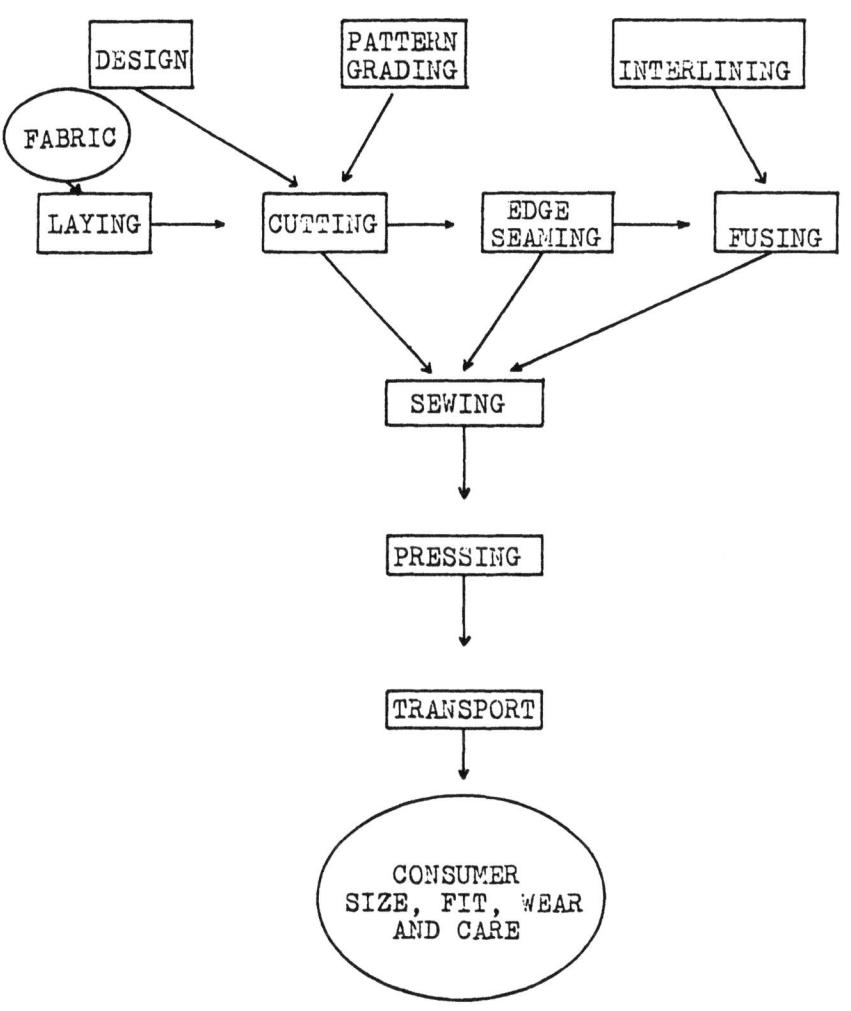

single or multiple layers of woven fabrics (sometimes made from horsehair) but recently the use of nonwoven fusible materials in interfacing and interlinings have showed promising results at a lower cost. The use of fusible nonwovens has created certain problems for the manufacturer as well as for the consumer. These include blistering, delamination, and wrinkles due to differential shrinkage and loss of some physical properties due to ageing.

When designing and manufacturing garments it is essential to keep in mind the end-use requirements and consequently modify the conditions that influence the manufacturing processes to meet those requirements. Some of the factors that should be considered are: functional requirements; normal clothing - work clothing - protective clothing; sportswear, outdoors, indoors, military; hospitals; ease of movement.

Dr. Shishoo then discussed the mechanics of seam puckering. He pointed out that in sewing of fabrics one must determine the compatibility of fabric and sewing threads in terms of extension and recovery and shrinkage behaviour of the composite. Type of needle, seam and the stitch type are other factors that need considerations. Seam puckering is affected by the sewing thread tension and by a number of sewing machine parameters (adjustment, stitch type, stitch length, needle size and design, sewing speed; presser foot; feed dog and pressure).

The fabric property that influences the seam puckering the most is the shrinkage of fabric. He introduced the concept of "Limit of Contraction (LOC)' of a fabric-seam system, defined as 'the decrease in length of a fabric before any visible puckering', as a parameter to determine the role of fabric properties in seam puckering. This concept helps to allow for the fabric to adjust in the seam before it puckers. The fabric properties that may influence seam puckering include: differential extension, weight, thickness, compressibility, resilience, shear, yarn to yarn friction, fabric to fabric friction, fabric bending stiffness, fabric extension at stresses less than 0.1 N/Cm, plate buckling, and shell buckling.

A concomitant effect of seam puckering is increase in thickness of the seam. Consequently, if one plots fabric contraction in seam against change in thickness, the point at which the slope of the curve changes is taken as the 'Limit of Contraction' for that particular system. This can be observed in the accompanying Figure 1 (b) where the LOC for a system for conditions of 'after sewing' and 'after steam pressing' are shown. Dr. Shishoo then described an experimental study that was carried out at TEFO. The experimental study involves the measurement of change in length (measured to 1/10th of a mm. accuracy) and thickness (determined at 0.001 N/cm² pressure) in a fabric-seam system on a sample of 30 cm x 5 cm dimensions as shown in Figure 2 (a). This figure shows a seam with (i) Severe and (ii) essentially no puckering. He showed that the LOC generally varies between a minimum of 0.10% for fabrics such as 100% light weight cotton and a maximum of 0.6% for 100% wool fabrics after sewing as shown in the accompanying diagram. However, the minimum and the maximum values of LOC for condition after pressing were found to be 0.8% and 4.1%, respectively. When the regression analysis was carried out to determine the influence of various physical properties of fabrics on LOC, it was observed that fabric bending stiffness, fabric weight and the coefficient of friction (static fabric-to-fabric friction) were found to be highly significant. The relationship can be expressed by the form:

$$LOC = 0.005 \frac{E_b \times W}{M_s} + 0.27$$

When E_b = bending stiffness
W = fabric weight
M_s = static coefficient of friction between fabrics.

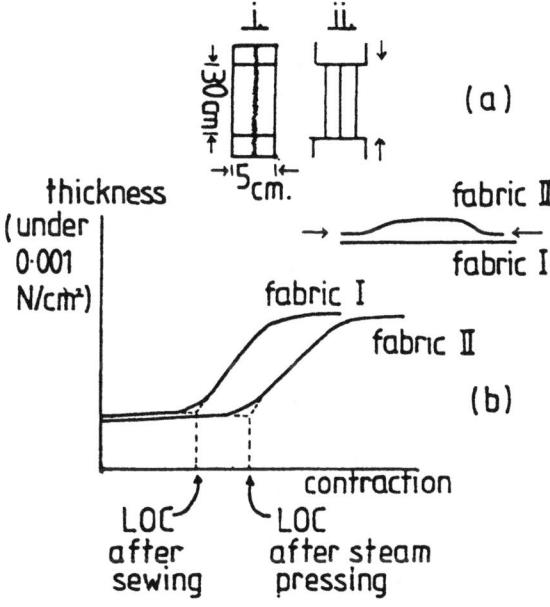

Figure 2
Limit of Contraction
100% Wool Fabric   - 0.45 - 0.60%
Wool/Polyester Blends - 0.10 - 0.45%
Cotton (Lightweight) - 0.10 - 0.30%

This type of analysis has been used to determine LOC values for a wide variety of fabric constructions, sewing thread tensions and the types of stitches used in the apparel industry.

The methods used in shaping garments comprise another area which needs attention by textile scientists. There are essentially three methods that are used in shaping garments: pattern design and sewing; adhesive based materials; and by the moulding process. Of these methods the moulding process which is used for thermoplastic materials requires critical conditions of temperature and pressure. In addition, the time of heating before moulding coupled with the moulding pressure has a profound effect on the shrinkage and the other aesthetic properties of the end-use material. Yet another critical factor is the heating time immediately after loading.

Dr. Shishoo then addressed the question of performance and the fitting requirements for knitwear. He suggested that for best garment performance one needs approximately 8% extension in the fabric. However, for best fit and performance one needs to establish a size system that is based on a relationship between critical body measurements and the corresponding garment dimensions. The setting of control dimensions of the garment should take the following points into account: it should represent one or more body measurements; should either represent certain critical body measurement as regards garment fit or should be well related to the critical measurements; dimensions should either be known or be easily measureable by the customer; in addition, the distribution of dimensions should also be known. The development of a size system for garments requires the acquisition of knowledge from an extensive analysis of garment fit and the body measurements. For example, for ladies panties one would require such critical body measurements as (a) thigh circumference (b) waist (c) waist back-to-crotch-to-waist front, and (d) hip circumference. Figure 3 shows the fitting requirements for various types of knitted under- and outer-wear garments (P_L and P_W are the loads in the length and width directions, respectively). Dr. Shishoo demonstrated how one can determine minimum and maximum dimensions for various body measurements. Figure 4 (a) shows the minimum and maximum extension of a knitted fabric at loads of 2 newtons and 10 newtons, respectively. From this information one can then determine the minimum (L_1) and maximum (L_2) waist dimensions from the plot between body weight and waist circumference dimensions. These dimensions would cover approximately 96% of the population between body weight of 50 and 60 kilograms as shown in Figure 4 (b). Yet another information one must have is the relationship between the lengthwise extension for widthwise contraction and this is shown in Figure 5. One piece of information which one can obtain from this relationship is the level of contration that can be expected

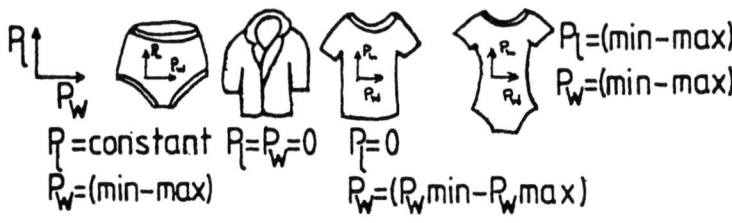

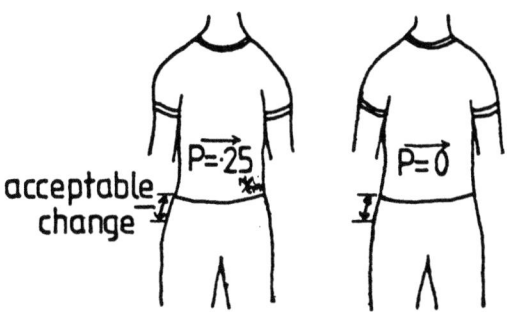

Fig. 3. Fitting Requirements of Knitwear.

for the maximum widthwise extension. This analysis is important in knitted T-shirts where the length contraction can make a difference between an uneasy, unpleasant appearance and fit and proper and comfortable wear performance of the garment. It is also important to have this type of information for unset (unwashed) and set or washed garments. In his discussion of the subject of the mechanics of textile clothing design, structure and performance, Dr. Shishoo tried to point out the importance of the application of applied mechanics to such structures in improved design of fabric systems and improved performance of apparel and household textiles.

The Use of the Computer in Optimizing Cloth Manufacturing (Konopasek)

The conversion of flat fabric into a shell covering particular parts of the human body is a fairly complicated, expensive and inefficient process. Two major sources of inefficiency are:
(a) considerable cutting waste accompanying the cutting of fancy shaped panels out of sheet material, and
(b) labour-intensive manually controlled operations of sewing the panels together.

Many current or potential innovations in apparel technology depend on better understanding and control over mechanical response and behaviour of the fabrics. Most of the problems in apparel making are similar or identical to problems encountered in other areas of use of fibres and fibre assemblies, and there is ample space and need for exchange of theoretical and engineering know-how.

The problem of cutting waste may be seen from a broader viewpoint of material utilization in the production of apparel and other fibre-based substrates or products. We can identify a whole hierarchy of optimization problems in this area:

1st level: Optimum marking or pattern layout making, i.e. nesting and packing a given set of panels on a rectangular strip of fabric of a given width so as to secure minimum fabric consumption per garment.

2nd level: Cut order planning, i.e. selection and break-down of the fabric supply and cutting schedule (i.e. widths and lengths of fabric rolls, size combinations on markers, table allocation etc.) in a way which would match fabric manufacture and cutting technology constraints, and secure the best possible results when solving the problems at the 1st level.

3rd level: Garment design and pattern engineering, i.e. design or modification of such a set of panel shapes which, besides satisfying the aesthetic and functional requirements of the garments, and technical and economical standards of sewing operations, would yield the best possible results when subjected to solving of problems at the 2nd and 1st level.

4th level: Design of fabrics with such mechanical and other properties within technical and economical standards of fabric manufacture which, when involved subsequently in solving of problems at 3rd, 2nd and 1st level, would require minimum amount of material and other resources per garment/service unit.

5th level: Design or selection of fibre/textile production processes which would ease the above constraints and secure optimum environment for solving the problems at the 4th, 3rd, 2nd and 1st level.

There is a good deal of optimization going on at each of these levels separately, mainly through intuitive decision making or through spontaneous effect of the market forces. The use of more rigorous optimization methods and simultaneous coverage of two or more levels is expected to bring about deeper insight and better solutions. The amount of information and the mathematical complexity of the problems necessitates the use of computers.

A word of caution: even the most powerful computer systems may not be capable of generating the optimum solution, not just

through the whole hierarchy, but even at the first level alone.
It may be shown that the problem of a true optimum packing of
a set of panels on a piece of sheet material belongs to the class
of so called NP-complete problems of combinatorial programming.
Optimization in cases like that requires an exhaustive search
through the solution space and consequently the time required is
proportional to the factorial of the number of panels n. This
gets out of hand very quickly for n>10. However, it also means,
that there is an open space and unlimited opportunity for
improving the solution methods and the solutions themselves.

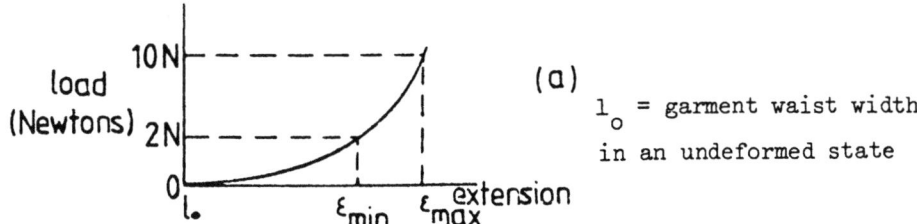

Fig. 4(a).   Knitted fabric extension curve

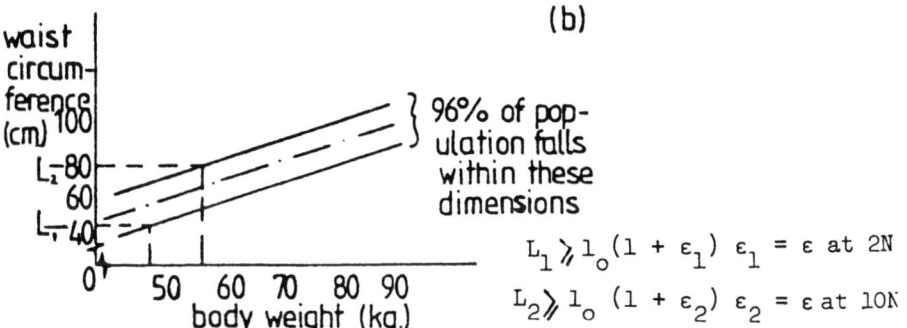

Fig. 4(b)   Relationship between body weight and waist circumference

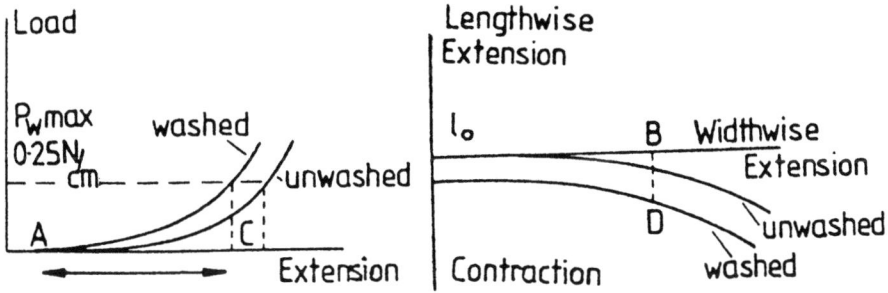

Fig. 5. Knitted fabric extension and contraction curves

DISCUSSION XIV

COMPUTATIONAL METHODS

Leader   : Dr. W. Schaffers
Rapporteur : Mr. A. Uttley

On the Use of Finite Element Methods

In studying the effects of ballistic impact on a network Lloyd uses a finite element method (FEM) to solve equations (42) and (49) (p. 311) repeatedly for each node of the network. The choice of FEM is inevitably an expensive one because of an effect described as the "computational black hole power law", by which the number of operations, and hence the cost of the analysis, increases as a power function. The terms in this function are as follows:

(i)   Interpolation of mechanical properties at each node.
(ii)   The linear solution for this interpolation.
(iii)   The constraint equations and boundary conditions.
(iv)   Equilibrium iterations.
(v)   Incremental solutions for time or for displacement.

When using a FEM the expense usually dictates compromise and promotes efficiency. The analysis may specify 18 degrees of freedom at each node and so the computer must have a large store with not only fast access but also fast system operation. Accuracy requires a good word length which also adds computational time. For example, the CDC 7600 uses a 60 bit word length. This effectively rules out the use of minicomputers.

The cost of the analysis can be reduced and even minimised by efficient programming. The large matrices involved must be stored compactly and concisely. For example, if a number is known to be zero it need not be stored or transferred. Overlay techniques which place only the most currently used data into computer memory can reduce the time taken reading from peripheral memory such as disc or tape. It is important to know the computer system's own overlay methods and to organise data transfer which complements the machine's operation. The management of matrix order can influence program speed; for example moving non-zero numbers to the matrix diagonal can improve the overall speed of matrix arithmetic.

It must be remembered that programming that attempts to organise resources efficiently does itself incur direct overhead cost in both machine and operator time. A trade-off must be sought which balances improved operation with cost effectiveness. Above all, the interpolation solution must be efficient because this is the basic unit of cost. All subroutines that are used frequently must be fast and written in a fast language, using efficient procedures. Accuracy can be checked by changing the nodes but this is yet another term for the power function.

The expertise and sophistication required for programming proficiency rules out DIY programs and implies the use of the large versatile FEM packages. Universal use of these may lead to more uniform presentation of result.

An alternative view (Konopasek) is that new computer technology makes the FEM a more attractive numerical technique by reducing the cost of the procedures exploited by the FEM. Fast processors, bubble memory, array/factor processes all speed up the number manipulation. Interactive programming becomes possible and allows the analyst to give intelligent guidance during iterative calculations.

The composite problem reported in Discussion Session 9 (Curiskis) is an example of the application of a 3-D FEM. The isoparametric compatible displacement model is utilized for the continuum discretization and the initial stress method is employed for the non-linear solution scheme. The discretized FE problem is defined in terms of constraint equations on the nodal displacements of the boundary surfaces of the repeat unit which is representative of the idealized multi-fibre composite geometry.

In the program package developed for this problem the constraint equations are enforced during assembly and the global stiffness equations are inverted by special procedures [1]. The equation solver utilizes a direct elimination method based on FESS, developed at Swansea. For large non-linear problems, a split package technique is used to monitor the iterative solution.

The difficulties encountered with the project were chiefly associated with computer availability in an academic environment (for example, core and time limitations). It is expected that computer times may be reduced by the adaptation of a frontal solver and of some acceleration scheme for the iterative process. The program has been run on an IBM 370/165 and on a CYBER 76-26. Core limitations on the latter machine required an overlay organisation. The problem discussed required about 70 minutes of CPU time on the IBM 370/165.

FE analysts have long recognised the need to check computer solutions by running test cases with known analytic solutions or experimental results, or by varying the number and spacing of nodes (convergence study). For the composite problem, the stress-strain curve obtained from a 20-element model compared favourably with that obtained from the 84-element model.

Dr. Blum took up this last point by emphasising the potential pitfalls of using FEM. In Stuttgart the biggest users of FEM also have the biggest experimental facilities. This is a consequence of modelling - to gain faith in the model we must test its predictions. But there is much to be said for exploring the mathematics of the model first. With deeper understanding of the mathematics it is often possible to simplify the model and to be more certain of its predictions. This is especially true for non-linear problems. Then is the time to go to the computer.

Use of large packages such as NONSAP can be dangerous because it is easy to get out of touch with what the computer is doing and what assumptions are made in the program. Writing one's own program can teach a lot and allow one to keep in touch.

These points are made by Olden [2] who uses non-linear FEM and stresses the need for an understanding of the physics first of all. He remarks that there is a need, not for FEMs, but for effective methods of solving non-linear equations. Perhaps the physics is even more important than the mathematics. A secure understanding of the physics may allow direct computation of a model with suitable experimental testing.

Computation of Fabric Mechanics

Although the computational aspects of modelling the plain weft-knitted structure are fairly fully covered in the paper (p 175) there is an aspect, mentioned by Konopasek in another context, that is worth commenting upon (Hepworth). It is that once a solution has been obtained for one set of data, it can be used as a "pilot solution", a starting point for further solutions with different data. For the relaxed or unloaded fabric a solution obtained for one value of the ratio d/l leads to solutions for a range of values. Next, taking a fixed value of d/l, the solution obtained for zero load leads to solutions for increasing uniaxial loads, first parallel to the courses and then parallel to the wales and subsequently for a whole range of biaxial loads.

Difficulties encountered include the general one when finding any minimum, the danger of converging on a local minimum instead of a global one. This is usually avoided by a good choice of starting point and the use of small enough step size. In the present case, it is helpful to know that the required minimum is $F = 0$. Powell's method is very efficient for the small number of variables involved (≤ 7), but unfortunately, it has now been removed from standard libraries of routines in favour of methods which are more efficient for a large number of variables.

The alternative computational method, using the energy optimization technique, requires just such a routine (de Jong). For the plain-woven fabric for example, neglecting yarn torsion, discretization of the ordinary differential equations over n points results in a set of $4n$ equations (equation (14) for both yarns involved). The boundary conditions provide two more and, if friction is included, this results in a total of 31 equations with appropriate residuals. The routine used is QNWT (CDC Mathematics Science Library).

Difficulties include unstable integration routines (Simpson's rule as used) and an inherent one in that the gradient is calculated by $\partial H/\partial m_i$ is not the true gradient, so that the convergence of the process is slower than might be expected.

Monte Carlo Methods for Strength Computation (Popper)

Monte Carlo (or Direct Simulation) methods offer a convenient method for solving stocastic problems which occur in Textile Technology but cannot be solved readily by classical probability theory. This applies to problems with complex relations between many stocastic variables.

The basic technique consists of the following steps:
(1) Generate a random number (from the appropriate distribution) for each stocastic variable; (2) solve the deterministic physical problem by conventional methods; and (3) repeat the process to determine the average value of the output or its functional relation with some input parameter.

The major advantage of the technique is simplicity - especially when using modern computers. By contrast, classical probablistic methods often lead to horrendous formulations even in relatively simple problems. The major disadvantages of the Monte Carlo method are the large number of replicate runs needed for good accuracy and the poor estimates of variance of the computed outputs.

One unique feature of the method is the possibility of varying a
parameter in a system in which the random variables remain fixed.
This makes it possible to isolate the effect of a single parameter
which would normally be hard to detect if many sources of variability
exist. For example, in computing the strength of a yarn, it
is possible to generate a unique irregular fiber array and evaluate
it with varying levels of friction. In that way the effect of
friction can be determined without averaging results of many
repeated runs.

Example of Monte Carlo Simulation in Computing Yarn Strength (Popper)

The force-displacement relation for a yarn can be computed by this
technique. With it, the complex effect of variable properties and
frictional interactions can both be included. This was done by
modeling the yarn as a parallel array of fibres, each of which
consists of a number of segments. The properties of each segment
(including the frictional force between them) were generated as
the first step of the computation. Then, the appropriate force
equilibrium and property equations were solved numerically to
find the yarn tension at any elongation.

Effectively, the computer generated a random sample, and then
tested it. Each run represents a new sample with slightly
different response. The results were computed through the failure
point to obtain strength and elongation data. In addition, all
internal stresses and strains are computed and retained.

Figure 1 shows the distribution of tensile stress in a broken
fiber and its neighbours. This illustrates the additional load
buildup on the nearest neighbours near the break point; and, it
shows how interfiber shear stresses act to increase tension away
from the break.

Figure 2 shows typical break patterns observed in simulations with
low and high frictional interactions. This effect is also shown
on Figure 3, by the force-elongation response. For the yarn
simulated, increased interactions improved the yarn strength and
changed the break characteristics from gradual failure to sudden
break.

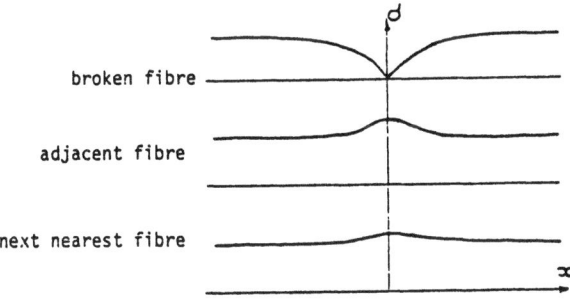

Fig. 1. Tensile stress distribution in fibres near a break.

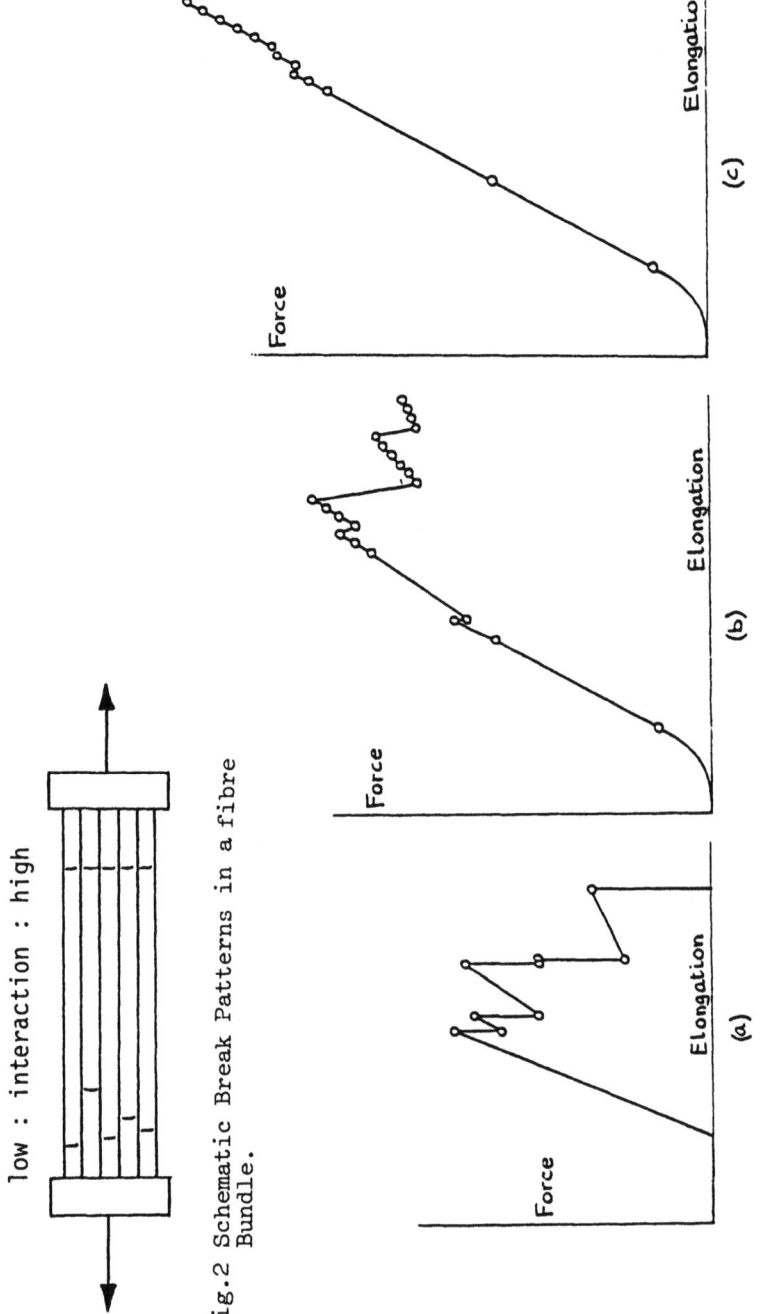

low : interaction : high

Fig. 2 Schematic Break Patterns in a fibre Bundle.

Fig. 3. Simulated force/elongation curves for a fibre bundle: (a) low interaction (b) increasing interaction (c) high interaction.

Future Developments: Interactive Computation (Konopasek)

The development of theoretical and applied work in our field is seen as involving an almost total switch to interactive computation, extensive use of computer graphics and increasing use of library sub-programs and program packages. A great deal of computation will be performed on a more routine basis. Consequently, more attention must be given to human engineering aspects of the programs and packages.

The user of an interactive system should be able to:

(i)   assign or change at any time the input data from the terminal keyboard (and enquire about its status) without being forced to key in redundant information;

(ii) store a current status of conversation and recall it when needed;

(iii) get sufficient feed-back information about any inconsistencies and errors in data assignment or computation, including prompts on the fixing of the errors;

(iv) enquire of a "HELP" file about the basics as well as the details of the program operation.

Almost all the programming in our field is currently done in FORTRAN. The use of this language conceived over 25 years ago under completely different circumstances is becoming anachronistic. Some newer languages either do not offer much of an advantage (e.g. BASIC) or did not gain popularity (e.g. PL/1) or are designed for a different purpose (e.g. PASCAL). However, the quickly spreading APL language seems to fit our needs. Among the attractive features of APL are its rich variety of some 40 operators which simplify programming chores, its structured build-up of programs using simple functions as building blocks, its variety of array functions obviating DO loops, its incremental program development and debugging, and its implicit array declarations.

All these and other features improve the productivity of an applications programmer; this is a substantial advantage in an era of rapid redistribution of computation costs involving a reduction in hardware and increase in software development. The effort put into the development of finite element and other packages written in FORTRAN may not be wasted in transition to APL or other languages. FORTRAN subroutines may be used within an APL framework.

There are two other general purpose systems and/or languages which can be recommended.

MACSYMA [3] (Project MAC's SYmbolic MAnipulation system) offers

capabilities which one should, but usually does not, expect from
the computer: it performs symbolic differentiation and integration,
it takes limits, solves systems of linear of polynomial equations,
factors polynomials, expands functions in Laurent or Taylor series,
solves differential equations, plots curves and manipulates matrices
and tensors. In addition to symbolic manipulation MACSYMA returns
results in terms of FORTRAN statements or numerical values when
requested. The benefits from using MACSYMA when the mathematical
problems are too involved either in a qualitative or a quantitative
sense (or both) are tremendous.

QAS [4] (Question Answering System on mathematical models and
related data bases) is a very high level computer language
conceived as a problem solving tool for an engineer, scientist
or economist. The QAS

(i) accepts, stores and allows access to pieces of knowledge defined in terms of mathematical equations and empirical data;

(ii) obviates writing computer programs in the traditional sense;

(iii) offers an ultimate flexibility in computer assisted problem solving by allowing for any feasible combination of input data and output information;

(iv) helps the user to overcome inconsistencies in problem formulation by indicating the trouble and prompting an alternative course of action;

(v) may be taught in a couple of hours and implemented on portable microcomputers.

The most useful feature of QAS for applications in our field are
built-in procedures for solving sets of non-linear algebraic
equations, unconstrained as well as constrained optimization,
two-way interpolations of empirical relationships, evaluation of
derivatives based on automatically performed symbolic differentiation
and evaluation of definite integrals.

REFERENCES

1. J.I. Curiskis and S. Valliappan, Computers and Structures, 8, 117 (see also Discussion Session 9).
2. J.T. Olden, Applied Mechanics Reviews
3. MACSYMA Reference Manual. Laboratory for Computer Science, MIT
4. M. Konopasek and C. Papacoustadopoulos, Computer Language, 3, 145.

DISCUSSION XV

MECHANICAL PERFORMANCE THROUGHOUT THE LIFE OF A PRODUCT

Leader   : Dr G.A. Carnaby
Rapporteur : Dr S.K. Batra

Carpet piles (Carnaby)

This topic encompassed various subjects in textile mechanics which had in themselves been studied in some detail. However it was necessary to make a synthesis of the results of this work in order to deal with a practical problem such as carpet wear. It was suggested that the carpet wear process could be characterized by changes in carpet thickness and that the main factors responsible for the reduction in thickness with wear were fibre shedding, frictional locking effects, viscoelastic behaviour of the fibres, and abrasion array of the pile.

It was then shown how Chapman's work on repeated wrinkling of fabrics could be adapted to describe the viscoelastic behaviour of a similar model for the pile and backing of a carpet. It was shown that the reduction in specific volume of the carpet could be expressed through an inverse cubic relationship, as the summation of a geometric progression relating to the number of treads on the carpet. To express this relationship in terms of pile thickness it was then necessary to consider the rate of mass loss of the carpet pile. Experimental evidence regarding the abrasion process and carpet wear were shown to indicate that a linear rate of pile mass loss would be a reasonable first approximation.

The speaker then derived an expression for the total carpet thickness as a function of the number of treads. Other terms related to the initial carpet properties and the observed changes in specific volume of both pile and backing. The expression given was:

$$h_{t_n} = m_{p_0}(1+c.n)\left\{\left(\frac{1}{v_{p_\infty}^{1/3}} - \frac{1}{v_{p_0}^{1/3}}\right)\left[1 - \left(1 - \frac{\frac{1}{v_{p_1}^{1/3}} - \frac{1}{v_{p_0}^{1/3}}}{\frac{1}{v_{p_\infty}^{1/3}} - \frac{1}{v_{p_0}^{1/3}}}\right)^n\right]^{-1/3} + \frac{1}{v_{p_0}^{1/3}}\right\} +$$

$$m_b\left\{\left(\frac{1}{v_{b_\infty}^{1/3}} - \frac{1}{v_{b_0}^{1/3}}\right)\left[1 - \left(1 - \frac{\frac{1}{v_{b_1}^{1/3}} - \frac{1}{v_{b_0}^{1/3}}}{\frac{1}{v_{b_\infty}^{1/3}} - \frac{1}{v_{b_0}^{1/3}}}\right)^n\right]^{-1/3} + \frac{1}{v_{b_0}^{1/3}}\right\}$$

for $n \geq 1$.

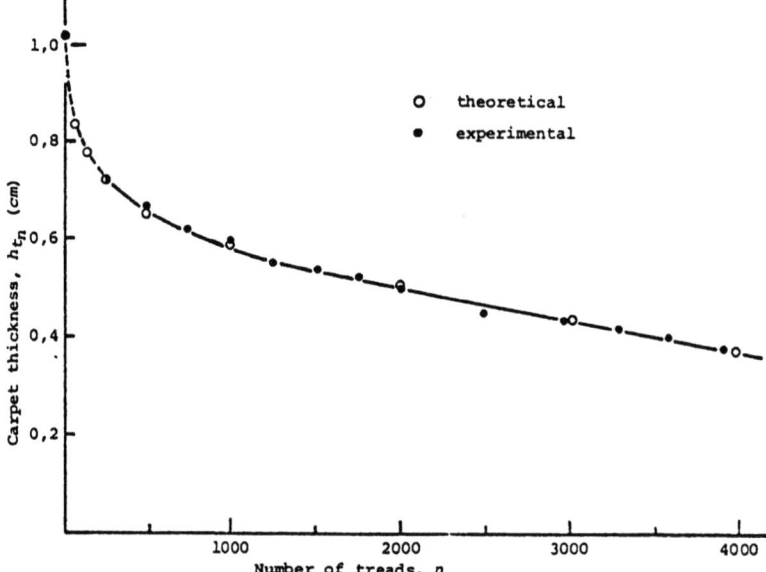

Variation in carpet thickness with actual number of treads (corr.)
Measured parameters: $m_{p_0} = 0{,}0665$ g/cm^2; $m_b = 0{,}129$ g/cm^2;
$v_{p_0} = 9{,}50$ cm^3/g; $v_{b_0} = 3{,}05$ cm^3/g
Fitted parameters: $v_{p_1} = 9{,}46$ cm^3/g; $v_{p_\infty} = 5{,}0$ cm^3/g;
$v_{b_1} = 3{,}00$ cm^3/g $v_{b_\infty} = 2{,}5$ cm^3/g; $c = -2{,}1315 \times 10^{-4}$

Fig. 1.

This equation was then shown fitted to experimental points taken from a typical wear curve as obtained on the WRONZ Carpet Wear Simulator. The use of this relationship for end point prediction of carpet wear tests using the microprocessor and other sophisticated monitoring systems in this new testing machine were discussed.

Aspects of Viscoelasticity (Chapman)

The following is a collection of simple bits of information which provide a useful guide to using linear viscoelasticity and avoiding pitfalls.

(1) (a) The stress strain curve of the LVE material is non-linear and always convex upwards Fig. 2(a) determined by the equation

$f(t) = \rho \int_0^t G(c) \, dc$ where F - stress, ρ - strain ratio, t - time.

Differentiation shows the slope of the stress-strain curve is equal to the stress relaxation modulus

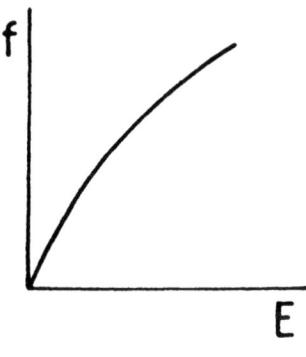
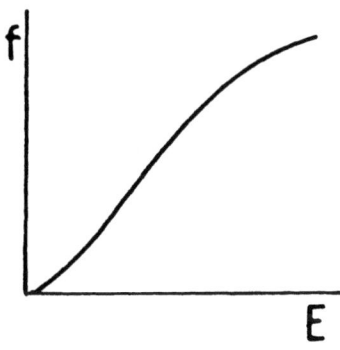

Fig. 2a.   Fig. 2b.

(b) If there is any slight crimp in a fibre or non-alignment of jaws a toe-in Fig. 2(b) will result with opposite curvature to the viscoelastic curvature resulting in an inflexion region through which we are tempted to draw a tangent and consider it to have fundamental physical significance (ref. Bendit CSIRO).

(c) When testing for time effects with stress-strain curves, the rate of strain needs to be changed by orders of magnitude to observe large effects.

(d) Hysteresis loss is not necessarily changed by doing experiments much faster or much slower.

(2)

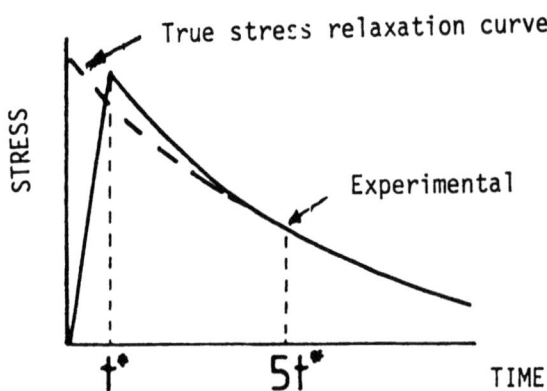

Fig. 3.

(a) When measuring stress relaxation or creep behaviour a finite time t* is required to achieve maximum strain or stress. Zero time should be taken from the time that extension is commenced and the experiment will not represent the true stress relaxation behaviour until at least time 5t* (Fig.3).

(b) Tensile fibres in stress relaxation and creep experiments are usually extremely sensitive to RH changes. Very accurate control is necessary on RH and temperature. (Air conditioned rooms are not good enough).

(c) Recovery from deformation usually takes roughly 10 times the length of time a fibre has been deformed.

(d) LVE fibres given sufficient time always recover completely.

(3) The term internal friction to describe viscoelastic loss should be avoided. Viscoelastic and frictional dissipation are completely different in their action.

(4) Remember the Denby formula theory works well.

Prediction of the Compression Property of Carpets (Kawabata)

Let us consider straight rods as shown in Fig. 4. When the rods are compressed by a plate, some modes of deformations will occur with the friction between the plate and the tip of the rods as shown in Fig. 5.

In this time, let us consider a simple case shown in (1) Fig. 5. The experiment to examine the prediction for this type of deformation is possible by using the apparatus shown in Fig. 6.

where the horizontal force component is eliminated. The compressional force is estimated as follows:-

1st stage of deformation: Bending of rod -----(elastica)
2nd   "   "   "   "   "   Overlap zone appears as shown in Fig. 8.
3rd   "   "   "   "   "   Shortening of the rod length with increasing of the overlap zone, and slippage between fibres in the overlap zone will occur. These lead to the increasing of compressional force as well as hysteresis behaviour as shown in Fig. 9, Fig.10.

In addition, we have to add the deformation of the base fabric to the compressional deformation of the rods zone as shown in Fig.11. and Fig.12. Carpet deformation is a typical fibre assembly problem having

(1) large deflection of fibre
(2) space between fibres
(3) friction between fibres

Studies of Fibre Fracture and Fatigue (Hearle)

The availability of scanning electron microscopy has opened up[1-18] the study of fibre failure. Considerable progress has been made in research at UMIST over the last 12 years in the classification of forms of fracture and the study of fatigue; but much interesting work remains to be done in comparative studies and the concept of fracture maps, and in the study of fracture mechanics and the relation to structure in materials subject to large strains, with non-linear, inelastic, anisotropic mechanical properties, and special geometries.

Figure 13 illustrates ten forms of fracture. There are six forms of tensile failure:

(1) Classical brittle fracture, occurring in glass and elastomeric fibres; (2) ductile (stable) crack propagation in a V-notch or diamond followed by catastrophic failure, occurring in thermoplastic fibres; (3) mushroom breaks at high rates when there is adiabatic heating; (4) axial splitting in highly oriented fibres with weak lateral cohesion, such as Kevlar; (5) granular fracture, when some individuality of structural elements blocks crack propagation, but local load sharing leads to localisation of the cross-section of failure in acrylic fibres, rayon and wool; (6) separate breakage of individual elements over a length of

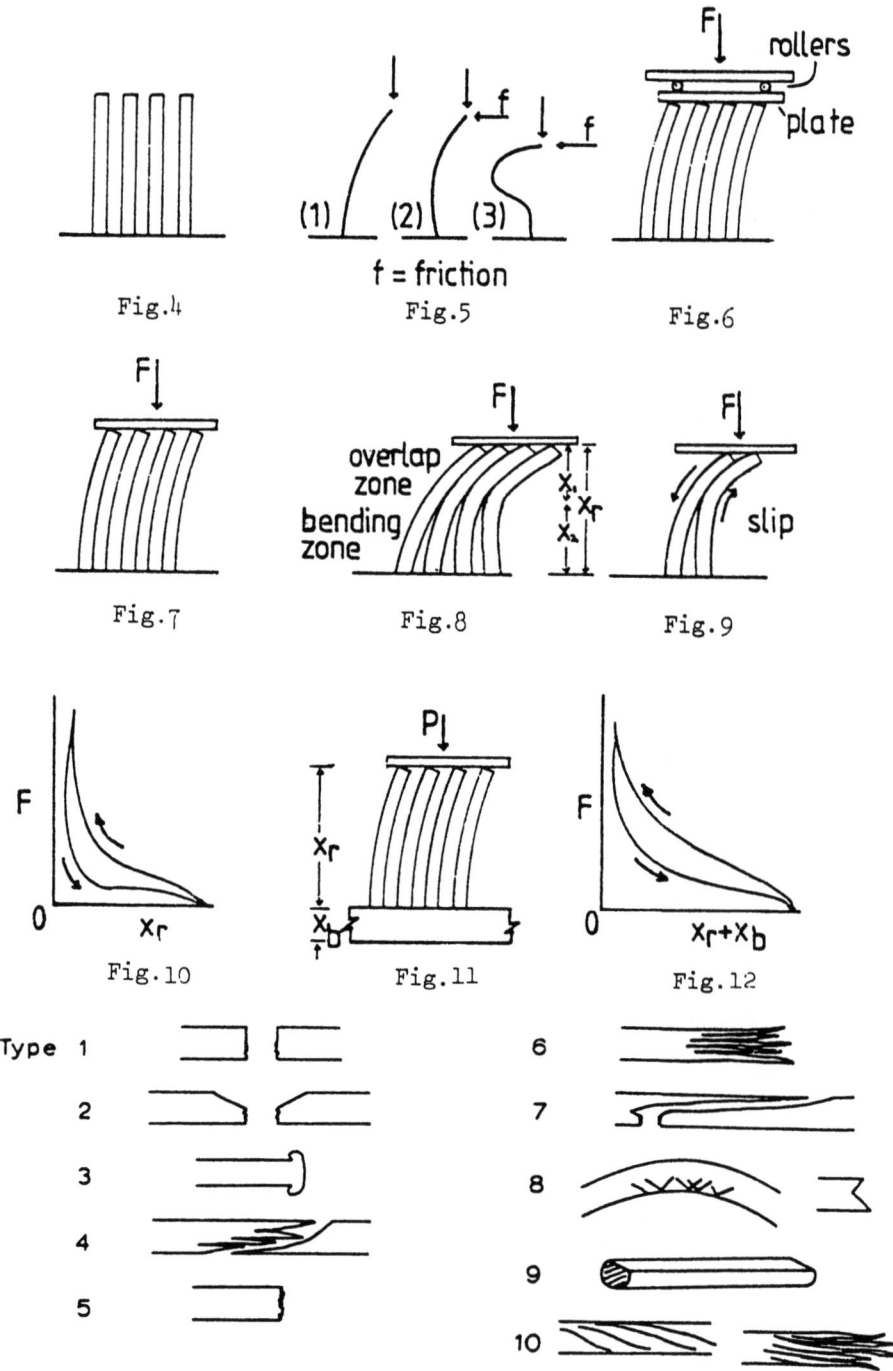

Fig.4

Fig.5 f = friction

Fig.6

Fig.7

Fig.8

Fig.9

Fig.10

Fig.11

Fig.12

Fig. 13

fibre, in wet cotton. None of these failures resemble typical wear in use, though they may be found in special circumstances.

In use fibres are subject to repetitive loading and so it is appropriate to use fatigue testing giving four forms of break: (7) cyclic tensile loading from zero to about half the normal breaking load in nylon and some other fibres; (8) flex fatigue by oscillation backwards and forwards over a pin causing breakdown in nylon and polyester along kink-bands; (9) direct surface on rubbing over a pin, in Kevlar and wool; (10) biaxial rotation over a pin. Breaks of type (7) have been found in parachute cords, and of type (8) in tyres, but the commonest mode of failure in use is type (10).

When a bent fibre is rotated, material goes alternately into tension and compression, weakening the structure through the formation of kink-bands. However the energy loss due to bending hysteresis must be overcome by the rotational drive. Consequently there must be torque, and resulting twist, present and the shear stresses cause multiple splitting and ultimate failure.

The motivation for a concentration of research effort on the laboratory test method by biaxial rotation over a pin came from the observations of failures on worn products. In shirts, trousers, socks and other products the usual sequence of fibre failure is multiple splitting, leading to a brush-like fracture, and then a subsequent rounding of the free ends: this has been found with many fibre types. We have not studied carpet wear, but I would expect the same effects to occur, since the sources of damage will be the buckling of fibres in three-dimensionas with a consequent twisting or the rolling of bent fibres over one another.

REFERENCES

1. J.W.S. Hearle in Contributions of Science to the Development of the Textile Industry, edited by M. Cordelier and P.W. Harrison, Published by Institute Textile de France and Textile Institute, 1975, p.60.

2. A.R. Bunsell and J.W.S. Hearle, Fracture and Fatigue of Fibres, Rheol. Acta., 13, 711, 1974.

3. B.C. Goswami and J.W.S. Hearle, Text. Res. J., 46, 1, 55, 1976.

4. J.W.S. Hearle and B.S. Wong, J.Text.Inst., 68, 89, 1977.

5. J.W.S. Hearle and B.S. Wong, J.Text.Inst., 68, 127, 1977.

6. S.F. Calil and J.W.S. Hearle, ICF4 Conf., Canada, p.1267-1271, 1977.

7. J.W.S. Hearle and B.S. Wong, J.Text.Inst., 4, 155, 1977.

8. J.W.S. Hearle and B. Lomas, J. Applied Pol. Sci., 21, 1103, 1977.

9. J.W.S. Hearle and B.S. Wong, J.Materials Sci., 12, 2447, 1977.

10. L. Konopasek and J.W.S. Hearle, J. Applied Pol. Sci., 21, 2791, 1977.

11. N.E. Dweltz, J.W.S. Hearle, G.E. Cusick, J.Text.Inst., 69, 9, 294, 1978.

12. J.W.S. Hearle and N. Hasnain, Annual Conf., of the Textile Institute, New Delhi, 163, 1979.

13. J.W.S. Hearle and J.T. Sparrow, Text.Res.J., 49, 268, 1979.

14. J.W.S. Hearle and J.T. Sparrow, Text.Res.J., 49, 242, 1979.

15. I.A.O, Bahari, I.E. Clark, M.Sc. theses, S.F. Calil, R. Mandal Ph.D. theses, University of Manchester.

16. S.F. Calil, B.C. Goswami and J.W.S. Hearle, in press

17. J.W.S. Hearle and B. Lomas, Clothing Res.J., 5, 47, 1977.

18. I.E. Clark and J.W.S. Hearle, J.Phys.E., to be published.

The Compression of Pile Carpets (Leaf)

The analysis of the compressive behaviour of pile carpets has occupied a number of workers, and I should like to describe two abortive attempts carried out at Leeds University.

When a pile carpet is deformed by forces P per tuft the result is usually as shown in figure 1.

To try and explain this behaviour we have considered two models.

Model I[1]

The tufts bend according to the law

$K = 0 \quad (M < M_o)$
$BK = M - M_o \quad (M \geqslant M_o)$

where M is the bending moment, K the curvature, B flexural rigidity and M_o a coercive couple. The yarns were assumed to be in

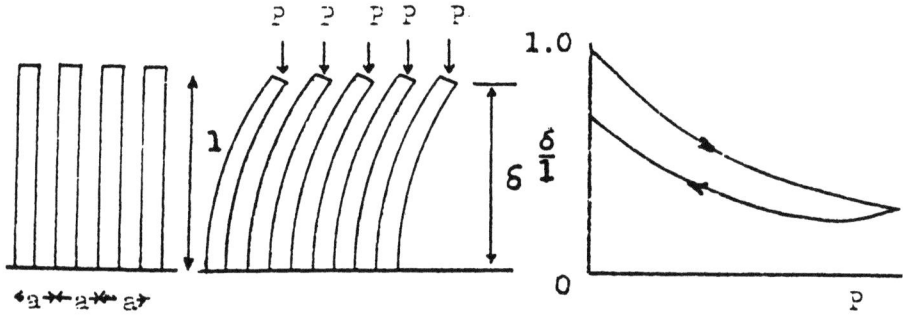

Figure 14

compressible laterally and inextensible. The resulting relation between P and δ/ℓ is as shown in Figure 15, and

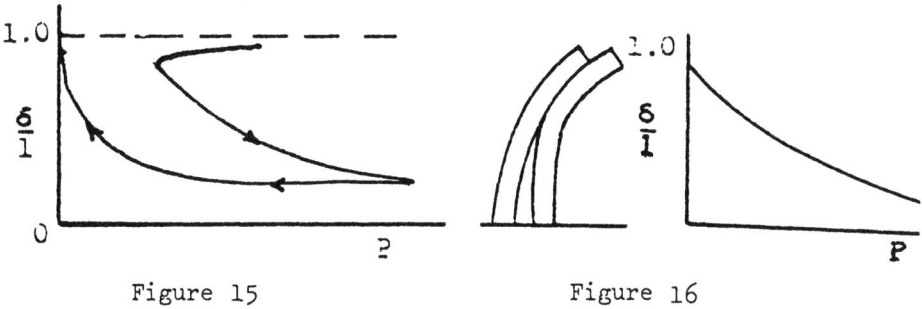

Figure 15   Figure 16

agreement with experiment is not good.

Model II[2]

The tufts bend according to the law

$$B_k = M$$

for all values of M, but friction between the tufts was allowed for and lateral yarn compression was assumed to take place according to a relation of the form.

$$t = a + b\, e^{-cN}$$

where t is the tuft diameter, N is the normal force per unit length, and a, b, c are constants for a given yarn. The theoretical pile compression curve is shown in Figure 16, i.e. roughly similar in form to the experimental curve but agreement with experimental data was still poor.

It is apparent that a more satisfactory analysis is still required.

References

1. O.H. Culcuoglu, Ph.D. thesis, Leeds University, 1971.
2. B.L. Thomas, Ph.D. thesis, Leeds University, 1973.

DISCUSSION XVI

COMMUNICATION OF RESEARCH RESULTS

Leader   : Mr C.R. Williams
Rapporteur : Dr. R.W. Rennell.

Before inviting a number of prepared contributions the session
leader introduced a framework for the discussion by posing the
following questions:

    1. What needs to be communicated? 2. Why? 3. By whom?
    4. To whom? 5. How do we obtain the feedback and iteration
    needed for the communication to be successful. 6. By what
    methods can the results of research be communicated?

Information retrieval (Backer)

Professor Backer became concerned some 10 years ago that a large
amount of information was potentially available, but no efficient
information retrieval system existed and so he set out to develop
such a system.

The first requirement is for some means of language control for
indexing. The indexing system developed allowed 8,500 words
specific to textiles and also allows 8,500 relationships between
these key words. This basic framework was developed in Europe
and at present 25,000 articels/year are abstracted and stored
in two main computer data bases. The TITUS system developed in
France and the world textile abstract is developed at Shirley
Institute. Both these systems are available through the Lockheed
DIALOG system.

These information retrieval systems have improved the flow in information but their effects are limited. The main problem is that few people in industry have time to read the literature. This is where a "gatekeeper" who can read, understand, translate and communicate information is valuable. The "gatekeeper" could also survey conferences etc. to gain early knowledge of information which may take some time to be published in full.

Industrial Relevance (Popper)

Speaking from an industrial viewpoint there appears to be a lack of feedback concerning which areas are of real interest and what use is made of results of research by industry. In order to provide some feedback the following areas were suggested. These represented a personal view based on Dr Popper's experience of real problems from Du Pont operating divisions.

1. Uniformity - general problems such as distortion, irregularity, barré etc. There are many other forms of non-uniformity but they can be roughly classified:
    (a) sheet buckling, e.g. transport belts - limited amount of theory available,
    (b) complex deformations, e.g. laying sheets - moving materials through machinery,
    (c) self excited disturbances e.g. drafting waves - no characterization of the disturbance available.

2. Analysis of intermediate or temporary fibre assembly structures, e.g. tow, sliver, webs, bales, packages etc.
   Here thermomechanical models are needed to characterize such things as shrinkage and set. None of the existing viscoelastic models appear to be valid here.

3. Dynamics of high speed processes.
    (a) transport stability - high air drag etc.
    (b) limiting speed in false-twisting - is there a limiting speed?

4. Relationship between complex end-use response and engineering variables. This covers the translation of textile terms into engineering terms and related to this is Kawabata's work on comfort.

5. Mechanics of new structures.
   e.g. geotextiles, high speed rotating structures utilizing high strength-to-weight ratio fibres (fly wheels).

6. Instrumentation.
   (a) Tracer and marking techniques – tracking of fibres.
   (b) Measurement of microgeometry – no simple techniques.
   (c) Development of a fibre-like transducer for strain, temperature etc.

Dr. Popper then described how feedback is ensured within du Pont between the Engineering R & D division who provide research for the rest of Du Pont on a contract basis. These contracts range from 2 hour consultations to 2 year contracts. All work is funded and an idea or concept must be sold to the user. This guarantees a meaningful research area and automatically ensures the user's interest in the results. He also mentioned that du Pont fibres could be obtained free of charge for research purposes.

General discussion

Contributors commented on the value of using author indexes rather than subject indexes (Stevens), on tracing back from a recent relevant paper and on using keywords for searching new areas (Dent), on citation indexes for searching forward by means of an author's name (Backer), on how to secure cross-fertilisation of ideas from related fields when occasions like this Kilini meeting were rare (Shephard), on the problems of industrial secrecy (Ansell).

Shishoo described the operating system at TEFO. Prof. Lindberg re-organised TEFO when it appeared to be drifting away from the main problems of the industry. They now have seven target groups:

1. Knitted fabrics; 2. Nonwovens; 3. Industrial;
4. Finishing; 5. Clothing; 6. Woven fabrics;
7. Carpet and wallcoverings.

Each group consists of users, manufacturers and management. The groups decide on 3 year programmes which are part-Government, part-industry supported. Twenty percent of the total budget is set aside for fundamental research.

Their literature sources therefore effectively include information from industry and users who appear willing to finance any work which can be shown to have tangible objectives.

van Harten suggested that asking industry if they had problems would reveal none. However if you investigate you find problems exist.

The application of the subject (Hearle)

Intercommunication among the experts in the subject of the mechanics of flexible fibre assemblies, apart from the publication of research results in the literature or at conferences, would be useful in standardisation of terminology, in exchange of computer programs, and in comparative studies both in terms of explicit comparison of the predictions of different theories on the same input data and in the accumulation of a large set of test results on standard fabrics.

However, it is more important to consider how the research results can be communicated to potential users. The attendance at the ASI is predominantly academic, with some industrial research contractors, some specialists in military or engineering applications, some representatives of fibre producers, but little representation of the textile industry and none from any of the major textile companies.

The state of the art, to date, has led to qualitative understanding, which is helpful in solving industrial problems, but hardly to any quantitative use in the engineering design of textile fabrics. The adoption of modern interactive computational methods could provide a tool of practical day-to-day value for textile technologists, but in order to achieve such use we have to break into the chicken-and-egg situation. Until programs are readily available, technologists will not use the methods; and until there is a demand, there is no motivation for carrying out the development and marketing which is needed. Through education and other forms of publicity, we must slowly chip away at the problem.

In many important areas of the subject, the basic research has been done, although there are other basic problems still to be solved. The temptation for the academic is to move into these new areas, and this should be done. But it is more necessary to develop a dialogue between the academic researchers and those in industry who can formulate the special problems, where there would be value in working out detailed solutions by the methods currently available, or moderate developments of them. This might then lead to a more widespread routine use, similar to the use of mechanics in other branches of engineering.

This meeting has brought together, for the first time, specialists in the mechanics of flexible fibre assemblies. It will be of benefit to the advance of the subject if we, and others, can continue to communicate with one another. I recommend the formation of an informal group for this purpose: it might be called the Kilini Research Group.

APPENDIX

SOME TEXTILE QUANTITIES

Fineness

expressed by:

linear density = $\frac{\text{mass}}{\text{length}}$ count = $\frac{\text{length}}{\text{mass}}$

denier = g/9000m   cotton count = 840 yard hank per lb.

tex = g/km   worsted count = 560 yard hank per lb.

    ETC   ETC

Twist Factor

expressed by:

twist x (linear density)$^{\frac{1}{2}}$ twist/(count)$^{\frac{1}{2}}$

e.g. 10 tex$^{\frac{1}{2}}$ cm^{-1} $\doteq$ 1 t.p.i./(cotton count)$^{\frac{1}{2}}$

"Fabric Weight"

expressed by:

mass/area

100 g/m^2 $\doteq$ 3 oz per sq. yd.

Specific Stress

expressed by:

force/linear density (text of first chapter gives alternatives)

See Table for unit conversions

CONVERSION: SPECIFIC STRESS

1 ——— N/tex, kJ/g, GN m^{-2}/g cm^{-3}, (km/s)^2

— 10.2gf/dtex, 11.3gf/den

— 102gf/tex, kmf, kgf mm^{-2}/g cm^{-3}
   239 cal/g, 430 Btu/lb

10^3 ——— mN/tex, J/g, MPa/g cm^{-3}

— 145,000 psi/g cm^{-3}

*10^6 ——— N/kg m^{-1}, J/kg, Pa/k gm^{-3}, m^2 s^{-2}

— 3.94 x 10^6 inchf, psi/(lb/cu.in)

10^9 ———

— 10^{10} dyn/g cm^{-1}, erg/g

*Strict SI

Notes: (a) Other multiples also used

(c) Gravitational force units are often written as g, g-wt or pond; lb or lb-wt; km or Rkm.

STRESS

Density in g/cm^{-3} times:

1 ——— GN/m^2, J/mm^3

— 102 kg/mm^2
  145 ksi

10^3 ——— N/mm^2, MPa

— 10^4 bar, 9869atm, 1.02 x 10^4 kg/cm^2

— 145,000 psi, lbf/in^2

10^6 ———

— 7.5 x 10^6 mm Hg

*10^9 ——— Pa, N/m^2, J/m^3, kg m^{-1} s^{-2}

— 10^{10} dyn/cm^2

(b) Nm^{-2} = Pa

NATO ASI - MECHANICS OF FLEXIBLE FIBRE ASSEMBLIES

19 August - 2 September 1979

List of Participants

J. Amirbayat
Textile Technology Dept.
U.M.I.S.T.
P.O. Bix 88
Sackville St.
Manchester M60 1QD
U.K.

A. Ankara
Met. E. Dept.
Middle East Technical University
Ankara
Turkey.

M.P. Ansell
School of Materials Sci.
University of Bath
Bath
U.K.

J. Ari-Gur
Aeronautical Engineering Dept.
Israel Inst. of Technology
Haifa
Israel

P. Arghyros
Textile Technology Dept.
U.M.I.S.T.
P.O. Box 88
Sackville Street
Manchester M60 1QD
U.K.

S. Arghyros
Preveza Mills S.A.
Paleologou Benizelou Street
Athens
Greece

M.E. Arponen
Tampella Textile Div.
P.O. Box 266
33101 Tempere
Finland

S. Backer
5 Irving Road
Waban, Massachussetts 02168
U.S.A.

M. Bakioglu
I.T.U.
Insaat Fakultesi
Tekisla-Taksim
Instanbul
Turkey

J. Banbaji
Israel Fibre Institute
P.O. Box 80001
Jerusalem
Israel

S. Batra
Textile Materials & Management
                               Dept.
N.C. State University
Raleigh, North Carolina
U.S.A.

H. Behery
Textile Dept.
Clemson University
Clemson, S.Carolina 29631
U.S.A.

O.G. Bilir
Mechanical Engineering Dept.
Middle East Tech. University
Ankara
Turkey

R. Blum, Lecturer
Melonen Strasse 42
Stuttgart 75
Germany

C.P. Buckley, Lecturer
Textile Technology Dept.
U.M.I.S.T.
P.O. Box 88
Sackville Street
Manchester M60 1QD.
U.K.

A.R. Bunsell
Ecole des Mines de Paris
P.O. Box 87
91003 Evry Cedex
France

H.W. Busching
Civil Engineering Dept.
Clemson University
Clemson, S.Carolina
U.S.A.

G. Carnaby, Lecturer
Wool Res. Org. of New Zealand.
Private Bag
Christchurch
New Zealand

C.D. Cassolas
School of Textiles
Philadelphia School of Textiles & Sci.
Philadelphia
U.S.A.

B.M. Chapman, Lecturer
C.S.I.R.O.
338 Blasland Road
Ryde, New South Wales
Australia

T. Chytiris
Hellenic Cotton Board
150 Syngrou Ave.,
Athens 404
Greece

S.K. Clark, Lecturer
Applied Mechanics Dept.,
University of Michigan
Ann Arbor, Michigan
U.S.A.

A. Collios
Inst. de Recherches Interdisciplinaires
des Geologie et de Mecanique
Boite Postale 53,
38041 Grenoble Cedex
France

C. Delides
Physical Chemistry Dept
Strathclyde University
194 Cathedral Street
Glasgow
Scotland

R.W. Dent
2327 Oriole Drive
Durham N. Carolina,
U.S.A.

J.H. Dorr
Dosbouw
Postbus 5003
Burgh-Haamstede
Holland

J.E. Fitzgerald
School of Civil Engineering
Georgia Inst. of Technology
Georgia, Atlant
U.S.A.

M. Frias
University of Lisboa
P.Jose Fontana 16-4
Lisboa
Portugal.

V. Friedrich
Sommeracher Str. 19
D8712 Volkach/Main
Germany.

B.C. Goswami
Textiles & Clothing Dept.
University of Tennessee
Knoxville, Tennessee
U.S.A.

P. Grosberg, Lecturer
Textiles Industries Dept.
University of Leeds
Leeds
U.K.

J.P. Giroud, Lecturer
University of Grenoble
Grenoble
France

K. van Harten
Mechanical Engineering Dept.
University of Delft
P.O. Box 5036
Delft
Holland

J.W.S. Hearle, Lecturer, Director
Textile Technology Dept.
U.M.I.S.T.
P.O. Box 88
Sackville Street
Manchester M60 1QD
U.K.

A.D. Hearle
St. John's College
University of Cambridge
Cambridge
U.K.

R.B. Hepworth, Lecturer
Textile Industries Dept.
University of Leeds
Leeds
U.K.

H.C. Huang
Mechanical Engineering Dept.
M.I.T.
Cambridge, Mass.
U.S.A.

S. de Jong, Lecturer
School of Textile Technology,
University of New South Wales
Kensington N.S.W.
Australia

B. Kaftanoglu
Middle East Tech. University
Ankara
Turkey

K. Kawabata, Lecturer
Polymer Chemistry Dept.
Kyoto University
·Kyoto
Japan

F. Ko
Philadelphia Coll. of Textiles
           & Sci.
Philadelphia 30319
U.S.A.

M. Konopasek, Lecturer
1289 West Nancy Creek Drive
Atlanta, Georgia
U.S.A.

V. Kyriazis
Hellenic Cotton Board
150 Syngrou Ave.,
Athens
Greece.

G. Leaf, Lecturer
Textile Industries Dept.
University of Leeds
Leeds
U.K.

C. Leech, Lecturer
Mechanical Engineering Dept.
U.M.I.S.T.
P.O. Box 88
Sackville Street
Manchester M60 1QD
U.K.

M. Levinson
Civil Engineering Dept.
McMaster University
Hamilton, Ontario
Canada

D.W. Lloyd, Lecturer
Textile Industries Dept.
University of Leeds,
Leeds
U.K.

S. Luong
Laboratoire de Mechanique de Solids
Ecole Polytechnique
91128 Palaiseau Cedex
France

J. Mackeprang
Textile Department
Denmarks Tekniske Hojskole
Bygning
Denmark

S.J. Van der Meer
Peter van Anvooy
Laan 5
Diereh
Holland

H. Meyer
Kaj Neckelmann
Keilstruprej
Silkeborg 8610
Denmark

S. Moghe, Lecturer
B.F. Goodrich & Co.
9921 Bresksville Road
Brecksville, Ohio
U.S.A.

C.P. Morgado
Institute Polytechnico da Covilha
Covilha
Portugal

K.A. Mouw
Bykswaterstaat Deltadien Str.
P.O. Box 5002
Busch Haamstede
Holland

G. Nemoz
Institute Textile de France
Section Lyon 69130
Eculy, B.P. 60
France

A. Newton
Textile Technology Dept.
U.M.I.S.T.
P.O. Box 88
Sackville Street
Manchester M60 1QD
U.K.

A. Nijhof
Mechanical Engineering Dept.
University of Delft
P.O. Box 5036
Delft
Holland

G.C. Papanicolaou, Lecturer
Theoretical & Applied Mechanics
                              Dept.
National Tech. University
Athens
Greece

S.L. Phoenix, Lecturer
248 Upson Hall
Cornell University
Ithaca, New York,
U.S.A.

P. Popper
E.I. du Pont de Nemours
Exp. Station, Bldg, 304
Wilmington, Delaware
U.S.A.

N. Raagard
Fluid Mechanics Dept.
Bldg. 404
D.T.H.
Lyngby
Denmark

E. Racah
School of Architecture & Building
University of Bath
Bath
U.K.

A.K. Ray
Suite No. 202
585 King Edward Ave.,
Ottawa, Ontario
Canada

R.W. Rennell
Shirley Institute
Didsbury, Manchester
U.K.

W.J. Schaffers
2408 Dorval Road
Chalfonte
Wilmington, Delaware
U.S.A.

A. Schenek
Halderlinstr 11,
D7445 Bemflingen
Germany

R.S. Seth Lecturer
Pulp & Paper Research Inst.
570 St. John's Blvd.
Pointe Claire
Canada

K.M. Shanks
36 Kingfield Drive
Didsbury, Manchester
U.K.

R.G. Shephard
Ministry of Defence
S.C.R.D.E.
Flagstaff Road
Colchester, Essex
U.K.

R. Shishoo
Svenska Textileforschunge Inst.
S-402 Gothenberg
Sweden

J. Skelton Lecturer
Fabric Res. Laboratories
Route 128 at U.S.1.
Dedham, Mass.
U.S.A.

K. Slater
Textile Sci. Division
University of Guelph
Guelph, Ontario
Canada

G. Stevens
159 Mytchett Road
Mytchett, Camberley
U.K.

J.J. Thwaites, Lecturer,
             Assoc. Director
Engineering Department
University of Cambridge
Cambridge
U.K.

A. Uttley
30 Moorfield Road,
W. Didsbury, Manchester
U.K.

C.R. Williams
U.S. Army
Natick Research & Development
Natick, Mass.
U.S.A.

G. Yazicioglu
E.U. Tekstil Fakultesi
Bornovo, Izmir
Turkey

T. Yazicioglu
E.U. Tekstil Fakultesi
Bornovo, Izmir
Turkey

L.K. Yu
Corporation Research
International Paper Co.
Tuxedo Park
New York
U.S.A.

The participants at the ASI during the excursion to Delphi

MIX
Papier aus verantwortungsvollen Quellen
Paper from responsible sources
FSC® C105338

If you have any concerns about our products,
you can contact us on
ProductSafety@springernature.com

In case Publisher is established outside the EU,
the EU authorized representative is:
**Springer Nature Customer Service Center GmbH
Europaplatz 3, 69115 Heidelberg, Germany**

Printed by Libri Plureos GmbH
in Hamburg, Germany